The Fungi

To the memory of two distinguished mycologists,

S. D. Garrett FRS (1906–1989)

and

P. H. Gregory FRS (1907–1986)

The Fungi

Second Edition

Michael J Carlile
Imperial College
University of London

Sarah C Watkinson
University of Oxford

Graham W Gooday
University of Aberdeen

ACADEMIC PRESS

A Harcourt Science and Technology Company

San Diego San Francisco New York Boston
London Sydney Tokyo

Academic Press
A Harcourt Science and Technology Company
Harcourt Place, 32 Jamestown Road, London NW1 7BY, UK
http://www.academicpress.com

Academic Press
A Harcourt Science and Technology Company
525 B Street, Suite 1900, San Diego, California 92101-4495, USA
http://www.academicpress.com

ISBN 0-12-738445-6 (Hardback)
0-12-738446-4 (Paperback)

Library of Congress Catalog Number: 00-109116

A catalogue record for this book is available from the British Library

Typeset by Phoenix Photosetting, Chatham, Kent
Printed and bound in Great Britain by Bath Press, Avon

01 02 03 04 05 06 BP 9 8 7 6 5 4 3 2 1

Contents

3 Fungal cells and vegetative growth   85

Spores, dormancy and dispersal   185

4

Fungi and biotechnology   461   8

Colour plates appear between pages 300 and 301.

Foreword

Fungi: The threads that keep ecosystems together

When people ask what I do for a living, and I tell them I'm a mycologist, they usually react with surprise. Often they don't know what a mycologist is, but when I tell them, the next question is "why?" Why study fungi?

When someone mentions "fungi" you may think immediately of mushrooms on pizza or maybe moldy food in your refrigerator or the fungus growing on your toes – But in fact fungi are everywhere and affect our lives every day, from mushrooms to industrially important products to plant helpers to plant pathogens to human diseases.

Fungi affect human lives in many and varied ways, so it is important to know something about fungal biology in order to be able to control or exploit them for our own purposes. The study of fungi has increased exponentially in the past 100 years, but they are still being ignored or neglected in many fields of study. For example, more than 90% of fungal species have never been screened for antibiotics or other useful compounds. Many ecologists do not even think about fungi when doing their experiments or observations. However fungi play very important roles in the ecosystem. They are a vital part of the links in the food web as decomposers and pathogens and are important in grassland and forest ecosystems alike. Fungi have many different kinds of associations with other organisms, both living and dead. Since all fungi are heterotrophic, they rely on organic material, either living or dead, as a source of energy. Thus, many are excellent scavengers in nature, breaking down dead animal and vegetable material into simpler compounds that become available to other members of the ecosystem. Fungi are also important mutualists; over 90% of plants in nature have mycorrhizae, associations of their roots with fungi, which help to scavenge essential minerals from nutrient poor soils. Fungi also form mutualistic associations with algae and cyanobacteria in the dual organisms known as lichens.

On the other hand, many fungi are detrimental, inciting a large number of plant diseases, resulting in the loss of billions of dollars worth of economic crops each year, and an increasing number of animal diseases, including many human maladies. Fungi can cause human disease, either directly or through their toxins, including mycotoxins and mushroom poisons. They often cause rot and contamination of foods – you probably have something green and moldy in the back of your refrigerator right now. They can destroy almost every kind of manufactured good – with the exception of some plastics and some pesticides. In this age of immunosuppression, previously innocuous fungi are causing more and more human disease.

There are many ways in which people have learned to exploit fungi. Of course, there are many edible mushrooms, both cultivated and collected from the wild. Yeasts have been used for baking and brewing for many millennia. Antibiotics such as penicillin and cephalosporin are produced by fungi. The immunosuppressive anti-rejection transplant drug cyclosporin is produced by the mitosporic fungus *Tolypocladium inflatum*. Steroids and hormones – and even birth control pills – are commercially produced by various fungi. Many organic acids are commercially produced with fungi – e.g. citric acid in cola and other soda pop products is produced by an *Aspergillus* species. Some gourmet cheeses such as Roquefort and other blue cheeses, brie and camembert are fermented with certain *Penicillium* species. Stone washed jeans are softened by *Trichoderma* species. There are likely many potential uses that have not yet been explored.

Fungi are also important experimental organisms. They are easily cultured, occupy little space, multiply rapidly, and have a short life cycle. Since they are eukaryotes and more closely related to animals, their study is more applicable to human problems than is the study of bacteria. Fungi are used to study metabolite pathways, for studying growth, development, and differentiation, for determining mechanisms of cell division and development, and for microbial assays of vitamins and amino acids. Fungi are also important genetic tools, e.g. the "one gene one enzyme" theory in *Neurospora* won Beadle and Tatum the Nobel prize for Physiology or Medicine in 1958. The first eukaryote to have its entire DNA genome sequenced was the bakers' and brewers' yeast *Saccharomyces cerevisiae*.

Mycologists study many aspects of the biology of fungi, usually starting with their systematics, taxonomy, and classification (you have to know "what it is" before you can work effectively with it), and continuing on to their physiology, ecology, pathology, evolution, genetics, and molecular biology. There are quite a few disciplines of applied mycology, such as plant pathology, human pathology, fermentation technology, mushroom cultivation and many other fields.

Fungi never fail to fascinate me. They have interesting life cycles and occupy many strange, even bizarre, niches in the environment. Take for example *Entomophthora muscae*, a fungus that infects houseflies. The spores of the fungus land on the unfortunate fly and germinate, then penetrate the exoskeleton of the fly. The first thing the fungus does, according to reports, is grow into the brain of the fly, in order to control its activities. The mycelium of the fungus grows into the particular area of the brain that controls the crawling behavior of the fly, forcing the fly to land on a nearby surface and crawl up as high as possible. Eventually the hyphae of the fungus grow throughout the body of the fly, digesting its guts, and the fly dies. Small cracks open in the body of the fly and the *Entomophthora* produces sporangia, each with a single spore, which are then released in hopes of landing on another fly.

Other fungi, such as the dung fungus *Pilobolus*, produce spore "capsules" that are shot off with great force, up to 3 meters away from their 1 cm sporulating structure. Some fungi are "farmed" by Attine ants and by termites. Some fungi can actually trap and eat small worms called nematodes. Known for their diverse and amazing physiology, fungi can grow through solid wood, and in lichen associations can even break down rocks. Fungi have intriguing and captivating sex lives, some species with thousands of different sexes. Tetrad analysis in the Ascomycetes has helped to solve some fundamental mysteries about genetics in eukaryotic organisms.

I am pleased to introduce you to THE book for teaching and for learning fungal biology. Michael Carlile, Sarah Watkinson, and Graham Gooday have produced an eminently readable book to introduce students to all aspects of the biology of fungi, including physiology and growth of hyphae and spores, fungal genetics, fungal ecology and how these aspects of the fungi can be exploited in biotechnology. The authors cover many of the topics I have alluded to above in great depth, as well as thoroughly explaining the mostly hidden lives of fungi.

For new students of the fungi, I know you will enjoy learning about these amazing organisms. For those of you who are already mycophiles, this book will serve as a handy reference to fungi and their activities.

Thomas J. Volk
Department of Biology
University of Wisconsin – La Crosse
http://www.wisc.edu/botany/fungi/volkmyco.html

Preface to the
Second Edition

The preface to the first edition of this book, which follows, discusses the significance of mycology for various branches of science and its value for students of different biological disciplines, and explains the approach used and rationale behind the arrangement of chapters and the selection of topics. All the principles there discussed apply to the second edition, but the passage of six years has inevitably led to changes in the material presented. Many of these changes result from progress in molecular biology and its application to the fungi. When the writing of the first edition was completed in 1993, it was clear that molecular methods, which were already having revolutionary consequences for bacterial classification, identification and ecology, would have equally profound effects in mycology. However, at that time, apart from the use of molecular methods in biotechnology (such as yeast as a host for cloned genes) and population genetics of plant pathogens, rather little had been published. We therefore had to limit ourselves very largely to outlining molecular methods and stating their potential for mycology. The situation had changed sufficiently by the time of publication for some otherwise enthusiastic reviewers to regret the paucity of molecular material. We are now able to make proper use of molecular insights in discussing fungal development, classification, ecology and pathogenicity, as indicated below in a consideration of changes in each chapter.

Chapter 1 now has an improved presentation of the place of fungi in the major groups of organisms, made possible by progress in molecular phylogeny. This applies also to the consideration of the major groups of fungi in Chapter 2, although the terminology used by practising mycologists is emphasized. The publication of the 8th edition of the authoritative *Ainsworth & Bisby's Dictionary of the Fungi* has facilitated a revision of Appendix 2 to provide an up-to-date classification. Chapter 3 includes new material on the way in which hyphae and yeast cells grow, and Chapter 4 on mating in fungi, much of which results from the application of molecular methods. Whereas in 1993 there had been a very limited application of molecular cladistics to fungi, now every issue of leading mycological journals has phylogenetic trees for further fungal groups. Molecular methods are also giving an increased insight into the extent of genetic recombination in nature, especially among apparently asexual fungi. These developments have necessitated considerable revision of Chapter 5. A major problem in fungal ecology has been a very limited ability to identify fungi in nature unless they are sporulating, a problem that is diminishing through the application of molecular methods, as described in Chapter 6, which also includes an account of the recently established threat to fungal biodiversity along with approaches to fungal conservation. Chapter 7 includes new material on the molecular basis of fungal pathogenicity, and has more material on medical mycology than did the first edition. A major recent development in medicine has been the rise of immunosuppressive and cholesterol-lowering drugs of fungal origin into the category of best-selling pharmaceuticals. These and other new products of fungal origin are considered in Chapter 8. Chapters now end with questions, with answers at the end of the book.

We wish to renew our thanks to our friends who read or commented upon chapters or sections and provided illustrations for the first edition of the book. We now add our gratitude to those who have provided similar help in the preparation of the second edition, including Professor Joan Bennett, Professor Tom Bruns, Professor Mark Seaward, Professor Nick Talbot, Professor Tom Volk, those whose names appear in the legends to the Figures that they provided, and Lilian Leung of Academic Press. Finally, as before, we thank members of our families, Elizabeth Carlile, Margaret Gooday and Anthony, Charles and Ruth Watkinson, for their help.

Dr Michael J. Carlile
42 Durleigh Road
Bridgwater
TA6 7HU

Dr Sarah C. Watkinson
Department of Plant Sciences
University of Oxford
South Parks Road
Oxford OX1 3RB

Professor Graham W. Gooday
Department of Molecular & Cell Biology
Institute of Medical Sciences
University of Aberdeen
Foresterhill
Aberdeen AB25 2ZD

Preface to the First Edition

The study of fungi, mycology, is of importance for students of many branches of the life sciences. Fungi are of major significance as mutualistic symbionts and parasites of plants, so their study is an important part of courses in *plant sciences*, and essential for students of *plant pathology*. Fungi are a major component of the microbial world, and it is increasingly being recognized that they should receive proper consideration in *microbiology* courses. Yeasts, filamentous fungi or both are important in brewing, in the preparation of many foods, in biodeterioration and in the fermentation industry, and hence need consideration in the context of *biotechnology*. Yeasts have also had a major role in the development of *biochemistry*, and filamentous fungi in that of *genetics*, especially biochemical genetics, and both are now of major importance in *molecular biology* as hosts for gene cloning. Fungi are of crucial importance in the breakdown of the vast amounts of organic carbon produced annually by photosynthesis, and thus are important in *ecology* and *environmental science*. Like bacteria, plants and animals, the fungi are one of the great groups of living organisms, whether considered in terms of numbers of species, biomass or role in the environment. Fungi therefore deserve a place in the curriculum for degrees in *biology*, and in introductory courses for any branch of the life sciences. The present book is intended to provide an account of the fungi useful for students of all the above mentioned disciplines, as well as to others who need information about the fungi.

The present authors have taught general courses in mycology to first and second year undergraduates, and more specialized fungal topics to senior students including those on MSc courses. The students were specializing in varied aspects of biology, including biochemistry, biotechnology, microbiology, plant sciences and plant pathology. Most students, prior to taking a course in mycology, had acquired some knowledge of biochemistry, genetics, microbiology and molecular biology and were interested in these subjects. Their knowledge of the structure and classification of organisms, and of the procedures involved in identification and morphology was however limited, and they needed to be convinced of the importance of these topics and the extent to which they had been revitalized by new, often molecular, methods. Colleagues teaching mycology in other universities in both the United Kingdom and overseas report similar situations. They agree that a book that covered aspects of the fungi of importance in a wide range of disciplines, and which took into account the strengths and limitations of present day students, would be a useful one. We have aimed to produce such a book.

The perspective that we have adopted is a microbiological one. Most students now come to mycology with some microbiological knowledge and, without the application of the microbiological techniques of isolation and growth in pure culture, most of our present understanding of the fungi could never have been gained. We hence concentrate attention on fungi that can be grown in pure culture, while maintaining an interest in their performance in nature. The opening chapter introduces the fungi by reference to the cultivated mushroom, considers their status as one of the major groups of organisms, and indicates the ways in which varied disciplines have contributed to knowledge of the fungi, and the study of fungi to fundamental biological discoveries. The second chapter surveys fungal biodiversity, describing in some detail a well-studied representative of each major group before discussing variety within the group, a didactic approach pioneered by the great nineteenth century biologist and educationalist Thomas Henry Huxley. A practical attitude is adopted with respect to classification and nomenclature, with the terms and groupings used by mycologists in their work and conversation, rather than in their taxonomic papers. There are sections on yeasts and lichens, since books, journals, conferences and scientific societies are devoted to them even though, on the basis of phylogeny, they should be distributed among other groups of fungi. A formal classification and nomenclature is provided in Appendix 2. High growth rates and yields are essential for success in the fermentation industry, and valuable for the microbial biochemist. Chapter 3 hence deals in some detail with the growth of fungi in pure culture and the conditions that influence growth. Spores – which seem to be able to get nearly everywhere and to survive almost everything – are crucial to the success of most fungi. Their formation, in both yeasts and filamentous fungi, also provides promising systems for fundamental studies on the control of developmental processes – fungi develop rapidly, are amenable to genetic manipulation, and have some of the smallest genomes among eukaryotes. Chapter 4 hence includes both classical material on the production, dispersal, survival and germination of fungi spores and an introduction to some recent approaches to the control of sporulation. Variability within fungal species is an important topic. For example, variability in culture concerns fermentation technologists, interested in the stability or improvement of their strains, and variability in nature, plant pathologists assessing whether plants may succumb to more virulent strains. Chapter 5 deals with this topic, as well as the principles involved in classification and in understanding fungal evolution. Fungi are the organisms mainly responsible for the breakdown of the most abundant form of organic carbon, lignocellulose, and so have a crucial role in the ecosystem. This, and other saprotrophic activities, are dealt with in Chapter 6. Since conclusions about the presence and activities of microbes in the environment are highly dependent upon which of a range of techniques are employed, this chapter has a substantial section on methods, some of which are revolutionizing microbial ecology as well as other aspects of mycology. Chapter 7 concentrates on the relationships, mutualistic and parasitic, between fungi and plants, which have interacted with each other throughout their evolution, although the relations between fungi and other groups of organisms are also considered. The final chapter on fungal biotechnology is concerned with both traditional and novel ways in which fungi are exploited by man. Fundamental principles are stressed throughout, rather than details likely to be modified by future research, or matters best taught by observation and experiment in the laboratory.

We wish to thank friends who have read and commented upon chapters or sections. These include Dr Ken Alvin, Dr Simon Archer, Mr Paul Browning, Professor Ken Buck, Professor Keith Clay, Dr Molly Dewey, Dr John Gay, Professor Graham Gooday (who read the whole book), Dr Paul Kirk, Dr Bernard Lamb, Dr Peter Newell, Dr Nick Read, Dr Tim Taylor and Professor Tony Trinci. We are also grateful to Dr Maureen Lacey, who drew Fig. 4.11 illustrating the air spora especially for the book, Dr Jeff Smith and Professor David Wood who helped obtain mushroom photographs for Chapters 1 and 8, Mr Frank Wright who prepared many of the prints, John Baker who took the cover photograph, those who are named in the legends to the Figures that they provided and finally, the many colleagues who in conversation have corrected errors or introduced us to new developments. Finally, we thank members of our families, Elizabeth Carlile and Anthony and Charles Watkinson, who helped in many ways.

Dr Michael J. Carlile
Department of Biology
Imperial College at Silwood Park
Ascot SL5 7PY

Dr Sarah C. Watkinson
Department of Plant Sciences
University of Oxford
South Parks Road
Oxford OX1 3RA

The Fungi as a Major Group of Organisms

1

A fungus familiar to almost everyone is the cultivated mushroom, *Agaricus bisporus* (Fig. 1.1), which is grown commercially on a very large scale, and also survives in nature. The role of the edible **fruit bodies** (or **fruiting bodies**) is the production of large numbers of **spores** by means of which dispersal occurs. The spores are borne on the gills below the cap, and a stalk raises the fruit body above the ground to facilitate spore dispersal by air currents.

Examination of the stalk, cap and gills with a microscope shows that the fruit body is composed of long, cylindrical branching threads known as **hyphae** (sing., **hypha**). The hyphae are divided by cross-walls into compartments which typically contain several nuclei. Such compartments, together with their walls, are equivalent to the cells of other organisms. The spores are borne on specialized cells termed **basidia** (sing., **basidium**). In most mushrooms each basidium bears four spores, but in *Agaricus bisporus* only two – hence the specific epithet *bisporus*.

If a spore of a cultivated mushroom is placed on a suitable substratum, such as a nutrient agar medium, it may germinate, a slender hypha or **germ-tube** emerging from the spore. Hyphal growth is apical, wall extension being limited to the roughly hemispherical apex of the hypha. Nutrients are absorbed from the substratum, and growth, nuclear division and hyphal branching occur to give an approximately circular **colony** which increases in diameter at a uniform rate. Similar colonies can be established by excising tissue from the fruiting body and placing on a suitable medium. The hyphae of a growing colony are termed a **mycelium** (pl., **mycelia**). Although the fruiting body is the spectacular feature of the cultivated mushroom and related fungi, it is entirely dependent for its nutrition on an extensive mycelium penetrating the substratum.

Different kinds of large fungi or **macrofungi** have been recognized for thousands of years. In current English edible ones are often called mushrooms and poisonous ones toadstools. During the eighteenth century, botanists made considerable progress in the recognition and classification of the fungi, and early microscopists observed and described hyphae and spores. In the early nineteenth century it was established that many serious plant diseases were caused by infection of the plant by minute living organisms. These organisms were found to be composed of hyphae and to produce spores, so many of the causal organisms

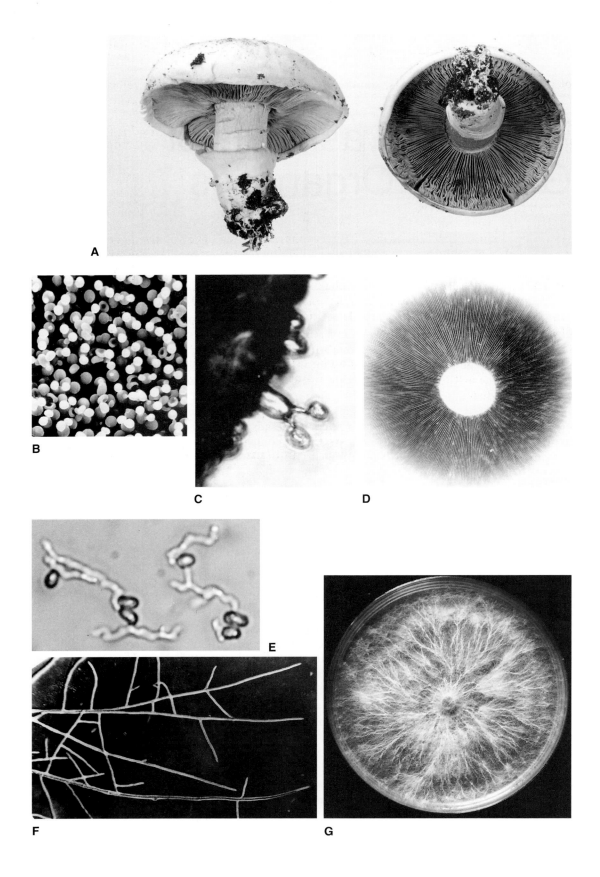

A

B

C

D

E

F

G

of plant diseases, such as rusts, smuts and mildews, were recognized as being microscopic fungi (**microfungi**). Microfungi were also found attacking dead organic materials. Such microfungi were termed **moulds**, spelt **molds** in the USA. The concept developed of fungi as non-photosynthetic plants composed of hyphae, depending for their nutrition on the absorption of organic materials and producing a variety of spores.

Meanwhile, studies on alcoholic fermentation established that the yeast responsible for the process was a microscopic organism reproducing by budding. Although the cells of brewer's yeast are ellipsoidal in form and do not produce hyphae, they were regarded as fungi on the basis of being plants that live by absorbing organic materials rather than by photosynthesis. The discovery of yeasts capable of producing a hyphal phase and of moulds able to produce a yeast phase confirmed the soundness of classifying yeasts as fungi, even though some of the best-known species did not form hyphae.

In the mid-twentieth century electron microscopy showed that all cellular organisms, that is all organisms other than viruses, could be classified on the basis of cell structure into two groups, the **prokaryotes** and the **eukaryotes**, as discussed below. It was found that fungi had a cell organization that was clearly eukaryotic. The fungi can hence be defined as non-photosynthetic hyphal eukaryotes and related forms. Their status as one of the major groups of living organisms is considered below.

The Classification of Organisms into Major Groups

Man, faced with the diversity of living things, has classified them in a variety of ways on the basis of their more striking features. Traditionally, the most fundamental distinction is between **animals**, motile and food-ingesting, and **plants**, static and apparently drawing their nourishment from the soil or in some

---

Figure 1.1 The mushroom. A, Fruit bodies of *Agaricus bitorquis*, a close relative of the cultivated mushroom. On the left note the stalk and the gills below the cap. At the right is a fruit body inverted to show the gill pattern more clearly (T. J. Elliott). B–G, The cultivated mushroom, *Agaricus bisporus*. B, Scanning electron micrograph of the surface of a gill, showing basidia each bearing two spores, interspersed with sterile spacer cells, paraphyses (P. T. Atkey). C, Light micrograph of a basidium bearing two spores (T. J. Elliott). D, Spore print, made by slicing off the stalk of a fruit body, and laying the cap with gills facing downwards on a surface. The discharged spores reproduce the pattern of the gills (M. P. Challen). E, Germ-tubes emerging from spores (From Elliott, T. J. (1985). The general biology of the mushroom. In Flegg, P. B., Spencer, D. M. & Wood, D. A., eds., *The Biology and Technology of the Cultivated Mushroom*, pp. 9–22. Reprinted by permission of John Wiley & Sons Ltd, Chichester.) F, Branching hyphae at the edge of a colony on agar medium (T. J. Elliott). G, A colony covering a Petri dish. Prominent multihyphal strands, radiating from the centre, are developing (Reproduced with permission from Challen, M. P. & Elliott, T. J. (1987). Production and evaluation of fungicide resistant mutants in the cultivated mushroom, *Agaricus bisporus. Transactions of the British Mycological Society* 88, 433–439. (A–G reproduced by permission of Horticulture Research International.)

instances from other plants. This concept of two kingdoms, animals and plants, has dominated scientific classification from ancient times until quite recently.

At first it seemed that the fungi could be assigned without question to the plant kingdom, since they are non-motile and draw their nourishment from the substratum. During the nineteenth century it was realized, however, that the most fundamental features of green plants are that they are **phototrophs**, utilizing energy from light, and **autotrophs**, synthesizing their organic components from atmospheric carbon dioxide. Animals on the other hand are **chemotrophs**, obtaining energy from organic materials, and **heterotrophs**, utilizing the same materials as the source of carbon for the synthesis of their own organic components. On these fundamental metabolic criteria it is clear that fungi, although non-motile, resemble animals rather than plants. Further problems were created by studies on unicellular organisms, which revealed the existence of numerous photosynthetic but motile forms, and of species which were obviously closely related but differed from each other in that some ingested food and some were photosynthetic. These and other problems led to criticism of the two kingdom scheme and to various proposed alternatives, but a deep attachment to the traditional idea of living organisms being divisible into plants and animals dominated biology until a couple of decades ago.

A willingness to consider alternative schemes can be traced to progress in knowledge of cell structure that resulted from electron microscopy in the period 1945–1960. It became clear that at the most fundamental level there are two types of organism – not animals and plants, but those with cells that have a true nucleus (**eukaryotes**) and those with cells that do not (**prokaryotes**). Differences in cellular organization are so profound as to indicate a very early evolutionary divergence of cellular organisms into prokaryotes and eukaryotes. Fungi, in their cellular organization, are clearly eukaryotes.

Whittaker in 1969 proposed a five kingdom classification of organisms. The prokaryotes were accepted as constituting one kingdom, the Monera, but the eukaryotes, within which there is a far greater number of species and structural diversity, were divided into four groups on nutritional and structural criteria. Unicellular eukaryotes (protozoa and unicellular algae) were considered as a single kingdom, the Protista. The multicellular eukaryotes, however, were subdivided on the basis of nutrition into three kingdoms, the photosynthetic plants (Plantae), the absorptive fungi (Fungi) and the ingestive animals (Animalia).

This classification of fungi as one of five kingdoms of living organisms, all with equal taxonomic status, was until recently a useful one. However, new molecular and cladistic approaches (pages 281–286) have yielded a wealth of new information about fundamental similarities and differences between organisms, and these new approaches have been recognized as providing valid evidence for interpreting evolutionary relationships. This has led to the discovery that at the molecular level, life on earth can be classified into three groups, called domains, of which two are prokaryotic and the third eukaryotic (Fig. 1.2). **Fungi** are now recognized as one of five eukaryotic kingdoms, the others being **Animalia** (animals), **Plantae** (plants), **Chromista** (corresponding roughly to the algae and also known as Stramenopila) and **Protozoa**, which contains a wide variety of mainly phagotrophic unicellular organisms.

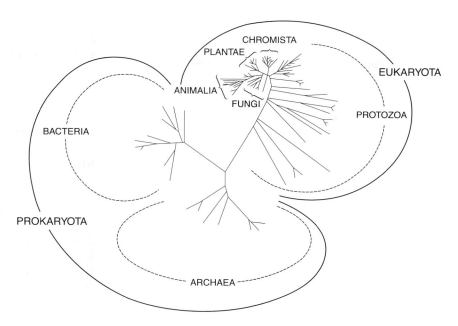

Figure 1.2 A phylogenetic tree showing the relationships between the two Prokaryote and five Eukaryote kingdoms. The kingdom Fungi consists solely of organisms regarded as fungi, but there are phyla within the Chromista and Protozoa that either resemble fungi (such as the Oomycota) or have been studied by mycologists (such as the Myxomycota). The tree is an unrooted one (page 282), involving no assumptions about the point where the common ancestor is situated, but indicating the amount of evolutionary change and pattern of divergence. It is based on the extent of differences between the small sub-unit ribosomal RNA sequences (page 326) for over 70 species. Note that the Fungi, Animals (Animalia), Plants (Plantae) and Chromista form compact groups indicating relatively close relationships compared with much greater evolutionary distances in the protozoa. After Hawksworth, D. L. *et al.* (1995) *Ainsworth & Bisby's Dictionary of the Fungi*, 8th edn. CAB International, Wallingford.

The Study of Fungi

Mycology, the study of fungi, arose as a branch of botany. As indicated earlier, fungi were at one time considered to be members of the plant kingdom, and their structure, life cycles and dispersal have received a great deal of attention from scientists initially trained as botanists.

The microfungi are, however, microorganisms (microbes). Colonies of microfungi in nature are usually of microscopic dimensions, and such organisms can only be studied in detail by the methods of microbiology, separating them from all other organisms and growing them in pure culture. The techniques necessary for achieving and maintaining pure culture were developed by Robert Koch in the late nineteenth century for the study of pathogenic bacteria, but were soon applied to both micro- and macrofungi and were indeed essential for the further development of mycology. Like most bacteria, the nutrition of fungi is heterotrophic and absorptive, and in many environments microfungi and yeasts are closely associated with bacteria and compete with them. Hence in many investigations in microbial ecology it is essential for the activities of bacteria and of fungi to receive equal attention. The similarities between bacteria and fungi as regards the techniques needed for their study, their physiology and their ecology are such that mycology can be considered as a branch of microbiology, and major contributions to the study of fungi are now being made by microbiologists.

The fungi are relatively simple eukaryotes, and many species can be grown easily in pure culture, with high growth rates and if necessary in large amounts. These features have made them attractive research material for scientists whose interest lies not in any specific group of organisms but in fundamental biological processes such as the generation of energy, the control of metabolism and the mechanisms of inheritance. Fungi have been the material with which many fundamental biological discoveries have been made. For example, at the end of the nineteenth century Buchner showed that yeast extracts could perform the conversion of sugar into alcohol, a process previously known only as an activity of the intact cell. The elucidation of the pathways involved was a major activity of biochemistry during the first quarter of the twentieth century. During the 1940s studies on nutritional mutants of the mould *Neurospora crassa* by Beadle and Tatum established the concept that an enzyme is specified by a gene, and founded biochemical genetics. In the 1950s work by Pontecorvo on *Aspergillus nidulans*, another mould, showed that genetic analysis could be carried out in the absence of the sexual process, and the methods of genetic analysis developed with this organism have subsequently been of great value in mapping human chromosomes. Currently fungi are used as model organisms to study the structure and function of genes. The sequencing of the genome of the yeast *Saccharomyces cerevisiae,* completed in 1996, contributed to a recognition that not only many genes, but also their cellular functions, were common to animals and fungi. The application of recombinant DNA technology to fungi, and their commonly haploid state, in which a change in a gene is not concealed by the activity of a dominant allele, increases the value of fungi for the analysis of fundamental cellular processes. Fungi have hence been of great value to biochemists and geneticists, who have in turn made important contributions to the study of fungi.

In addition to having a role in fundamental biological research, fungi are of great practical importance. In most natural ecosystems there are fungi associated with the roots of plants which help to take up nutrients from soil, and the decomposition of plant litter by fungi is an essential part of the global carbon cycle. Fungi cause some of the most important plant diseases, and hence receive much attention from plant pathologists. Some cause disease in man and domestic animals, so there are specialists in medical and veterinary mycology. Many cause spoilage of food, damage manufactured goods or cause decay of timber. These attract the attention of food microbiologists, experts in biodeterioration, and timber technologists, respectively. Fungi also fulfil many roles beneficial to humans. The larger fungi have been gathered for food from ancient times, but now *Agaricus bisporus* and a variety of other species are cultivated, and a branch of mycology termed mushroom science is seeking to improve the strains and methods used. Yeasts have been used for thousands of years in brewing and baking and the preparation of a variety of foods, and their study is a major aspect of research in brewing science and in food technology. The metabolic versatility of fungi is exploited by the fermentation industry, to make antibiotics and other high value substances of interest to medicine, agriculture and the chemical industry, to produce enzymes and to carry out specific steps in chemical processes. Recent developments in recombinant DNA technology (genetic manipulation or gene cloning) have led to fungi being used to produce hormones and vaccines hitherto available only from mammalian sources. Fungi are likely to remain of great practical as well as academic interest, and to attract the attention of scientists trained in a variety of disciplines.

Further Reading and Reference

General Works on Fungi

Alexopoulos, C. J., Mims, C. W. & Blackwell, M. (1996). *Introductory Mycology*, 4th edn. Wiley, Chichester.

Deacon, J. W. (1997). *Introduction to Modern Mycology*, 4th edn. Blackwell, Oxford.

Esser, K. & Lemke, P. (1993–). *The Mycota*. Springer-Verlag, Berlin.

Gow, N. A. R. & Gadd, G. M., eds. (1994). *The Growing Fungus*. Chapman & Hall, London.

Gravesen, S., Frisvad, J. C. & Samson, R. A. (1994). *Microfungi*. Munksgaard, Copenhagen.

Griffin, D. H (1994). *Fungal Physiology*, 2nd edn. Wiley-Liss, New York.

Hawksworth, D. L., ed. (1990). *Frontiers in Mycology*. CAB International, Wallingford.

Hawksworth, D. L., Kirk, P. M., Pegler, D. N. & Sutton, B. C. (1995). *Ainsworth & Bisby's Dictionary of the Fungi*, 8th edn. CAB International, Wallingford.

Hudson, H. J. (1986). *Fungal Biology*. Arnold, London.

Ingold, C. T. & Hudson, H. J. (1993). *The Biology of the Fungi*, 6th edn. Chapman & Hall, London.

Jennings, D. H. & Lysek, G. (1999). *Fungal Biology: Understanding the Fungal Lifestyle*, 2nd edn. Bios, Oxford.

Moore, D. (1998). *Fungal Morphogenesis*. Cambridge University Press, Cambridge.

Moore-Landecker, E. (1996). *Fundamentals of the Fungi*, 4th edn. Prentice-Hall, New Jersey.

Oliver, R. P. & Schweizer, M., eds. (1999). *Molecular Fungal Biology*. Cambridge University Press, Cambridge.

Webster, J. (1980). *Introduction to Fungi*, 2nd edn. Cambridge University Press, Cambridge.

Prokaryotes, Eukaryotes and Major Groups of Microorganisms

Barr, D. J. S. (1992). Evolution and kingdoms of organisms from the perspective of a mycologist. *Mycologia* **84**, 1–11.

Carlile, M. (1982). Prokaryotes and eukaryotes: strategies and successes. *Trends in Biochemical Sciences* **7**, 128–130.

Gooday, G. W., Lloyd, D. & Trinci, A. P. J., eds. (1980). *The Eukaryotic Microbial Cell. Symposium of the Society for General Microbiology*, Vol. 30. Cambridge University Press, Cambridge.

Gouy, M. & Wen-Hsiung Li (1989). Molecular phylogeny of the kingdoms Animalia, Plantae and Fungi. *Molecular Biology and Evolution* **6**(2), 109–122.

Lederberg, J., ed. (2000). *Encyclopedia of Microbiology*, 2nd edn. Academic Press, London.

Madigan, M. T., Martinko, J. M. & Parker, J. (2000). *Brock's Biology of Microorganisms*, 9th edn. Prentice-Hall, New Jersey.

Margulis, L. & Schwartz, K. V. (1998) *Five Kingdoms: an Illustrated Guide to the Phyla of Life on Earth*, 3rd edn. Freeman, New York.

Margulis, L., Corliss, J. O., Melkonian, M. & Chapman, D. J., eds. (1989). *Handbook of Protoctista: the Structure, Cultivation, Habitats and Life Cycles of Eukaryotic Microorganisms and their Descendants*. Jones & Bartlett, Boston.

Postgate, J. (2000). *Microbes and Man*, 4th edn. Cambridge University Press, Cambridge.

Roberts, D. Mc. L., Sharp, P., Alderson, G. & Collins, M., eds. (1996). *Evolution of Microbial Life*. Cambridge University Press, Cambridge.

Tudge, C. (2000). *The Variety of Life*. Oxford University Press, Oxford.

Whittaker, R. H. (1969). New concepts of kingdoms of organisms. *Science* **163**, 150–160.

The Study of Fungi: Methodology

Hawksworth, D. L. & Kirsop, B. E., eds. (1988). *Living Resources for Biotechnology: Filamentous Fungi*. Cambridge University Press, Cambridge.

Kirsop, B. E. & Doyle, A., eds. (1991). *Maintenance of Microorganisms and Cultured Cells*, 2nd edn. Academic Press, London.

Kirsop, B. E. & Kurtzman, C. P., eds. (1994). *Living Resources for Biotechnology: Yeasts*, 2nd edn. Cambridge University Press, Cambridge.

Paterson, R. R. M. & Bridge, P. D., eds. (1994). *Biochemical Techniques for Filamentous Fungi*. IMI Technical Handbooks 1. CAB International, Wallingford.

Smith, D. & Onions, A. H. S. (1994). *The Preservation and Maintenance of Living Fungi*, 2nd edn. Commonwealth Mycological Institute, Kew.

Stamets, P. (1993). *Growing Gourmet and Medicinal Mushrooms*. Ten Speed Press, Berkeley.

The Study of Fungi: History

Ainsworth, G. C. (1976). *Introduction to the History of Mycology*. Cambridge University Press, Cambridge.

Ainsworth, G. C. (1981). *Introduction to the History of Plant Pathology*. Cambridge University Press, Cambridge.

Ainsworth, G. C. (1986). *Introduction to the History of Medical and Veterinary Mycology*. Cambridge University Press, Cambridge.

Sutton, B. C., ed. (1996) *A Century of Mycology*. Cambridge University Press, Cambridge.

Journals and Serial Publications on Fungi and other Microbes

Advances in Microbial Ecology
Advances in Microbial Physiology
Annual Review of Microbiology
Annual Review of Phytopathology
Applied and Environmental Microbiology
Archives of Microbiology

Current Opinion in Microbiology
FEMS Microbiology Ecology
FEMS Microbiology Letters
FEMS Microbiology Reviews
Fungal Genetics and Biology
Journal of Bacteriology
Medical Mycology
Microbiological Reviews
Microbiology
Mycologia
Mycological Research
Mycologist
Mycorrhiza
Mycoses
Phytopathology
Studies in Mycology
Symposia of the British Mycological Society
Symposia of the Society for General Microbiology
Trends in Biochemical Sciences
Trends in Ecology and Evolution
Trends in Microbiology

Web sites

http://phylogeny.arizona.edu/tree/phylogeny
This site maintained at the University of Arizona incorporates molecular data as they become available to present a continually developing phylogenetic tree for the diversification of life throughout evolution. The Ascomycete and Basidiomycete branches of the tree are illustrated by pictures of fungi in different groups, and references are given to the scientific literature supporting the divergences within the tree.

http://mycology.cornell.edu/
This site at Cornell University provides an enormous collection of mycological links.

http://www.wisc.edu/botany/fungi/html
This is maintained by T. Volk of the University of Wisconsin and contains pictures of hundreds of different fungal fruiting bodies with accompanying details of special points of interest and biological significance of each.

www.THEFUNGI.com
A website maintained in connection with this book. Contains mycological links, images and list of book contents.

Fungal Diversity

2

The Classification of Fungi and Slime Moulds into Major Groups

For an organism to be formally recognized by taxonomists, it must be described in accordance with internationally accepted rules and given a **Latin binomial**, namely a generic name followed by a specific epithet. Man, for example, is *Homo sapiens*, and the cultivated mushroom, as indicated in the previous chapter, *Agaricus bisporus*. It was estimated in 1983 that about 64 000 fungal species were known, and in 1995 the estimate was 72 000, suggesting that about 700 new species are discovered each year. The number of species so far discovered, however, is probably only a small proportion of those that exist, as few habitats and regions have been intensively studied. Various approaches have been used in trying to estimate the number of fungal species in the world. For example, in well-studied regions fungal species can be six times as numerous as those of flowering plants. On this basis, since about 270 000 flowering plants are known, there may be about 1.6 million fungal species. Other approaches suggest similar or larger numbers.

A genus may contain one, a few or many species. Since there are such a large number of fungal species, there are also a large number of genera, most of which will be unfamiliar to any one mycologist. Hence categories intermediate in status between genus and kingdom are needed. Genera can be grouped into **families**, families into **orders**, orders into **classes**, and classes into **phyla** (sing., **phylum**) (Table 2.1). *Mucor mucedo*, for example, is in the class Zygomycetes, which is in the phylum Zygomycota. In a formal classification the ending -**mycetes** indicates a class, and -**mycota** a phylum. Standard endings are also used for most of the other formal taxonomic categories. Informal terms are also used. The Zygomycota, for example, are often referred to informally as **Zygomycetes**, although formally the Zygomycetes are the larger of two classes within the Zygomycota. In addition there are informal groupings, such as the yeasts, which do not correspond to any formal grouping.

Formal classifications and nomenclature change as new information becomes available, and at any one time there are differences of opinion as to classification and the names of categories. There is greater stability as regards informal nomenclature which will therefore generally be used in this book, although formal categories will be mentioned as necessary, and an outline formal classification is provided in Appendix 2.

Table 2.1 Categories in classification illustrated by reference to two Zygomycetes, *Mucor mucedo* (page 39) and *Glomus macrocarpum* (page 397)

Kingdom Fungi		
	Phylum Zygomycota	
	Class Zygomycetes	
Order Mucorales[a]		Order Glomales[b]
Family Mucoraceae		Family Glomaceae
Genus *Mucor*		Genus *Glomus*
Species *mucedo*		Species *macrocarpum*

[a] There are several families in the order Mucorales, and many genera in the family Mucoraceae. The genus given illustrates the use of the same name with different endings at several levels in a hierarchy.

[b] The order Glomales has only one family, but the family has several genera.

It is now clear that the fungi studied by mycologists include organisms from three Kingdoms, the **Chromista**, the **Protozoa** and the **Fungi**. The features that distinguish these three Kingdoms are set out in Table 2.2. Only the Kingdom Fungi consists exclusively of fungi. The Chromista are aquatic and mainly photosynthetic organisms. The latter, the algae, are studied by plant biologists and include the seaweeds, as well as filamentous and microscopic unicellular forms. The group of fungi informally called **Oomycetes** or water moulds have structural, genetic and biochemical features that have now established them as a phylum, the Oomycota, within the Chromista. They are common microfungi with many species and include important plant pathogens such as the organism that causes potato blight. They have motile spores which swim by means of two flagella, and grow as hyphae with cellulose-containing walls. In addition to the Oomycetes there are two other Chromistan groups of fungi which are also aquatic or parasitic and with motile spores; the Hyphochytriomycota and the Labyrinthulomycota. They have few species and are dealt with briefly in Appendix 2. The Protozoa are diverse unicellular organisms with separate lines of descent from a unicellular ancestor. The fungi known as slime moulds all belong to this kingdom. They do not form hyphae, lack cell walls during the phase in which they obtain nutrients and grow, and are capable of ingesting nutrients in particulate form by phagocytosis. The slime moulds hence fail to meet any normal definition of the fungi, and are now established as members of the Kingdom Protozoa, but they produce fruiting bodies which have a superficial resemblance to those of fungi. This resulted in slime moulds early attracting the attention of mycologists, and to their inclusion in most textbooks on mycology. Two of the best studied groups of slime moulds are the **cellular slime moulds** and the **plasmodial slime moulds** or **Myxomycetes**.

The Kingdom Fungi consists solely of species that are hyphal, or clearly related to hyphal species. Throughout most or all of their life cycle they possess walls which normally contain chitin but not cellulose, and they are exclusively absorptive in their nutrition. They are divided into four divisions as shown in Fig. 2.1. The Chytridiomycetes are the only group with motile cells (known as **zoospores**). The form taken by the sexual phase of the life cycle is an important criterion in the classification of fungi that lack zoospores. The sexual process leads to the production of characteristic spores in the different groups. The fungi

Table 2.2 Important groups of fungi

Features that may be shared by several groups	Group	Kingdom
Trophic phase[a] lacking cell walls. Able to ingest particulate food. **The slime moulds**	1. **Cellular slime moulds.** Amoebae aggregate to form a 'slug' which gives rise to a fruit body. Example: *Dictyostelium* 2. **Plasmodial slime moulds (Myxomycetes).** Amoeboid phase followed by a multinucleate plasmodial phase. Example: *Physarum*	Protozoa
Trophic phase with cell walls; nutrition exclusively absorptive. The fungi (in the informal sense)	*Many species produce zoospores* 3. **Oomycetes.** Zoospores biflagellate[b], sexually produced spores are oospores. Walls contain cellulose. Example: *Saprolegnia* 4. **Chytridiomycetes.** Zoospores uniflagellate[c]. Sexual process may involve fusion of motile gametes, walls contain chitin. Example: *Allomyces*	Chromista
	Zoospores not produced 5. **Zygomycetes.** Sexually produced spores are zygospores. Example: *Mucor* 6. **Ascomycetes.** Sexually produced spores are ascospores. Example: *Pyronema* 7. **Basidiomycetes** (mushrooms, toadstools, rusts and smuts). Sexually produced spores are basidiospores. Example: *Agaricus* 8. **Mitosporic fungi.** No sexually produced spores. Example: *Penicillium*	Fungi

[a] The phase concerned with nutrition and growth.
[b] Having two flagella.
[c] Having a single flagellum.

that form **zygospores** are classified as **Zygomycetes**, those that form **ascospores** as **Ascomycetes**, and those forming **basidiospores** as **Basidiomycetes.** The hyphae of Ascomycetes and Basidiomycetes have numerous cross-walls. Another feature widespread in Ascomycetes and Basidiomycetes is that when hyphae within a fungal colony come into contact they may fuse with each other. This hyphal anastomosis, if frequent, can convert the radiating hyphae of a colony into a three-dimensional network. Hyphal anastomosis, as indicated later, may be a major factor in permitting the mycelium of some Ascomycetes and Basidiomycetes to produce large fruit bodies. Cross-walls and hyphal anastomoses are largely lacking in the Zygomycetes and Chytridiomycetes. These organisms are sometimes termed the '**lower fungi**', in contrast to the '**higher fungi**', the Ascomycetes, Basidiomycetes and related forms. There is some justification for a loose distinction of this kind, in that the potentialities of hyphal and mycelial organization have been more fully exploited in the latter groups.

Many Ascomycetes and Basidiomycetes, in addition to producing spores by a sexual process, form other types of spores asexually. There are also many species, recognizable as higher fungi through the presence of cross-walls in their hyphae, that produce asexual spores but lack a sexual phase. These are known as **mitosporic fungi,** as all their spores are produced following mitosis but none by meiosis. They were formerly termed the **Deuteromycetes** or **Fungi Imperfecti**, and a Deuteromycete was reclassified as an Ascomycete or Basidiomycete if a sexual phase was discovered. However, analysis of DNA sequences now allows these asexual fungi to be classified with their closest sexual relatives, and it appears that they have arisen from many different groups of fungi by the loss of sexuality. They thus do not constitute a natural group, and ultimately they could all be assigned to Ascomycete or Basidiomycete groups.

Some important features of the fungal groups mentioned above are indicated in Table 2.2. Fig. 1.2 shows the relationships between the Kingdoms to which fungi belong, and other living organisms. The way in which the four major divisions of the Kingdom Fungi are thought to have diverged is shown in Fig. 2.1.

These groups also form the subject of the rest of the chapter, along with two additional groups, the yeasts and the lichens. **Yeasts** are fungi that are normally unicellular and reproduce by budding, although some will, under appropriate conditions, produce hyphae, just as some normally hyphal fungi may produce a yeast phase. Many yeasts have a sexual phase that enables them to be classified as Ascomycetes or Basidiomycetes, although some do not. It is useful, however, to deal with the yeasts, which have much in common with each other with respect to form, habitat, practical importance and methods of identification as a group, and many books are devoted to the subject of yeasts, as are some institutes. The **lichens** are intimate symbiotic associations of a fungus, nearly always an Ascomycete, with an algal or a cyanobacterial species. The fungal components of lichens can be assigned to order within the Ascomycetes, but it is often useful, on the basis of morphology, physiology and ecology, to consider the lichens as a group.

The eight groups listed in Table 2.2 along with the yeasts and lichens will now be considered in more detail.

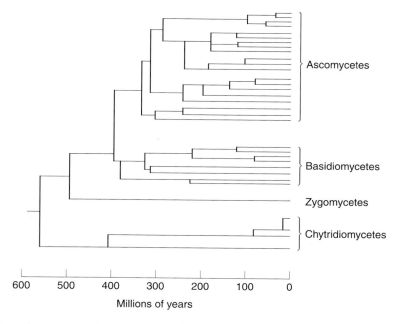

Figure 2.1 The pattern of divergence of fungal phyla. The tree was constructed from comparisons of 18S RNA gene sequences of over 30 species of fungi. The dates of divergences are based on the earliest appearance of key structures in the fossil record. For example, the oldest fossils showing clamp connections, diagnostic of Basidiomycetes, are almost 400 million years old. Dates of divergence available from the fossil record were used to deduce that the average rate at which the rDNA base sequence changes is 1% per hundred million years. Hence divergence times could also be theoretically assigned to branches in the phylogenetic tree for which there was no fossil evidence available. Both fossil and molecular data indicate that the diversification of Ascomycetes and Basidiomycetes has occurred in parallel with the diversification of a land flora during the last 500 million years, and that the first terrestrial fungi evolved about 550 million years ago. Fossil Ascomycete fruiting bodies resembling present day Pyrenomycetes have been found on a 400 million year old fossil plant. Based on Berbee, M.L. & Taylor, J.W. (2000), *The Mycota*, Springer-Verlag, Berlin; Taylor,T.N., Hass, H. & Kerp, H. (1999) *Nature* **399**, 648.

The Cellular Slime Moulds

The slime moulds are Protozoa that have been much studied by mycologists. The **cellular slime moulds** are so designated to contrast them with slime moulds that produce plasmodia. Two phyla are recognized, the **Acrasiomycota** and the **Dictyosteliomycota**. Members of the Acrasiomycota occur mainly on dung and decaying vegetation and will not be considered further. Members of the Dictyosteliomycota occur in soils throughout the world, especially in the surface soil and leaf litter of deciduous forests. Although they are common, only about fifty species are known. One of these, *Dictyostelium discoideum*, has been studied intensively by biologists and biochemists interested in cellular interaction and developmental processes. The life cycle of *D. discoideum* (Figs. 2.2, 2.3) will now be described.

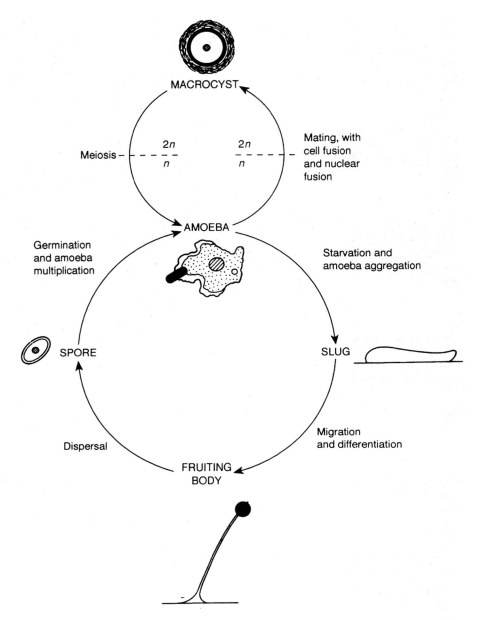

Figure 2.2   The life cycle of the cellular slime mould *Dictyostelium discoideum*. An amoeba is shown with a central nucleus and a contractile vacuole, and is ingesting a rod-shaped bacterium. When starvation occurs in a population of amoebae, aggregation gives slugs, which may consist of thousands of amoebae and be 1 mm long. After migration to a suitable site, slugs differentiate into a fruiting body consisting of a basal disc, a stalk, and a sporangium containing spores. If a spore reaches a suitable substratum, it germinates and an amoeba emerges. Amoebae that differ in mating type can mate to give diploid ($2n$) macrocysts, with loosely textured primary walls and more dense, secondary, inner walls. Macrocysts are capable of prolonged survival, and on germination undergo meiosis to give haploid (n) amoebae once more.

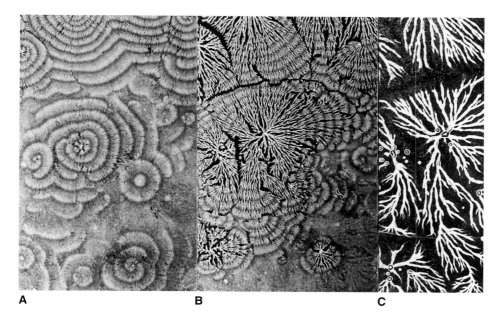

A   B   C

Figure 2.3 Amoeba aggregation in the cellular slime mould *Dictyostelium discoideum*. A, Centres of attraction have formed and are surrounded by bright zones of elongated amoebae moving towards centres, and dark zones of roughly circular, temporarily stationary amoebae. B, After about an hour the zones are breaking up into streams moving towards the centres. C, After a further hour all the amoebae have joined streams. Dark-field microscopy. (Reproduced with permission from Newell, P. C. (1981). Chemotaxis in the cellular slime moulds. In Lackie, J. M. & Wilkinson, P. C., eds, *Biology of the Chemotactic Response*, pp. 89–114. Cambridge University Press, Cambridge.)

The amoebae of *D. discoideum* can be grown readily in two-membered culture with a variety of bacteria such as *Escherichia coli*. An agar medium that will permit growth of the chosen bacterial species is spread with bacteria and inoculated with *Dictyostelium* amoebae. The bacteria multiply and the amoebae feed on them by phagocytosis, taking the bacteria into food vacuoles within which the bacteria are digested. Under suitable conditions nuclear division followed by cell division occurs about every 3 hours, so a large amoeba population is soon produced. Mutant strains of *D. discoideum* have been obtained that can be grown in pure culture on a complex soluble medium, but growth is slower with a generation time of about 9 hours. The nutrition of cellular slime mould amoebae is, however, mainly ingestive, particulate food normally being taken by phagocytosis.

The amoebae of *D. discoideum* are about 10 μm in diameter. In addition to food vacuoles a contractile vacuole is present, expelling excess water that has entered the cell by osmosis. There is a single haploid nucleus with seven chromosomes. The haploid DNA content is low in comparison with most eukaryotes, being about 12 times that of *E. coli*, or about 50 million base pairs. Genetic studies can be carried out by making use of the **parasexual cycle**.

Occasional cell fusion followed by nuclear fusion produces diploid amoebae. This is a rare event, occurring once in a population of 10^5 to 10^6 amoebae, but selective procedures have been devised for the ready isolation of such diploids. Such diploids can lose chromosomes one at a time to give aneuploids and finally the haploid condition. Such haploidization can be encouraged by selective procedures. The loss of whole chromosomes during haploidization makes the parasexual cycle very useful for assigning genes to linkage groups. Rather infrequent mitotic crossing over within linkage groups can also occur.

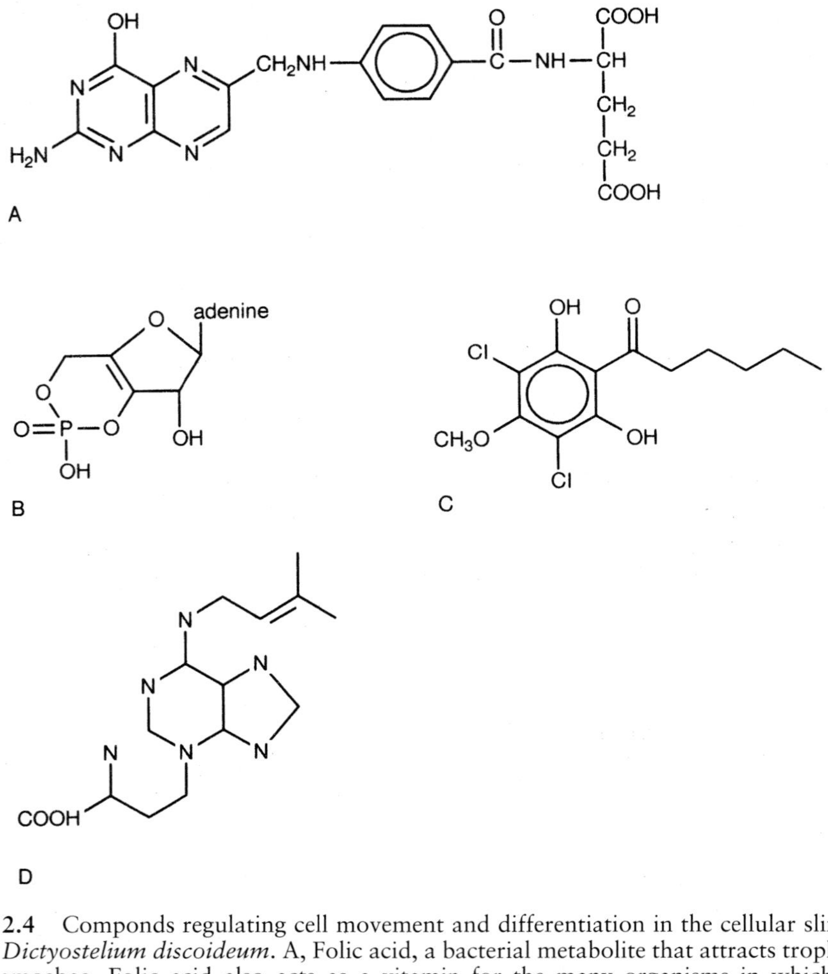

Figure 2.4 Componds regulating cell movement and differentiation in the cellular slime mould *Dictyostelium discoideum*. A, Folic acid, a bacterial metabolite that attracts trophic phase amoebae. Folic acid also acts as a vitamin for the many organisms in which a derivative of folic acid, tetrahydrofolate, is an enzyme co-factor in the metabolism of C_1 componds. B, Cyclic AMP ($3',5'$-cyclic adenosine monophosphate) is emitted by aggregating amoebae, attracting further amoebae into the aggregate. It also has an important role in metabolic regulation in both prokaryotes and eukaryotes. C, DIF, differentiation inducing factor, brings about the differentiation of amoebae at the anterior end of the slug into the stalk cells of the fruiting body. D, Discadenine prevents the premature germination of *D. discoideum* spores.

Efficient location of bacteria by amoebae is facilitated by chemotactic responses. Amoebae repel each other by a factor, as yet unidentified, that they release. They hence avoid high concentrations of amoebae, where there will be few surviving bacteria, and are dispersed to areas where bacteria are more likely to be present. They also show positive chemotaxis to a product released by bacteria, folic acid (Fig. 2.4A), and themselves release an enzyme which destroys folic acid. This presumably prevents the building up of a uniform background concentration of folic acid, which would not give any indication as to the direction of the folic acid source and the bacteria.

Ultimately amoebae exhaust the supply of bacteria in their vicinity. Their behaviour then changes. They cease to repel each other and cease to respond to folic acid. Instead some begin to emit cyclic AMP (3′,5′-cyclic adenosine monophosphate (AMP), Fig. 2.4B) and others respond to this substance by positive chemotaxis. Attractant centres are formed. These centres emit pulses of cyclic AMP every few minutes. Nearby amoebae respond by moving towards the centre for about 100 s, covering about 20 μm and also releasing a pulse of cyclic AMP which attracts amoebae further from the centre. After a refractory period amoebae recover cyclic AMP sensitivity and become ready to respond to a further pulse of cyclic AMP. Thus, a relay system operates that can attract amoebae a centimetre or more from a centre. Each centre becomes surrounded by a field of amoebae moving towards the centre. With dark field microscopy alternate zones of moving amoebae, which are elongated and bright, and stationary amoebae, roughly circular and dark, can be recognized (Fig. 2.3). Subsequently, the field breaks up into streams moving towards the centre. Finally, all the amoebae within range of a centre's influence reach the centre to produce an aggregate which, depending on the amoeba population at the time food was exhausted, may contain from a few hundred to a few hundred thousand cells.

The aggregation process in *D. discoideum* has been the subject of intensive study, and the sensory transduction path, from the binding of cyclic AMP at the surface of the plasma membrane to the movement of the amoeba in the direction from which a cyclic AMP pulse originated, is gradually being elucidated. Amoebae produce phosphodiesterases which destroy cyclic AMP. One is released into the medium and presumably prevents the build-up of a uniform background of cyclic AMP, the other is membrane bound and perhaps frees receptors from cyclic AMP thus permitting response to further pulses.

In some cellular slime moulds attractants other than cyclic AMP are responsible for aggregation. Cyclic AMP is, however, the attractant for several *Dictyostelium* spp. other than *D. discoideum*. There is hence the possibility that a single centre may attract more than one species. If this occurs a sorting process takes place, resulting in aggregates consisting of cells of only one species. This is the result of species specificity in cell adhesion, resulting from the release of species-specific proteins. A molecule of such a protein has two sites able to bind to surface polysaccharides on the amoebae of the producer species, and can thus cause adhesion between such cells but not those of other species. The specific proteins involved in cell adhesion in *D. discoideum* are called **discoidins**.

The mass of cells resulting from aggregation develops into a slug-like organism (sometimes termed a grex or pseudoplasmodium) which is enclosed in a slime sheath. The '**slug**', depending on the number of cells in the aggregate from which it originated, may be minute or as much as 1 mm in length. It can migrate for several days, and is positively thermotactic and phototactic, moving towards warmth and light. In nature this will help the slug to move through leaf litter or soil to a site on the surface suitable for the development of a fruiting body from which spores can readily disperse.

The **fruit body** consists of a basal disc (hence the specific epithet, *discoideum*), a multicellular stalk and a roughly spherical mass of spores, the sporangium. The stalk consists of cell wall materials, largely cellulose, secreted by the stalk cells before they die. During slug migration the cells that will become the stalk (the pre-stalk cells) are at the tip of the slug. The conversion of amoeboid pre-stalk cells into the vacuolate, walled, stalk cells is brought about by a **differentiation inducing factor**, DIF (Fig. 2.4C) produced at the tip of the slug. The fruiting bodies, as they rise from the substratum, avoid colliding with each other. This, and the adequate spacing of fruiting bodies, is due to the emission by the developing fruiting bodies of a volatile factor, ammonia, which repels other fruiting bodies. The spores can remain viable for several years, their premature germination either within the sporangium or in a dense mass being prevented by a germination inhibitor, discadenine (Fig. 2.4D). When spores are well dispersed the inhibitor is lost by diffusion. Germination is stimulated by amino acids, which will be encountered if a spore arrives in the vicinity of bacteria. A germinating spore swells, the spore wall ruptures, and an amoeba emerges and begins to feed on bacteria. Under unfavourable conditions the amoebae of some cellular slime moulds, but not *D. discoideum*, can develop cell walls to become **microcysts**. These cells, more resistant than amoebae, germinate when favourable conditions return.

The life cycle so far described can be accomplished by amoebae that remain haploid and constitute a clone, that is have originated from a single haploid cell. A sexual process, however, can be initiated by bringing together amoebae that differ in mating type. The mating type of a cell is determined by which of two alleles, *mat A* or *mat a*, is present. Cell clumping is brought about by a volatile factor, ethylene, released by *mat A* cells and acting on *mat a* cells. Cyclic AMP is then released attracting more cells into the clump. Within the clump, cell and nuclear fusion occurs between two cells of different mating type. The resulting zygote ingests and digests many of the surrounding cells to produce a large cell which develops a thick wall to become a **macrocyst**. Under suitable conditions the macrocyst germinates with meiosis occurring and haploid amoebae being released.

The Plasmodial Slime Moulds (Myxomycetes)

Another protozoan phylum studied by mycologists is the **Myxomycota** (**plasmodial slime moulds**). Two classes are recognized, the Protosteliomycetes, which will not be considered further, and the **Myxomycetes**. Members of the latter, much larger, class produce fruit bodies visible to the naked eye. They are usually found on dead plant materials and have been collected by naturalists for over a century. Although not obtrusive, they are common, and an observant collector can obtain a dozen or more species in a visit of a few hours to woodland. The fruiting bodies are made of durable materials, and museum specimens can remain in good condition for many years.

The special feature of the Myxomycete life cycle is the **plasmodium**, a multinucleate mass of protoplasm not subdivided into cells. It is a transient stage in the life cycle, so is less often seen in nature than are the fruiting bodies. Some species have only minute plasmodia, but in others the plasmodium can reach the size of a dinner plate; such plasmodia are occasionally seen in nature as a slimy yellow mass on decaying wood. As with the cellular slime moulds, there is an amoeboid phase in the life cycle. Myxomycete amoebae, however, differ from those of the cellular slime moulds in that they can produce flagella and swim. The amoebae of Myxomycetes are common in soil and in decaying timber. Both amoebae and plasmodia are phagotrophic, in this resembling protozoa rather than true fungi.

About 700 Myxomycete species are known, and of these about 300 are in the order Physarales, which have large plasmodia. A few isolates of one species, *Physarum polycephalum*, have been extensively used for research in cell and molecular biology. Studies on genetic variability in nature have been carried out mainly with a second species, *Didymium iridis*. An account of the life cycle (Figs. 2.5, 2.6) of *P. polycephalum* follows.

The **amoebae** of *P. polycephalum* feed on bacteria, and like those of *Dictyostelium* are readily grown in two-membered culture with *Escherichia coli*. They have been grown in pure culture on a complex medium, but multiply more slowly. The nucleus has a prominent nucleolus, and food vacuoles and a contractile vacuole are present. If a culture is flooded with water, the amoebae elongate and turn into **flagellates**. The flagella are smooth and emerge from the anterior end of the cell. There are usually two flagella, but whereas one, pointing forward, is active, the other commonly points backward, is held close to the cell surface, and is inactive. The flagellates neither feed nor undergo cell division, and when free water disappears, revert to the amoeboid state. When all the bacteria present have been consumed, the amoebae turn into thick-walled **cysts**, which can survive for a long period in the absence of nutrients. In the presence of bacteria a cyst will germinate and an amoeba emerge. The amoeboflagellate phase is hence able to multiply in the presence of bacteria, swim when flooding occurs, and survive periods without nutrients. It can be maintained indefinitely in two-membered culture and could probably similarly persist in nature; amoeboflagellate protozoa are common in soil and water.

Initiation of the **plasmodial phase** usually requires mating between two genetically different strains. Plasmodium initiation occurs with the highest frequency if the two strains differ at two loci, designated *mat A* and *mat B*. At each locus many alleles have been found so the number of possible mating types is large. If two amoebae differ at the *mat B* locus the probability that they will fuse to give a diploid cell is greatly increased. The probability that a diploid cell (zygote) will develop into a plasmodium is greatly increased if it contains two different *mat A* alleles. Although plasmodium formation is most likely if amoebae differing at both mating type loci are brought together, plasmodium formation can occur at a much lower frequency when there are differences at only one of the

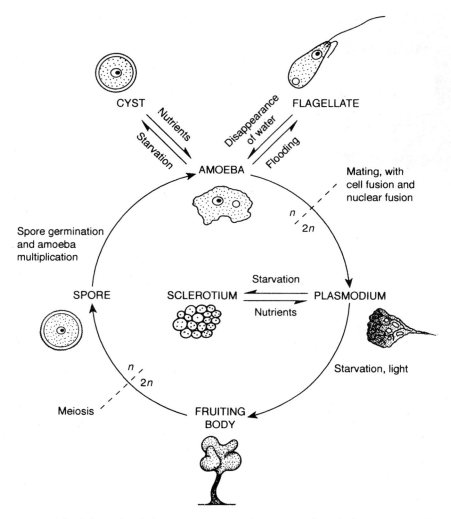

Figure 2.5   The life cycle of the Myxomycete *Physarum polycephalum*. An amoeba is shown with a contractile vacuole and a nucleus with a prominent nucleolus, typical of *Physarum*. Starvation results in cyst formation, and free liquid in conversion to flagellates. Cell and nuclear fusion between haploid (n) amoebae of different mating type initiates the formation of diploid ($2n$) plasmodia, which grow large and show rapid protoplasmic streaming in prominent veins. Starvation of plasmodia may cause the formation of sclerotia consisting of multinucleate spherules, but accompanied by exposure to light, results in fruit body development. Meiosis occurs during spore formation, and under suitable conditions spores germinate to give haploid amoebae. The life cycle of the apomictic strain is identical except that there is no mating or ploidy change.

loci, and very occasionally, within a clone. *P. polycephalum* is hence normally **self-sterile** (heterothallic), and has diploid plasmodia, since zygote production is involved in their origin. There are also, however, facultatively **apomictic** strains. In these a mutation at the *mat A* locus enables amoebae to give rise, without mating, to plasmodia which are haploid. The amoebae of these strains can also

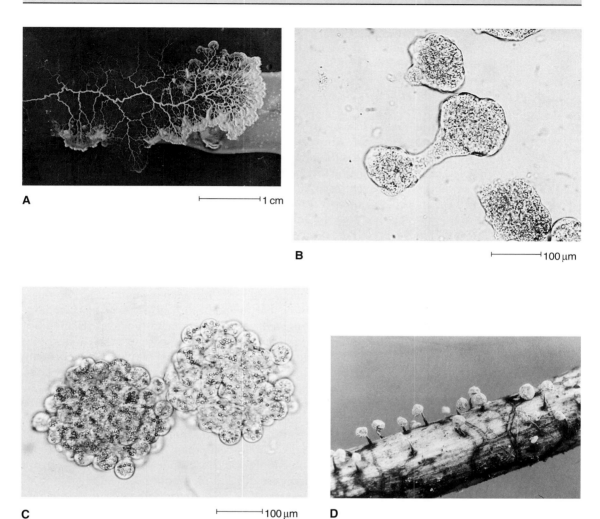

A ⊢————————⊣1 cm

B ⊢————————⊣100 μm

C ⊢————————⊣100 μm **D**

Figure 2.6 Some stages in the Myxomycete life cycle. A–C, *Physarum polycephalum*. A, Plasmodium migrating along and ingesting a streak of washed yeast cells on water agar. Plasmodia in pure culture on an agar medium are shown in Fig. 5.5. B, Microplasmodia in shaken liquid culture. C, Microsclerotia, consisting of numerous multinucleate spherules, formed in liquid culture after nutrient exhaustion. Phosphate storage granules are present in the spherules. D, Fruit bodies of *Physarum viride* on dead plant material. Tracks left by plasmodia as they migrated to exposed sites suitable for fruit body formation and spore dispersal are visible. (A–C from Carlile M. J. (1971). Myxomycetes and other slime moulds. In Booth, C., ed., *Methods in Microbiology* vol. 4, pp. 237–265. Academic Press, London. D, John and Irene Palmer.)

mate with other strains to produce diploid plasmodia. Studies on other Myxomycetes have demonstrated **self-fertile** (homothallic) species, in which mating occurs within a clone. The sexual behaviour of Myxomycetes is thus very flexible. Not only are self-sterile, self-fertile and apomictic species known but all three forms of behaviour have been found among strains of a single species, *Didymium iridis*.

A zygote grows and undergoes mitosis without cell division occurring. Since nuclear division is synchronous, the young plasmodium has in turn 2, 4, 8 and 16 nuclei. Uncountably large numbers are soon reached and plasmodia may cover many square centimetres and contain millions of nuclei. Plasmodial nuclear divisions differ from those of amoebae in that the nuclear membrane remains intact. Perhaps in a multinucleate cell the open type of mitosis might risk 'mix-ups' between nuclei. Plasmodia increase in size by fusion with each other as well as by growth. As the size of plasmodia increases protoplasmic streaming becomes more pronounced, and ultimately develops into the shuttle streaming in well-defined channels ('veins') characteristic of the mature plasmodium. Observation of a 'vein' with the microscope shows a torrent of protoplasm moving in one direction at speeds of up to 1 mm s^{-1} for about a minute, after which flow occurs in the opposite direction for about a minute. Presumably it is this shuttle streaming in the branching pattern of veins, partly overcoming the limitations of diffusion as a means of transporting oxygen, nutrients and wastes, which makes possible the development of a 'cell' as enormous as the plasmodium.

Plasmodia can be grown in pure culture on a soluble medium containing a suitable carbohydrate (e.g. glucose, starch), nitrogen source (e.g. peptone), mineral salts and vitamins (thiamine, biotin and haem), either on an agar medium or in shaken liquid culture. Provided with adequate nutrients a plasmodium grows, with a doubling time of about 8 hours with respect to mass, and slowly spreads. If starved, a plasmodium migrates, and is attracted by various nutrients including many carbohydrates. In nature plasmodia can surround and digest large fungus fruit bodies, and in the laboratory phagocytosis of smaller organisms can be demonstrated. Shaken liquid culture, however, shows that plasmodia are capable of absorbing nutrients and in nature plasmodial nutrition is probably partly ingestive and partly absorptive. The plasmodium, as a result of its size, is able to move further and attack far larger organisms than can amoebae. Plasmodia are surrounded by a mucopolysaccharide slime sheath which probably gives some protection from desiccation and aids locomotion.

When genetically identical plasmodia meet, they fuse to form a single plasmodium. If, however, the plasmodia belong to different strains, then fusion may not occur, or if it does, the nuclei of one strain may be eliminated, sometimes with considerable destruction of protoplasm. This vegetative incompatibility and its genetic basis, is considered in more detail later (page 259, Fig. 5.5).

Starvation of plasmodia can have two consequences, depending on whether or not light is present. In darkness a starved plasmodium becomes a **sclerotium**, which consists of numerous spherules, each thick-walled and containing several nuclei and protoplasm. The sclerotia can survive for some years, but in the presence of nutrients the spherules germinate and the protoplasts merge to form a plasmodium. A starved plasmodium in light gives rise to a **fruit body** containing spores. Meiosis occurs during sporulation to return the organism to the haploid state; in apomictic strains there is a pseudomeiosis which does not involve a reduction in ploidy. Spores are capable of prolonged survival and are readily dispersed, and if they arrive at a suitable site germinate to release amoebae.

The remarkable properties of the Myxomycete plasmodium have been extensively exploited in fundamental biological research. The synchronous division of millions of nuclei provides unequalled opportunities for studying

DNA, RNA and protein changes throughout the nuclear cycle, and the control of mitosis. The rapid protoplasmic streaming has led to investigations on the molecular basis of cell motility, and the complex life cycle to a variety of studies on genetics and developmental biology.

The Oomycetes

The group of organisms informally known as **Oomycetes** are now recognized as constituting a phylum, the **Oomycota**, in the kingdom Chromista. A remarkable product of parallel evolution, they so resemble the true fungi in structure and life style that they have always been studied by mycologists and were until recently regarded as fungi. About 700 species are known. Their sexual phase has a clear differentiation into large female and small male structures, termed **oogonia** (sing. **oogonium**) and **antheridia** (sing. **antheridium**), respectively. Within an oogonium meiosis occurs and, depending upon the species, one or a few **oospheres**

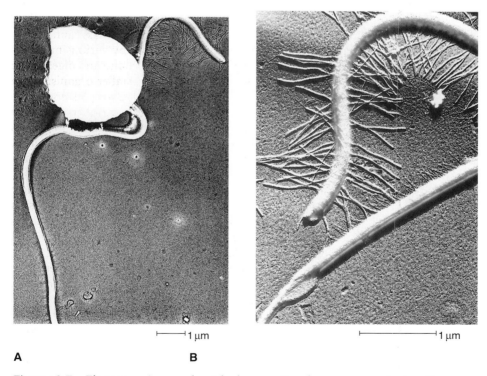

├────── 1 μm   ├────── 1 μm

A   **B**

Figure 2.7 Electron micrographs (shadow cast) of zoospores of the Oomycete *Phytophthora palmivora*. A, Zoospore with a short anterior flagellum bearing mastigonemes (stiff lateral hairs) and a long, nearly smooth posterior flagellum. B, Details of flagella. The anterior bears prominent mastigonemes and the posterior has just perceptible fine hairs. (Reproduced with permission from Desjardins, P. R., Zentmyer, G. A. & Reynolds, D. A. (1969). Electron microscopic observations of the flagellar hairs of *Phytophthora palmivora* zoospores. *Canadian Journal of Botany* 47, 1077–1079).

Table 2.3   Important differences between the Oomycetes and Fungi

	Oomycetes	The Fungi
Zoospores	Biflagellate; an anterior tinsel and a posterior smooth flagellum	Uniflagellate; a posterior smooth flagellum in the Chytridiomycetes
Lysine biosynthesis	Via diaminopimelic acid	Via α-aminoadipic acid
Mitochondria	Cristae tubular	Cristae plate-like
Wall polysaccharides	Cellulose present; chitin also in some species	No cellulose; chitin usually present
Wall proteins	Hydroxyproline present	Proline present

(unfertilized eggs) are produced. Each contains, when mature, a single haploid nucleus. Meiosis also occurs in antheridia, and the antheridia grow towards oogonia. From the antheridia **fertilization tubes** penetrate oogonia, and a single fertilization tube enters each oosphere. A single haploid nucleus passes from the antheridium through the fertilization tube, and fuses with the haploid nucleus in the oosphere. The oosphere then develops into the **oospore** (fertilized egg) characteristic of the class. Each oospore has a single diploid nucleus. When the oospore germinates it gives rise to a mycelium that is diploid, in contrast to the haploid mycelium of most other fungi.

Another characteristic of the Oomycetes that distinguishes them from the Fungi is the **biflagellate zoospore** (Figs. 2.7, 4.15). This has a forwardly directed **tinsel flagellum** – a flagellum with a row of hairs on each side when viewed with the electron microscope – as well as a posteriorly directed **smooth flagellum**. Ultrastructural and biochemical studies have shown that the Oomycetes differ from the Fungi in further fundamental ways (Table 2.3).

Three Oomycete orders, the **Saprolegniales**, **Pythiales** and **Peronosporales**, have attracted considerable attention, and will be considered further.

The Saprolegniales (Water Moulds)

Members of the order Saprolegniales, often referred to as the water moulds, are common in fresh water and also occur in damp soil. The life cycle (Fig. 2.8) of the common genus *Saprolegnia* will be described, along with features of interest from other genera.

Saprolegnia is common in soil and fresh water. Many species are saprotrophic, living on animal and plant remains, but a few are parasites, killing fish and destroying fish eggs. Hyphae of *Saprolegnia* are sometimes seen emerging from

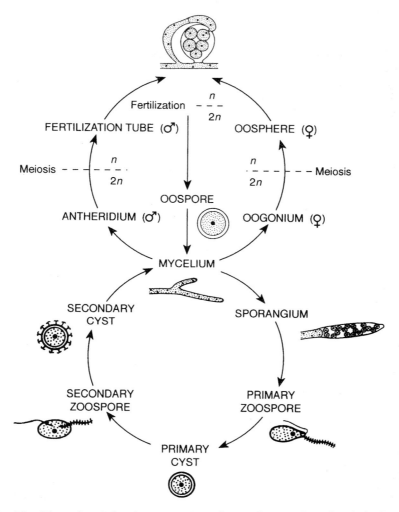

Figure 2.8 The life cycle of the Oomycete *Saprolegnia ferax*. When the diploid (*2n*) mycelium ceases growth through nutrient exhaustion, septa form delimiting sporangia at hyphal tips. A sporangium is shown just prior to the breakdown of its tip and the discharge of zoospores. Both primary and secondary zoospores have a smooth and a tinsel flagellum, but the point of insertion differs. Secondary cysts have hooks which probably facilitate attachment to suitable surfaces. Meiosis occurs in antheridia and oogonia to give haploid (*n*) nuclei, and oospheres are delimited within oogonia. Antheridia grow on to the surface of oogonia, and produce fertilization tubes that penetrate oogonia and fuse with oospheres. Here a single fertilization tube is shown contributing a nucleus to an oosphere. Fusion between antheridial and oosphere nuclei results in the formation of diploid (*2n*) oospores.

sick or dead goldfish in aquaria. Crude cultures of *Saprolegnia* and other water moulds can be obtained by placing boiled split hemp seeds, ants' 'eggs' or other forms of 'bait' in pond water samples; the bait soon supports growth of a water mould. After contaminating bacteria are eliminated, most water moulds grow readily in pure culture on common liquid or agar media, such as glucose–yeast extract–peptone–mineral salts agar. Water moulds cannot reduce sulphate, so in defined media sulphur is best supplied in a reduced organic form such as methionine. Glucose is a good carbon source, and nitrogen can be supplied as an ammonium salt or as amino acids. Vitamins need not be supplied.

The hyphae of *Saprolegnia* are, compared with most fungi, large and extend rapidly. They taper towards the tip, and the maximum width is only attained at some distance behind the tip. The hyphae lack cross-walls, as do those of other Oomycetes, and nuclei are scattered through the cytoplasm. Prominent protoplasmic streaming in the direction of the tip occurs. The number of branches depends on nutrition, high nutrient concentrations increasing branch frequency. Old cultures may develop resting structures termed **gemmae** (sing. **gemma**) which germinate under suitable conditions.

Starvation conditions can initiate **sporangium** production, and ultimately release of zoospores. In the laboratory this can be achieved by replacing a liquid nutrient medium by distilled water or a dilute salts medium; after sporangia have developed zoospore discharge can often be brought about by a further replacement of the distilled water or salts medium. A sporangium is a swollen hyphal tip, separated from the rest of the hypha by a cross-wall; zoospores form inside the sporangium. Zoospores when first discharged from *Saprolegnia* sporangia are pear-shaped with flagella emerging at the front. These **primary zoospores** swim rather sluggishly and soon develop thin cell walls to become spherical **primary cysts**. The cysts germinate to release kidney-shaped **secondary zoospores** with laterally emerging flagella. These swim much more vigorously and may form **secondary cysts** only after some hours. Secondary zoospores show a chemotaxis towards salts which is enhanced by the presence of amino acids. Perhaps the role of the primary zoospore is to swim away from the sporangium and the static water adjacent to the fungus, so that the primary cyst may be dispersed by currents. The chemotaxis of the secondary zoospore may assist in reaching hosts or nutrient materials and it appears that the secondary cysts, copiously provided with spines and hooks as revealed by electron microscopy, may be well adapted for attachment to a host. Secondary cysts normally germinate by producing a slender hypha, the **germ-tube**, which shows chemotropism (oriented growth) towards amino acids, facilitating penetration of nutritious substrata.

There is much variation in the Saprolegniales, both within and between species and genera, in the pattern of zoospore activity and encystment indicated above. In *Saprolegnia* the secondary cyst may sometimes germinate to produce a zoospore instead of a germ-tube. Such a 'tertiary' zoospore resembles the secondary zoospore in form and activity. The primary zoospores of *Achlya* commonly encyst immediately on discharge, forming at the exit pore of the sporangium a cluster of cysts from which secondary zoospores emerge. In *Thraustotheca* the primary zoospores encyst within the sporangium and are released by sporangium rupture. In *Dictyuchus* the primary zoospore phase is suppressed; the developing zoospores encyst within the sporangium and the cysts germinate to discharge secondary zoospores, each through a separate pore in the sporangium wall. In *Pythiopsis* it is the secondary zoospore that is suppressed, the primary cyst germinating to give a germ-tube. Some fungi, known to be Saprolegniales from their sexual morphology, do not produce sporangia, and others produce within their sporangia non-motile spores that have presumably evolved from cysts. The motile phase of the zoospore life cycle is clearly one that shows flexibility within a species according to requirements, and in the course of evolution is eliminated or modified in response to natural selection.

Most water moulds are **hermaphrodite** or **monoecious**, bearing both male and female reproductive structures on a single diploid mycelium. Oogonia are usually sited singly, often at the tips of hyphal branches, and depending on species may contain one or a few oospheres. Antheridia may develop on the same hypha as an oogonium and close to it, or on other hyphae. Monoecious species are **self-fertile** (**homothallic**). The oospores that develop from fertilized oospheres are thick walled and can survive for long periods. They germinate by means of a germ-tube, which may develop into a mycelium, or may terminate in a small sporangium from which zoospores are discharged.

There are a few species that are **self-sterile** (**heterothallic**), with the cooperation of two mycelia being needed for the sexual process to occur. When two compatible mycelia are brought together, one produces male, and one female structures. Such species are referred to as **dioecious**. The dioecious condition facilitates research on the sexual process, and two dioecious species, *Achlya bisexualis* and *Achlya ambisexualis*, have been employed in such studies. In these species mycelium that has arisen from a single oospore will not form sexual structures. If, however, culture filtrate from a 'female' strain is added to a 'male' strain, **antheridial initials** (incipient antheridia) develop. This is due to a steroid sex hormone, hormone A or antheridiol (Fig. 2.9A) which is produced by the female strain and is active at very low concentrations. When antheridial initials have been induced in the male strain, it in turn releases a steroid sex hormone, hormone B or oogoniol (Fig. 2.9B), which induces **oogonial initials** in the female strain. Both antheridiol and oogoniol are synthesized by the fungus from fucosterol, the most abundant sterol in *Achlya* and other Oomycetes. Antheridial initials grow towards oogonia by chemotropism up a concentration gradient of antheridiol. At one time the occurrence of hormones additional to hormones A and B and acting later in the sexual process was postulated; the occurrence of such hormones is now thought unlikely. The distinction between male and female strains is not absolute. A comparison of many strains demonstrates the existence not only of 'strong males' that act only as males, and 'strong females' that act only as females, but also of intermediate weak male and weak female strains. Such strains will act as male when paired with a strong female, or female when paired with a strong male. The physiological basis for this series of strains seems to be in the amounts of hormone A produced, with a strong female producing most hormone A and a strong male the least. Relative sensitivity to hormone A may also be important.

Monoecious species of *Achlya* and *Thraustotheca* have been shown to respond to hormones from *Achlya ambisexualis* and *A. bisexualis*, and to produce factors initiating antheridium development in dioecious species. It is likely, therefore, that sexual development in monoecious species is controlled by the same (or similar) hormones as in dioecious species. Under some conditions of nutrition and temperature a normally dioecious species may develop both oogonia and antheridia on a single mycelium. It seems likely that monoecious and dioecious species are closely related and that the evolution of one type from the other could

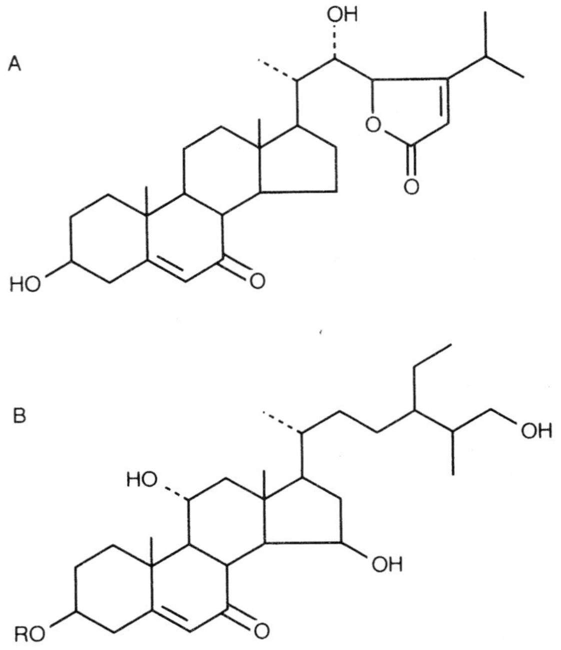

Figure 2.9   The sexual hormones of the Oomycte *Achlya*. A, Antheridiol ($C_{29}H_{42}O_5$: MW 470) is a sterol produced by the female strain. It induces antheridial initials in the male strain at concentrations as low as 2×10^{-11} M. B, The sterol oogoniol and its esters (R = H, acetate, propionate or isobutyrate residues) are produced by a male strain that has developed antheridial initials in response to antheridiol. The major component of the hormone mixture is the isobutyrate ester ($C_{33}H_{54}O_6$: MW 546). Oogoniol and its esters induce oogonial initials in the female strain. They are less potent than antheridiol, with the most active component of the mixture inducing sexual development in the female at concentrations of ca 10^{-7} M.

readily occur. Although good progress has been made in understanding hormonal and physiological aspects of sexuality in water moulds, the genetic basis of their sexuality remains to be clarified.

The Pythiales

One of the families in the order Pythiales, the Pythiaceae, includes two of the best known fungal genera, *Pythium* and *Phytophthora*.

Species of *Pythium* occur in fresh water and soil and can live as saprotrophs on plant debris, but some species under suitable conditions will attack plants, and are hence facultative parasites. *Pythium ultimum* and *Pythium debaryanum* can attack a wide variety of seedlings if they are overcrowded and under too moist conditions, causing 'damping-off' and extensive death. The hyphae of *Pythium* in their invasion of plant tissues are both intercellular and intracellular, passing between and penetrating through plant cells, and pectic enzymes diffuse ahead of the hyphae, dissolving the middle lamella of cell walls and softening the tissues. The life cycle of *Pythium* is similar to that of *Saprolegnia*. Sporangia are formed, commonly at the tips of hyphal branches, and zoospore maturation occurs in a thin-walled vesicle which emerges from the sporangium under appropriate conditions. The zoospores resemble the secondary zoospores of *Saprolegnia* and in some species have been shown to be attracted by root exudates. After a period of swimming, or on arrival at a suitable substratum, encystment and germ-tube emergence occurs. Many *Pythium* species are self-fertile, forming antheridia and oogonia and successfully producing oospores in cultures that have been derived from a single zoospore. Others, however, tend to be self-sterile. In *Pythium sylvaticum* there are predominantly male and predominantly female strains and the sexual process occurs most readily when the two types are brought together, although most of the male and a few of the female strains can if necessary be self-fertile.

Phytophthora (Greek: *phyton*, plant; *phthora*, destroyer) is a genus of outstanding importance to plant pathologists, killing or otherwise damaging a wide variety of economically important plants. There are some aquatic species, but it is the terrestrial species that are of economic importance. Many species can attack a broad range of hosts. *Phytophthora cinnamomi*, for example, can infect many woody plants, including conifers, and in recent years has devastated the native eucalypt forests of Australia. At the other extreme *Phytophthora infestans* is limited to the Solanaceae (the potato and tomato family) and is best known as the cause of late blight of potato, a disease responsible for a famine in Ireland in the 1840s and still of economic importance. *Phytophthora* hyphae differ in their behaviour from those of *Pythium* in that they are intercellular, passing between plant cells but producing short specialized branches, *haustoria*, which penetrate the cells. The sporangia of *Phytophthora* are borne on specialized branched hyphae, the sporangiophores, which typically emerge from the infected plant and extend *ca* 200 µm into the air. This facilitates the dispersal of the sporangia which are detachable in terrestrial species when mature. Such sporangia may behave as conidia and produce a germ-tube (direct germination) or may release zoospores (indirect germination). Which occurs depends on environmental conditions, low temperatures favouring zoospore release. Thus in *P. infestans* zoospores are produced at 15°C and germ-tubes at 20°C, behaviour which is understandable as at the lower temperature, free water – from dew or rainfall – in which zoospores can swim will persist longer. Chemotaxis of zoospores to plant exudates has been shown in several *Phytophthora* species. In some species the sexual phase can be produced in cultures derived from single zoospores whereas other species are normally self-sterile and it is necessary to bring together two mating types, designated A_1 and A_2. Oospores are thick-walled and probably are capable of prolonged survival in soil at times when there are no suitable hosts for infection. Mycelium too has some capacity for survival on plant debris.

Pythium and *Phytophthora*, like members of the Saprolegniales, can be grown in pure culture, although they tend to be nutritionally more demanding. A remarkable feature of the Pythiaceae, and one in which they are apparently unique among eukaryotic organisms, is that although they cannot synthesize sterols, vegetative growth can occur in the absence of added sterols. There is, however, a requirement for sterols for both sporangium production and the sexual process. Sterols are thought to endow plasma membranes with greater strength; perhaps this is essential for the zoospore membrane but not for the membranes of hyphal cells that are protected by a tough cell wall. The role of sterols in the sexual process may be as precursors for sex hormones similar to those known in *Achlya*.

The Peronosporales

This order includes two families, the Peronosporaceae (downy mildews) and the Albuginaceae (white rusts), that are obligate parasites of vascular plants. Examples of downy mildews are *Plasmopara viticola*, which causes an important disease of the grape vine, and *Peronospora parasitica*, which attacks members of the Cruciferae (the cabbage family). The white rust *Albugo candida* also attacks Cruciferae. The downy mildews and white rusts are biotrophs, obtaining their nourishment only from living cells, in contrast to the Pythiaceae, which are usually necrotrophs, killing cells and then obtaining nourishment from them. Although downy mildews and white rusts cannot be grown in pure culture, some species have been grown in association with cultured cells of the relevant host plant. Zoospores remain of importance in some species, but in others, such as *P. parasitica*, the sporangia function solely as conidia.

     The Peronosporales can be arranged as a series with respect to their parasitic activity. *Pythium* has considerable saprotrophic ability, and as a parasite has a wide host range and causes rapid death. *Phytophthora* has little saprotrophic ability, but as a parasite has a greater degree of host specialization and some capacity for obtaining nourishment from living cells. Finally, the downy mildews and white rusts are obligately biotrophic parasites, usually with a limited host range and rarely killing host plants. A parallel series can be traced from aquatic to fully terrestrial species. These series probably also reflect an evolutionary trend from aquatic saprotrophs, through facultative necrotrophs with a considerable dependence on damp conditions, to fully terrestrial biotrophs.

The Chytridiomycetes

The **Chytridiomycota**, more informally the **Chytridiomycetes**, are a phylum in the Kingdom Fungi. The main feature that occurs in Chytridiomycetes, and is absent from the other fungi that remain to be considered, is the **zoospore**, which has a single smooth posterior flagellum. This could well be a primitive feature of the fungi, lost in the groups that adopted a more terrestrial life style.

     The Chytridiomycota, of which about 800 species are known, are classified into five orders. Three of these orders, the **Blastocladiales**, the **Chytridiales** (chytrids) and the **Neocallimastigales** (anaerobic rumen fungi) will be considered further.

The Blastocladiales

Members of the order Blastocladiales are mainly saprotrophs living on plant or animal debris in fresh water, mud or soil. Most produce true mycelium, which branches dichotomously and is divided into compartments by occasional partitions, **pseudosepta**, differing in chemical composition from the hyphal wall. The sexual process consists of the fusion of a pair of uninucleate gametes, both of which are motile. Two genera, *Allomyces* and *Blastocladiella*, which have been used extensively in fundamental research, are considered in more detail below. Another genus, *Coelomomyces*, which will not be considered in detail, is an obligate parasite of arthropods. In the course of its life cycle it has to infect in turn mosquito larvae and adult copepods; it may thus prove useful for the biological control of mosquitoes and hence malaria.

Allomyces macrogynus, which occurs on organic debris in ponds and in soil, is readily grown in agar or liquid culture on a yeast–peptone–soluble starch medium or on suitable defined media. The life cycle (Figs. 2.10, 2.11C) has both haploid and diploid mycelial phases. The latter is a convenient point at which to begin a description of the life cycle. The diploid mycelium is often termed the **sporophyte** ('spore-producing plant'), since when nutrients are exhausted it develops sporangia. Zoospores are formed within the sporangia, which hence may be termed **zoosporangia**. The zoosporangia are of two types. In one type, the thin-walled **mitosporangia**, all nuclear division is by mitosis, and diploid zoospores (**mitospores**) are produced. Zoospore release can be obtained by immersing sporangia in pond water or in very dilute mineral salts solution – distilled water is harmful to the zoospores, which are not protected by a cell wall. The zoospores, which swim by means of a single posterior flagellum, have a **nuclear cap**, consisting of a store of ribosomes enclosed within a membrane. This feature, confined to the Blastocladiales, presumably facilitates prompt protein synthesis and rapid growth when germination occurs. The zoospores show chemotaxis to amino acids, and on reaching a suitable substratum immediately encyst and germinate. Germination is bipolar. A rhizoid emerges, penetrates the substratum and then repeatedly branches giving finer rhizoids. Then from the opposite side of the cyst a stout hypha emerges and by repeated dichotomous branching produces a new sporophyte mycelium.

The other type of zoosporangia produced by the sporophyte mycelium is thick-walled **meiosporangia**. Within these meiosporangia, which remain dormant for some weeks, meiosis occurs, and when germination takes place, haploid zoospores (**meiospores**) are released. These behave similarly to mitospores, but following encystment give rise to haploid mycelium which is referred to as the **gametophyte** ('gamete-producing plant') because, on nutrient exhaustion, it gives rise to **gametangia** within which gametes develop. The male gametangia occur at the end of hyphae, with the female gametangia immediately below, whereas in another species, *Allomyces arbusculus*, the opposite arrangement occurs. The female gametangia are colourless, and the female gametes (Fig. 2.11C) resemble zoospores in size and form. The male gametangia and the male gametes, which are smaller than the female gametes, are bright orange with γ-carotene. Exposure to pond water or very dilute mineral salt solutions brings about gamete discharge.

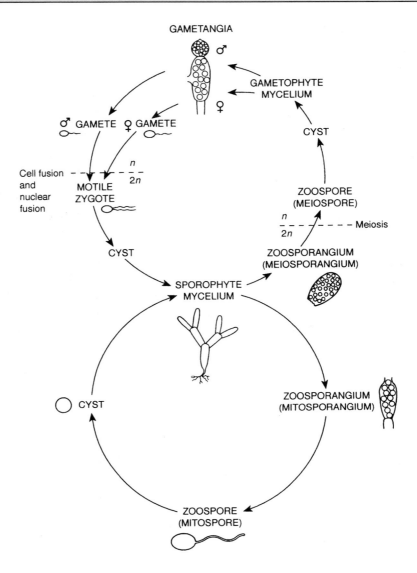

Figure 2.10 The life cycle of the Chytridiomycete *Allomyces macrogynus*. The diploid (2*n*), sporophyte mycelium branches dichotomously, and on nutrient exhaustion bears thin-walled sporangia (mitosporangia) one of which is shown at the beginning of zoospore discharge. Zoospores are propelled by a single posterior flagellum, and after a period of motility form cysts. These germinate to give further diploid sporophytes. The sporophyte mycelium bears thick-walled resting sporangia (meiosporangia), as well as thin-walled sporangia. A detached meiosporangium, with its ornamented, nearly opaque wall is shown. Meiosis occurs in the meiosporangium. The zoospores and cysts of the haploid (*n*) phase closely resemble those of the diploid phase. Haploid mycelium, however, bears male and female gametangia. The female gametangium shown has begun to discharge female gametes, but the papilla on the male gametangium which normally discharges first is still unopened. The large female and small male gametes are similar to each other and to zoospores, and fuse to give motile zygotes. These, after encystment, germinate to produce diploid mycelium.

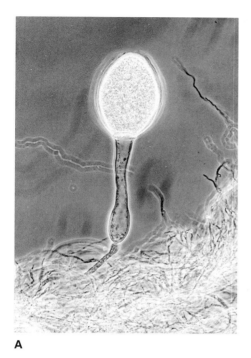

A

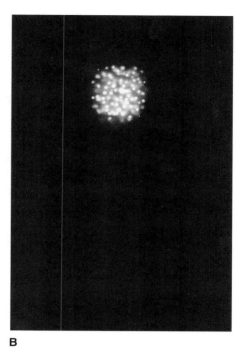

B

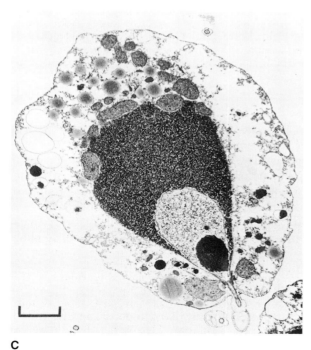

C

Figure 2.11 Stages in the life cycle of Chytridiomycetes. A, Phase contrast photomicrograph of a zoosporangium (width ca 100 μm) of an anaerobic rumen Chytridomycete, *Neocallimastix* sp., isolated from faeces of a Malaysian water buffalo. B, The same zoosporangium viewed by fluorescence microscopy after treating with a stain (DAPI) specific for nuclei. The nuclei are confined to the zoosporangium and are absent from the extensive rhizomycelium. C, Electron micrograph of a female gamete of *Allomyces macrogynus*. The nucleus contains a single electron-dense nucleolus, and is surrounded by a massive nuclear cap, consisting of ribosomes. The cytoplasm contains lipid droplets (spherical, grey) and mitochondria. The base of the single posterior flagellum is visible below the nucleus. The zoospores of *Allomyces* closely resemble the female gamete. Scale bar, 1 μm. (A, B from Trinci, A. P. J., Davies, D. R., Gull, K., Lawrence, M. I., Nielsen, B. B., Rickers, A. & Theodorou, M. K. (1994). Anaerobic fungi in herbivorous animals. *Mycological Research* **98**: 129–152. C, reproduced with permission from Pommerville, J. & Fuller, M. S. (1976). The cytology of the gametes and fertilization of *Allomyces macrogynus*. *Archives of Microbiology* **109**, 21–30.)

The male gametes, which are the first to be released, swarm around the female gametangia. This is due to the sex attractant, **sirenin** (Fig. 2.12), which is released by the female gametangia and gametes and which is effective at attracting male gametes over a wide range of concentrations. Sirenin was named after the sirens, females of classical Greek mythology having an attractant power over navigators. The male gametes fertilize the female gametes as they emerge from the gametangia, nuclear fusion rapidly following cell fusion. The **zygote** formed by fertilization retains the flagella from both gametes and is hence the sole biflagellate phase in the life cycle. The zygote differs from the male gamete and resembles zoospores in being insensitive to sirenin but being attracted by amino acids. On arrival at a suitable substratum it encysts and germinates to give rise to the sporophyte mycelium.

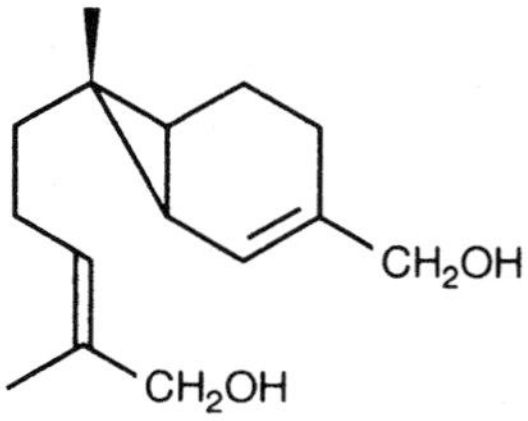

Figure 2.12   Sirenin, the attractant released by female gametangia and gametes of the Chytridiomycete *Allomyces*, and to which the male gametes respond. It is a bicyclic sesquiterpene ($C_{15}H_{24}O_2$: MW 236). It is active over a wide range of concentrations, from 10^{-10} to 10^{-5} M. In addition to being highly active, it is highly specific, most analogues and the D-isomer being without activity. Sirenin is destroyed by the male gametes after it is taken up. There is also evidence for a complementary sex hormone, parisin, released by male gametes and attracting female gametes.

A. macrogynus, which has a basic haploid chromosome number of 14, and *A. arbusculus* which has one of 8, are both hermaphrodite but self-fertile. There are polyploid strains in both species. The two species can hybridize, and a third species, *Allomyces javanicus*, is a naturally occurring hybrid between them. Artificial hybrids between *A. macrogynus* and *A. arbusculus* have been produced, and include some strains that are effectively female (i.e. almost wholly lack male gametangia) and others that are effectively male. The isolation of sirenin was achieved by using a female hybrid for sirenin production, and a male, yielding only male gametes, for assay.

Zoospores of *Blastocladiella emersonii* (Fig. 4.15) closely resemble those of *Allomyces*. On arriving at a suitable surface, a zoospore encysts, and soon produces a rhizoid that penetrates the substratum and repeatedly branches. Synthesis of cell constituents and repeated nuclear division occur, and what had been the cyst enlarges and becomes an ovoid **thallus**, from which more rhizoids emerge. The thallus then usually differentiates into a **thin-walled zoosporangium** from which zoospores are discharged. Under optimal conditions the entire cycle takes about 20 hours, during which time up to eight successive nuclear divisions have occurred to yield about 256 (i.e. 2^8) zoospores. This cycle has been the subject of extensive physiological and biochemical research. The thallus can also develop into a **thick-walled zoosporangium** which is capable of prolonged survival under adverse conditions. This is often termed the RS or resistant sporangium. Such sporangia develop in the presence of high bicarbonate concentrations, produced naturally by overcrowding and high carbon dioxide production. Within resistant sporangia synaptonemal complexes, indicative of meiosis, have been demonstrated by electron microscopy, and the zoospores released from RS sporangia have been found to have half the DNA content of those from the thin-walled OC (ordinary colourless) sporangia. This suggests that the intensively studied cycle involving OC sporangia corresponds to the diploid cycle in *Allomyces*, but the significance of the haploid zoospores is less clear. Two other types of zoosporangia have been observed in *B. emersonii*, one of which is orange from the presence of γ-carotene. Perhaps this corresponds to the male gametangium of *Allomyces*. It is surprising that the life cycle of *B. emersonii*, parts of which have been so intensively studied, has not yet been elucidated fully. The life cycle of another species, *Blastocladiella variabilis*, has been fully elucidated, and in this sexuality has been clearly demonstrated, with two types of gametangia, discharging gametes which although of equal size, are orange and colourless, respectively.

The Chytridiales

Members of the order Chytridiales, often referred to as chytrids, lack a true mycelium but may produce slender rhizoids penetrating the substratum or very slender hyphae (the **rhizomycelium**) linking sporangia. They are mainly aquatic, with some species saprotrophs on plant or animal debris, and others parasites on algae or small aquatic animals. There are also species that live in soil. Some of these are saprotrophs, but others are parasites of vascular plants, in a few instances causing economically important diseases. The saprotrophic aquatic species can be obtained in crude culture by baiting water samples with suitable substrates such as cellulose (e.g. boiled grass leaves), chitin (e.g. shrimp skeletons) or keratin (e.g. hair or moulted snake skin!). Many of the saprotrophic species have been grown in pure culture but those attacking living organisms are obligate parasites. The Chytridiales include a wide diversity of forms, and the more readily handled species should provide interesting material for future physiological, biochemical, genetical and ecological research.

The Anaerobic Rumen Fungi

Ruminant animals, such as sheep and cattle, are highly effective consumers of plant biomass. This effectiveness is due to the activity of microorganisms in the rumen, a specialized region of the gut. Conditions in the rumen are virtually anaerobic, and the study of rumen microbes requires stringent exclusion of oxygen, both in the sampling and processing of rumen contents and in the subsequent culture of the rumen organisms. Obligately anaerobic bacteria and protozoa have long been known to be abundant in the rumen, but the presence of fungi was unsuspected until 1974. It was then realized that some organisms which had been regarded as protozoa were in fact Chytridiomycete zoospores. Attention which had hitherto been directed towards the liquid from strained rumen contents was then switched to the residual plant debris, where abundant obligately anaerobic Chytridiomycetes were found.

The anaerobic rumen fungi are now regarded as constituting the order Neocallimastigales. The life cycles of several rumen fungi have been studied, and that of one species, *Neocallimastix hurleyensis*, followed in axenic culture. Zoospores of *Neocallimastix* are unusual in being multiflagellate, with 8–17 flagella. Zoospores are attracted to, and encyst upon, fragments of plant material. Cyst germination follows, with the germ-tube penetrating the substratum to become a rhizoid that branches profusely. As with *Blastocladiella* (page 36) a thallus develops and becomes a zoosporangium (Fig. 2.11A), which in *N. hurleyensis* yields an average of 88 zoospores. When this fungus is grown at 39°C, the temperature of the rumen, about 30 hours elapse between a zoospore encysting and the release of further zoospores from the mature sporangium that develops.

The activities of anaerobic Chytridiomycetes in the herbivore rumen are considered further in Chapter 7.

The Zygomycetes

The phylum **Zygomycota** consists of two classes, the **Zygomycetes** and the **Trichomycetes**. Within these two classes the sexual process consists of the fusion of two gametangia to give a resting spore, the **zygospore**. Whether the two classes are closely related is not clear. The Trichomycetes, of which about 200 species are known, are obligate parasites that live in the gut of insects and other arthropods, and will not be considered further. There are about 900 species of Zygomycetes, and members of one order, the **Mucorales**, are very widespread and abundant. The Mucorales will now be considered, after which brief reference will be made to other orders.

The Mucorales

Most members of the Mucorales are saprotrophs, and are common in soil and on the droppings of rodents and large herbivores. Others cause rots of fruits and some occur on the decaying fruiting bodies of mushrooms and toadstools. The saprotrophic members of the Mucorales usually have little ability to attack refractory substrates such as cellulose or chitin, but have large hyphae that spread rapidly, and soon sporulate to produce spores that germinate easily. They hence 'get there first' and can exploit readily assimilable nutrients such as sugars before other fungi arrive, and hence have been termed 'primary saprotrophic sugar fungi'. Others, including *Mucor hiemalis*, have been found on rotting wood, apparently obtaining sugars released from wood by Basidiomycete enzymes, and hence are 'secondary saprotrophic sugar fungi'. A few members of the Mucorales are **mycoparasites**, attacking other fungi, particularly saprotrophic Mucorales growing on dung. Their hyphae are generally small and many are obligate parasites and cannot be cultured. The largest genus in the Mucorales is *Mucor* itself, many species of which are common in soil and on decaying plant materials. The life cycle (Fig. 2.13) of *Mucor* and similar members of the Mucorales will now be described.

Sporangiospores of the Mucorales may contain one or several nuclei. Such a spore, placed on a suitable agar medium, will germinate. The spore swells, often to many times the original volume, and a new inner wall is synthesized beneath the original spore wall. After a few hours a hypha, the germ-tube, breaks through

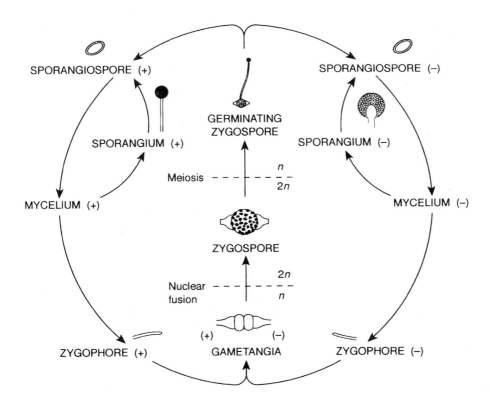

Figure 2.13 The life cycle of the Zygomycete *Mucor mucedo*. Multinucleate sporangiospores germinate to give mycelia of the same mating type (+ or −). Mycelia bear sporangia on sporangiophores, as shown on the left. On the right a sporangium in section shows the columella, sporangiospores, and oxalate crystals on the sporangium surface. If colonies of different mating type come into proximity, zygophores (stout aerial hyphae) differing in mating type grow towards each other and form gametangia when they come into contact. Cell fusion and nuclear fusion occur to give thick-walled diploid ($2n$) zygospores. Meiosis precedes germination, which gives a sporangiophore terminating in a sporangium containing haploid (n) sporangiospores.

the old spore wall. The cell wall of the germ-tube is continuous with the new inner wall of the spore. Sometimes several germ-tubes are produced. These hyphae grow and branch, and within a day a circular colony of vegetative mycelium is established. The vegetative hyphae spread upon and penetrate the substratum, and in some species aerial hyphae rising above the substratum are also produced. Some genera, such as *Rhizopus* and *Absidia*, achieve rapid spread by means of stout rapidly growing aerial hyphae known as **stolons**. When these touch a suitable substratum slender hyphae, **rhizoids**, develop and penetrate the substratum. The hyphae of the Mucorales are coenocytic with many nuclei and few cross-walls.

Some members of the Mucorales are capable of growth under anaerobic conditions (page 147). In these circumstances metabolism is fermentative and alcohol is produced from sugar. In the absence of air and in the presence of high carbon dioxide concentrations sporangiospores of some *Mucor* species, such as *Mucor rouxii* and *Mucor racemosus*, germinate by budding, and a form resembling yeast is produced. This **mould–yeast dimorphism**, controlled by environmental conditions, has been studied in *Mucor* and *Mycotypha*. Some *Rhizopus* species, such as *Rhizopus oryzae*, will grow under anaerobic conditions but remain hyphal. Some other members of the Mucorales, such as *Phycomyces blakesleeanus*, are incapable of anaerobic growth.

The most widespread mode of asexual sporulation in the Mucorales is the production of sporangiospores. The details of **sporangium** (pl. **sporangia**) form and their arrangement on the erect hyphae that bear them, the **sporangiophores**, vary and are used in defining genera and species. Detailed studies of sporangiophore growth have been carried out in a genus, *Phycomyces*, in which the sporangiophores – and also the mycelial hyphae, sporangia and zygospores – are very large. The *Phycomyces* sporangiophore is at first merely a stout hypha rising above the substratum and elongating by means of wall extension in the apical region. Then elongation ceases and the apex swells to form a spherical sporangium. The sporangium is at first bright yellow from the presence of β-carotene, but then turns black as the carotene undergoes oxidative polymerization to form **sporopollenin**, a substance very resistant to chemical and biological degradation. Within the sporangium the protoplasm is cleaved and rounds off to give about 100 000 sporangiospores, each containing a few nuclei. The walls of the sporangiospores contain sporopollenin. Most of the sporangium is occupied by sporangiospores, but the sporangiophore projects into the sporangium as a **columella**. After the sporangium is formed, elongation of the sporangiophore is resumed, with extension occurring in a growth zone just below the sporangium. The sporangiophore, both in the early and late phases of elongation, displays a range of sensory responses. It shows positive phototropism (Fig. 4.5), growing towards the light over a very wide range of light intensities. Other sensory responses that enable the sporangiophore to rise above the substratum (in nature usually dung) while avoiding obstacles are negative geotropism (upward growth), negative hydrotropism (growth towards dry conditions), an avoidance response when approaching obstacles including other sporangiophores, and growth into air currents (positive anemotropism). The sporangial wall of *Phycomyces* is thin, and readily ruptured when the sporangium is mature, to expose the spores which are contained in a mucilaginous matrix. It is probable that such 'slime spores' are dispersed by rain splash.

Mucor, like *Phycomyces*, has an approximately spherical sporangium borne on an erect sporangiophore, but both sporangium and sporangiophore are much smaller. In some species the sporangium wall dissolves at maturity leaving a 'stalked spore drop'. The sporangium of *Rhizopus* resembles that of *Mucor* but there is no mucilage in the sporangium; on rupture 'dry spores' are released which are readily dispersed by air currents. In other genera of Mucorales there is a wide variety of sporangium form. In *Thamnidium* a sporangium similar to that of *Mucor* occurs at the end of the sporangiophore, but about halfway up the sporangiophore there are short branches carrying tiny sporangia (sporangioles) each containing only a few sporangiospores. *Thamnidium* may, depending on conditions of light, temperature and humidity, produce a sporangium only, sporangioles only, or both. *Cunninghamella* apparently produces conidia, but developmentally these, and similar structures in some other genera, appear to be sporangioles containing a single spore. The sporangiospores of the Mucorales are mostly classifiable into dry spores, dispersed by air currents, and slime spores, dispersed by rain splash. The genus *Pilobolus*, however, members of which grow on dung, has an active discharge mechanism. The sporangiophores are phototropic and when mature, the osmotic rupture of a subsporangial swelling 'shoots' the sporangium a metre or so away from the dung, to alight on vegetation and perhaps to be consumed by a herbivore and thus to arrive in fresh dung.

Some members of the Mucorales produce, in addition to sporangiospores, a further type of asexual spore, the **chlamydospore**. These thick-walled spores are produced within hyphae and have no dispersal mechanism, so probably they remain at the site of production until conditions favourable for growth recur. *Mucor racemosus* produces copious chlamydospores both in vegetative hyphae and within sporangiophores. The sexual process has received detailed study in *Mucor mucedo*. When vegetative growth brings two colonies that differ in mating type into close proximity, both produce **zygophores**, specialized aerial hyphae. Zygophores of differing mating type grow towards each other through the air, a form of positive autotropism termed zygotropism and effective over several millimetres. When two zygophores of different mating type come into contact the walls firmly fuse to each other and zygophore elongation ceases. The two zygophores then swell in the region immediately adjacent to the area of contact to give two multinucleate **progametangia**. Each progametangium develops into a **gametangium** by the production of a cross-wall which delimits it from the adjacent region of the zygophore which is then termed the **suspensor**. The cross-wall separating the two gametangia then breaks down and the fused gametangia develop into a zygospore. The development of a thick zygospore wall containing the black pigments melanin and sporopollenin soon renders the observation of cytological events difficult. Limited cytological observations supplemented by genetic analysis suggest that nuclei of different mating type pair and fuse, and that unpaired nuclei degenerate. Meiosis occurs, but in *M. mucedo* only one of the four recombinant types from a single diploid nucleus survives, and subsequently multiplies. Zygospores do not readily germinate. In *M. hiemalis* there is no germination for 30 days, and 1% of the zygospores germinate in the following 90 days. A sporangiophore emerges from the zygospore and terminates in a sporangium. In *M. hiemalis* and *M. mucedo* all the resulting sporangiospores, representing one of the four products of a single meiosis, are of the same mating type. Many members of the Mucorales other than *M. mucedo* are self-sterile, with two mating types designated plus (+) and minus (–), respectively. There are, however, self-fertile species, such as *Rhizopus sexualis*, in which colonies developing from a single uninucleate sporangiospore will produce zygospores. Other details of the sexual process also vary.

In *Phycomyces blakesleeanus* parental haploid nuclei degenerate, and usually only a single diploid nucleus survives and undergoes meiosis, to yield four recombinant haploid nuclei which then multiply. Sometimes, however, more than one diploid nucleus undergoes meiosis and hence a larger variety of recombinant types are produced. Up to 80% of the zygospores have been shown to germinate, from 80 to 120 days after their production.

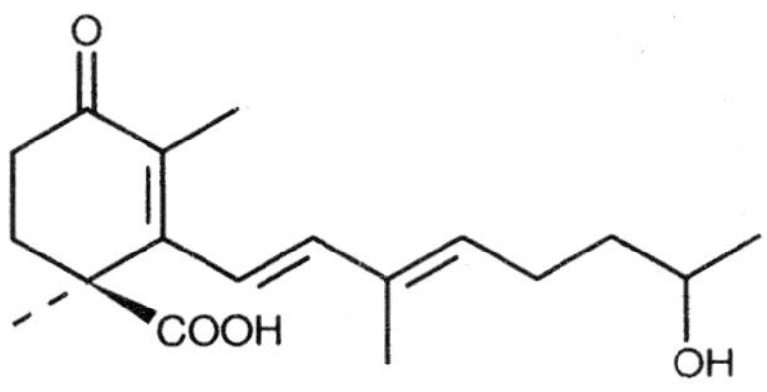

Figure 2.14   The main zygophore-inducing hormone in *Mucor mucedo*, trisporic acid C ($C_{18}H_{26}O_4$: MW 306). It is active at about 10^{-8} M. *M. mucedo* also produces small amounts of a second zygophore-inducer, trisporic acid B, which carries an oxygen instead of a hydroxyl group.

The hormonal control of the sexual process in *Mucor mucedo* has been studied in detail. It has been found that a mixed culture of a plus and a minus strain will produce **trisporic acid** (Fig. 2.14), a substance which is not produced by either plus or minus strains when grown alone. Trisporic acid, applied to a plus or minus strain, induces zygophores. The production of trisporic acid is the result of a remarkable collaborative biosynthesis involving both plus and minus strains (Fig. 2.15). The pathway is known in outline although the identity of some intermediates and steps are uncertain. Both plus and minus strains are able to cleave ß-carotene to yield retinal. The minus strain can convert this compound by a series of steps into trisporol, but trisporol can only be converted into trisporic acid by the plus strain. The plus strain on the other hand, can convert retinal into 4-dehydrotrisporic acid, which can only be converted into trisporic acid by the minus strain. Thus, the production of trisporic acid is dependent upon the diffusion of trisporol and 4-dehydrotrisporic acid between plus and minus strains either through the substratum or through air – the two compounds are both water soluble and volatile. Although trisporic acid is the factor responsible for zygophore induction it is not volatile and hence cannot account for the attraction through air of the plus and minus zygophores to each other. This mutual attraction appears to be due to the diffusion of volatile strain-specific precursors.

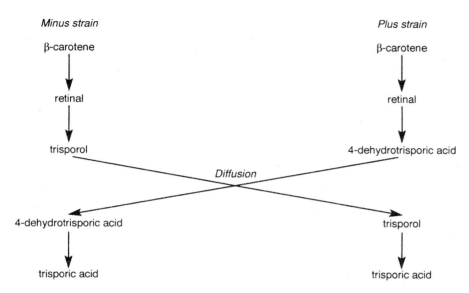

Minus strain

β-carotene

retinal

trisporol

4-dehydrotrisporic acid

Diffusion

trisporic acid

Plus strain

β-carotene

retinal

4-dehydrotrisporic acid

trisporol

trisporic acid

Figure 2.15   The collaborative synthesis of trisporic acid by plus (+) and minus (−) strains of *Mucor mucedo*. There are two possible routes for the synthesis of trisporic acid from retinal, the trisporol and the 4-dehydrotrisporic acid pathways. The minus strain has the enzymes required for the first part of the trisporol and the second part of the 4-dehydrotrisporic acid path. Conversely the plus strain has the enzymes for the first part of the 4-dehydrotrisporic acid and the second part of the trisporol path. Thus trisporic acid synthesis is accomplished only when both strains are present.

   In addition to inducing zygophores, trisporic acid stimulates the production of ß-carotene and other intermediates in its own biosynthetic path, thus greatly increasing its own rate of biosynthesis, an example of **positive feedback** or **metabolic amplification**. Trisporic acid stimulates carotenoid production and induces zygophores in a wide range of Zygomycetes, including self-fertile species. It is likely, therefore, that the hormonal control of the sexual process demonstrated in *M. mucedo* is widespread and perhaps universal in the Zygomycetes.

Other Zygomycete Orders

The Mucorales are so widespread and abundant that they are frequently encountered by almost all mycologists. The other Zygomycete orders are of a more specialist interest, although in some environments they are of considerable significance. The **Glomales** are important since they form a characteristic mutualistic association with the roots of a wide variety of vascular plants (page 397). Their hyphae penetrate between the cells of roots where they send haustoria, which branch like trees (**arbuscules**; from *arbor*, Latin for tree), into cells, and in some groups also form swellings (**vesicles**) inside cells. Such fungi are hence known as **arbuscular** or **vesicular–arbuscular** fungi, and the associations as **arbuscular** or **vesicular–arbuscular mycorrhizas** (*mycorrhiza*; Greek, fungus-root). The **Zoopagales** are predators or parasites of small animals such as amoebae and nematodes (eelworms). The predatory forms have a sticky mycelium which traps the prey which is then invaded by hyphae, and the parasites infect their host by means of spores, which germinate after being ingested by the host or sticking to its surface. The **Entomophthorales** include the large genus *Entomophthora* (Greek, insect destroyer) whose members infect and kill a wide variety of insects including house flies (page 428). Some fungi in this group are used in the biological control of insect pests.

The Ascomycetes

Members of the phylum **Ascomycota**, commonly referred to as the **Ascomycetes**, are those fungi in which the sexual process involves the production of haploid **ascospores** through the meiosis of a diploid nucleus in an **ascus** (pl. **asci**). Most Ascomycetes also carry out asexual sporulation, **conidiospores** (**conidia**) being

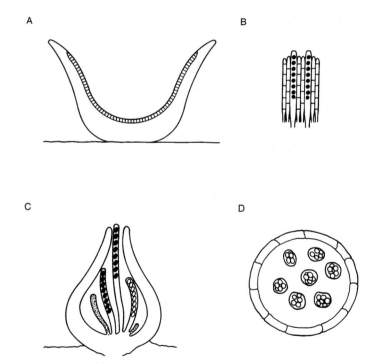

Figure 2.16   Diagrams of Ascomycete fruit bodies. A, Apothecium of *Aleuria vesiculosa*. The upper surface is lined with asci and spacer hyphae, paraphyses. B, Asci and paraphyses of *A. vesiculosa* (A, B after Buller, A. H. R., *Researches on Fungi*, vol. 6, Longman, Green, London, 1934). C, Perithecium of *Sordaria fimicola*. An ascus is shown protruding from the opening (ostiole) of the perithecium prior to discharging its ascospores and collapsing. It will be replaced by a sequence of other asci, shown in various stages of development (after Ingold, C. T. (1971). *Fungal Spores: their Liberation and Dispersal*, Clarendon Press, Oxford): D; Cleistothecium of *Eurotium repens*. The ascus walls lyse and finally the perithecium ruptures to release ascospores (after Webster, J. (1980). *Introduction to Fungi*; 2nd edn, Cambridge University Press, Cambridge).

A

B

C

D

Figure 2.17 Fruit bodies (ascocarps) of Ascomycetes. A–C, Diverse forms in the order Pezizales. A, Apothecia of *Aleuria aurantia*. Asci line the inside of the cups. B, The stalked apothecium of the morel, *Morchella hortensis*. The pits are lined with asci. C, Excavated fruit bodies of the truffle *Tuber aestivum*, one intact and one sectioned. Animals dig up and eat the fruit bodies, and in so doing, scatter the spores that they contain. *Tuber* is thought to have evolved from Pezizales with more usual fruit bodies. D, Several stalked perithecial stromata (two in focus) have developed from a buried sclerotium of *Claviceps purpurea*. The exits to the perithecia (ostioles) appear as dots on the heads of the stromata. See also Fig. 8.17. (A–C, John and Irene Palmer, D, Stephen Shaw and Peter Mantle.) Further illustrations of fruit bodies are available at http://www.wisc.edu/botany/fungi/html

produced on specialized aerial hyphae, the **conidiophores**, that rise above the substratum. The sexual phase of an Ascomycete is now termed the teleomorph, and the asexual phase the anamorph, the concept of 'perfect' and 'imperfect' states having been abandoned. Isolates are often obtained that fail to produce a teleomorph, but which from their asexual sporulation are clearly identical with a known Ascomycete. Such anamorphic isolates are assigned a separate Latin binomial. For example, strains of *Eupenicillium brefeldianum* that fail to produce ascocarps are designated *Penicillium dodgei*. When both anamorphic and teleomorphic phases are present, the teleomorph name is used.

Many Ascomycetes produce their asci in complex fruiting bodies termed ascocarps (Figs. 2.16, 2.17). Such Ascomycetes were formerly regarded as Euascomycetes ('true Ascomycetes'), and classified on the basis of ascocarp form. The 'Discomycetes', for example, were those which had a disc-shaped ascocarp, the **apothecium**, on which asci are exposed. The 'Pyrenomycetes' were those that produced asci within a flask-shaped ascocarp, the **perithecium**. The 'Plectomycetes' were those in which the asci developed inside an approximately spherical ascocarp, the **cleistothecium**. Ascocarp form is crucial in relation to spore dispersal. An apothecium is ideal for the discharge of ascospores into the air, but the asci are ill-protected during development. A perithecium gives some protection, but limits the rate at which ascospores can be discharged. In a cleistothecium, the asci are well protected, but can only be released by rupture of the cleistothecium.

There are a very large number of saprotrophic and parasitic Ascomycetes, at least 18 000, which have ascocarps. In addition there are many more such Ascomycetes, perhaps a further 14 000, that have a mutualistic association with phototrophic microorganisms and constitute the lichens (see page 76). Other Ascomycetes, formerly termed the Hemiascomycetes ('Half Ascomycetes'), do not have ascocarps, solitary asci being produced. Such Ascomycetes are not numerous, but include many important yeasts (page 71).

There has been a major reclassification of the phylum Ascomycota. Grouping into such classes as the Discomycetes, Pyrenomycetes and Plectomycetes on the basis of ascocarp form is no longer accepted. This is because it brings together fungi that on a range of other criteria are dissimilar, and separates ones that are similar. Instead a wide range of characters, including details of ascus structure, are being used to assemble species into genera, genera into families, and families into orders. Some of the orders now recognized contain not only saprotrophic and parasitic species but also lichens.

A few well-studied Ascomycetes, selected on the basis of importance, or illustrative of Ascomycete diversity, will now be considered.

The Pezizales: *Pyronema, Ascobolus*, Morels and Truffles

The Pezizales commonly produce their ascocarps, which are often quite large, on the surface of forest soil, dead wood or dung. A common species is *Pyronema omphalodes*, formerly known as *P. confluens*, which is found on bonfire sites and on sterilized soil in greenhouses. It is readily grown in pure culture on defined media containing a sugar and mineral salts. Ascospores placed on the medium germinate to produce a rapidly spreading and branching mycelium of large hyphae. The hyphae show features which are lacking in the 'lower fungi' discussed in earlier pages but are usual in 'higher fungi' – Basidiomycetes, Ascomycetes and mitosporic fungi. The hyphae have numerous cross-walls (**septa**, sing. **septum**) but these are perforated by **septal pores** which permit rapid protoplasmic streaming and even the passage of nuclei. A hypha may also undergo fusion (**anastomosis**) with a neighbouring hypha. Usually each hypha produces a slender side branch. These grow towards each other and fuse. Hyphal anastomosis can convert a mycelium of radiating hyphae (as is seen in the lower fungi) into a three-dimensional network characteristic of the higher fungi. Such a network can facilitate the transport of protoplasm and nutrients to any point on a mycelium and facilitate the production of the large fruiting bodies which are widespread in the higher fungi.

The life cycle of *Pyronema* is outlined in Fig. 2.18. There is no asexual sporulation. Induction of sexual sporulation requires carbohydrate limitation. One procedure is to grow the fungus in a small Petri dish of glucose mineral salts agar placed in a large Petri dish of water agar. The dense mycelium in the inner dish spills over to form a thin mycelium with **apothecia** on the outer dish. This procedure, which produces copious apothecia, also demonstrates the efficiency with which nutrients and protoplasm are translocated from the inner dish. Alternatively both moderate vegetative growth and moderate apothecium production can be obtained in a single Petri dish by providing the slowly utilized sugar lactose instead of the readily utilized glucose. Exposure to light is essential for apothecium production, and apothecia and mycelium produced in daylight are pink from the presence of carotenoids. It is, however, the ultraviolet component of daylight that is responsible for apothecium formation and the visible component for carotenoid induction, so it is possible, by exposure to suitable ultraviolet radiation, to produce white apothecia.

Pyronema is self-fertile and apothecium production begins with the formation of clusters of the male structure, the slender **antheridium**, and the female structure, the approximately globose **ascogonium**. Both are multinucleate. The ascogonium carries a projection, the **trichogyne**, which grows and curves to make contact and fuse with an antheridium. Nuclei then pass from the antheridium into the ascogonium. **Ascogenous hyphae** grow out from the fertilized ascogonium, and nuclei enter these hyphae. It is likely that the nuclei pass into the ascogenous hyphae in pairs, one member of each pair being derived from the antheridium. In *Pyronema* there is no proof of this, but in Ascomycetes in which genetic analysis has been carried out the results of hybridization experiments establish firmly that pairing must occur. The growth of an ascogenous hypha is terminated by the tip curving sharply back on itself to form a **crozier**. The two nuclei nearest the hyphal tip then divide, thus giving a sequence of four nuclei, two of which will have come from the antheridium and two from the ascogonium. Septa then form, isolating one nucleus in a terminal cell, two nuclei, one 'male' and one 'female', in the penultimate cell, and leaving the fourth nucleus in a stalk cell. Usually the penultimate cell then elongates to form a tubular **ascus**. The two haploid nuclei fuse to give a diploid nucleus which undergoes meiosis to yield four haploid nuclei. These in turn undergo mitosis to form eight nuclei. Walls develop around these nuclei and associated cytoplasm to yield eight **ascospores**.

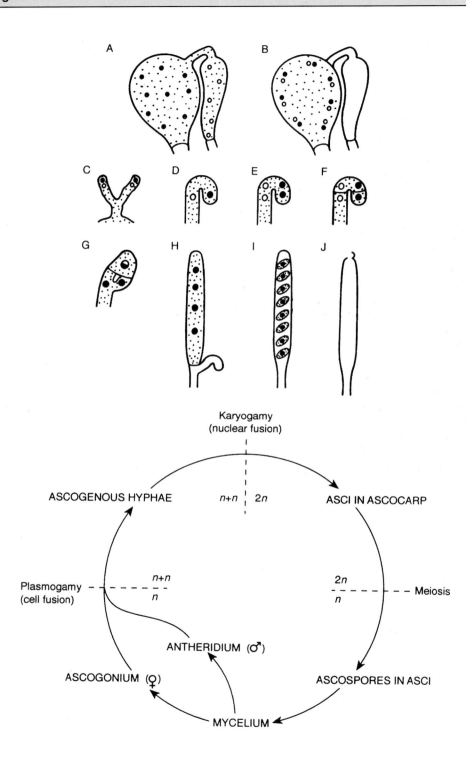

Meanwhile the terminal and stalk cells may have fused to bring together a male and female nucleus to initiate the development of another crozier, and sometimes penultimate cells will develop into croziers instead of asci. So a cluster of ascogonia will produce many ascogenous hypha, and an ascogenous hyphae (which sometimes may branch) can produce many croziers and many asci. The resulting apothecium will contain a large number of parallel tubular asci interspersed with other hyphae, the **paraphyses**, which perhaps can be regarded as spacers, keeping the asci apart and allowing their optimal orientation. Finally the asci discharge their ascospores. A small lid, the **operculum**, at the tip of the ascus is forced open, and the ascospores are shot into the air where they can be dispersed by air currents.

An important feature of Ascomycete meiosis is that the **tetrad** of four nuclei produced in a single meiosis all survive and for a time are kept together in the form of four or more, usually eight, ascospores in a single ascus. This permits **tetrad analysis**, the isolation and genetic study of all four recombinant types formed in the meiosis of a single hybrid nucleus. Moreover in Ascomycetes with narrow tubular asci the nuclei cannot slip past each other so that the four products of meiosis are represented by four successive pairs of ascospores arranged along the ascus. These can be dissected out in order and cultured. Such an **ordered tetrad analysis** yields information on whether two alleles segregated at the first or second meiotic division. In those Ascomycetes in which broad asci allow nuclei to slip past each other, ordered tetrad analysis cannot be carried out. However, an **unordered tetrad analysis** in which the eight ascospores are separately cultured without attempting to keep them in order is less laborious and still yields much useful genetic information. *Pyronema omphalodes* has not been utilized in genetic work, but the potentialities of tetrad analysis have been exploited in several other Ascomycetes.

Another genus in the Pezizales, the coprophilous *Ascobolus*, has been the subject of considerable study. *A. crenulatus*, formerly termed *A. viridulus*, like *Pyronema omphalodes* is self-fertile. *A. immersus*, however, which has been extensively used for genetical research, is self-sterile, and the sexual process only occurs if the two mating types are brought together. Each strain then produces ascogonia, and fertilization occurs through hyphae of the other strain acting as

Figure 2.18 The life cycle and stages in ascus production in the Ascomycete *Pyronema omphalodes*. Here *n* + *n* indicates the association of as yet unfused nuclei from the antheridium and ascogonium. In self-sterile Ascomycetes, however, *n* + *n* represents a dikaryotic phase, with nuclei of different mating types within the same cell. Stages in ascus production are illustrated as follows. A, A trichogyne growing from an ascogonium has just fused with an antheridium. Antheridial nuclei are shown white and ascogonium nuclei black. B, The contents of the antheridium have passed into the ascogonium and pairing of nuclei has occurred. C, Ascogenous hyphae are developing at the surface of the ascogonium. Paired nuclei move into the hyphae and divide repeatedly. D, Crozier formation at the tip of an ascogenous hypha. E, Nuclear division has occurred in the crozier. F, Septa have formed in the crozier. G, Nuclear fusion has occurred in the penultimate cell, initiating ascus development, and the tip cell has fused with the stalk cell. H, Meiosis has occurred in the developing ascus, and the fused tip and stalk cell have given rise to a further crozier. I, Mature ascus with eight ascospores. J, Discharged ascus, showing operculum still attached at ascus tip.

antheridia. The broad asci of *A. immersus* only allow the analysis of unordered tetrads (Fig. 5.8C, D). The occurrence of a mutation affecting spore colour, however, permits the analysis of the contents of very large numbers of asci without having to culture the ascospores. When mutant and wild-type are hybridized, the asci produced would be expected to contain four wild-type and four mutant ascospores. Most asci do, but occasional asci are found with different frequencies of wild-type and mutant. This results from the process of **gene conversion**, which may be a very widespread phenomenon but is most readily demonstrated in Ascomycetes in which tetrad analysis can be carried out. *A. fufuraceous* (formerly called *A. stercorarius*) is another self-sterile species, requiring the interaction of the two mating types for the sexual process to occur. It also produces spores asexually by the fragmentation of aerial hyphae. These spores, termed **oidia** or **arthrospores** (Fig. 2.19A), germinate to give a mycelium if they fall on a suitable substratum. If they fall on a mycelium of the same mating type they do not germinate, but on the mycelium of the opposite mating type an oidium will induce the formation of an ascogonium. The trichogyne of the ascogonium grows towards the oidium and fuses with it. Apothecium development follows. The asci of *Ascobolus* and probably most Pezizales are phototropic, curving to point in the direction of maximum light intensity and hence shooting their ascospores in the direction with least obstruction.

The Pezizales also include the morels, members of the genus *Morchella* (Fig. 2.17B). Morels grow in woodlands and their fruiting bodies, edible and much prized, are stalked and up to 15 cm high. Some success has been achieved both in the field cultivation of morels, and the production of mycelium, but not fruiting bodies, in pure culture. Other Pezizales are known as cup-fungi, because the disc bearing the asci is curved into a saucer or even a cup-shape (Fig. 2.17A). In Pezizales with subterranean fruiting bodies, the truffles (Fig. 2.17C), this tendency has gone further so that the asci line the interior of a roughly spherical fruiting body. Here there is no discharge mechanism, and it is likely that spores are dispersed by rodents which, attracted by the smell of mature fruiting bodies, dig them up and eat them. A few species, such as *Tuber melanosporum*, are highly prized by gourmets and are found with the help of trained dogs or pigs. They form mycorrhizal associations with the roots of beech and oak, and recently some success has been achieved in promoting the mycorrhizal association between *T. melanosporum* and the oak.

The Sordariales: *Neurospora, Podospora, Sordaria* and *Chaetomium*

The Sordariales include one of the most intensively studied of all fungi, *Neurospora crassa*. It is readily grown in pure culture and has large hyphae which spread rapidly. Two types of asexual sporulation occur. The aerial mycelium rising above the substratum consists largely of branched chains of **macroconidia** (Figs. 2.19B, 4.1B), roughly ellipsoidal cells containing several nuclei. Macroconidia are pink due to the presence of carotenoids, and are produced in enormous numbers. They are readily detached and dispersed by air currents and on landing on a suitable substratum germinate readily. In the tropics burnt vegetation is often festooned with the pink mycelium and macroconidia of *Neurospora*, and in the laboratory care is needed to avoid the massive release of macroconidia with resulting contamination of other cultures when Petri dishes are opened. Tiny uninucleate **microconidia** (Figs. 2.19C, 4.1C) are also produced, but do not germinate so readily as macroconidia and do not survive so long. It is probable that the role of microconidia is participation in the sexual process.

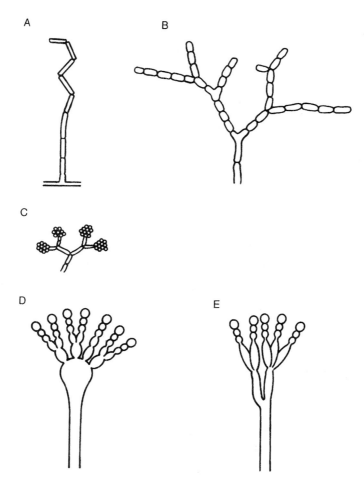

Figure 2.19 Asexual sporulation in Ascomycetes. A, Formation of arthrospores in *Ascobolus furfuraceus*, by the separation of the constituent cells of an aerial hypha. B, Macroconidia of *Neurospora crassa*. Alternate bulging and constriction occurs throughout the aerial hypha, septa develop at the constrictions and the resulting macroconidia detach at the slightest disturbance. C, Microconidia of *Neurospora crassa*. The tiny microconidia are budded from the conidiophore and form sticky clusters. D, Conidiophore of *Aspergillus (Emericella) nidulans* bearing primary and secondary sterigmata (phialides). From the latter conidia are being produced, the basal ones not yet swollen to full size or separated from the phialides. E, Conidiophore of *Penicillum expansum* bearing metullae on which phialides are producing conidia. D and E are diagrammatic, phialides and conidia being more numerous than shown.

N. crassa is self-sterile with two mating types designated A and a. Mating type is controlled by a single locus with the alternative alleles *A* and *a*. A single strain of either mating type will produce protoperithecia consisting of ascogonia bearing trichogynes but further development occurs only if a macroconidium or microconidium from the other mating type reaches the trichogyne. Then a **perithecium** is produced containing numerous asci. Each ascus contains eight initially uninucleate ascospores within which mitosis occurs to give a multinucleate condition. The asci elongate through the neck of the perithecium to discharge their ascospores. *Neurospora* ascospores are dark and thick-walled and can survive for many years. They germinate only if dormancy is broken by heat shock (60°C for 20 minutes) or treatment with furfural.

Genetic studies on *N. crassa*, including analysis of ordered tetrads and chromosome mapping, were already being carried out in the 1930s. At the end of the decade the organism was chosen by Beadle and Tatum to examine the possibility that genes act by controlling the production of enzymes, a view that can be precisely formulated as the one-gene one-enzyme hypothesis. They showed that wild *Neurospora* strains could be grown on inorganic media containing a sugar and the vitamin biotin. Conidia were then treated with X-rays, mated with wild-type cultures, and single ascospores isolated with a view to obtaining mutant cultures derived from single haploid nuclei. Some cultures obtained in this way were shown to have nutrient requirements additional to those of the wild type, and it was established that such a deficiency could result from the mutation of a single gene. The study of a range of mutations resulting in the same nutritional deficiency led to the elucidation of the biosynthetic pathway for tryptophan and for ornithine. *Neurospora crassa* thus had a major role in the foundation of biochemical genetics and is now genetically and biochemically one of the most intensively studied organisms.

Hyphal anastomosis occurs readily in *N. crassa*, as in many Ascomycetes. It is therefore possible to obtain **heterokaryons**, mycelia that contain genetically different nuclei. This can be done by bringing together two mutant strains each with a different nutritional deficiency on a medium that lacks both nutrients. Under these circumstances heterokaryosis leads to an organism with a selective advantage as compared with either constituent strain. Heterokaryosis has been extensively utilized in genetical research, but the formation of heterokaryons through hyphal anastomosis in nature may be rare. This is due to **vegetative incompatibility**. Isolation of strains from nature has demonstrated the occurrence of many incompatibility groups, heterokaryon formation being able to occur within a group but not between groups. A genetic difference at any one of a large number of loci, termed *het* loci, results in somatic incompatibility.

Neurospora sitophila and *N. intermedia* resemble *N. crassa* in being self-sterile and having eight-spored asci. *N. tetrasperma* has, as the name indicates, four spores in each ascus. Each spore contains two nuclei, one carrying the *A* and one the *a* mating type allele. This is the result of the precise way in which meiosis and ascospore wall delineation occur in this species. The presence of both *A* and *a* alleles in single ascospores, yields a culture capable of producing perithecia. The consequences of possessing a mating system are thus evaded and apparent self-fertility results. These four species with a mating system are sufficiently closely related to be self-fertile. There are also five *Neurospora* species known which have eight-spored asci, are self-fertile without any indications of a mating system, and lack conidia.

The genus *Podospora*, which grows on herbivore dung, is closely related to *Neurospora*. In some species repeated mitosis following meiosis results in large numbers of ascospores in the ascus. *P. decipiens*, for example, has strains characterized by 8, 16, 32 or 64 ascospores per ascus. The most intensively studied species, *P. anserina*, resembles *N. tetrasperma* in normally having four binucleate ascospores per ascus, with the two nuclei in each ascospore carrying nuclei of opposite mating type. The precise meiotic manoeuvres and spore wall delineation that result in this condition are, however, different from those in *N. tetrasperma* and must have evolved independently. Some asci of *P. anserina* contain both large binucleate and small uninucleate ascospores. Germination of the small ascospores yields self-sterile cultures. *P. anserina* does not produce macroconidia, and the microconidia, although capable of fertilizing ascogonia of the opposite mating type, hardly ever germinate to yield a mycelium. *P. anserina* has been employed extensively in studies on vegetative incompatibility.

Sordaria (Fig. 5.8A, B) is another genus that is closely related to *Neurospora* and is found on dung. In the self-fertile species *S. fimicola* extensive studies on the physiological control of perithecium formation have been carried out and mutants have been obtained that block various steps in perithecium production. Ascospore discharge and its control have been studied in detail. Each ascus in turn elongates through the neck of the perithecium. The ascus then bursts, discharging the ascospores through the ruptured tip, and collapses. The perithecium neck is capable of phototropic orientation to point in the direction of maximum illumination. Ascospore discharge is stimulated by warmth, light and low relative humidity and hence tends to occur during daytime. Ascospore colour mutants have been used for the analysis of linear tetrads by direct spore counting and have been valuable for studies on gene conversion.

Another genus, *Chaetomium*, is common in soil and on dead vegetation. It produces cellulase which enables it to attack cellulose-rich substrates, causing deterioration of paper, straw and wood under damp conditions. The perithecia lack any obvious neck and are hairy. The asci lyse outside the perithecium liberating the ascospores which ooze through the aperture of the perithecium like toothpaste from a tube as a 'spore tendril'. Dispersal is probably by rain splash.

Ophiostoma and *Claviceps*

Ophiostoma ulmi (formerly *Ceratocystis ulmi*), a member of the order Ophiostomatales, causes Dutch elm disease. The epidemic that has devastated the elms of Western Europe in the last few decades is due to an aggressive strain that has recently been given specific status as *Ophiostoma novo-ulmi*. The fungus produces asexual spores (**conidia**) by budding at the tips of single hyphal branches (**conidiophores**), at the tips of bundles of such hyphae, known as **coremia** (sing. **coremium**) or **synnemata** (sing. **synnema**) or by the budding of hyphae growing within an older hypha. The conidia themselves may subsequently multiply by yeast-like budding. As in *Chaetomium* the asci lyse within the perithecium, releasing ascospores. These ooze out of the tiny perithecium through its long (*ca* 1 mm) slender straight neck to give a stalked spore drop. This, as well as the droplets of conidia produced at the tips of conidiophores and coremia, are viscous and are probably carried from tree to tree by the bark-boring beetles which by tunnelling beneath the bark initiate infections. Spread within a tree occurs by mycelial growth and the transport of conidia through the water-conducting vessels of the sapwood. The fungus produces toxins and also causes blocking of the vessels by gum production.

The ergot fungi, forming the genus *Claviceps* in the order Hyopocreales, are parasites of grasses and cereals. Infection is limited to the ovary, which is penetrated by hyphae from germinating ascospores or conidia. The ovary is extensively colonized by hyphae, which produce large numbers of conidia at the surface of the ovary. The infected ovary produces an exudate rich in sugars and amino acids, 'honeydew', which accumulates in the floral cavity and may even overflow. The exudate supports further fungal growth and attracts insects which feed on honeydew and the conidia it contains. The conidia are dispersed to other florets and plants by contact, rain splash or on insects. Sporulation and honeydew secretion ceases in a few weeks and the infected ovary develops into a hard black mass of hyphae, the **sclerotium** (pl. **sclerotia**) (Fig. 8.17) instead of the seed that it should have become. Sclerotia, which are capable of prolonged survival, are widespread in higher fungi. Those of *Claviceps* are conspicuous and have been known for centuries as ergots. The ergots fall to the ground in autumn and survive the winter. In early summer one or a few fruit bodies emerge (Fig. 2.17D) and on them numerous perithecia form. Long needle-shaped ascospores develop, are discharged into the air, dispersed by air currents and initiate a fresh season's round of infections.

The best known ergot fungus is *Claviceps purpurea* (Figs. 2.17D, 8.17), the ergot of rye. In past centuries the contamination of rye bread with ergot was responsible for horrifying outbreaks of ergotism involving gangrene, loss of limbs and death. The disease is now of mainly veterinary interest, as cattle or sheep eating ergot-infected grasses may suffer serious effects including abortion. Ergot contains a wide variety of alkaloids that stimulate the central and sympathetic nervous systems in various ways producing a range of secondary effects. Ergot preparations have been used for centuries for hastening childbirth and controlling subsequent bleeding, and in the twentieth century chemists and pharmacologists identified many active components and obtained safer and more reliable preparations. Therapeutically useful ergot alkaloids are now produced on a large scale by the fermentation industry (page 521).

The Eurotiales: *Aspergillus*, *Penicillium* and their Teleomorphs

Members of the order Eurotiales produce their asci within cleistothecia, approximately spherical closed ascocarps. The ascus walls lyse, releasing ascospores (Fig. 4.1F) into the ascocarp cavity. Ultimately the ascocarp is ruptured, but there is no specific mechanism for ascospore dispersal. The Eurotiales are of interest because of the abundance and importance of the anamorphic states of some genera. Several teleomorphic genera, including *Eurotium* and *Emericella*, have their anamorphic states in the genus *Aspergillus*, and the teleomorphic genera *Eupenicillium* and *Talaromyces* in the genus *Penicillium*. Teleomorphic states have not been demonstrated for many species of *Aspergillus* and *Penicillium*, and they are usually infrequent even in those species where they do occur. The designations *Aspergillus* and *Penicillium* are hence much more widely used than those of the corresponding teleomorphic genera.

The generic name *Aspergillus* is derived from the Latin *aspergillum*, a mop for distributing holy water, and refers to the appearance of the mature conidiophores which are the means of asexual sporulation in the genus. A conidiophore (Fig. 2.19D) rises perpendicularly from the substrate hyphae and at its tip develops a swelling, the **vesicle**. From the vesicle arise numerous short hyphae. These may bear conidia, or they may branch to give further short hyphae which carry the conidia (Fig. 4.1E). The hyphae which actually carry the conidia are termed **phialides**, and the hyphae which bear phialides are termed **metullae** (sing. **metullus**). Some authors term the hyphae that arise directly from the vesicle **primary sterigmata**, regardless of whether they bear conidia or a second set of hyphae, and the second set of hyphae, if present, **secondary stigmata**. The conidia that are produced on the phialides are not immediately shed, so the production of a succession of conidia from a single sterigma results in a chain of conidia. It is the numerous chains of conidia arising from the vesicle that produce the mop-like appearance. The conidia are very hydrophobic and thus not readily wetted. On maturity, however, they separate readily and are dispersed by air currents. *Aspergillus* is tolerant of low water activity, being able to grow on substrates of high osmotic potential and to sporulate in an atmosphere of low relative humidity.

Aspergillus is divided into subgenera and sections, often informally termed groups. Each group is named from one of the better known species within it. Assignment of an isolate to a group is usually easy but the determination of the precise species more difficult. Some groups are of great practical significance, but the *Aspergillus nidulans* group is also of interest because *A. nidulans* (for sporulation, see Figs. 2.19 and 4.1D–F) has had an important role in the development of genetics. A common soil organism, it produces cleistothecia (often called perithecia by geneticists), on the basis of the morphology of which the teleomorph is designated *Emericella nidulans*. It was utilized by Pontecorvo to establish that genetic analysis could be carried out in self-fertile organisms. Mutagenic treatment of a single isolate resulted in strains with a range of nutritional deficiencies, and bringing together such strains on minimal medium forces heterokaryon formation. Examination of the progeny of such heterokaryons established that recombination had occurred during meiosis and that linkage data could be obtained. Following chromosome mapping by meiotic recombination it was demonstrated in addition that genetic analysis could be carried out without recourse to the sexual process. In heterokaryons some nuclear fusion to give diploids occurs, and within these diploids crossing over between homologous chromosomes sometimes takes place during mitosis. The diploids may revert to the haploid state giving nuclei which will be genetically different from those of the parental strains. The frequency of diploidization, mitotic crossing over and haploidization was low but means were found to increase the frequency and to detect recombination efficiently. The construction of linkage maps by mitotic recombination in *A. nidulans* showed that genetic mapping in systems lacking a sexual phase, such as mitosporic fungi and mammalian tissue cultures, was possible. *A. nidulans* has become genetically and biochemically one of the most intensively studied of all organisms. Another member of the *A. nidulans* group, *Aspergillus heterothallicus* (teleomorph *Emericella heterothallica*) is so far the only known *Aspergillus* species showing sexual sporulation in which the cooperation of two mating types is needed.

Aspergillus glaucus (teleomorph *Eurotium herbariorum*) and other members of the *A. glaucus* group are exceptional, even among *Aspergillus* species, for their tolerance of low water activities. This renders them important agents in biodeterioration, attacking foods such as jam that have high sugar contents and products such as textiles and leather under even moderately damp conditions. The *A. restrictus* group also tolerate low water activities, and in the humid tropics will damage camera and microscope lenses, living on traces of organic materials and producing acids that etch the glass. Members of the *Aspergillus niger* group have long been of importance in the fermentation industry as the principal source of citric acid for soft drinks (page 509), and more recently for the production of enzymes (pages 476–481). *Aspergillus oryzae* and other members of the important *A. flavus-oryzae* group have an important role in the production of many Asian foods and beverages (pages 489, 503). On the other hand, some strains of *A. flavus* can attack stored groundnuts and other foods, contaminating them with aflatoxin, a highly toxic and carcinogenic mycotoxin (pp. 440–441). Members of the *Aspergillus fumigatus* group are serious pathogens of birds and occasionally humans, causing the lung infection aspergillosis. Various other species can cause infection of individuals with impaired immune responses, and the inhalation of *Aspergillus* spores should be avoided – even if infections do not result, allergic responses (pp. 442–443) may be produced.

The genus *Penicillium* is of comparable importance to *Aspergillus*. The name is derived from the Latin *penicillus*, an artist's brush, and refers to the branching conidiophores (Fig. 2.19E) on which chains of conidia are borne. Taxonomically, it is a very difficult genus, species being distinguished by subtle and frequently variable differences in the details of asexual sporulation. In some species conidia are borne on phialides which arise at the apex of the conidiophore. In others the conidiophore bears metullae on which the phialides are borne, and in some the conidiophore itself may branch to varying extents prior to bearing metullae. In some species branching is symmetrical, in others irregular. The spores as in *Aspergillus* are hydrophobic, difficult to wet, but easily dispersed by air currents. Many members of the genus are important in causing biodeterioration, especially in conditions of low relative humidity. Many species are involved in food spoilage. *Penicillium digitatum* attacks citrus fruits. It produces ethylene which accelerates ripening and renders the fruit (and adjacent ones in storage) more susceptible to attack. Finally, the fruit is covered in masses of olive green conidia and shrivels and dries. *Penicillium italicum*, on the other hand, causes a nauseous slimy rot of citrus fruit and produces blue–green conidia. *Penicillium expansum* causes a storage rot of apples.

In both *P. italicum* and *P. expansum* conidiophores may adhere to each other to give **synnemata** (sing. **synnema**), a tendency reaching a spectacular climax in

Penicillium claviforme, which can produce highly differentiated synnemata centimetres long. Some members of the genus are useful to humans. *Penicillium camembertii* and *P. roquefortii*, for example, have a role in the ripening of soft and blue-vein cheeses, respectively (page 505). The most famous contribution of *Penicillium* to human welfare, however, was its role as the source of penicillin, the first clinically useful antibiotic (page 516–517).

The Basidiomycetes

Members of the subdivision Basidiomycota, commonly referred to as **Basidiomycetes**, are those fungi in which the sexual process involves the production of haploid **basidiospores** borne on a **basidium** in which a diploid nucleus undergoes meiosis. They are an important group with about 22 000 species known. Most of the conspicuous fungi of fields and woods are Basidiomycetes. At the microscopic level there are two features that are characteristic of Basidiomycetes and very widespread in the group. One such feature is the presence of 'clamp connections' (page 61) apparently linking adjacent cells of hyphae. The other is that the basidiospores are usually **ballistospores** (page 217) which are actively launched from the basidium. The classification of Basidiomycetes is controversial. Traditionally, four major groups have been recognized. Two of these, the Hymenomycetes and Gasteromycetes are, with a few exceptions, macrofungi with large fruit bodies. The other two, the smuts and rusts, are microfungi and parasites of plants. The Hymenomycetes have a hymenium, an extensive area of basidia exposed in such a way that basidiospores can be launched efficiently into the air by a process of active discharge. The Gasteromycetes have basidiospores that are not actively launched. Instead there are a wide variety of fruit body forms and dispersal methods. It would seem that the only common feature of the Gasteromycetes is that they lack ballistospores and that the group is not a natural one. A recent view, therefore, is that they should not be formally treated as a separate group, and that within the Basidiomycota three classes should be recognized – the **Basidiomycetes** (Basidiomycete macrofungi and closely related forms), the **Ustomycetes** (with a single order, the **Ustilaginales** or smuts) and the **Teliomycetes** (with a single order, the **Uredinales** or rusts).

The Basidiomycete Macrofungi – the Class Basidiomycetes

There are many orders within the Basidiomycete macrofungi. Here a few intensively studied species and the features of some orders will be considered.

The Agaricales: Coprinus *and* Agaricus
Members of the Agaricales, often referred to as the agarics, include the cultivated mushroom, *Agaricus bisporus*, already mentioned and illustrated in Chapter 1.

Other agarics have similar fruiting bodies with stalk, cap and gills, although many are much smaller and a few larger. They are the fungi commonly referred to as mushrooms if edible and toadstools if not. Not only *A. bisporus* but also several other species are cultivated for food (Chapter 8). Although the cultivated mushroom is the species best known to the public, it has proved a difficult subject for physiological and genetical research, and many aspects of its biology are ill understood. The life cycle of another species, *Coprinus cinereus* (Fig. 2.20), will hence first be described. This species, which in nature grows on horse dung, can be grown easily in the laboratory, readily produces fruiting bodies in pure culture, and is amenable to genetic analysis. Until recently most publications on this species purported to be on *Coprinus lagopus*, which grows on sticks and on soil in woodlands, leading to some confusion.

A single basidiospore of *C. cinereus* placed on a suitable substratum will produce a branched septate mycelium. The septa divide the hyphae into compartments each of which contains a single nucleus. The septa are of a type limited to Basidiomycetes and known as **dolipore septa** (Fig. 3.1C). There is a pore at the centre of the septum, but the septum is greatly thickened adjacent to the pore, so that the pore resembles a tube rather than a simple opening. On each side of the pore modified endoplasmic reticulum forms a **septal pore cap**

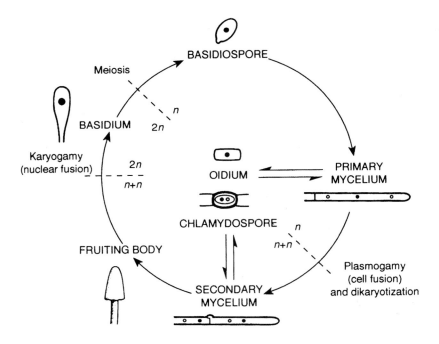

Figure 2.20 The life cycle of the Agaric *Coprinus cinereus*: Haploid (*n*) basidiospores germinate to give primary mycelium which can produce oidia, and fuse with oidia or mycelium of compatible mating type to give dikaryotic (*n* + *n*) secondary mycelium with clamp connections. Secondary mycelium can produce chlamydospores and give rise to fruit bodies. Fusion of two haploid nuclei to give diploid (2*n*) nuclei occurs in basidia. Meiosis follows and each basidium bears four haploid basidiospores.

(sometimes called a **parenthosome**) with a number of perforations. This complex septal apparatus permits cytoplasmic streaming between compartments but not the passage of nuclei. Asexual sporulation takes the form of the production of **oidia** on short hyphal branches to form stalked spore drops. *C. cinereus* is self-sterile and the initiation of the sexual process requires an encounter between two strains that differ in mating type.

The mycelium that arises from a single basidiospore is known as the **primary mycelium**. It is a **homokaryon**, having only one type of nucleus, and a **monokaryon**, with only one nucleus per cell. Mating type is determined by two unlinked genetic factors, A and B. Such a mating system is known as **bifactorial**. A survey of natural populations of *C. cinereus* demonstrated 36 alternative factors at the A locus and 32 at the B locus, but further sampling would undoubtedly reveal higher numbers. Even on the basis of 36 A factors and 32 B factors, 1152 (i.e. 36×32) mating types are possible. An encounter between two strains of identical mating type (e.g. $A_1 B_1$ and $A_1 B_1$, a situation referred to as A= B=) will lead to no further developments, but if the strains differ with respect to **both** mating factors (A $\neq$ B $\neq$, e.g. $A_1 B_1$ meets $A_2 B_2$) the sexual process goes to completion and fruit bodies are formed (Table 2.4). The sexual process can be initiated as a result of encounters classifiable as A=B$\neq$ or A$\neq$B= but will subsequently fail. Full sexual compatibility between two strains requires differences with respect to both A and B factors, and hence not all possible pairs of different mating types are fully compatible.

Encounters between two strains can occur as a result of two basidiospores germinating near each other or of the spreading of two adjacent primary mycelia. The encounter between the two strains is followed by hyphal anastomosis. Then, if the two strains are fully compatible (A$\neq$B$\neq$), nuclei from each strain migrate rapidly through the mycelium of the other. This migration requires that the dolipore septa are converted into simple perforate septa, a process that is controlled by the B factor, so nuclear migration is possible only when the pairs of

Table 2.4 The consequences of encounters between primary mycelia of *Coprinus cinereus* of different mating type

Type of encounter	Example of mating type	Consequence
A = B =	$A_1 B_1 \times A_1 B_1$	Dikaryon formation not initiated
A = B $\neq$	$A_1 B_1 \times A_1 B_2$	Dolipore septa are converted into simple perforate septa and nuclear migration occurs, but no synchronization of nuclear division or formation of clamp connections
A $\neq$ B =	$A_1 B_1 \times A_2 B_1$	Only a few dikaryotic cells and clamp connections are formed, since nuclear migration cannot occur and most cells remain monokaryons
A $\neq$ B $\neq$	$A_1 B_1 \times A_2 B_2$	Nuclear migration and dikaryon formation occur, the secondary mycelium develops clamp connections, and fruiting bodies are formed

strains differ with respect to the B factor (i.e. A≠B≠ or A=B≠). Nuclear migration in *C. cinereus* occurs at a rate of up to 1 mm h⁻¹. The result of migration is that each compartment in both mycelia contains a nucleus of each mating type. The mycelia have hence become **dikaryons**, and the migration process is termed **dikaryotization** or, in some of the older literature, diploidization. Migration is accompanied by nuclear division, so as to permit the doubling of the number of nuclei in both mycelia. Dikaryotization can occur by the deposition of an oidium from one strain close to the mycelium of another. A hypha from the mycelium shows chemotropic curvature towards the oidium and fuses with it and, if the strains are compatible, dikaryotization follows. Dikaryotization can also occur if a monokaryon encounters a dikaryon carrying nuclei of a compatible mating type. Under these circumstances nuclear migration will be unilateral instead of reciprocal.

The processes of nuclear migration and dikaryon development are controlled by proteins encoded by the A and B factors. The B factors encode pheromones and their receptors which control interactions between the nuclei of the two

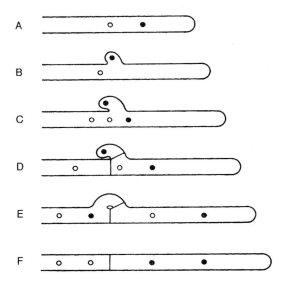

Figure 2.21   Diagram illustrating how the dikaryotic state is maintained by means of clamp connections. A, A dikaryotic hyphal tip, with two nuclei of different mating types. B, The development of a backwardly growing side branch, into which one of the nuclei has moved. C, Synchronous division of the two nuclei has occurred, and one of the daughter nuclei from the side branch nucleus has moved back into the main hypha. D, A septum has developed between the daughter nuclei derived from the nucleus that remained in the main hypha, and a second septum cutting off the side branch. Septal pores are not indicated. E, The tip of the side branch has fused with the main hypha, giving the clamp connection typical of the Basidiomycete dikaryon, and the side branch nucleus has moved into the main hypha. A new compartment with two nuclei of differing mating type has thus been formed, with the apical compartment remaining dikaryotic. F, An illustration of the probable consequence of nuclear division is a dikaryotic hyphal tip as illustrated in A, but without clamp connection development, resulting in hyphal compartments reverting to the monokaryotic state.

parent strains in the mycelium. The A factors encode transcription factors which regulate the development of the dikaryotic mycelium.

The dikaryotic mycelium produced by the fusion of two fully compatible mycelia will continue to grow, but the new growth will differ in form and constitute a **secondary mycelium**. The most striking feature of the secondary mycelium is the presence of **clamp connections** (Fig. 2.21), devices for maintaining the dikaryotic condition as growth continues. Consider a dikaryotic apical cell with two nuclei of differing mating type. If these undergo nuclear division the most likely consequence is two nuclei of one mating type near the apex and two nuclei of the other mating type towards the rear of the cell. Cross-wall formation which restores the binucleate state would then result in two homokaryotic cells. What in fact happens is that the apical cell produces a short backward pointing side branch and one nucleus moves into it, the other nucleus remaining in the hypha but close to the branch. Simultaneous nuclear division then occurs, and a nucleus from within the side branch and one from within the hypha move forwards towards the hyphal apex. The formation of a cross-wall at the base of the side branch and across the main hypha close to it results in a new apical cell containing two nuclei, one of each mating type. The side branch curves in towards the hypha and fuses with the sub-apical cell which already contains one nucleus. The nucleus in the side branch, which is of the complementary mating type, migrates into the sub-apical cell. Thus one cell has given rise to two, still dikaryotic, cells.

The proper development of secondary mycelium requires the participation of nuclei that differ with respect to both the A and B factors. The cooperation of complementary A factors is needed for the properly synchronized division of the two nuclei in apical cells, and for the production of clamp connections, and the cooperation of complementary B factors for the fusion of the clamp cell with the sub-apical cell. The secondary mycelium (dikaryon) differs from the primary mycelium (monokaryon) not only in having binucleate cells and clamp connections but in other ways. The hyphae are wider (about 7 μm instead of 4 μm), advance faster, giving a less dense growth, and have branches at a more acute angle with respect to the parent hypha. Oidia are not formed; instead thick walled chlamydospores may be produced.

Fruiting bodies (Fig. 2.22, for a related species) are formed only on A≠B≠ secondary mycelia. The production of primordia can be initiated by light, even a short exposure at low intensity being effective. The primordia develop in about 5 days from a tiny tangle of mycelium into a 'button' about 1.5 cm high in which the various parts of the mature fruiting body are present in miniature. Then, over a period of about 6–9 hours, elongation of the stipe to a length of about 10 cm occurs. The stipe at first grows towards the light (positive phototropism) and then upwards (negative geotropism). This sequence of sensory responses is probably the most effective for placing the gills, which meanwhile are exposed by the expansion of the cap, in a position optimal for the discharge of basidiospores. Stipe extension and cap expansion is due almost entirely to cell elongation and enlargement, there being very little cell division at this stage. Glycogen accumulated in the primordium disappears, and a high chitin synthase activity is observed. It is likely, therefore, that the glycogen is the source, via glucose, of the N-acetylglucosamine from which the chitin of new wall material is synthesized.

Figure 2.22   The fruit bodies (basidiocarps) of Basidiomycetes. A, A young fruit body of the ink cap *Coprinus comatus*, with older fruit bodies above and to the side. These are undergoing autodigestion, which follows a wave of basidiospore production moving upwards along the gills, which are concealed from view. B, The stipe and under surface of the fruit body of *Boletus tomentosus*, showing the openings of the pores which are lined with basidia. C, View from below of the fruit body of *Hericium crinaceus*, showing spines which are covered with basidia. (A–C, John and Irene Palmer.) See also Fig. 4.10. Further illustrations of fruit bodies are available at http://www.wisc.edu/botany/fungi/html

The surface of a gill is covered by basidia (Fig. 2.23) and by **paraphyses,** cells which act as spacers, keeping the basidia separate. Nuclear fusion and meiosis take place during the final stages of button development but basidiospore production only occurs after the gills are exposed. Each basidium develops four projections, sterigmata, the tip of each of which swells to become a basidiospore. The four nuclei resulting from meiosis move into the four developing basidiospores. The basidiospores are discharged violently from sterigmata in a manner which is characteristic of all actively discharged basidiospores. A droplet

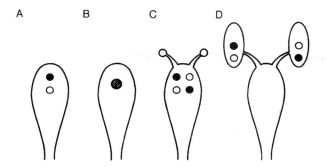

Figure 2.23 Diagram of basidiospore development in *Agaricus bisporus*. A, A basidium with two haploid nuclei of compatible mating types. B, A diploid nucleus has resulted from nuclear fusion. C, Meiosis has occurred, and since *A. bisporus* is bipolar, two nuclei are of one parental mating type and two of the other. Sterigmata have been formed and the basidiospore initials have begun to swell. D, Two nuclei have moved into one spore and two into the other. The basidia of other Agarics (e.g. *Coprinus cinereus*) are very similar, but bear four spores each of which receives a single nucleus.

appears at the base of the basidiospore and in the course of a few minutes expands. The spore is then shot off, by a 'surface-tension catapult' mechanism, described in detail later (pages 217–219). Although the initial launching speed is high, basidiospores are small so the momentum is such that they are discharged only a short distance. This is essential, otherwise they would be deposited on the opposite gill. Instead, gravity results in the basidiospores falling clear of the gills to be dispersed by air currents. Actively discharged basidiospores, which occur throughout the Hymenomycetes, are termed ballistospores (page 217).

Basidium maturation commences at the base of the gills and moves upwards. Following basidiospore discharge, the exhausted area of the gill is lysed by chitinases. Its removal eliminates an obstacle to falling basidiospores, enabling gills to be close to each other and not absolutely vertical in orientation. Gill autolysis produces black droplets, hence the popular name for members of the genus *Coprinus*, ink caps. A large ink cap, *Coprinus comatus*, is estimated to produce a total of about nine thousand million basidiospores in its active period of 2–3 days, or about $30\,000$ s^{-1}.

Since fruiting bodies of *Coprinus cinereus* are produced only on A≠B≠ mycelium, the diploid basidium nucleus is heterozygotic for both mating type factors, having a genetic constitution with respect to mating type of, for example, $A_1A_2B_1B_2$. This means that a single basidium can produce four basidiospores each of which has a different mating type specificity, e.g. A_1B_1, A_1B_2, A_2B_1, A_2B_2. Such a bifactorial mating system is hence also termed tetrapolar. The way that it promotes outbreeding is discussed in Chapter 5 (Table 5.2).

The large ink cap *Coprinus comatus* also has many mating types, although they result from the occurrence of many alleles of one instead of two mating type genes. This means that a basidium nucleus with, for example, the mating genotype A_1A_2, can give rise to only two basidiospore mating types, A_1 and A_2. Such mating type determination is termed unifactorial or bipolar. Its effectiveness in promoting outbreeding is discussed in Chapter 5 (page 250, Table 5.1). Another ink cap, *Coprinus sterquilinus*, is self-fertile. A single basidiospore gives a mycelium that after a short period of growth produces cells with two nuclei and clamp connections and is able to give rise to fruiting bodies.

The genus *Agaricus* is a large one, with many edible species. *Agaricus bisporus*, relatively uncommon in nature, is well known as the cultivated mushroom. Basidiospores of the cultivated mushroom are difficult to germinate, but some germination can be obtained by placing spores on an agar surface in the same vessel as growing mycelium of the fungus, which produces a volatile factor stimulating germination of spores of the same species. A single spore that germinates will usually yield a mycelium capable of producing fruiting bodies. The mycelium has many nuclei per cell, and nuclear migration and clamp connections do not occur. The existence of a mating factor that occurs in two forms, A_1 and A_2 has been established. In a basidium an A_1 and an A_2 nucleus fuse to give a diploid nucleus that undergoes meiosis to yield two A_1 and two A_2 nuclei. The two basidiospores commonly receive a nucleus of each type and hence a single spore can usually produce a self-fertile mycelium. *A. bisporus* thus seems to have evolved to a self-fertile state from a self-sterile ancestor with a bipolar mating system. Bipolar Basidiomycetes commonly have a large number of A factors, and the occurrence of only two A factors in cultivated strains of *A. bisporus* suggests that the cultivated mushroom arose from a single wild source rather than having been repeatedly introduced into cultivation. The origin of the cultivated mushroom and its cultivation are considered further in Chapter 8.

The gills of *Agaricus* are wedge shaped, broad at the base and tapering downwards, and are precisely oriented in the vertical plane by geotropism. This permits discharged spores to fall clear of the gills. Such an arrangement is usual in the agarics, the wave of spore maturation and gill autolysis in *Coprinus* being exceptional.

The Boletales
Fruit bodies in this order have a central stalk like those in the Agaricales, but the hymenium lines vertically oriented pores instead of gills (Fig. 2.22B). Some genera, such as *Rhizopogon* and *Suillus*, are important as mycorrhizal fungi (Figs. 7.13, 7.14).

The Poriales
Members of this order are important agents of timber decay (Chapter 6). Some, such as *Coriolus versicolor* (Fig. 6.19A) and *Fomes fomentarius* (Fig. 6.19C), have leathery or woody fruit bodies with the hymenium lining vertical pores. Others, such as the oyster fungus, *Pleurotus ostreatus* (Fig. 6.19A) and shiitake, *Lentinus edodes* (Fig. 8.11C), have gills, and an attractive texture and flavour and are grown commercially for food.

The Schizophyllales
This order includes an intensively studied species, *Schizophyllum commune*, which is common on fallen branches. It produces basidia on gill-like lamellae on

the undersurface of small fruit bodies. Under dry conditions the fruit bodies curl up, but even after a couple of years in the dry state will uncurl in humid conditions and start discharging basidiospores within a few hours. *Schizophyllum* is easily grown in pure culture on defined media and readily produces fruit bodies under such conditions. It is self-fertile with dikaryon formation and fruiting controlled, as in *Coprinus cinereus*, by two genetic factors, A and B. A study of natural populations identified 96 A and 56 B factors. Both the A and B factors have been shown to consist of two closely linked genes, $A\alpha$ and $A\beta$, and $B\alpha$ and $B\beta$. Many alleles have been found at each locus, and recombination between the α and β loci results in new specificities. Nine $A\alpha$ and 32 $A\beta$ alleles are known, and these recombined in different ways would give 9×32 or 288 different A factors. Similarly, the 9 $B\alpha$ and 9 $B\beta$ alleles known would allow 81 B factors. With many more α and β alleles likely in nature, the probable number of A and B factors existing is likely to be very large. In *Coprinus cinereus* too both the A and B factors are made up of closely linked genes. Both *S. commune* and *C. cinereus* have been used in studies on the mode of action of the A and B factors.

The Lycoperdales (Puff-balls), Nidulariales (Bird's Nest Fungi) and Phallales (Stinkhorns)

Members of these orders have basidiospores that are not ballistospores, and which detach from the basidia in a non-violent manner. In puff-balls (Fig. 4.10A, B) they are released into a cavity in the fruit body to form a powdery mass. Drops of rain striking the fruit body, or other pressures on it, can then eject puffs of spores through an aperture. Bird's nest fungi (Fig. 4.10C) have splash cups from which rain drops fling spore packages. In the stinkhorns (Fig. 4.10D) the spores form a sticky layer which attracts flies and other insects which effect dispersal. All these and other orders which lack ballistospores were formerly included in the Gasteromycetes.

The Auriculariales and Tremellales

Members of these orders have phragmobasidia, in which septa develop in the basidium after meiosis, rendering it multicellular. This contrasts with other orders, members of which have holobasidia, in which the basidium does not undergo septation, and is a single club-shaped cell which when mature carries the basidiospores – almost always four – at one end. A member of the Auriculariales, *Auricularia polytricha*, and a member of the Tremellales, *Tremella fusiformis*, are cultivated on a large scale for food in the Far East. Other species of *Tremella* have been used for studies on sex hormones (page 205) and mating systems (page 208). Many members of the Tremellales readily form a yeast phase, and some Basidiomycete yeasts (page 75) have been assigned to the Tremellales.

The Class Ustomycetes and Order Ustilaginales (smuts)

The largest order in the class Ustomycetes is the Ustilaginales or smuts, important plant pathogens. The smuts, of which about 1000 species are known, are biotrophic (page 391) pathogens of flowering plants, and cause economically important diseases of cereals. Masses of black sooty spores are formed, hence the popular name, smuts. The part of the plant which is commonly the most susceptible to infection and damage is the flower, so smuts can cause loss of seed, such as cereal grain. The smut of maize, *Ustilago maydis*, results in galls (hypertrophied host tissue containing fungus mycelium) that replace kernels in the maize cob. These galls are a Mexican delicacy, *huitlacoche*. The life cycle (Fig. 2.24) of this well studied species will be described.

The galls produced by *Ustilago maydis* on maize kernels break open to expose the sooty spores characteristic of smuts. The spores when mature are uninucleate and diploid and are known as **teliospores**, **teleutospores**, **chlamydospores** or sometimes **brandspores**. These spores, produced on the maturing maize cobs in the autumn, are both a means of dispersal and the way in which the fungus

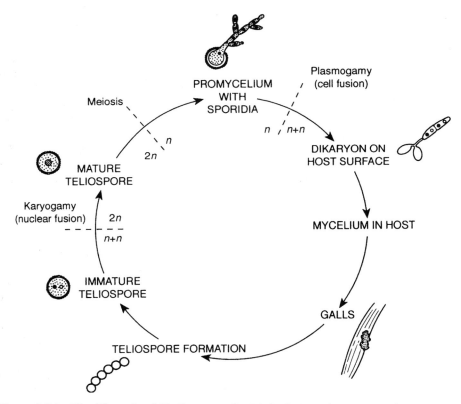

Figure 2.24   The life cycle of *Ustilago maydis*. Diploid ($2n$) teliospores undergo meiosis when they germinate to give haploid (n) promycelia. The promycelium bears sporidia, equivalent to basidiospores. Sporidia can bud like yeasts, and probably multiply in this way on plant debris and other organic matter at the soil surface. If two sporidia of compatible mating type germinate in proximity on the surface of a maize plant, they mate to give a dikaryon ($n + n$) able to initiate infection. Galls composed of mycelium and host tissue arise on stems, leaves and corn cobs. Mycelial cells finally round off to give teliospores, during the maturation of which nuclear fusion gives the diploid ($2n$) state.

survives the winter. Since mature teliospores are diploid they constitute a stage comparable to the basidia of other Basidiomycetes. Meiosis occurs in the germinating spore and a **promycelium** with four cells each containing a haploid nucleus is produced. Each cell produces a bud, the **sporidium** (pl. **sporidia**) which is regarded as equivalent to a basidiospore. The cells of the promycelium, however, are capable of repeated sporidium production, and the sporidia are able to multiply by budding. The haploid sporidial phase grows readily in pure culture on nutritionally simple media, forming yeast-like cells and compact colonies on agar. Mating is controlled by two genetic factors, a and b.

Only two alleles occur at the a locus, a_1 and a_2. Fusion between haploid cells only occurs if they differ at the a locus. The a alleles each have two regions. One region specifies a diffusible sex hormone that attracts the conjugation tube that emerges from a nearby cell of the opposite mating type. The other region specifies the receptor for the sex attractant produced by the opposite mating type. Hence, with two compatible cells close to each other, the conjugation tubes home in on each other and fusion results. There are at least 30 alleles at the b locus. *U. maydis*, with two alleles at one mating type locus and many at a second, thus displays what has been termed modified tetrapolar incompatibility (page 251). The development of a dikaryotic mycelium and growth within the host plant takes place only if fusion was between haploid cells that carried different b factors. Mating can be carried out on agar media but attempts to grow the dikaryotic mycelial phase in pure culture have had very limited success. The dikaryotic mycelium that grows within the plant may send hyphae to the surface where dikaryotic conidia are produced and dispersed to initiate further infections. Nuclear fusion occurs in the developing brandspore to give the diploid condition. Since mating is controlled by two genetic factors, sporidia of four different mating types can be produced by meiosis during the germination of a single brandspore.

Another smut that has received considerable study is *Ustilago violacea*, the anther smut of the Caryophyllaceae (carnations and campions). Infection results in the production of brandspores instead of pollen in the anthers. Two hosts, the campions *Silene dioica* and *Silene alba*, have separate male and female plants. Infection of female plants results in partial suppression of the female parts of the flower and the production of anthers within which brandspores develop. *Ustilago violacea* has a unifactorial mating system with two alleles, a_1 and a_2. This is more common in the Ustilaginales than is the bifactorial system of *Ustilago maydis*. Mating in *U. violacea* has received considerable study. Contact is first established between compatible cells by means of fimbriae, long proteinaceous hairs produced by both mating types and visible only by electron microscopy. A conjugation tube grows from the a_2 cell and fuses with a peg it induces in the a_1 cell.

Another order, the **Sporidiales**, includes some important yeasts, such as *Rhodosporidium* (page 75).

The Class Teliomycetes and Order Uredinales (rusts)

The larger of the two orders in the class Teliomycetes is the Uredinales or rusts, so called because of the rust-coloured masses of spores produced by many species on the plants that they infect. About 7000 species are known. All are biotrophic parasites, attacking ferns, conifers and especially flowering plants. Some, for example *Puccinia graminis* (Fig. 2.25), which attacks cereals and grasses, are economically important. The variety of *P. graminis* that attacks wheat has been intensively studied. The diploid stage, the **teliospore** or **teleutospore**, develops in the autumn on wheat plants and survives the winter in stubble. Germination and meiosis occur in the spring and four basidiospores are produced on sterigmata. These are of two mating types, designated plus (+) and minus (−). The actively discharged basidiospores cannot infect wheat plants but will infect a second host, the barberry, in which a haploid monokaryotic mycelium is formed. Pustules

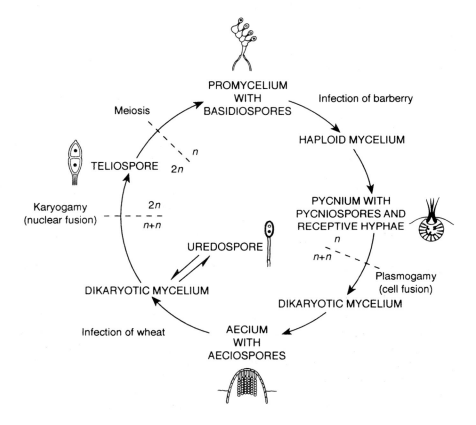

Figure 2.25   The life cycle of the rust *Puccinia graminis*. Teliospores (teleutospores) develop on wheat plants during the autumn, with nuclear fusion occurring during their formation to give the diploid ($2n$) state. The teliospores survive the winter in stubble. Meiosis to give the haploid (n) state occurs during germination, and a promycelium which bears four basidiospores emerges from each of the two compartments of the teliospore. The basidiospores infect barberry plants, and pycnia develop on the upper surface of leaves. These contain pycniospores (spermatia), which ooze from the pycnia, and receptive hyphae. Insects can carry pycniospores to the receptive hyphae of pycnia of the opposite mating type, resulting in the establishment of dikaryotic ($n + n$) mycelia. Dikaryotic aeciospores formed in aecia on the lower surface of barberry leaves are dispersed, and may infect wheat plants. Uredospores spread the infection among wheat plants, but in early autumn their production is gradually replaced by that of teliospores.

develop at the leaf surface. These have **pycniospores** (spermatia) and **receptive hyphae** and produce a sweet-smelling sugary solution that attracts insects. The insects may carry pycniospores between pustules formed by strains of different mating type. If a pycniospore of one mating type reaches a receptive hypha of the other, cell fusion occurs and a dikaryotic mycelium is established. The dikaryotic mycelium produces binucleate, dikaryotic **aeciospores** that can infect wheat but not barberry. In the wheat plant binucleate dikaryotic **uredospores** are produced, which spread the infection among wheat plants, and finally teliospores in the production of which nuclear fusion occurs to give the diploid condition once more. Some rusts have a less complex life cycle and only one host. About 30 species have been grown in pure culture, although growth under such conditions is very slow.

The Mitosporic Fungi

Mitosporic fungi were at one time referred to as Fungi Imperfecti or imperfect fungi, and later as Deuteromycetes. The 'imperfection' that leads to classification as a mitosporic fungus is the absence of any sexual stage in the life cycle. Not all fungi, however, that are permanently anamorphic are assigned to the mitosporic fungi. For example, anamorphic fungi that have sporangia resembling those of Zygomycetes are assigned to Zygomycete genera, and those that are clearly anamorphic rusts, to rust genera. Most fungi classified as mitosporic fungi are likely to have arisen from Ascomycetes through loss of the sexual phase, although some are anamorphic Basidiomycetes. About 15 000 mitosporic fungi are known. Mitosporic fungi can be divided informally into three classes, the Hyphomycetes, the Agonomycetes and the Coelomycetes. Mitosporic fungi that usually multiply by budding can be included in the Hyphomycetes but are here dealt with as anamorphic yeasts (page 75).

The Hyphomycetes

Hyphomycetes are abundant in the soil, and many are of importance as plant pathogens, as agents of biodeterioration, and in fermentation technology. Their accurate identification is hence of importance, and there is an extensive literature on the taxonomy of *Aspergillus*, *Penicillium* and other common genera. Most Hyphomycetes bear conidia on separate conidiophores, but in some the conidiophores adhere to give synnemata. Spore form and the way in which the spores are arranged on the conidiophores has been extensively used in classification, but increasing use is being made of the details of the way in which the conidia are produced.

The Agonomycetes

Agonomycetes are 'sterile' in the sense of lacking spores and so are sometimes termed Mycelia Sterilia. Some species produce sclerotia (page 171), and some, predators on eelworms (page 429), the traps that entangle their prey, but many have few features that aid identification.

The Coelomycetes

The Coelomycetes, many of which infect plants, are distinguished from the Hyphomycetes by the production of conidiophores and conidia within **conidiomata** (sing. **conidioma**), structures which with plant pathogenic species may consist of plant as well as fungus tissue. A common type of conidioma is the **pycnidium** (pl. **pycnidia**), which is a flask-shaped structure resembling the perithecium of Ascomycetes. Classification of Coelomycetes is based mainly on conidium and conidioma form, but, as with the Hyphomycetes, increasing use is being made of the details of conidiophore and conidium development.

The Yeasts

Yeasts are fungi which occur predominantly or exclusively in a unicellular state. Many are found on the surfaces of plants where they are able to exploit naturally exuded nutrients and more copious exudates that may follow injury or senescence of plant parts. They are abundant on leaves and fruits and in the nectaries of flowers. Other yeasts are found on the surface and in the gut of animals, especially insects, and a few are pathogens of humans and warm-blooded animals (pages 433–438). The surfaces of plants are, however, the most common habitat. Here yeasts have to compete with fast-growing bacteria in the exploitation of the nutrients present. Droplets or films of liquid on the surface of plants and animals will be liable to evaporation, leading to very high solute concentrations and even to desiccation. Alternatively, rain or dew may result in dilution. Hence microorganisms on plant and animal surfaces are liable to experience violent fluctuations in ambient water potential. As will be discussed later (page 163), an approximately spherical form, and perhaps especially the cell multiplication by budding characteristic of yeasts, seems to be more appropriate than the hypha and its apical growth for coping with such fluctuations in water potential.

The production of a yeast form is very widespread in fungi. Relatively few species, however, have given up the advantages of mycelial growth that permit success in many environments. There are only a few hundred species that, on the basis of an exclusively or predominantly unicellular form, are regarded as yeasts. Their growth in the natural environment seems to be restricted to the limited class of habitats already discussed, with a frequent occurrence in soil and water reflecting a capacity for prolonged survival during the course of dispersal. A few are pathogens of plants or animals, but the great significance of yeasts for humans lies in their ability to grow readily and with high metabolic rates in nutrient rich solutions and especially in sugary liquids. This has resulted in their importance in the production or spoilage of a variety of foods and beverages.

Some yeasts produce ascospores, some basidiospores and some are anamorphic, lacking a sexual phase.

Ascosporogenous or Ascomycete Yeasts

The ascospore-producing (ascosporogenous) yeasts are by definition Ascomycetes. They do not, however, form ascocarps, and so were formerly included in the Hemiascomycetes (page 46). Almost all are assigned to the order Saccharomycetales. Although most members of the order can bud and are regarded as yeasts, there are a few, such as *Dipodascus candidum*, which cannot do so and are hence not yeasts. A few species constitute a second order, the Schizosaccharomycetales. These increase in numbers by not by budding but by

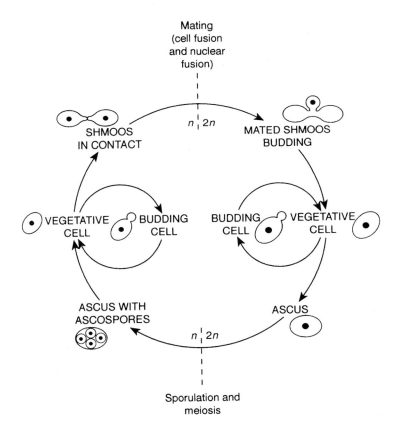

Figure 2.26 The life cycle of the Ascomycete yeast *Saccharomyces cerevisiae*. Diploid (*2n*) vegetative cells multiply by budding, but on nitrogen starvation may give rise to asci in which meiosis occurs to give four haploid (*n*) ascospores, two of each mating type. Ascospores germinate to give haploid vegetative cells which are slightly smaller than diploid cells. Although they can multiply by budding, proximity of cells of the opposite mating type leads to shmoo formation, mating and return to the diploid phase. In some yeasts ('haploid yeasts') budding occurs only in the haploid phase, and mating is followed by meiosis and ascospore formation. In others ('diploid yeasts'), ascospore germination is immediately followed by fusion between cells of opposite mating type and return to the diploid state. *S. cerevisiae* is commonly regarded as a diploid yeast, since mating usually soon follows ascospore germination; single cells can, however, be used to establish permanently haploid cultures.

the formation of a cross-wall followed by fission, so are informally known as the fission yeasts.

Saccharomyces cerevisiae

The best known of all yeasts is *Saccharomyces cerevisiae* (Figs. 2.26, 8.7); its natural habitat is the surface of fruit, but it has been used by man for thousands of years to produce alcoholic beverages (page 481) and bread (page 500). Biochemically and genetically it is one of the most intensively studied of all organisms, and is widely used in genetic manipulation as a host for the genes of other organisms, both for basic research and in biotechnology (page 533).

Saccharomyces – the name means 'sugar fungus' – metabolizes glucose via the glycolytic (Embden–Meyerhof) pathway to pyruvate. If oxygen is present the pyruvate can be oxidized via the tricarboxylic acid cycle to carbon dioxide and water. In the absence of oxygen, or at high sugar concentrations (see below), alcoholic fermentation occurs, with ethanol and carbon dioxide being produced. Alcoholic fermentation by *Saccharomyces* is responsible not only for the production of beer and wine but also, through carbon dioxide formation, for the raising of the dough in bread making.

The cells of *Saccharomyces cerevisiae* are normally oblate spheroids – they have two axes of equal length and the third longer. Sometimes the third axis is only a little longer and hence the cell is nearly spherical, and sometimes it is much longer giving an elliptical outline. Cells usually have a prominent central vacuole. The nucleus is small with a haploid DNA content about four times that of *Escherichia coli*. Since the haploid chromosome number is 16, on average the chromosomes are smaller than the chromosome of *E. coli*. In 1996 *S. cerevisiae* became the first eukaryote to have its DNA sequence fully determined. About 6000 genes are present. Mitochondria vary in number and shape, but several are normally present with one larger and more branched than the rest. Cells increase in number by budding (Figs. 3.6 and 8.7). A bud is at first small but rapidly swells to reach the same size as the mother cell. Meanwhile nuclear division has occurred and one nucleus has passed into the daughter cell. Finally the two cells separate. The process of budding leaves a birth scar on the daughter cell and a bud scar on the mother cell. Birth and bud scars can be distinguished by scanning electron microscopy. A mother cell will carry a single birth scar and bud scars corresponding in number to the daughter cells that it has produced. A non-growing population of yeast consists of single cells, but if rapid growth is occurring many buds and pairs of cells will be seen. Sometimes separation of daughter cells occurs more slowly than their production and clusters result, mother cells being attached to several daughter cells, some of which may themselves have become mother cells and carry buds or daughter cells.

S. cerevisiae is able to grow rapidly in rich media under anaerobic conditions, a cell population doubling in about 90 minutes. However, since the energy yield from fermentation is low, about 98% of the glucose present is metabolized to provide energy and only 2% is incorporated into cell materials. Cell yield per gram of sugar metabolized is hence low. A high level of glucose in the cell will suppress aerobic metabolism even if oxygen is present, a phenomenon known as **catabolite repression** or the **Crabtree effect**. Hence on a rich medium with abundant sugar, fermentation of glucose to ethanol and carbon dioxide occurs even in aerobic conditions. However, when the glucose supply is exhausted, ethanol is converted into pyruvate. Some is then metabolized to carbon dioxide and water via the tricarboxylic acid cycle to yield energy. In addition some is converted back into the sugars needed for wall synthesis by a reversal of the glycolytic pathway, a process known as **gluconeogenesis** (gluco-neo-genesis). Since aerobic metabolism yields much more energy than fermentation, approximately 10% instead of 2% of the glucose supplied is converted into cell material. This increased cell yield under aerobic conditions per gram of sugar supplied is known as the **Pasteur effect**.

The life cycle of *S. cerevisiae* is illustrated in Fig. 2.26. The strains employed in baking and brewing are commonly diploid. Sporulation can usually be induced in such diploid strains. Aerobic conditions are essential, and a suitable medium will lack assimilable nitrogen and sugars and contain acetate or glycoxylate as the carbon source. Under these circumstances vegetative growth cannot occur and the operation of the glyoxylate cycle is favoured. A proportion of the vegetative cells will then act as asci. Within an ascus meiosis occurs and four ascospores are formed. The ascus wall is not readily lysed, except by treatment with the enzyme glucuronidase, but presumably in nature breakdown ultimately occurs to release ascospores. The ascospores are resting cells, more resistant to adverse conditions than vegetative cells. An ascospore will germinate on a suitable medium containing assimilable sugar to yield a haploid vegetative cell. Such a cell is smaller than a diploid cell but has a similar metabolism and can bud and multiply in the same way. The isolation and germination of single ascospores will give haploid strains that can be propagated indefinitely in the vegetative state. Return to the diploid condition requires mating between two haploid cells differing in mating type. Two mating types occur and are designated **a** and α. Cells of mating type α produce α-factor, a peptide sexual hormone (Fig. 4.7). This acts on cells of mating type **a** causing their cell cycle to be arrested in phase G1 (page 106) and their surface properties to change so that they readily adhere to cells of mating type α. Cells of mating type **a** produce a-factor, another peptide sexual hormone (Fig. 4.7). This acts in a similar way to α-factor but on cells of mating type α. The hormones also cause the yeast cells to change in shape, becoming 'shmoos' (Fig. 4.8). These have projections which fuse with corresponding projections from cells of the opposite mating type. Cell fusion is followed by nuclear fusion, initiating the diploid phase (Plate 7). The control of the mating process in *S. cerevisiae* is considered further in Chapter 4 (page 207) and the basis for self-fertility in some strains in Chapter 5 (page 252).

Other Ascomycete yeasts

The grouping of Ascomycete yeasts into genera is based largely on ascus and ascospore morphology, but nutritional and biochemical tests are important in delimiting species. Some yeasts differ from *Saccharomyces* in being exclusively oxidative in their metabolism, and others are only weakly fermentative. Vegetative form is also of importance; some yeasts are able to produce mycelium and some are not. *Schizosaccharomyces* (Fig. 2.27C, D), which like *Saccharomyces* has been intensively studied, is known as the fission yeast because

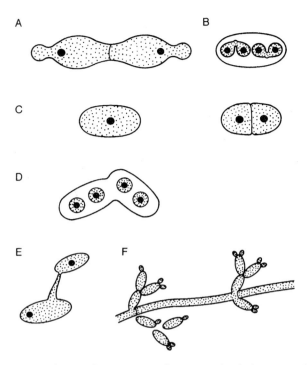

Figure 2.27   Diversity of yeast form. A, *Saccharomycodes ludwigii*. Budding is from a broad base and from the ends of cells. The two mother cells have not yet separated. B, *S. ludwigii*. Ascospores germinate and mate within the ascus to restore the diploid state. C, *Schizosaccharomyces pombe*. Cell division is by binary fission. D, *S. pombe*. Ascospore production immediately follows mating to restore the haploid state. E, *Sporobolomyces* sp. Production of a ballistospore, later forcibly discharged, on a sterigma. F, *Candida albicans*. Production of budding yeast cells, from a hypha.

its rod-shaped cells increase in number by fission, instead of budding. After nuclear division a cross-wall is formed halfway along the cell; then the two daughter cells so formed separate.

The life cycles of Ascomycete yeasts appear varied and complicated. They are, however, basically similar to that of *Saccharomyces*, except that in some, diploid vegetative multiplication is lacking and in some, haploid multiplication. In *Schizosaccharomyces*, for example, mating is immediately followed by meiosis. It is hence regarded as a haploid yeast. In *Saccharomycodes ludwigii* (Fig. 2.27A, B), however, ascospores of different mating type fuse within the ascus to restore the diploid state. It is hence a diploid yeast. Although *Saccharomyces* is capable of both haploid and diploid budding, it is commonly regarded as a diploid yeast. This is because, unless cultures have been directly or indirectly derived from a single ascospore, encounters between cells of different mating type and the restoration of the diploid state will soon occur, and because the strains used in baking and brewing are diploid.

Basidiosporogenous or Basidiomycete Yeasts

There are many Basidiomycetes that have been shown to produce a yeast phase in which cells multiply by budding. This does not usually result in such fungi being referred to as yeasts. This term is usually restricted to fungi long recognized as yeasts and subsequently found to produce a sexual phase with Basidiomycete features. Some Basidiomycete yeasts have been assigned to the Tremellales (class Basidiomycetes) and others to the Sporidiales (class Ustomycetes).

A well-studied Basidiomycete member of the Sporidiales is *Rhodosporidium toruloides*, long known in its haploid anamorphic state as *Rhodotorula glutinis*. This species has two mating types, and following conjugation between two cells differing in mating type, a dikaryotic mycelium with clamp connections is formed. The mycelium bears teliospores, which contain single diploid nuclei formed by nuclear fusion during their development. When these thick-walled resting spores germinate, meiosis occurs, and the short promycelium, the equivalent of a basidium, carries sporidia (basidiospores). These, on dispersal, bud to give the yeast phase so completing the life cycle. An example of a yeast assigned to the Tremellales is *Filobasidiella neoformans*, the teleomorph of *Cryptococcus neoformans*, a pathogen of humans responsible for cryptococcosis (page 436). Basidiomycete yeasts, like other Basidiomycetes, have a variety of mating systems. *F. neoformans*, like *R. toruloides*, is bipolar with two mating types. There are also species that have many mating types, some being bipolar and some tetrapolar, and species that are self-fertile.

Asporogenous or Anamorphic Yeasts

'Asporogenous' means 'not forming spores' and asporogenous yeasts are those which do not undergo sexual sporulation. Many, however, do produce spores asexually, so the term 'anamorphic yeasts', referring to the absence of a sexual cycle, is perhaps preferable.

An example of an anamorphic yeast that produces spores is *Sporobolomyces* (Fig. 2.27E), the anamorph of *Sporidiobolus* (class Sporidiales). It is a pink yeast that is very common on the surface of leaves. Cells of *Sporobolomyces*, as well as multiplying by budding, can produce a single ballistospore on a sterigma. Inversion of a Petri dish in which cells have been streaked on an agar medium results in a faint reproduction of the pattern of streaking on the Petri dish lid. This 'shadow' or 'reflection' on the lid is the result of ballistospore discharge and fall on to the lid. The ballistospore-producing yeasts have hence been termed the shadow or mirror yeasts.

Most anamorphic yeasts do not form ballistospores. Many species are of importance in food spoilage, as pathogens of humans or animals or in the fermentation industry. The identification of species is mainly on the basis of nutritional and biochemical tests, and species are grouped into genera using such features as the pattern of budding, whether the cells are pigmented, whether mycelium is formed and the kinds of spores produced, if any. The genera are often artificial, in the sense of bringing together species that are only distantly related.

Teleomorphs have not yet been found in some genera, but increasingly molecular and ultrastructural detail will make it possible to judge whether purely anamorphic species are related to Ascomycetes or to Basidiomycetes.

The Lichens

Lichens (Figs. 2.28, 2.29, Plate 1) can be loosely grouped into crustose, foliose and fruticose forms. **Crustose** species form a crust so firmly attached to rock or bark as to make removal difficult. **Foliose** species (*folium*, Latin, a leaf) are leaf-like and more loosely attached to the substratum. **Fruticose** species (*fruticosus*, Latin, like a bush) are varied in form and are free from the substratum except at the point of attachment; some occur on the ground and others festoon trees. About 14 000 species are known. They are world wide in distribution but are most striking, both in abundance and variety, in humid areas free from human disturbance, particularly atmospheric pollution. Moist conditions are needed for growth but many species can survive prolonged drought. Lichens are found in both hot and cold deserts and dominate about 8% of the earth's surface.

A lichen species is an intimate symbiotic association of a fungus – nearly always an Ascomycete but sometimes a Basidiomycete – and a photosynthetic microorganism. The fungal component of a lichen (the **mycobiont**) is unique to that species and otherwise unknown in nature. The photosynthetic component (the **photobiont**) is an alga or a cyanobacterium. Common photobionts are the green alga *Trebouxia*, which is uncommon except as a lichen component, and the green alga *Trentepohlia* and the cyanobacterium *Nostoc*, which are widespread in the free-living state. The cells of the photobiont are in close contact with those of the mycobiont. They may be limited to a distinct layer or, more rarely, distributed throughout the lichen. The photobiont provides the association with the products of photosynthesis, and, if a cyanobacterium, also those of nitrogen fixation. The benefits it receives are less obvious, but may include protection from desiccation, screening from excessive light and the provision of mineral nutrients extracted from the substratum by the fungus. Lichens grow at a very slow rate – often only a few millimetres per year – and may be very long-lived. The situations in which they are found are often too poor in nutrients or too exposed to be colonized by other organisms. Some even occur within the translucent outer layers of porous stones in the cold deserts of Antarctica. These slow-growing cryptoendoliths (Greek, hidden in stones) may be the longest-lived organisms on earth. Lichens have provided chemists with an immense variety of novel substances, including one of the most useful of pH indicators, litmus.

Lichens may be dispersed in a variety of ways. Fragments of dried lichen may get scattered by wind. Some species release **soredia**, small clumps of algal cells and fungal mycelia, whereas others produce **isidia**, in effect miniature lichens, easily detached and dispersed. Most species have perithecia or more commonly apothecia containing asci from which ascospores are discharged, usually forcibly. These fruiting bodies are long-lived and may continue to discharge ascospores for

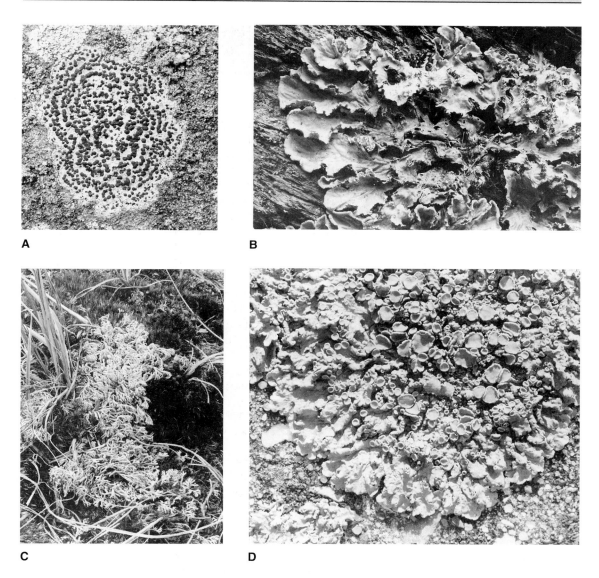

Figure 2.28 Diversity of form in lichens. A, *Rhizocarpon concentricum*, a crustose (encrusting) lichen, with concentric rings of black apothecia. It is widespread on slightly basic, siliceous rocks. B, *Peltigera membranacea*, a foliose (leaf-like) lichen, here on bark, but common also in grass in damp meadows and even by roadsides. C, *Cladonia arbuscula*, a fruticose (bush-like) lichen, among sedges and mosses, it is common on acid heathlands and peat moors in upland areas. D, *Xanthoria parietina*, a placodioid lichen–crustose at the centre but lobed and detachable at the margin. There are numerous nearly circular apothecia. This lichen is unusual in favouring nutrient-rich sites, such as roofs below TV aerials on which birds perch, and in being resistant to all but the most extreme air pollution. (A–D reproduced with permission from Dobson, F. S. (1992). *The Lichens: an Illustrated Guide to British and Irish Species*, 3rd edn. Richmond Publishing Company, Slough.)

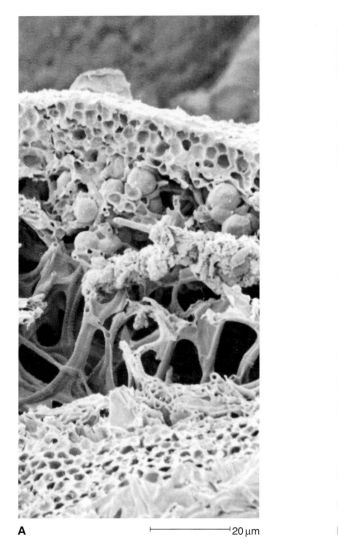

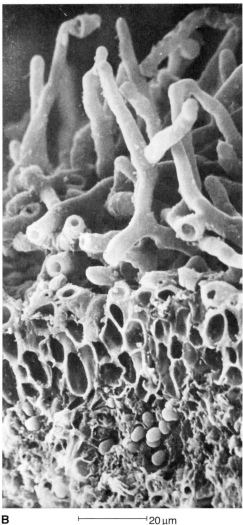

A   ⊢————————⊣ 20 μm   B   ⊢————————⊣ 20 μm

Figure 2.29   Scanning electron micrographs of sections across lichen thalli. A, *Xanthoria* sp. At the top there is a dense cortex of fungal cells, then a zone in which, in addition to fungal hyphae, there are spherical cells (diam. ca 7 μm) of *Trebouxia*, a green alga uncommon except in lichens. Fungal hyphae cross the medulla, the central area with air spaces, to reach the lower cortex. B, *Peltigera canina*. At the top are fine hairs (fungal hyphae), then the palisade-like cortex, also composed of fungal hyphae, and at the bottom the algal zone. This contains spherical cells (diam. ca 4 μm) of *Nostoc*, a nitrogen-fixing member of the cyanobacteria, a group formerly known as the blue–green algae. (A, B.E. Juniper. B, Bronwen Griffiths.)

years, unlike the short-lived fruiting bodies of non-lichenized Ascomycetes. Presumably if an ascospore germinates close to an appropriate alga the lichen is re-established. In a few species such proximity is guaranteed since the ascospores carry cells of the photobiont on their surface when they are discharged.

Many lichens are very sensitive to atmospheric pollutants, especially to sulphur dioxide. This has led to the disappearance of many species from industrialized countries over the last century, although some species are moderately and a few very tolerant of pollutants. These differences in sensitivity enable the severity of pollution in different areas to be evaluated by observing the presence or absence of lichen species with known degrees of tolerance.

Careful attention to environmental conditions is needed for lichens to be maintained in the laboratory, otherwise they will die or dissociate into photobiont and mycobiont. The germination of ascospores and isolation of single algal cells has permitted a number of photobionts and mycobionts to be established in pure culture. The growth of the mycobiont is commonly extremely slow, the production of a medium-sized colony taking many months instead of the few days characteristic of most free-living fungi. The early stages in the establishment of a lichen have been studied by bringing together an appropriate mycobiont and photobiont. Sparse nutrition is required, in order to prevent excessive growth of one of the partners, and the simulation of natural conditions, especially alternate wetting and drying. Careful attention to such factors has led to the establishment of several species in pure culture throughout their life cycles.

Further Reading and Reference

General Works on Fungal Diversity

Alexopoulos, C. J., Mims, C. W. & Blackwell, M. (1996). *Introductory Mycology*, 4th edn. Wiley, New York.

Hawksworth, D. L. (1991). The fungal dimension of biodiversity: magnitude, significance and conservation. *Mycological Research* **95**, 641–655.

Hawksworth, D. L., Kirk, P. M., Pegler, D. N. & Sutton, B. C. (1995). *Ainsworth & Bisby's Dictionary of the Fungi*, 8th edn. CAB International, Wallingford.

Isaac, S., Frankland, J. C., Watling, R. & Whalley, A. J. S, ed. (1993). *Aspects of Tropical Mycology. Nineteenth Symposium of the British Mycological Society*. Cambridge University Press, Cambridge.

McLaughlin, D. J., McLaughlin, E. G. & Lemke, P. A. (2000). *Systematics and Evolution*. Vol 7B, *The Mycota*, K. Essen & P. A. Lemke, eds. Springer-Verlag, Berlin.

Onions, A. H. S., Alsopp, D. & Eggins, H. O. W. (1981). *Smith's Introduction to Industrial Mycology*, 7th edn. Arnold, London.

Reynolds, D. R. & Taylor, J. W., eds. (1993). *The Fungal Holomorph: Mitotic, Meiotic and Pleomorphic Speciation in Fungal Systematics*. CAB International, Wallingford.

Taylor, J. W., Bowman, B., Berbee, M. L. & White, T. J. (1993). Fungal Model Organisms: Phylogenies of *Saccharomyces*, *Aspergillus* and *Neurospora*. *Systematic Biology* **42**, 440–457.

Webster, J. (1980). *Introduction to Fungi*, 2nd edn. Cambridge University Press, Cambridge.

Wessels, J. G. H. (1999). Fungi in their own right. *Fungal Genetics and Biology* **27**, 134–145.

Cellular Slime Moulds

Ashworth, J. M. & Dee, J. (1975). *The Biology of Slime Moulds*. Arnold, London.

Aubry, L. & Firtel, R. (1999). Integration of signalling networks that regulate *Dictyostelium* differentiation. *Annual Review of Cell and Developmental Biology* **15**, 469–517.

Kessin, R. H., Gundersen, G. G., Zaydfudim, V., Grimson, M. & Blanton, R. L. (1996). How cellular slime moulds evade nematodes. *Proceedings of the National Academy of Sciences of the United States of America* **93**, 4857–4861.

Loomis, W. F. (1975). *Dictyostelium discoideum: a Developmental System*. Academic Press, New York.

Newell, P. C. (1981). Chemotaxis in the cellular slime moulds. In Lackie, J. M. & Wilkinson, P. C., eds., *Biology of the Chemotactic Response*, pp. 89–114. Cambridge University Press, Cambridge.

Olive, L. S. (1975). *The Mycetozoans*. University of Iowa Press, Iowa.

Raper, K. B. (1984). *The Dictyostelids*. Princeton University Press, Princeton, New Jersey.

Plasmodial Slime Moulds (Myxomycetes)

Aldrich, H. C. & Daniel, J. W., eds. (1982). *Cell Biology of Physarum and Didymium*, 2 vols. Academic Press, New York.

Bailey, J. (1995). Plasmodium development in the myxomycete *Physarum polycephalum*: genetic control and cellular events. *Microbiology* **141**, 2355–2365.

Dove, W. F., Dee, J., Hatano, S., Haugli, F. B. & Wohlfarth-Bottermann, K. E., eds. (1986). *The Molecular Biology of Physarum polycephalum*. Plenum Press, New York.

Ing, B. (1999). *The Myxomycetes of Britain and Ireland*. Richmond Publishing Company, Slough.

Martin, G. W. & Alexopoulos, C. J. (1969). *The Myxomycetes*. University of Iowa Press, Iowa.

Sauer, H. (1982). *Developmental Biology of Physarum*. Cambridge University Press, Cambridge.

Stephenson, S. L. & Stempen, H. (1994). *Myxomycetes: A Handbook of Slime Moulds*. Timber Press, Portland, Oregon.

Oomycetes, Chytridiomycetes and Zygomycetes

Buczaki, S. T., ed. (1983). *Zoosporic Plant Pathogens: a Modern Perspective*. Academic Press, London.

Cerdá-Olmedo, E. & Lipson, E. D., eds. (1987). *Phycomyces*. Cold Spring Harbor Laboratory, New York.

Couch, J. N. & Bland, C. E., eds. (1985). *The Genus Coelomomyces*. Academic Press, Orlando, Florida.

Erwin, D. C., Bartnicki-Garcia, S. & Tsao, P. H., eds. (1983). *Phytophthora: its Biology, Taxonomy, Ecology and Pathology*. American Phytopathological Society, St. Paul, Minnesota.

Fuller, M. S. & Jaworski, A., eds. (1987). *Zoosporic Fungi in Teaching and Research*. Southeastern Publishing Corporation, Athens, Georgia.

Ingram, D. S. & Williams, P. H., eds. (1991). *Phytophthora infestans*, the cause of late blight of potato. *Advances in Plant Pathology*, volume 7, Academic Press, London.

Lucas, J. A., Shattock, R. C., Shaw, D. S. & Cooke, L. R., eds. (1991). *Phytophthora. Seventeenth Symposium of the British Mycological Society*. Cambridge University Press, Cambridge.

Mountford, D. O. & Orpin, C. G., eds. (1994). *Anaerobic Fungi: Biology, Ecology and Function*. Marcel Dekker, New York.

Spencer, D. M., ed. (1981). *The Downy Mildews*. Academic Press, London.

Trinci, A. P. J., Davies, D. R., Gull, K., Lawrence, M. I., Nielsen, B. B., Rickers, A. & Theodorou, M. K. (1994). Anaerobic fungi in herbivorous animals. *Mycological Research* **98**, 129–152.

Ascomycetes, Basidiomycetes and Mitosporic Fungi

Banuet, F. (1992). *Ustilago maydis*, the delightful blight. *Trends in Genetics* **8**, 174–180.

Barnett, H. L. & Hunter, B. B. (1998). *Illustrated Genera of Imperfect Fungi*, 4th edn. APS Press, St. Paul, Minnesota.

Braun, U. (1995). *The Powdery Mildews (Erysiphales) of Europe*. Gustav Fischer Verlag, Stuttgart.

Breitenbach, J. & Kranzlin, F. (1984–1996). *Fungi of Switzerland*. 4 vols. Verlag Mykologia, Lucerne.

Carmichael, J. W., Kendrick, B. W., Connors, I. L. & Sigler, L. (1980). *Genera of Hyphomycetes*, University of Alberta Press, Edmonton, Alberta.

Casselton, L. A. (1997). Molecular recognition in fungal mating. *Endeavour* **21**, 159–163.

Cole, G. T. & Kendrick, B., eds. (1981). *Biology of Conidial Fungi*, 2 vols. Academic Press, New York.

Cole, G. T. & Samson, R. A. (1979). *Patterns of Development in Conidial Fungi*. Pitman, London.

Coley-Smith, J. R., Verhoef, K. & Jarvis, W. R., eds. (1980). *The Biology of Botrytis*. Academic Press, London.

Cummins, G. B. & Hiratsuka, Y. (1983). *Illustrated Genera of Rust Fungi*, revised edn. American Phytopathological Society, St Paul, Minnesota.

Ellis, M. B. & Ellis, J. P. (1990). *Fungi Without Gills (Hymenomycetes and Gasteromycetes): An Identification Handbook*. Chapman & Hall, London.

Ellis, M. B. & Ellis, J. P. (1997). *Microfungi on Land Plants: An Identification Handbook*. 2nd edn, Richmond Publishing Co., Slough.

Ellis, M. B. & Ellis, J. P. (1998). *Microfungi on Miscellaneous Substrates: An Identification Handbook*. Richmond Publishing Co., Slough.

Ginns, J. & Lefebvre, M. N. L. (1993). *Lignicolous Corticioid Fungi (Basidiomycota) of North America: Systematics, Distribution and Ecology*. APS Press, St Paul, Minnesota.

Glass, N. L. & Kuldau, G. A. (1992). Mating type and vegetative incompatibility in filamentous Ascomycetes. *Annual Review of Phytopathology* **30**, 201-224.

Hanlin, R. T. (1990). *Illustrated Genera of Ascomycetes*. The American Phytopathological Society, St Paul, Minnesota.

Hawksworth, D. L. (1994). *Ascomycete Systematics: Problems and Perspectives in the Nineties*. NATO ASI Series A, Life Sciences, Vol. 269. Plenum, New York.

Hibbett, D. S., Pine, E. M., Langer, E., Langer, G. & Donoghue, M. (1997). Evolution of gilled mushrooms and puffballs inferred from ribosomal DNA sequences. *Proceedings of the National Academy of Sciences, USA*, **94**, 12002–12006.

Hyde, K., ed. (1997). *Biodiversity of Tropical Microfungi*. University of Hong Kong Press, Hong Kong.

Kinghorn, J. & Martinelli, S., eds. (1994). *Physiology and Genetics of Aspergillus nidulans*. Chapman & Hall, London.

Kubicek, C. P. & Harman, G. E. (1998). *Trichoderma and Gliocladium*. Vol I: *Basic Biology, Taxonomy and Genetics*; Vol II: *Enzymes, Biological Control, and Commercial Applications*. Taylor & Francis, London.

Minter, D. W. (1984). New concepts in the interpretation of conidiogenesis in Deuteromycetes. *Microbiological Sciences* **1**, 86–89.

Moore, D., Casselton, L. A., Wood, D. A. & Frankland, J. C., eds. (1985). *Developmental Biology of Higher Fungi. Tenth Symposium of the British Mycological Society*. Cambridge University Press, Cambridge.

Moss, M. O. & Smith, J. E., eds. (1984). *The Applied Biology of Fusarium. Seventh Symposium of the British Mycological Society*. Cambridge University Press, Cambridge.

Nelson, P. E., Tousson, T. A. & Cook, R. J., eds. (1981). *Fusarium: Diseases, Biology and Taxonomy*. Pennsylvania State University, University Park and London.

Peberdy, J. F., ed. (1987). *Penicillium and Acremonium*. Plenum Press, New York.

Pegler, D. N. (1998). *Field Guide to the Mushrooms and Toadstools of Britain and Europe*. Larousse, London.

Phillips, R. (1981). *Mushrooms and other Fungi of Great Britain and Europe*. Pan Books, London.

Pitt, J. I. (1988). *A Laboratory Guide to Common Penicillium Species*. CSIRO Food Research Laboratory, North Ryde, New South Wales.

Powell, K. A., Renwick, A. & Peberdy, J. F., eds. (1994). *The Genus Aspergillus: From Taxonomy and Genetics to Industrial Applications*. Plenum Press, New York.

Ryvarden, L. & Gilbertson, R. L. (1993–94). *European Polypores*. 2 vols. Fungiflora, Oslo.

Samson, R. A. & Pitt, J. I., eds. (1990). *Modern Concepts in Penicillium and Aspergillus Classification*. Plenum Press, New York.

Scott, K. J. & Chakravorty, A. K., eds. (1982). *The Rust Fungi*. Academic Press, London.

Spencer, D. M., ed. (1978). *The Powdery Mildews*. Academic Press, London.

Subramanian, C. V. (1983). *Hyphomycetes: Taxonomy and Biology*. Academic Press, London.

Yeasts

Barnett, J. A., Payne, R. W., Yarrow, D. & Barnett L. (2000). *Yeasts: Characteristics and Identification*, 3rd edn. Cambridge University Press, Cambridge.

Campbell, I. & Duffus, J. H., eds. (1988). *Yeasts: a Practical Approach*. IRL Press, Oxford.

De Hoog, G. S., Smith, M. Th. & Weijman, A. C. M., eds. (1987). *The Expanding Realm of the Yeast-like Fungi*. Elsevier, Amsterdam.

Feldmann, H. (1999). The yeast *Saccharomyces cerevisiae*: insights from the first complete eukaryotic gene sequence. In R. P. Oliver & M. Schweizer, eds., *Molecular Fungal Biology*, pp. 78–134. Cambridge University Press, Cambridge.

Forsberg, S. L. & Nurse, P. (1991). Cell cycle regulation in the yeasts *Saccharomyces cerevisiae* and *Schizosaccharomyces pombe*. *Annual Review of Cell Biology* 7, 227–256.

Herskowitz, I. (1988). Life cycle of the budding yeast *Saccharomyces cerevisiae*. *Microbiological Reviews* **52**, 536–553.

Kurtzman, C. P. & Fell, J.W. (1996). *The Yeasts, a Taxonomic Study*, 4th edn. North-Holland, Amsterdam.

Nasim, A., Young, P. & Johnston, B. F., eds. (1989). *Molecular Biology of the Fission Yeast*. Academic Press, New York.

Pringle, J., Broach, J. R. & Jones, E. W., eds. (1997). *The Molecular and Cellular Biology of the Yeast Saccharomyces*. Cold Spring Harbor Laboratory Press, Cold Spring Harbor, New York.

Rose, A. H. & Harrison, J. S., eds. (1987–1993). *The Yeasts*, 2nd edn, 5 vols. Academic Press, London.

Russell, P. & Nurse, P. (1986). *Schizosaccharomyces pombe* and *Saccharomyces cerevisiae*: a look at yeasts divided. *Cell* **45**, 781–782.

Skinner, F. A., Passmore, S. M. & Davenport, R. D., eds. (1980). *Biology and Activities of Yeasts*. Academic Press, London.

Tuite, M. F. & Oliver, S. G., eds. (1991). *Saccharomyces*. Plenum Press, New York.

Walker, G. M. (1998). *Yeast Physiology and Biotechnology*. Wiley, Chichester.

Lichens

Bates, J. W. & Farmer, A. M., eds. (1992). *Bryophytes and Lichens in a Changing Environment*. Oxford University Press, Oxford.

Dalby, D. H., Hawksworth, D. L. & Jury, S. L., eds. (1988). *Horizons in Lichenology*. Academic Press, London.

Dobson, F. S. (2000). *Lichens: an Illustrated Guide to British and Irish Species*, 4th edn. The Richmond Publishing Company, Slough.

Gargas, A., DePriest, P. T., Grube, M. & Tehler, A. (1995). Multiple origins of lichen symbioses in fungi suggested by SSU rDNA phylogeny. *Science* **268**, 1492–1495.

Gilbert, O. (2000). *Lichens*. Harper Collins, London.

Hale, M. E. (1983). *The Biology of Lichens*, 3rd edn. Arnold, London.

Hawksworth, D. L. & Hill, D. J. (1984). *The Lichen-forming Fungi*. Blackie, Glasgow.

Honegger, R. (1991). Functional aspects of the lichen symbiosis. *Annual Review of Plant Physiology and Plant Molecular Biology* **42**, 553–558.

Kershaw, K. A. (1988). *The Physiological Ecology of Lichens*. Cambridge University Press, Cambridge.

Lawrey, J. W. (1984). *The Biology of Lichenized Fungi*. Praeger, New York.

Nash, T. H., ed. (1996). *Lichen Biology*. Cambridge University Press, Cambridge.

Price, R. C. (1992). The Methuselah Factor: age in cryptoendolithic communities. *Trends in Ecology and Evolution* **7**, 21.

Purvis, O. W., Coppins, B. J., Hawksworth, D. L., James, P. W. & Moore, D. M. (1992). *The Lichen Flora of Great Britain and Ireland*. Natural History Museum, London.

Richardson, D. H. S. (1975). *The Vanishing Lichens*. David & Charles, Newton Abbot.

Richardson, D. H. S. (1992). *Pollution Monitoring with Lichens*. The Richmond Publishing Company, Slough.

Richardson, D. H. S. (1999). War in the world of lichens: parasitism and symbiosis as exemplified by lichens and lichenicolous fungi. *Mycological Research* **103**, 641–650.

Seaward, M. R. D. (1988). Contributions of lichens to ecosystems. In: M. Galun, ed., *CRC Handbook of Lichenology*, vol 2, pp. 107–129. CRC Press, Boca Raton, Florida.

Questions

With each question one statement is correct. Answers on page 561.

2.1 Organisms which belong to the Chromista, not true Fungi, but are important members of the fungi in the broad sense and include many important plant pathogens:
a) Chytridiomycetes
b) Zygomycetes
c) Oomycetes
d) Basidiomycetes

2.2 The number of fungal species so far described is about:
a) 700
b) 7000
c) 70 000
d) 700 000

2.3 Fungi which form extensive mycelial systems consisting typically of hyphae with two different nuclear genotypes are found in the:
a) Ascomycetes
b) Basidiomycetes
c) Oomycetes
d) Myxomycetes

2.4 Many plants have mutualistic symbioses with fungi in the genus:
a) *Glomus*
b) *Mucor*
c) *Erysiphe*
d) *Phytophthora*

2.5 Lichens are formed by the association of:
a)   Ascomycetes or Basidiomycetes with algae or bacteria
b)   Ascomycetes only, with algae or bacteria
c)   Basidiomycetes only, with algae or bacteria
e)   Ascomycetes only, with algae only

Fungal Cells and Vegetative Growth

3

The Structure and Composition of Fungal Cells

An animal cell typically consists of a nucleus with associated cytoplasm bounded by a cell membrane. Plant cells are similar except for the presence of a cell wall external to the membrane. Yeasts, approximately spherical in form and with a single nucleus, are readily recognized as unicellular organisms. It is not easy to decide what constitutes a single cell when examining a fungal hypha, but ultrastructural studies of filamentous fungi demonstrate a typical eukaryotic cellular organization. It is important to realize that hyphae and yeast cells are just two forms of fungal growth, differing in the polarity shown by the cell as it grows, and that many fungi, particularly animal and plant pathogens, can grow as either form, dictated by environmental conditions. This phenomenon is known as **dimorphism**. Fungal cell structure will now be considered, stressing special features of the fungi rather than attributes common to most eukaryotic cells.

Form and Dimensions of Fungal Hyphae and Yeast Cells

The hyphae of a single species can differ considerably in diameter, depending on environmental conditions, their position in a colony, and the age of the colony. There are, however, clear differences between species. Thus the hyphae at the margin of a mature colony of *Aspergillus nidulans* are about 3–4 µm wide, whereas comparable hyphae in *Neurospora crassa* can have a diameter of 10 µm or more. The hyphae of these and other higher fungi are divided into compartments by cross-walls, septa (Fig. 3.1). In *A. nidulans* the apical compartment, including the hyphal tip, commonly has a length of 300–400 µm, and subsequent compartments an average length of about 50 µm. The hyphae of most lower fungi are not subdivided into compartments. Hyphae only reach their maximum diameter at a considerable distance behind the tip, as much as 30 µm in *N. crassa*. This gradual tapering towards the tip is not always obvious to the eye, and hyphal tips may appear approximately hemispherical. Most yeast cells, exemplified by *Saccharomyces cerevisiae*, are oval-shaped, about 5–10 µm in diameter, replicating by budding, but others, exemplified by the fission yeast *Schizosaccharomyces pombe*, are short rods, about 5 µm in diameter, replicating by growing longitudinally and dividing in the centre to give two equally sized daughter cells.

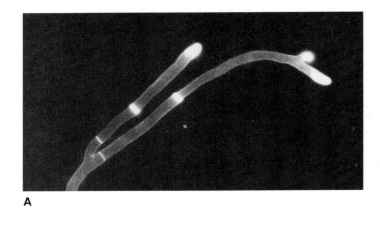

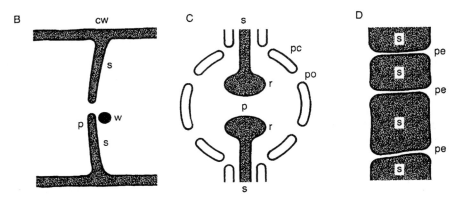

Figure 3.1   The hyphae and septa of higher fungi. A, Hyphae of *Botrytis cinerea* treated with the fluorescent stain Calcofluor White, and viewed by fluorescence microscopy. The stain is taken up most strongly in regions of active wall synthesis, about 10 μm at the hyphal tips, and developing septa. (Reproduced with permission from Gull, K. & Trinci, A. P. J. (1974). Detection of areas of wall differentiation in fungi using fluorescent staining. *Archives of Microbiology* **96**, 53–57.) B,C,D, Different types of septa, with increasing enlargement from B to D. Septa (s), pores (p), cell walls (cw). B, Simple septa with a single central pore occur in *B. cinerea*, most other mitosporic fungi and Ascomycetes and some Basidiomycetes. Cell organelles including nuclei and mitochondria can pass through the pore. If a compartment is damaged adjacent pores can be blocked rapidly by Woronin bodies (w). C, Dolipore septa are common in Basidiomycetes. Here only the area around the pore, which has a prominent rim (r), is shown. On either side of the pore is a membranous structure, the pore cap (pc), which is continuous with the endoplasmic reticulum and itself has pores (po). Cell organelles are able to pass through the septal pores. D, Multiperforate septa occur in some Ascomycetes and mitosporic fungi, such as *Endomyces* and *Geotrichum*. Only a small part of the septum, much enlarged, is shown. Very fine perforations (pe) cross the septum, through which protoplasmic strands (plasmodesmata) maintain cytoplasmic continuity between compartments. Organelles are unable to pass between compartments.

The Nucleus

The hyphal compartments of fungi may contain one or many nuclei. The apical compartment of *Aspergillus nidulans* can have about 50 nuclei and the smaller subsequent compartments each about four nuclei. In many Basidiomycetes the monokaryon has one nucleus per compartment and the dikaryon two, although some Basidiomycetes have many nuclei per compartment. Yeasts have a single nucleus per cell. The nuclei of the vegetative phase of most filamentous fungi are haploid, although those of Oomycetes, phylogenetically distinct from 'true' fungi, are diploid. Some yeasts including *Saccharomyces cerevisiae*, are capable of cell multiplication in both haploid and diploid phases of the life cycle, but in nature probably are diploid most of the time.

The nuclei of most fungi are small (*ca* 1–2 μm in diameter) compared with those of animals and plants and they have comparably small chromosomes. This renders cytological work difficult, but in a few instances cytological study, genetic mapping and pulsed field gel electrophoresis have resulted in firmly established haploid chromosome numbers and biochemical analysis has provided information on DNA content (Table 3.1). It emerges that the chromosomes of these species contain an amount of DNA similar to that of the prokaryotic

Table 3.1   Nuclear genome size and haploid chromosome number in fungi and other organisms

Group and species	Chromosome number	DNA, millions of base pairs	
		Nucleus	Average or range per chromosome
Prokaryotes			
Escherichia coli	1	4.6	4.6
Slime moulds			
Dictyostelium discoideum	7	50	7
Physarum polycephalum	*ca* 40	270	7
Oomycetes			
Achlya bisexualis	?	46	?
Zygomycetes			
Phycomyces blakesleeanus	?	31	?
Higher fungi			
Neurospora crassa	7	47	4–13
Aspergillus nidulans	8	25.4	3–5
Saccharomyces cerevisiae	16	12.8	0.2–2
Schizophyllum commune	6	36	1–5
Ustilago maydis	20	19	0.3–2
Tomato	12	2350	196
Human	23	3000	130

Compiled from various sources. For further data on nuclear genome size in filamentous fungi see Turner, G. (1993). Gene organization in filamentous fungi. In P. Broda, S. G. Oliver & P. F. G. Sims, eds. *The Eukaryotic Microbial Genome; Organization and Regulation. Fiftieth Symposium of the Society for General Microbiology*, pp. 107–125. Cambridge University Press, Cambridge.

chromosome and the haploid nucleus an amount only a few times greater than that of the prokaryote cell. Chromatin structure, however, resembles that of higher eukaryotes. Nucleosomes occur in which about 140 base pairs of DNA are wrapped around a disc consisting of four different histones (basic proteins) each present as two molecules. The histones show some differences in amino acid composition when compared with those of other eukaryotic organisms. The main differences are in the linking DNA between nucleosomes, which in fungi is shorter than the 60 base pair link of other organisms, and in the histone which is associated with the linking DNA, which throughout the eukaryotes is more variable than the disc histones. The chromosomes of *Aspergillus nidulans* and *Saccharomyces cerevisiae* contain a single linear double-stranded DNA molecule, and this is probably true of the chromosomes of other fungi. Little of this DNA (usually less than 10%) consists of repeated sequences, and it is likely that most of these sequences are accounted for by the repetition of the genes that code for ribosomal RNA. This contrasts with the situation in mammals and in the Myxomycete *Physarum polycephalum*, in which about one-third of the DNA is repetitive. About 60% of the genes from filamentous fungi that have been sequenced contain introns, DNA sequences that are transcribed but not translated. These sequences are short, generally less than 100 base pairs, as compared with those of mammals. In *S. cerevisiae* less than 5% of the genes sequenced have introns, but these introns on average are much longer. The gene density in *S. cerevisiae* is very high, with one gene being encountered for every two kilobases (kb) of DNA, in contrast to the human genome where the figure might be as low as one gene in 30 kb.

An enormous bank of information has been acquired by the complete sequencing of the genome of *Saccharomyces cerevisiae*, completed in 1996 – the first eukaryotic genome sequence. This has a total of 6000 genes, with open reading frames varying in size between 100 and more than 4000 codons, but most are less than 500 codons. About half of these have a known function or are homologous to genes with a known function. Unknown ones are being investigated systematically to determine their functions, using an increasingly powerful range of automated analytical microtechniques to investigate the expression of particular gene sequences under different conditions, and identification of expressed proteins to give the 'proteome' to accompany the genome. These techniques are also being used to investigate changes in total gene expression during cell development and during adaptations to different environmental conditions. Complete genome sequences are appearing for an increasing range of other fungi; including the fission yeast *Schizosaccharomyces pombe* and the human pathogen *Candida albicans*.

Mitochondria and Mitochondrial DNA

Mitochondria are the site of oxidative phosphorylation in eukaryotic cells. The enzymes concerned with electron transport and ATP production are located on the inner membrane of the mitochondrion, the area of which is increased by invaginations, the cristae. Slime moulds and Oomycetes resemble Protozoa in having tubular cristae, whereas those of other fungi are plate-like as in animals and plants. Mitochondria are derived from pre-existing mitochondria. In the sexual process, mitochondrial inheritance is often unisexual with the mitochondria being contributed by the larger cell. Thus in mammals the mitochondria come from the female gamete, and in *Neurospora crassa* from the proto-perithecium. In *Saccharomyces cerevisiae*, however, in which there is no obvious differentiation into male and female, both parents contribute mitochondria.

Mitochondria can vary in size, form and numbers during the cell cycle and in response to environmental conditions. Cells of *S. cerevisiae* may contain one or a few much-branched mitochondria or many, up to about twenty, small unbranched mitochondria. Fusion can occur between mitochondria, and fragmentation of a large mitochondrion can take place to yield several smaller ones. Under anaerobic conditions the membrane systems of *S. cerevisiae* mitochondria become less complex and electron-transfer enzymes such as cytochromes disappear. The mitochondria of *Physarum polycephalum* are simple in form and increase in numbers by binary fission. Much remains to be learnt about the population dynamics of mitochondria in yeasts and filamentous fungi.

Mitochondria contain DNA, which may form a nucleoid at the centre of the mitochondrion. The nucleoids in *P. polycephalum* mitochondria are particularly striking and can be demonstrated not only by electron microscopy but, with appropriate staining, by light microscopy. A nucleoid consists of one or several mitochondrial DNA (mtDNA) molecules, which can be extracted and examined by electron microscopy. An mtDNA molecule normally consists of a circle of double-stranded DNA, but in a few organisms, including the yeast *Hansenula mrakii*, the molecule is linear. The mtDNA constitutes the mitochondrial genome, and contains the genes for the transfer RNA (tRNA) and ribosomal RNA (rRNA) of the mitochondria and for some of the enzymes involved in oxidative phosphorylation. There are differences between organisms with respect to which enzymes or enzyme subunits are specified by mtDNA and which by the cell nucleus. Thus in humans subunit 9 of ATPase is specified by a nuclear gene, in *S. cerevisiae* by a mitochondrial gene, and in *N. crassa* the gene is present in both nucleus and mitochondrion. The mitochondrial genome of fungi differs in size between species (Table 3.2). The genome has been totally sequenced in several species, including a strain of *Schizosaccharomyces pombe* and in one of *Aspergillus nidulans*. In *S. pombe* the genome is only a little larger than the very compact human mitochondrial genome, but in *Saccharomyces cerevisiae* it is much larger, with considerable size differences between strains. In a strain of *S. cerevisiae* that has received detailed study, genes account for 16% of the mtDNA, introns for 22% and intergenic regions for 62%. Although the nuclear genome of fungi is several hundred times larger than the mitochondrial genome (compare Tables 3.1 and 3.2), mtDNA can be a considerable part of the total DNA in the cell, since cells contain many mitochondria, each perhaps with several mtDNA molecules. The proportion will vary, as mitochondrial frequency depends on the phase and conditions of growth. In *S. cerevisiae* mtDNA has been found to vary from 14 to 24% of the total DNA, and in *Schizosaccharomyces pombe* from 6 to 12%.

Saccharomyces cerevisiae, as a facultative anaerobe, can survive mutations that render the mitochondria non-functional. The resulting strains are dependent on

Table 3.2   Mitochondrial genome size in fungi and other organisms

Group and species	DNA (thousands of base pairs)
Slime moulds	
Dictyostelium discoideum	55–60
Physarum polycephalum	69
Oomycetes	
Achlya ambisexualis	50–51
Other lower fungi	
Phycomyces blakesleeanus	26
Higher fungi	
Neurospora crassa	62
Aspergillus nidulans	32
Saccharomyces cerevisiae	74–85
Schizosaccharomyces pombe	17–22
Schizophyllum commune	50
Ustilago cynodontis	77
Higher plants	160
Human	17

For some other fungal mitochondrial genome sizes see Hudspeth, M. E. S. (1992). The fungal mitochondrial genome. In D. K. Arora, R. P. Elander & K. G. Mukerji, eds., *Handbook of Applied Mycology*, vol. 4, *Fungal Biotechnology*, pp. 213–241. Marcel Dekker, New York.

glycolysis for their energy and as they produce only tiny colonies on agar media are known as 'petites'. A mitochondrial mutation in *N. crassa* results in a slow-growing strain known as 'poky'.

Hydrogenosomes

The anaerobic rumen chytrids (page 37) do not contain mitochondria, instead they have hydrogen-generating organelles, the **hydrogenosomes**, also found in anaerobic protozoa. They are a major site of fermentative metabolism in these organisms, enabling them to perform an oxidative metabolism following utilization of glucose by the glycolytic pathway, and leading to the production of hydrogen, carbon dioxide and acetate. These products are mainly formed from malate, the major substrate for the hydrogenosome. Hydrogenosomal metabolism also leads to substrate level ATP production. The hydrogenosomes are localized close to the flagellar bases in zoospores, as with mitochondria in zoospores of aerobic chytrids, consistent with a role in supplying energy for motility as well as for other cell activities. The evolutionary origin of hydrogenosomes was unclear for many years, but evidence suggesting that they are derived from mitochondria includes the following points.

1.   They are surrounded by two membranes, consistent with an endosymbiotic origin like that of mitochondria.

2. They may have inner membrane invaginations like the cristae of mitochondria.
3. Both organelles play a major role in the energy generation of the cell, with hydrogenosomes using protons as electron acceptors to give hydrogen, and mitochondria using oxygen to give water.
4. Their 'malic enzyme', a key component in their metabolism, is synthesized with an N-terminal 27 amino acid sequence homologous with mitochondrial targeting signals.

Unlike mitochondria, hydrogenosomes lack DNA, components of protein-synthesizing machinery and components of the mitochondrial electron transport chain. Some of the cytoplasmic enzymes, such as malate dehydrogenase, have amino acid sequences which are homologous to those of mitochondrial enzymes, but lack the mitochondrial targeting signals. Thus it is suggested that these fungi originally had mitochondria, and during the course of adaptation to an anaerobic lifestyle some mitochondrial enzymes have been re-targeted to the cytoplasm, whereas others, including those involved in mitochondrial energy metabolism, have been lost.

Microbodies

Microbodies are a family of organelles bounded by a single membrane. They carry out a range of functions in fungal cells. The major group are the peroxisomes, and this term is often used interchangeably with microbodies. A peroxisome contains at least one oxidase enzyme that produces hydrogen peroxide, together with peroxidase, which decomposes H_2O_2, either to produce oxygen or to oxidize a substrate. Thus it carries out respiration, but unlike a mitochondrion it does not conserve energy as ATP. Some peroxisomes have a second major function in the β-oxidation of fatty acids to give acetyl-CoA, for example in *Saccharomyces cerevisiae*. A third function is to provide a site for enzymes of the glyoxylate cycle, which are required for net conversion of two-carbon compounds into carbohydrate. These microbodies are sometimes known as glyoxysomes. *Neurospora crassa* contains two types of microbodies, one containing β-oxidation enzymes and no catalase, and the other catalysing peroxisomal respiration. Microbodies play an essential role in the utilization of a wide range of carbon and nitrogen sources. For example, peroxisomes of the yeast *Hansenula polymorpha* contain enzymes for the oxidation of methanol and methylamine, allowing growth on these substrates. Peroxisomal enzymes allow a range of yeast species to grow on other unusual substrates, such as alkanes, alkylamines, fatty acids, D-amino acids and uric acid. Growth on such compounds involves induction of the appropriate enzymes, and proliferation of peroxisomes, sometimes so massively that they pack the cell. It is thought that new microbodies arise by division of pre-existing ones. Electron micrographs of spores show that microbodies are commonly found in association with lipid storage droplets, so as to quickly mobilize these carbon sources when germination is triggered. A specialized group of microbodies in filamentous fungi are the Woronin bodies which contain hexagonal crystals, which have the role of plugging septal pores to seal off damaged hyphae (page 104).

Plasmids

Plasmids are pieces of DNA that are capable of replication independently of the replication of the chromosomes or mitochondrial DNA. Although plasmids are very common in bacteria, they are infrequent in fungi and other eukaryotes. An intensively studied fungal plasmid is the 2 μm DNA that occurs in the nucleus of *S. cerevisiae*. The 2 μm circular double-stranded DNA molecule consists of about 6200 base pairs. The plasmid can account for about 3% of the DNA in a cell and there can be up to 100 copies per cell. Whether the 2 μm DNA has any role in the life of the host is unknown, but it is proving valuable in the construction of vectors for the genetic manipulation of yeasts (Chapter 8). Another yeast, *Kluyveromyces lactis*, has two plasmids in the form of two linear dsDNA molecules of different lengths. Strains with the plasmid are 'killers', producing an extracellular protein toxin that kills a range of susceptible yeast in several genera, including *Saccharomyces*, *Candida*, and also *K. lactis* strains that lack the plasmids. The *K. lactis* toxin is composed of three subunits, γ, which enters the cell to cause permanent cell cycle arrest in the G1 phase, so that the cells cannot bud, α, a chitinase, which presumably aids entry of the γ subunit, and β, of unknown function. In *Podospora anserina* (page 53) plasmids that originate from mitochondrial DNA can spread through the mycelium and cause senescence, in which growth rate falls and vegetative propagation finally becomes impossible. The plasmids, termed 'senDNAs', can also be transmitted via the maternal parents through ascospores. Senescence in the progeny may be long delayed, allowing the continued propagation of the plasmids. It would seem that fungal plasmids are essentially 'selfish DNA', concerned with their own propagation.

Viruses

Viruses consist of nucleic acid capable of independent replication, with the nucleic acid usually being enclosed (encapsidated) in a coat (capsid) made of protein molecules genetically coded by part of the nucleic acid. The nucleic acid can be DNA or RNA, and single-stranded (ss) or double-stranded (ds). Viruses, known as mycoviruses, are of common occurrence in fungi. Unlike most viruses of animals, plants and bacteria, mycoviruses do not have an extracellular phase to their life cycle, and are transmitted only by intracellular routes. Most mycoviruses contain dsRNA and many are isometric (i.e. approximately spherical). Such viruses are widespread in Ascomycetes, Basidiomycetes, and related mitosporic fungi and yeasts. In some species, such as *Saccharomyces cerevisiae*, nearly all isolates contain virus, whereas in others, for example *Podospora anserina*, viruses have not been discovered. A single isolate may simultaneously contain viruses of more than one kind. Viruses, which multiply in the mycelium, can be transmitted between colonies by hyphal anastomosis, although transmission between strains may be limited by vegetative incompatibility. Virus transmission into asexual spores is usually efficient, but may vary in different strains of a fungus. For example, conidial inheritance of dsRNA virus in *Cryphonectria parasitica* can vary from 0 to 100% depending on the isolate. Transmission into basidiospores, for example of *Agaricus bisporus*, and into ascospores of yeasts such as *S. cerevisiae*, is generally efficient, but transmission into ascospores of filamentous Ascomycetes is often greatly restricted. In one study, only 10% of ascospores of *Magnaporthe grisea* contained dsRNA. Attempts to infect fungi with virus preparations have been unsuccessful except with protoplasts – cells artificially deprived of their walls. It is hence likely that in nature the transmission of viruses between fungi requires at least temporary protoplasmic continuity.

Extensive studies have been carried out on the isometric dsRNA viruses of *Aspergillus*, *Penicillium* and *Saccharomyces*. Each virus particle (or virion) has a capsid (protein coat) consisting of 60 or 120 copies of a single polypeptide. These polypeptides, singly or in pairs, form the 60 subunits (or capsomeres) required for a symmetrically organized, icosahedral capsid. Within each capsid is a single linear dsRNA molecule. These dsRNA molecules, however, can differ in size and base sequence. *Penicillium stoloniferum* virus S (Fig. 3.2), for example, consists of two types of virion with identical capsids but differing dsRNA molecules, and *Aspergillus foetidus* virus F of five virion types with identical capsids and differing dsRNA molecules. This separate encapsidation of different segments of the virus genome is common along the isometric dsRNA viruses of fungi, although there are also instances of viruses with a single type of virion containing an undivided genome such as the ScV-L virus of *Saccharomyces cerevisiae* (see below). The virus genome specifies, in addition to the capsid polypeptide, an RNA-dependent RNA polymerase which occurs inside each capsid bound to the RNA. The role of the polymerase is the replication of the genome and the transcription of a messenger RNA (mRNA) molecule from it.

Fungal viruses often have no recognizable phenotypic effects, and many were discovered only as a result of a search for dsRNA or virions in extracts of mycelium. An important exception is a virus which can cause severe malformation and reduced yields of fruiting bodies of the cultivated mushroom, *Agaricus bisporus*. Other exceptions are the viruses which result in the production by 'killer' strains of *Saccharomyces cerevisiae* or *Ustilago maydis* of proteins which kill sensitive strains of the same species. The viruses involved in the killing phenomenon in *S. cerevisiae* have been intensively studied. Most sensitive strains of *S. cerevisiae* carry virus ScV-L (Fig. 3.2C) with a genome of one dsRNA molecule (L dsRNA, with about 4500 base pairs) which encodes the capsid polypeptide and probably the RNA polymerase needed for dsRNA replication and transcription. Killer strains of *S. cerevisiae* carry an additional, 'satellite' dsRNA (M dsRNA, with 1500–2000 base pairs) which encodes a protein toxin, lethal to sensitive strains, and also a protein providing immunity to the toxin. Killer strains are classified into three main groups, K1, K2 and K28 on the basis of the molecular characteristics of their secreted protein toxins, which have differing structures and differing actions. The most studied toxin, K1, interacts with $\beta(1\rightarrow6)$ glucan receptors and is then transferred to form cation transmembrane channels in the cell membrane, leading to ion leakage and cell death. In contrast, the K28 toxin interacts with $\alpha(1\rightarrow3)$ mannose residues in a cell

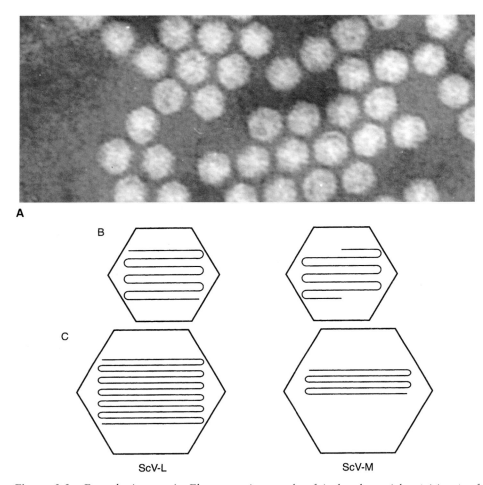

Figure 3.2 Fungal viruses. A, Electron micrograph of isolated particles (virions) of *Penicillium stoloniferum* virus S, negatively stained with potassium phosphotungstate. The virions (diam. *ca* 30 nm) are polyhedra, probably icosahedra with 20 faces, and are hexagonal in cross-section. A virion capsid consists of 60 capsomeres, just perceptible in the micrograph (Dr John Gay). B, Diagram showing the two dsRNA molecules of virus separately encapsidated. They are about 560 and 480 nm in length, equivalent to 1700 and 1450 base pairs. Their configurations in the virion are not known, and their lengths in the diagram in relation to virion size have been reduced to a quarter for clarity. C, A similar representation of *Saccharomyces cerevisiae* viruses ScV-L and ScV-M. The virions have a diameter of about 40 nm. The ScV-M virus occurs in killer and neutral strains of the yeast and its dsRNA specifies toxin production in killer strains and the protein that protects against the toxin in both killer and neutral strains. The dsRNA of the ScV-L specifies the requirements for its own replication and also the capsid polypeptide for ScV-M. Hence ScV-L is a 'helper' virus essential for the propagation of its 'satellite' ScV-M, and killer and neutral strains have to contain both viruses. ScV-L also occurs in most sensitive strains.

wall mannoprotein, and causes cell cycle arrest in the G2 phase and non-separation of mother and daughter cells. M dsRNA is encapsidated as a separate virion, ScV-M (Fig. 3.2C) the protein coat of which is identical to that of ScV-L and encoded by L dsRNA. Hence ScV-M can only replicate in the presence of ScV-L which acts as helper virus. The replication of M dsRNA also requires the products of several nuclear genes which are not needed for the replication of L dsRNA. In addition to killer strains and sensitive strains of *S. cerevisiae* there are also neutral strains, which are not killed by the killers and which do not kill the sensitive strains. In these the M dsRNA specifies the protective protein but not the toxin.

Some fungi contain dsRNA that is not encapsidated. An example is the linear dsRNA that occurs in strains of *Cryphonectria parasitica* (formerly *Endothia parasitica*), an Ascomycete that causes a devastating blight of the American chestnut tree (pages 208, 260). The virus is associated with lipid-rich cytoplasmic vesicles. Strains of the fungus having the dsRNA are hypovirulent – the disease that they produce is mild and damage to the host is slight. The dsRNA has been transmitted by hyphal anastomosis from hypovirulent to virulent strains of the fungus and a reduction in the pathogenicity of the latter demonstrated. This suggests that hypovirulent strains might be employed for biological control of the disease. Some strains of *C. parasitica* contain another dsRNA that is not encapsidated and which is located within their mitochondria. This dsRNA has evolved to use the fungal mitochondrial genetic code, in which UGA codes for tryptophan rather than acting as a termination codon.

Reserve Materials

It is rare that the environment will contain an ideal balance of the different nutrients required by a microorganism. The availability of assimilable nitrogen, for example, will often limit growth while utilizable carbon compounds are present in excess of immediate needs. Under such circumstances many microorganisms will take the carbon compounds into the cell and convert them into a form suitable for storage. Subsequently these carbon reserves may be mobilized and used as a source of energy and for biosynthesis. The nature of the carbon reserves of fungi will now be considered.

Fungi, like other eukaryotic organisms, can accumulate lipids as a carbon reserve; oil droplets are often observable by light microscopy in older cells. Another carbon reserve, widespread in fungi and slime moulds as well as in bacteria, protozoa and animals, is the polysaccharide glycogen. Although extracted glycogen has a molecular weight of 1–10 million and is water soluble, in the cell glycogen is in the form of insoluble granules of complex tertiary structure and has a molecular weight of over 100 million. Glycogen is a polymer of D-glucose molecules in the α-configuration and is hence an α-glucan. Most of the glucose molecules are linked to two others by $\alpha(1{\to}4)$ glycosidic bonds and thus form chains. Approximately every tenth glucose molecule, however, is linked also to a third glucose molecule by an $\alpha(1{\to}6)$ glycosidic bond, resulting in chain branching (Fig. 3.3A). Chains consisting of $\alpha(1{\to}4)$ linked glucose molecules

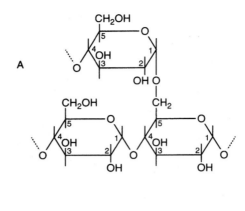

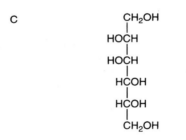

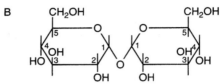

Figure 3.3 The molecular structure of four common reserve materials of fungi. A, Glycogen, an α-glucan. Three glucose molecules are represented, illustrating the α(1→4) linkage of the polymer chains and the α(1→6) linkage at a branching point. Carbon atoms C-1 to C-5 of the glucose molecules are represented by numbers only and C-6 by C. The H symbol is omitted from H atoms directly linked to C-1 to C-5 atoms. The continuation of the polymer chains is indicated by a dashed line. B, Trehalose, a non-reducing disaccharide. Symbols as with glycogen. C, Mannitol, a sugar alcohol. D, Polyphosphate. n represents a large, indefinite number of the groups indicated.

form a compact hollow helix stabilized by hydrogen bonding. Branching further facilitates the formation of compact macromolecules, and also results in large numbers of non-reducing end groups, facilitating rapid attack by the enzyme phosphorylase when the mobilization of the reserve is needed. Glycogen granules are observable by electron microscopy, and can constitute as much as 10% of the dry weight of fungi. The pathway and enzymes involved in the synthesis of glycogen from glucose-6-phosphate (G-6-P), a compound central to metabolism of most organisms, have been intensively studied. Details of this, and also of the different enzymes responsible for the breakdown of glycogen to G-6-P, are available in biochemistry textbooks. Extracellular $\beta(1\rightarrow3,1\rightarrow6)$ glucans can also serve as food reserves (see page 101).

Reserve materials also occur in fungi as soluble carbohydrates of low molecular weight. Monosaccharides are present only at very low concentrations and mainly in the phosphorylated state. The disacccharide trehalose (Fig. 3.3B), however, is a common reserve material in fungi. The trehalose molecule consists of two glucose molecules in the α configuration linked by a glycosidic bond between the two C-1 atoms. This means that it is a 'non-reducing' disaccharide. The enzyme trehalose phosphate synthase is responsible for the synthesis, from uridine diphosphate glucose (UDP-glucose) and G-6-P, of trehalose-6-phosphate which is then converted into trehalose by trehalose phosphatase. Reserves of trehalose can be converted directly into glucose by trehalase. As well as being a reserve material, trehalose has a major protective role in some spores against environmental stresses (page 233). Sugar alcohols (polyhydric alcohols or **polyols**) are also common reserves in fungi. They include three-carbon (glycerol), four-carbon (erythritol), five-carbon (arabitol and ribitol) and six-carbon (mannitol) compounds. Mannitol (Fig. 3.3C) and arabitol are particularly widespread. Trehalose and polyols can constitute up to 15% of the dry weight of fungi. Whereas lipids and high-molecular-weight polysaccharides are nearly universal carbon reserves, low-molecular-weight carbon compounds (i.e. sugars and polyols) as reserve materials are more limited in their distribution. Their common occurrence in this role in fungi and plants is probably a consequence of a further role for these materials, that of the regulation of water relations (page 161).

Other stored metabolites, basic amino acids and polyphosphate, are discussed in the following section.

Vacuoles

Vacuoles are often very abundant in fungal cells, and usually are the most obvious components seen with the light microscope. It has become increasingly clear that they are dynamic structures, playing multiple roles in hyphal and yeast cell growth, rather than being inert depositories for storage of water and nutrients or waste materials. Vacuoles do not occur in hyphal tips, but are characteristic of sub-apical regions. Tubular vacuolar projections, however, can often be seen within a few micrometres of the apex. In some hyphae, notably those of *Candida albicans* and *Basidiobolus ranarum*, the apical cells advance at the direct expense of cytoplasm in sub-apical cells, which end up containing a nucleus and a tiny amount of cytoplasm, seemingly pushed into a corner by a very large vacuole. Senescent cells in the centre of colonies become highly vacuolated.

In the growing fungus the vacuole plays three primary roles: (1) as a store for metabolites and cations; (2) as a regulator for pH and ion homeostasis in the cytoplasm; and (3) as an equivalent to an animal lysosome, with a range of lytic enzymes such as proteases, nucleases, phosphatases and trehalase in an acidic environment. The vacuolar membrane, sometimes termed the tonoplast, has a proton-pumping ATPase, similar to that of the plasma membrane, which energizes the transport of solutes and keeps the interior of the vacuole acidic. It hydrolyses ATP on its cytoplasmic face, pumping a proton into the vacuole for each molecule of ATPase hydrolysed, and generates a membrane potential of about 25–40 mV in *Neurospora crassa*, and a vacuolar pH about 6. Basic amino acids, in particular arginine, are pumped into the vacuole against their concentration gradient, by a proton/amino acid 'antiport system', with energy for the transport supplied by the simultaneous movement out of a proton via its electrochemical potential. Cations, such as K^+, Ca^{2+}, Mg^{2+}, Mn^{2+}, Fe^{2+} and Zn^{2+} are accumulated by a similar mechanism. The stored amino acids are mobilized to act as nitrogen sources under conditions of nitrogen starvation. The vacuole stores phosphate and polyphosphate (Fig. 3.3D). The polyphosphate, as well as acting as a supply of phosphate for metabolism, could act as a counter ion for the stored positively charged amino acids and cations, or as a pH regulator, as its hydrolysis releases protons, lowering the pH. Polyphosphate exerts a much lower osmotic pressure than an equivalent amount of phosphate, which is of importance to the water relations of the vacuole. The vacuolar hydrolytic enzymes may serve in the recycling of cell material during growth and development, and in the autolysis of cell components in senescent cells. An example of controlled autolysis resulting from the release of lytic enzymes from the vacuole is the progressive liquefaction of the tightly packed gills of fruit bodies of the ink cap mushrooms, *Coprinus* species, following spore release. This allows unimpeded release of the spores from higher up in the gills.

Vacuoles in living hyphae can be visualized by their accumulation of fluorescent dyes, such as 6-carboxyfluorescein. Fluorescence microscopy of treated hyphae of a range of species of Basidiomycetes, Ascomycetes, mitosporic fungi and Zygomycetes has shown that the vacuoles are interconnected by a pleiomorphic motile tubular system, that undergoes peristaltic-like dilations and contractions. This results in transport of vacuolar contents over considerable distances along hyphae. The most detailed observations have been with the ectomycorrhizal Basidiomycete, *Pisolithus tinctorius*, in which these tubules were seen transporting material across dolipore septa as well as within cells. Extension and retraction of the tubules is independent of cytoplasmic streaming, and transport can occur in either direction, covering distances up to 60 μm. It is suggested that this system could have particular importance in transporting phosphate in hyphae, especially those of mycorrhizal fungi. These vacuole movements probably involve microtubules, as they are inhibited by drugs that disrupt tubulin assembly. In yeast cells vacuoles are inherited, and during bud formation a tubule is seen to grow from the mother cell vacuole into the bud, and small vacuoles form in the growing daughter cell, to fuse as the daughter cell matures to form its vacuole.

The Cell Envelope

All cells are bounded by a **plasma membrane (plasmalemma)**. Information on the ways in which the plasma membranes of fungi differ from those of other organisms remains sparse. One important difference is that ergosterol is the major sterol of the plasma membranes of fungi, in contrast to cholesterol in membranes of animal and plants. This renders fungi more susceptible than animals and plants to some polyene antibiotics and to inhibitors of ergosterol biosynthesis (page 178), which are hence useful for treating fungal infections of animals and plants.

Cells, in the natural state, have a macromolecular coating, sometimes referred to as the glycocalyx, outside the plasma membrane. The glycocalyx of the plasmodium of the Myxomycete *Physarum polycephalum* is a sheath consisting largely of a slimy polysaccharide. This polysaccharide is a galactan, a polymer of galactose, in which some of the hydroxyl groups are replaced by phosphate or sulphate. In contrast to the slimy coatings of slime moulds, the glycocalyx in most other fungi is a firm cell wall. Some fungi, however, produce in liquid culture extracellular polysaccharides which pass into solution and make the medium viscous. Under natural conditions these polysaccharides probably constitute an outermost slimy coating to the hyphal wall, and hence can be considered as part of the cell envelope.

Hyphal and yeast walls

A satisfactory understanding of the structure and composition of fungal cell walls requires not only the isolation and identification of wall components but also the establishment of the location and arrangement of the components within the wall. This can be achieved by the treatment of isolated walls with a succession of enzymes together with electron microscopy to demonstrate the changes brought about by each enzyme treatment. Furthermore, it is important in dealing with hyphae to consider separately the structure and composition of hyphal tips, mature walls and septa if present. Such detailed studies have been carried out on only a small number of fungi, and it is becoming clear that there are considerable variations, both qualitative and quantitative, between different fungi and in different types of cells in individual fungi. The fungal wall is emerging as a dynamic structure rather than an inert coat to the cell. Its construction can change in response to environmental stresses such as osmotic shock. In all cases the major components are polysaccharides: chitin, glucans and mannoproteins in Ascomycetes, Basidiomycetes and mitosporic fungi; chitosan, chitin and polyglucuronic acid in Zygomycetes; cellulose and other glucans in Oomycetes (Table 3.3). There are also specific cell wall proteins and glycoproteins in walls of all of these groups.

An important component of the cell walls of most fungi is chitin, a polysaccharide that is also a major constituent of the exoskeleton of insects and other arthropods. Chitin is a linear polymer of the acetylated amino sugar N-acetylglucosamine, the subunits being linked by $\beta(1\rightarrow4)$ glycosidic bonds (Fig. 3.4). It is immensely strong, with a tensile strength much greater than artificial materials such as carbon fibres and steel. This strength is the result of extensive hydrogen bonding, along the chains, giving them rigidity as they are formed, and

Table 3.3 Major polymers occurring in fungal cell walls

Taxonomic group	Fibrous polymers	Gel-like polymers
Basidiomycetes, Ascomycetes, Mitosporic fungi	Chitin $\beta(1\rightarrow3),\beta(1\rightarrow6)$-Glucan	Mannoproteins $\alpha(1\rightarrow3)$-Glucan
Zygomycetes	Chitosan Chitin	Polyglucuronic acid Mannoproteins
Chytridiomycetes	Chitin Glucan	?
Hyphochytriomycetes[a]	Chitin Cellulose	?
Oomycetes[a]	$\beta(1\rightarrow3),\beta(1\rightarrow6)$-Glucan Cellulose	Glucan

Data from Gooday (1994) – see Further Reading and Reference.
[a]Phyla in the Kingdom Chromista.

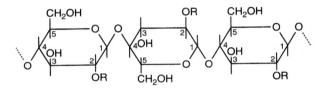

Figure 3.4 The structure of a linear glucan with $\beta(1\rightarrow4)$ glycoside linkages. Three such glucans – cellulose, chitosan and chitin – are important structural components of fungal cell walls. The symbol R represents -H in the cellulose molecule, $-NH_2$ in chitosan and $-NH-CO-CH_3$ in chitin. Other symbols as in Fig. 3.3A.

then between chains. In fungal walls the adjacent chitin chains aggregate in an anti-parallel fashion to give 'α-chitin'. This 'crystallization' of chains leads to the formation of microfibrils. The chitin microfibrils are randomly arranged in most hyphal walls, but are predominantly longitudinal in some germ tubes, transverse as shallow helices in the walls of stipe cells of some mushrooms, and tangential in septa. They range from 10 to 25 nm in diameter and may be over 1 μm long. In some fungi, they are much shorter, for example an average of 33 nm in hyphal walls of *Candida albicans*. They occur in the innermost region of the hyphal wall and mesh together to form a network embedded in an amorphous matrix. Such two-phase systems have great strength, as seen with fibre glass (glass fibres embedded in a resin) and similar reinforced plastics, and properties that can be varied by changing the proportions of the two materials. Zygomycete hyphal walls contain, in addition, chitosan, which is chitin that has had most of its acetyl groups removed enzymically while being synthesized, giving a polymer chiefly of $\beta(1\rightarrow4)$ glucosamine. Oomycete walls typically contain cellulose, the linear $\beta(1\rightarrow4)$ glucose polymer that is a major constituent of plant cell walls, instead of chitin as the primary skeletal wall, although some Oomycetes also contain small amounts of chitin together with the cellulose.

A second major fibrous component of fungal walls, except for those of Zygomycetes, is β(1→3) glucan. Frequently this is the most abundant wall component. Like chitin, chitosan and cellulose, long chains of it form fibrils. In walls of *Saccharomyces cerevisiae* its average degree of polymerization is 1500 glucose units, giving a molecular weight of 240 000 and a length of 600 nm. Its β(1→3) linkage gives it a helical structure, and three such helices readily form a triple helix held together by hydrogen bonds. β(1→6) Glucan is usually associated with the β(1→3) in the mature cell wall. In *Saccharomyces cerevisiae* β(1→6) glucan is about 10% of the total cell wall glucan, occurring as a highly branched polymer of about 350 glucose units.

A range of other glucans are found in fungi walls. α(1→3) Glucan is found as a fibrous polymer in walls of some fungi, such as *Schizosaccharomyces pombe*, *Schizophyllum commune*, and the pathogenic yeast cells of *Paracoccidioides brasiliensis*. Hyphae in vegetative colonies of *S. commune* become coated with a water-soluble mucilage termed schizophyllan, a β(1→3) glucan with single glucose units attached to approximately every third unit in the main chain by a β(1→6) linkage. An extracellular slime that has been extensively studied is pullulan, a linear glucan in which maltotriose units – three glucose molecules, β(1→4) linked – are joined by β(1→6) linkages. It is produced by the mitosporic fungus *Aureobasidium pullulans*. Some glucans have a dual role, serving both as a structural component of the wall and as a food reserve that may later be utilized. Thus when fruiting bodies are formed in *Schizophyllum commune* much of the glucan of the walls of the vegetative mycelium is metabolized. A hot-water soluble glucan with alternating α(1→3) and α(1→4) glucose unit, nigeran, is found in walls of some *Aspergillus* and *Penicillium* species.

Mannoproteins are major components of the outer wall of fungal cells, especially of yeasts. In *Saccharomyces cerevisiae* the cell wall is about 20% of the cell's weight, and almost half of it consists of mannoproteins. These macromolecules are elaborated in the endoplasmic reticulum and Golgi apparatus, and are secreted into the wall via the normal secretion pathway. There are two types. One is *N*-linked, in which a few long chains of α(1→6) mannose chains, with α(1→2) and α(1→3) side chains, some with phosphomannose units, are linked, via two *N*-acetylglucosamine units, to the nitrogen atom of the free amino group of asparagine residues in the protein. The other type is *O*-linked, in which many short α(1→2) and α(1→3)-linked mannose oligosaccharides are linked to the oxygen atom of the hydroxyl groups of serine and threonine residues. There is considerable variation in detail of the structures in different species. Small amounts of other sugars may be present, including glucose, galactose and glucuronic acid.

Two specific types of glycoproteins have been recognized as being covalently linked to glucans in cell walls of *S. cerevisiae*, and some other yeasts and hyphal fungi. Glycosylphosphatidylinositol (GPI) proteins are a class of proteins that have been modified in the endoplasmic reticulum by the addition of an inositol phospholipid to their C-terminal amino acids. This gives them a 'GPI-anchor' so that when they are secreted they are anchored by their phospholipid moieties to the lipid bilayer of the cell membrane, with the polypeptide chain extending into the cell wall. One such GPI-protein is α-agglutinin, the mating agglutinin on the surface of α-shmoos of *S. cerevisiae* (page 211). This exists on the surface of the cells in three forms: attached to the cell membrane via its GPI-anchor, free in the periplasmic space (i.e. within the matrix of the wall), and covalently linked to $\beta(1\rightarrow6)$ glucan in the outer part of the wall. Another class of proteins are covalently linked to $\beta(1\rightarrow3)$ glucan. These are proteins characterized by internal repeats of amino acid sequences, the 'Pir' cell wall proteins.

The hydrophobins are a class of hydrophobic proteins, first characterized from the Basidiomycete, *Schizophyllum commune*, by sequencing mRNA expressed during fruit body formation. They had not been noticed before, as they are too hydrophobic to run on electrophoretic gels. They have now been identified in other Basidiomycetes, and in Ascomycetes, mitosporic fungi and Zygomycetes. They are small secreted proteins with eight cysteine residues ordered in a particular manner in their amino acid sequences (C, cysteine; Xn, a variable number of amino acids):

$$X_n\text{-C-}X_n\text{-C-C-}X_n\text{-C-}X_n\text{-C-}X_n\text{-C-C-}X_n\text{-C-}X_n$$

Cerato-ulmin, a hydrophobin produced by *Ophiostoma ulmi* (page 53), has been analysed in detail. The eight cysteines form four disulphide bridges in order in the polypeptide chain, giving a protein with four loops, two of which are of predominantly hydrophobic amino acids. Hydrophobins form rodlets, 10–13 nm in diameter, by a process termed 'interfacial self-assembly' as it occurs in response to interfaces between water and air or between water and solid surfaces. The rodlets have a hydrophobic face and a hydrophilic face. Thus aerial hyphae that are secreting a hydrophobin become coated with a layer of it, as do any spores that they may be producing. This gives them a hydrophobic surface, visible by electron microscopy as an outer layer of ordered rodlets. Specific hydrophobins are also expressed during appressorium formation in the plant pathogen *Magnaporthe grisea* (page 378), and in fruit body formation in *S. commune* and *Agaricus bisporus*. In *M. grisea* the hydrophobin is necessary for appressorium development on the host plant leaves, suggesting a role in the interaction of fungus and substrate, and in the mushroom fruit bodies hydrophobins line internal air spaces and gaps between aggregated hyphae.

In addition to polysaccharide and protein components, some fungal walls contain melanins, dark pigments consisting of branched polymers derived from phenolic metabolites such as tyrosine, catechol and dihydroxynaphthalenes. Enzymes capable of degrading melanins have so far not been characterized, and it is probable that melanins endow the fungal wall with a high capacity to resist enzymic lysis, and may also have photoprotective and structural roles.

As mentioned earlier, there is considerable diversity in wall composition in different cell types in any fungus. Walls of hyphal and yeast cells of dimorphic fungi have differing compositions. The hyphae of the dimorphic Zygomycete, *Mucor rouxii*, have cell walls in which the chitosan and chitin account for 42% of the dry weight and mannoproteins for 2%. The cell walls in the yeast phase of this organism, however, have a chitin and chitosan content of 36% and a mannoprotein content of 9%. Similarly, the hyphal walls of *Candida albicans* have about three times as much chitin but a third as much protein as those of the yeast cells. However, in another human pathogen, *Paracoccidiodes brasilensis*, hyphal walls have less chitin (13 vs. 34%), no detectable $\alpha(1\rightarrow3)$ glucan, and more $\beta(1\rightarrow3)$ glucan (35 vs. 4%), galactomannans (6 vs. 1%) and protein (33 vs.

10%) than do the walls of the infective yeast cells, which also have 44% α(1→3) glucan. Spore walls may differ greatly from those of their parental hyphae or yeast phase cells. Thus sporangiospores, but not hyphae, of *Mucor rouxii* contain glucans and melanins, and ascospores of *Saccharomyces cerevisiae* contain considerable amounts of chitosan and an unusual dimerized amino acid, dityrosine, both of which are absent from the walls of vegetative cells. Different cell types have differing surface glycoproteins. In *Candida albicans*, for example, adhesins on hyphal and yeast phase surfaces differ, as do those (a- and α- sexual agglutinins) on the surface of the **a** and α mating cells of *S. cerevisiae*.

Architecture of the cell wall

It is clear from the above account that the fungal cell wall is a complex dynamic assemblage of many components. In thin sections of fungal cells viewed by electron microscopy the wall typically appears to be composed of layers (see Fig. 3.8), although it has recently become clear that the layers are not distinct, but interconnected as one massive multimolecular complex. Most work has been done with *Saccharomyces cerevisiae*, but there is considerable evidence that the model that has emerged can be extended to both yeast cells and hyphae of other fungi, with appropriate allowance being made for quantitative and qualitative differences in composition. The chitin microfibrils are in the innermost layer and appear to act as the primary scaffolding. Attached to them by glycosidic linkages are the β(1→3) glucan fibrils which are predominantly in the inner part of the wall. Moving out towards the cell surface are β(1→6) glucans covalently bound to the β(1→3) glucan fibrils. GPI-mannoproteins are either anchored in the membrane, extending out through the wall, or cleaved from their anchors and bound to β(1→6) glucans in the outer part of the wall, while Pir-cell wall proteins

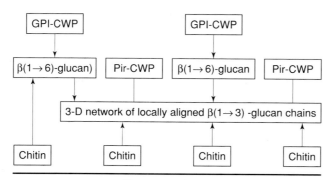

Plasma membrane

Figure 3.5 Diagram of suggested interrelationships of major wall components of *Saccharomyces cerevisiae*, but probably of applicability to cell walls of many fungi. There are covalent linkages between all components, to give one macromolecular structure, as described in text. Chitin microfibrils lie adjacent to the plasma membrane; β(1→3) and β(1→6) glucan chains are attached to the chitin; some of the cell wall mannoproteins (GPI-CWP) are attached to the β(1→6) glucan via remnants of glycosyl phosphatidylinositol anchors; cell wall proteins with internal repeats (Pir-CWP) are attached to β(1→3) glucan. Based on information from Smits, G. J. *et al.* (1999). *Current Opinion in Microbiology* **2**, 348–352; Kapteyn, J. C. (1999). *Biochimica et Biophysica Acta* **1426**, 373–383.

(page 102) appear to be attached to β(1→3) glucans. Some mannoproteins appear to be linked to chitin microfibrils, from where they extend out through the wall. Thus in summary the fibrous chitin and β(1→3) glucan are concentrated in the innermost part of the wall, and β(1→6) glucan, mannoproteins and proteins in the outermost part, but the components are cross-linked and interconnected (Figure 3.5). Aspects of the assembly of this mature structure in hyphal walls are discussed later (page 116).

Hyphal and yeast septa

The hyphae of Ascomycetes, Basidiomycetes and mitosporic fungi have frequent cross-walls (septa) dividing hyphae into compartments (Fig. 3.1). The septa are usually perforated. The septa of Ascomycetes and mitosporic fungi commonly have a single central pore, sometimes large enough to permit the passage of nuclei as well as protoplasmic streaming between cells. In over 50 species, one or more membrane-bound crystalline organelles termed **Woronin bodies**, about 0.1 μm in diameter, have been observed in the cytoplasm close to the pore. If a hyphal compartment is damaged and the protoplasm leaks out, a Woronin body in the adjacent cell moves and blocks the pore, stopping further leakage. A new hyphal tip can then grow through the septum. Woronin bodies of *Neurospora crassa* have been purified and characterized. They contain a major protein, Hex1 (for Hexagonal crystal) of molecular mass 19 000, which readily self-assembles to form crystals with a hexagonal cross-section. Expression of the gene *HEX1* in *Saccharomyces cerevisiae* resulted in the formation of Woronin bodies in the cytoplasm. Hex1 has a C-terminal tripeptide sequence that targets it to microbodies, so Woronin bodies, like peroxisomes, are a particular type of microbody (page 91). Deletion of the *HEX1* gene had no effect on growth, but when wounded by cutting with a razor blade or subjected to hypo-osmotic shock with water, the mutant strain leaked protoplasm uncontrollably, while the wild-type strain quickly recovered and re-grew, as its septal pores in undamaged cells had been sealed by Woronin bodies.

Septal perforations may be numerous and minute in the Saccharomycetales and some mitosporic fungi, such as *Geotrichum*, or there may be a single 'micropore' as in hyphae of *Candida albicans*. This micropore is 25 nm in diameter, and does not allow the passage of nuclei or mitochondria, but does give cytoplasmic continuity between adjacent compartments.

The most complex septa are the dolipore septa (page 86; Fig. 3.1) of some Basidiomycetes. After hyphal damage the septal pore channel is instantaneously sealed with material of unknown composition. Then the septal structure in the damaged cell becomes detached and replaced by a pad of material sealing off the damaged region.

The septa of *Neurospora crassa* consist of chitin covered with other glucans and protein. The most intensively investigated hyphal septa are those of *Schizophyllum commune*. A middle layer of chitin is covered on each side by a layer of chitin and glucan. The septal swelling which forms the rim of the septum consists of glucan. Dikaryotization involves the enzymic dissolution of the entire septum, whereas in *Coprinus* conversion of the dolipore septum into a simple septum is sufficient to permit nuclear migration. In hyphae of *Candida albicans*, the septum consists of two chitin-rich septal plates separated by a thin amorphous layer. There is a micropore in the centre, which is much too small to allow passage of organelles.

During budding of *Saccharomyces cerevisiae* a septum forms to allow separation of the bud from the mother cell. Initially a chitin-rich primary septum forms to separate the two cells. This is then overlain on both sides by glucan-rich secondary septa. The two cells separate, and the primary septum is left as a chitinous bud scar on the surface of the mother cell.

Isolated protoplasts of yeasts and filamentous fungi

Wall-free protoplasts have been isolated from a wide variety of yeasts and filamentous fungi. They are obtained by the treatment of yeast cells or fungus mycelium with wall-degrading enzymes in a medium having an osmolarity such that the released protoplasts will neither burst nor shrink. Inorganic salts, sugars and sugar alcohols have been used to give appropriate osmolarity. The digestive juices of the snail *Helix pomatia* can be used to produce protoplasts of yeasts, and a variety of enzyme preparations from soil bacteria (e.g. *Arthrobacter* and *Streptomyces*) and soil fungi (e.g. *Penicillium* and *Trichoderma*) to obtain protoplasts from filamentous fungi. The enzyme preparation from *Trichoderma* which releases *Schizophyllum commune* protoplasts contains glucanases and chitinases, which attack the major components of the *S. commune* wall. Wall-degrading enzyme preparations can be produced in the laboratory in which they are required. It is usually necessary to induce enzyme production by exposure of the soil fungus or bacterium to an appropriate substrate.

A range of metabolic activities, such as DNA, RNA and protein synthesis and respiration, have been demonstrated in isolated protoplasts, although rates are usually considerably lower than in the intact fungus. Transfer of protoplasts to media lacking wall-degrading enzymes results in wall synthesis and reversion to the yeast or hyphal form. Initially the spherical protoplast becomes coated with an open network of fibrous chitin and $\beta(1\rightarrow3)$ glucan, to slowly become an approximately spherical walled cell after which buds or hyphal outgrowths usually of abnormal form are produced. Soon, however, normal budding or hyphal growth is obtained.

Protoplasts are of value in many areas of research. The ease with which protoplasts are disrupted renders them more suitable than whole cells for the isolation of cellular organelles. This is particularly true for yeast cells, which have very tough cell walls, only ruptured by drastic procedures. Protoplasts are released more swiftly from the apical cells of hyphae than from older cells. This permits the separate isolation of the contents of the young and old cells of filamentous fungi and the comparison of their enzyme activities. Fusion of protoplasts of fungal cells can be obtained by treatment with polyethylene glycol and calcium ions. This enables fusions to occur between yeast cells of the same mating type or between, for example, cells of *Aspergillus nidulans* strains of different vegetative incompatibility groups, or even between filamentous fungi of different species. Hybridizations impossible with intact cells can thus be achieved. Protoplasts are also able to take up exogenous DNA. Hybrid plasmids carrying genes from *Escherichia coli* and *Saccharomyces cerevisiae* have been incorporated into *S. cerevisiae* cells. Both integration of DNA into chromosomes and cytoplasmic inheritance of introduced genes have been demonstrated. Isolated protoplasts are hence of considerable importance in genetic manipulation using fungal hosts.

The Growth and Form of Fungal Cells

The Yeast Cell Cycle

Eukaryotic cells, if supplied with adequate nutrients and if growth is not otherwise inhibited, increase in size continuously and undergo nuclear and cell division at intervals characteristic of the species and the conditions employed. Between successive cell divisions, events occur in a regular sequence, permitting the recognition of a cell cycle divisible into four phases, G1, S, G2 and M. The designations S and M indicate that these phases are characterized by major nuclear events, DNA synthesis (S) and mitosis (M), respectively. Phases G1 and G2 (G for gap) are the intervals between phases S and M. Cells in phase G2 have twice as much nuclear DNA as cells in G1, since cell division has not yet occurred but DNA replication has taken place. The amount of DNA representing a single copy of an organism's genome is sometimes referred to as the C value. Thus, in the G1 phase haploid cells have a C value of 1 and diploid cells of 2. During the G2 phase the C value of haploid cells is 2 and of diploid cells 4. Cell division occurs at the same time as mitosis in most organisms, but in the budding yeast *Saccharomyces cerevisiae* it takes place after the completion of mitosis, early in phase G1 (Fig. 3.6).

Populations of yeast cells of the same size and hence at the same point in the cell cycle can be obtained by centrifugation in a sucrose density gradient. All the cells in such a population will grow at approximately the same rate, so their cell cycles will be in synchrony for a few generations. Such synchronous populations of yeast cells have been useful for fundamental studies of the cell cycle, and the cell cycle of the budding yeast *S. cerevisiae* has been intensively studied. A range of temperature-sensitive cell cycle mutants have been isolated. These can be maintained in culture at a permissive temperature (e.g. 23°C) but at a restrictive temperature (e.g. 35°C) a specific step in the cell cycle fails. Such *cdc* (cell division cycle) mutants have permitted the elucidation of causal relationships between events in the cell cycle.

Studies on *S. cerevisiae* have identified a point in the cell cycle in the middle of phase G1 which has been designated 'Start'. It is at this point that gene *CDC28* acts and commits the cell to another round of nuclear and cell division. A cell can, however, only pass 'Start' if certain conditions are fulfilled. The previous nuclear division must have been completed, the cell must have reached a critical size, and there must be sufficient nutrients present for the completion of the cycle. Nutrient exhaustion hence results in a population of cells in phase G1. Haploid cells,

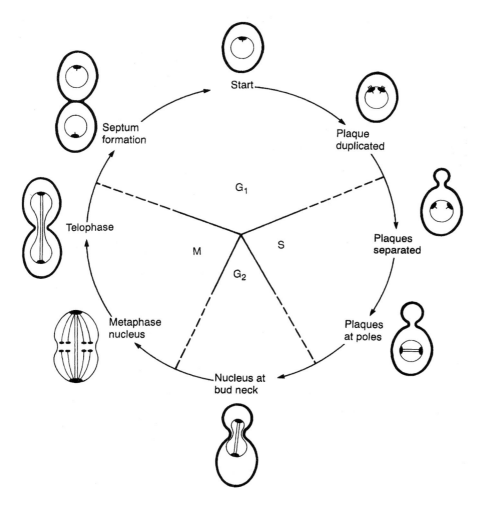

Figure 3.6   Life cycle of the budding yeast *Saccharomyces cerevisiae*. The phases of DNA synthesis (S), mitosis (M), and the intervening periods of growth (G1, G2) are indicated at the centre. The four phases are of approximately equal duration for rapidly growing cells (i.e. with generation time of under 2 h), but for slow-growing cells G1 is much extended. The cycle starts in G1 with spindle plaque duplication. Microtubules that extend into the cytoplasm are shown as these mark the site of future bud emergence. The plaques then separate, the microtubules extending from them into the nucleus (only one shown per plaque) extend, DNA synthesis is initiated, and a bud appears. By the end of the S phase the plaques are at opposite poles of the nucleus and are linked by microtubules that have grown from the plaques. During G2 the nucleus moves to the constriction (neck) between mother cell and bud. At the metaphase of mitosis each chromosome has divided into two chromatids, each linked to a spindle plaque by a microtubule. By the telophase of mitosis the chromatids, now daughter chromosomes, have been drawn to the poles of the dividing nucleus. Septum formation between mother and daughter cell occurs early in G1, and cell separation follows. (Diagram based on various sources.)

exposed to the mating pheromone produced by the opposite mating type, are arrested in G1 and diverted to conjugation. Diploid cells provided with conditions appropriate for sporulation also fail to pass 'Start' and undergo meiosis.

The master cell cycle protein is a regulatory kinase Cdc28, which is activated by binding to any one of at least nine regulatory proteins known as **cyclins**, as they accumulate in abundance and disappear at different times in the cell cycle. Activation of Cdc28 by cyclins Cln1, 2 and 3 in late G1 triggers passage through Start. During S phase, activation of Cdc28 by cyclins Clb3, 4, 5 and 6 promotes DNA replication and separation of spindle pole bodies, described below.

The nuclear envelope in *S. cerevisiae*, as in other eukaryotes, consists of two double unit membranes. Prior to 'Start' there is a single dense plaque about 150 nm in diameter lying between the nuclear membranes. This plaque, which has an important role in mitosis, is termed the **spindle plaque** or **spindle pole body**. The spindle plaques of *S. cerevisiae* have been isolated and it has been shown that they can act as centres from which microtubules will grow by polymerization of tubulin sub-units, each a dimer of α-tubulin and β-tubulin. Terms applied to structures in other organisms that are analogous to the spindle plaque of *S. cerevisiae* include **centrosome, microtubule organizing centre** and **nucleus-associated organelle**. The first event indicating that the cell is committed to a further cycle is the duplication of the spindle plaque, resulting in a double plaque connected by a bridge. Microtubules extend from the plaque for short distances both into the nucleus and into the cytoplasm. Those projecting into the cytoplasm indicate the point at which bud emergence subsequently occurs. Following spindle plaque duplication, the two plaques start to migrate, still within the nuclear membrane, to opposite poles of the nucleus. This marks the end of the G1 phase and the commencement of the period of nuclear DNA replication, the S phase.

The DNA in each chromosome of *S. cerevisiae*, like the chromosomal DNA of other eukaryotic cells, consists of several replicating units (replicons). Each replicon has an origin, where replication commences, and two termini, where the process ceases having progressed in both directions from the origin (Fig. 3.7). Replicons can be observed in gently extracted DNA as loops (often misleadingly termed bubbles) and, at the end of the chromosomes, as forks. Replicons in *S. cerevisiae* vary in length from 3 to 86 nm, but lengths of ca 20 or 30 nm are most common. The termini of adjacent replicons coincide so when DNA synthesis is completed in all the replicons in a chromosome, the duplication of the chromosomal DNA is complete, paving the way for the separation of each chromosome into two chromatids at the beginning of mitosis.

Four cell cycle genes, *CDC3, CDC10, CDC11* and *CDC12* encode **septins** which self-assemble as a complex to form filaments about 10 nm in diameter and 32 nm long. Once a yeast cell commits to division by activating the G1 cyclin–Cdc28 complex, it polarizes its cytoskeleton to the site of budding and actin, septin and tubulin polarize virtually simultaneously, well before a bud is visible. The septins form neck filaments which encircle the neck of the bud as it forms during the S phase, and appear to act as a scaffold for the localization of a number of activities, including chitin synthesis. At the site on the cell wall where bud emergence occurs lytic enzymes have rendered the wall thinner and softer.

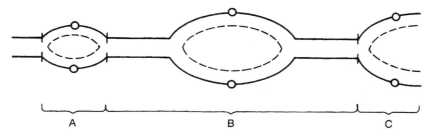

Figure 3.7 DNA replication in three adjacent replicons, A, B and C. Origins are indicated by -O-, termini by -I- and newly synthesized DNA is represented by the broken line. Replication is complete in the short replicons A and C but still in progress in the larger replicon B. Replicons A and B would be seen as loops (often referred to as 'bubbles') in electron microscopy of extracted DNA and the terminal replicon C as a fork.

Then synthesis of wall components and the action of the cell's hydrostatic pressure allow outgrowth of the wall in the softened area to give a bud. By the end of the G2 phase and the beginning of mitosis the bud has attained a considerable size and the nucleus has migrated into the neck that connects mother cell and bud. Meanwhile, some of the microtubules that project from the spindle plaques into the nucleus have extended, presumably by means of tubulin polymerization, to span the nucleus from pole to pole between the two plaques. Other microtubules then extend from the spindle plaques to the chromatids, which in metaphase have become arranged at the equator of the nucleus. Counts of microtubule numbers indicate that a single microtubule extends to each chromatid. Contraction of the pole-to-chromatid microtubules in anaphase, probably by depolymerization of tubulin, results in the movement of each set of homologous chromatids towards opposite poles of the nucleus. In telophase, rapid lengthening of the nucleus occurs, probably brought about by the elongation of the pole-to-pole microtubules. Mitosis ends with the nucleus, already dumb-bell shaped, separating into two nuclei, one in the mother cell and one in the bud. With mitosis complete, entry into the G1 phase occurs, with cell division involving the development of a cross-wall between the mother cell and the bud (now a daughter cell) and finally cell separation.

Throughout the cell cycle the nuclear membrane remains intact. This is true also of the other yeasts and of most of the filamentous fungi, but contrasts with most other eukaryotes, in which the nuclear membrane disappears for the duration of mitosis. When growth rates are maximal, the four phases in the cell cycle of *S. cerevisiae* are of approximately equal length. With slower growth rates, induced by nutrient limitation, that part of the G1 phase following cell division is expanded, but the length of the rest of the cycle, from the initiation of DNA synthesis to cell division, is unchanged. In synchronous cultures of *S. cerevisiae* there is a stepwise increase in cell numbers and DNA content, doubling occurring at an appropriate point in the cell cycle. Total protein and RNA, however, both increase exponentially throughout the cycle.

Studies on cell division cycle (*cdc*) mutants have demonstrated two causal chains of events in the cell cycle (Fig. 3.6). The blocking of budding while nuclear division continues can hence result in a cell with two nuclei, and the blocking of nuclear division but not budding, in a uninucleate cell with a bud. Blocking of the cell division process results in a budded, multinucleate cell.

Considerable work has also been carried out on the control of the cell cycle in the fission yeast, *Schizosaccharomyces pombe*, revealing similarities and differences (for example, a G1 phase that is much shorter and that may in some strains and conditions be absent) as compared with *S. cerevisiae*.

The Cell Cycle in Slime Moulds

Limited studies have been carried out on the cell cycle of the cellular slime mould *Dictyostelium discoideum*, utilizing a mutant *AX2* (AX for axenic) able to grow in pure culture. Growth in axenic conditions is much slower than growth in two-membered culture in this organism and the interval between successive cell divisions is about 9 hours instead of 2 hours. The G1 and S phases each account for about 20% of the cell cycle and G2 for about 60%. Mitosis is a rapid process, lasting about 8 minutes (about 1% of the cell cycle) and is open, the nuclear membrane disappearing. The amoeboid phase in the life cycle of the Myxomycete *Physarum polycephalum* also has an open mitosis (the nuclear membrane disappears) but in the multinucleate plasmodial phase the mitosis is closed, the nuclear membrane remaining intact.

Large plasmodia of *P. polycephalum* contain millions of nuclei which undergo mitosis with a high degree of synchrony. Such plasmodia hence provide a naturally synchronous cell cycle useful for a range of biochemical studies. Nuclear division in rapidly growing plasmodia occurs at intervals of about 10 h. Mitosis occupies about 3% of the cycle and DNA synthesis (the S phase) starts towards the end of mitosis. There is thus no G1 phase in growing plasmodia. The S phase occupies about 30% of the cell cycle and G2 about 66%. The plasmodium grows continuously through the cell cycle. Cell division does not occur so with the completion of each cell cycle the number of nuclei per plasmodium is doubled. Although a G1 phase is lacking in growing amoebae or plasmodia of *Physarum*, the nuclei of cysts and sclerotia, induced by starving amoebae and plasmodia, respectively, are in G1 phase.

The Growth of Hyphae

Vegetative growth of filamentous fungi in nature takes the form of hyphae spreading upon or penetrating a substratum. If extension of such hyphae occurred far behind the apex then friction between hyphae and substratum would make advance difficult and lead to buckling, distortion and damage to the hypha. Such problems do not occur, as extension is confined to the apical region, ceasing when the hypha attains its full width. An understanding of hyphal growth hence requires a consideration of the form and ultrastructure of the extension zone and of events occurring in the apical region.

Form and ultrastructure of the extension zone

The apical region of a hyphal tip approximates to a half ellipsoid of revolution, although for small hyphae the hemiellipsoid may not differ greatly from a hemisphere. The hemiellipsoid form is explicable in terms of the effect of the internal hydrostatic pressure acting upon a hyphal wall that diminishes in plasticity from the tip to the base of the extension zone. The extending hyphae of fungi have at or near the tip an approximately spherical region that stains with iron-haematoxylin and is known either as the **Spitzenkörper**, the term bestowed by the German discoverer, or by the English translation, the **apical body**. The region is dark under phase-contrast microscopy. Electron microscopy shows that the Spitzenkörper includes a mass of apical vesicles surrounding a vesicle-free core (Fig. 3.8). Fluorescent antibody staining of this core in a hypha of the chytrid *Allomyces macrogynus* shows that it is rich in the cytoskeletal components, actin and γ-tubulin, a form of tubulin associated with microtubule organizing centres and with centrosomes, i.e. sites for nucleation of microtubules (Plate 6). Staining for α-tubulin in the same hypha shows a mass of longitudinal microtubules from the sub-apical region converging on the Spitzenkörper.

Longitudinal microtubules are also seen in electron micrographs of hyphal tips (Fig. 3.8) The Spitzenkörper has a central position when a hypha continues to

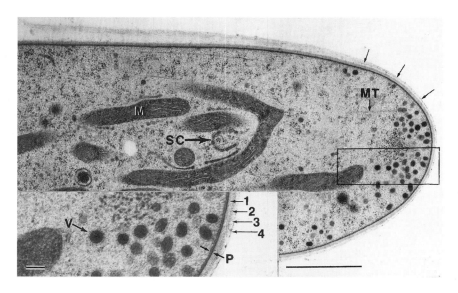

Figure 3.8 Electron micrograph of longitudinal section of a hyphal tip of *Fusarium acuminatum*, fixed by freeze substitution. Note the apical vesicles, longitudinal microtubules (MT), the close association between mitochondria (M) and smooth Golgi cisternae (SC) and the increase in wall thickness from the apex to the subapical region (arrows in upper right). The Spitzenkörper appears as a region surrounded by vesicles containing many small particles. The insert is an enlargement of the marked area, showing apical vesicles (V), the smooth profile of the plasma membrane (P), and the four-layered appearance of the cell wall (1, 2, 3, 4). Scale bars represent 1.0 and 0.1 μm respectively. Reproduced with permission from Howard, R. J. & Heist, J. R. (1979). Hyphal tip cell ultrastructure in the fungus *Fusarium*: improved preservation by freeze-substitution. *Journal of Ultrastructural Research* **66**, 224–234. See also Plate 6.

grow in the same direction but a change in direction of growth is preceded by a displacement of the Spitzenkörper to the side of the hyphal tip, followed by its re-establishing its central position. Displacing the Spitzenkörper with a laser beam results in redirection of apical growth. A new branch is preceded by the formation of a new Spitzenkörper at that site. Cessation of growth is accompanied by the disappearance of the Spitzenkörper. It is hence concluded that the Spitzenkörper has an important role in hyphal extension. Its role has been called that of a 'vesicle supply centre', for the vesicles required for apical extension.

The cytoplasmic vesicles associated with the Spitzenkörper are bounded by single-layered membranes, and commonly fall into two distinct classes, large vesicles, some of which appear to have contents, and microvesicles. In *Neurospora crassa* the large vesicles have a diameter of about 120 nm and the microvesicles about 40 nm. Cytoplasmic vesicles are associated not only with the tips of established hyphae but with the emergence of germ tubes from spores, the branching of hyphae and the development of clamp connections, all processes involving wall extension. Detailed studies on the distribution of cytoplasmic vesicles and other organelles within hyphae have been carried out with a strain of *Neurospora crassa*. Vesicles occupy about 80% of the volume within 1 μm of the hyphal tip, about 25% at 4–6 μm, and at the base of the extension zone, 25 μm from the tip, only 5%; endoplasmic reticulum and mitochondria are absent from the tip, appearing respectively at about 3 μm and 6 μm from the tip; and nuclei do not occur within 40 μm of the hyphal tip, so are absent from the extension zone.

Wall extension and biosynthesis

Experiments were carried out over 100 years ago in which particles were dusted onto the sub-apical region of growing hyphae. Particles that adhered to the surface of the hyphae did not change their position in relation to the substratum, but were left behind as the hypha extended. Axial extension of the wall must hence be limited to the extreme tip, and wall growth through the rest of the extension zone confined to circumferential extension. The picture that emerges of fungal growth is one of prodigious activity at the hyphal apex, with numerous vesicles delivering lytic and biosynthetic enzymes to the plasma membrane at a high rate. When cytoplasmic vesicles fuse with the plasma membrane they also contribute their own membrane components, thus permitting plasma membrane extension. An intensively studied strain of *Neurospora crassa* has a mean hyphal extension rate of 18.5 μm min^{-1}. It has been calculated that the necessary extension of the plasma membrane would require the fusion of about 40 000 cytoplasmic vesicles per minute with the plasma membrane. The same strain of *N. crassa* has an extension zone length of about 25 μm. This means that the entire extension zone is relocated, with the base of the extension arriving at the point previously occupied by the apex, in 80 seconds. 'Relocation times' ranging from 30 seconds to a few minutes have been found in other fungi. The ways in which such rates of hyphal advance, and necessary biosynthetic activity, are sustained will now be considered.

The growth of hyphae involves the extension of the wall and hence the biosynthesis of wall components. Extensive studies have been carried out on chitin biosynthesis and more limited investigations on the formation of other wall components. Studies on a variety of fungi have demonstrated the enzymes and biosynthetic steps involved in the conversion of glucose-6-phosphate into the chitin precursor uridine diphosphate N-acetylglucosamine, UDP-GlcNAc (Fig. 3.9). A single enzyme activity is involved in the final step, chitin synthase:

$$2UDP\text{-}GlcNAc + (GlcNAc)_n \rightarrow (GlcNAc)_{n+2} + 2UDP$$

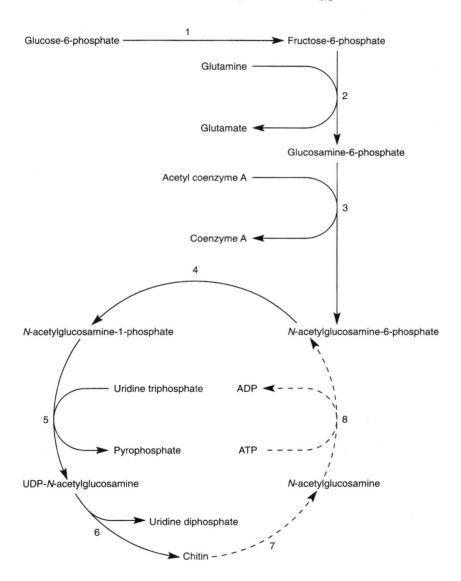

Figure 3.9 Pathway of chitin synthesis ($\rightarrow$) and N-acetylglucosamine cycle ($---\!\!>$). Enzymes as follows: 1, phosphoglucoisomerase; 2, glucosamine-6-phosphate synthase; 3, glucosamine-phosphate-acetyltransferase; 4, N-acetylglucosamine phosphomutase; 5, UDP-N-acetylglucosamine pyrophosphorylase; 6, chitin synthase; 7, chitinase and N-acetylglucosaminidase; 8, N-acetylglucosamine kinase.

In all fungi so far examined, however, there are multiple isozymes of chitin synthase. *Saccharomyces cerevisiae* has three structural chitin synthase genes, *Aspergillus fumigatus* has at least 7, *Phycomyces blakesleeanus* has at least 10. Studies with deletion mutants of the different genes in a range of fungi show that each gene product has a distinct role in the growth of walls of hyphae and yeast cells, of septa and of spores. Biochemical studies have nearly all used mixtures of isozymes, because there is as yet no technique to obtain active pure preparations of single isozymes. There is, however, no reason to suppose that there are fundamental differences between their biochemical properties, so the following generalizations can be made. Chitin synthases are intrinsic transmembrane proteins, requiring a phospholipid environment for activity, and they span the cell membrane, accepting substrate from the cytosol and extruding nascent chains of chitin into the growing wall. They require a divalent cation for activity, probably Mg^{2+} *in vivo*, are activated by the chitin monomer, GlcNAc, and inhibited by their product UDP. Enzyme activity in membrane preparations is usually activated, sometimes many-fold, by treatment with proteases such as trypsin, i.e. as isolated the enzyme is a 'zymogen', but the significance of this observation to the regulation of chitin deposition in the cell wall remains unclear. There is also growing evidence for regulation by post-translational modification of the enzyme by phosphorylation. Autoradiographic studies on the incorporation of tritium-labelled GlcNAc show that its incorporation into chitin is highly polarized at the hyphal tip, at branch points and at growing septa, with very little incorporation into lateral walls. The density of labelling over hyphal tips closely parallels the distribution of cytoplasmic vesicles, further implicating their role in wall biosynthesis. These distinct localizations show that the enzyme must be active at these points and inactive elsewhere. There is evidence that it can be almost instantly activated by hypo-osmotic stress, i.e. by membrane stretch, and this may be the situation at hyphal tips and branch points. There is also genetic evidence, however, that individual chitin synthases require associated proteins, including septins (page 108), to anchor them in the appropriate site in the cell membrane.

Chitosomes, a specific class of microvesicles containing chitin synthase in inactive zymogen form, have been identified by differential centrifugation of cell homogenates from a range of fungi. Synthesis of chitin microfibrils by appropriately treated chitosome preparations has been demonstrated, each chitosome producing a microfibril (Fig. 3.10). In the living cell it is probable that chitosomes fuse with the plasma membrane, at which stage the zymogen is enzymically activated. In addition to chitin synthase, the lytic enzyme chitinase is produced by growing fungi. It is probable that this enzyme is involved in a controlled breaking of the chitin microfibrillar network when a lateral branch is produced (page 122) and perhaps also in modelling of the antiparallel arrangement of the α-chitin microfibrils and in cross-linking chitin to other wall components. An *N*-acetylglucosamine cycle in which monomers released by chitinase and *N*-acetylglucosaminidase are reincorporated into chitin has been envisaged (Fig. 3.9). In the Zygomycetes, such as *Mucor* species, nascent chitin chains are immediately deacetylated to give chitosan by the enzyme chitin deacetylase.

β(1→3) Glucan fibrils are also synthesized in a highly polarized fashion in growing hyphae, but apparently somewhat less so than those of chitin. They are

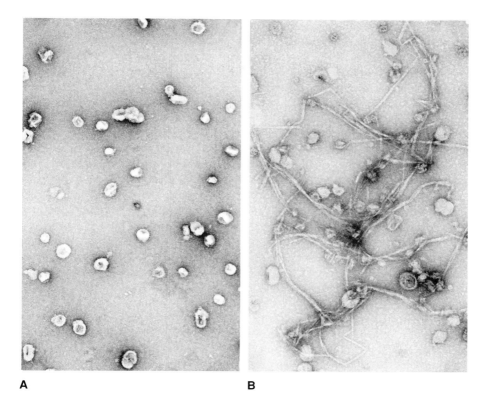

Figure 3.10 Electron micrographs of negatively stained preparations of chitosomes from *Mucor rouxii*. A, Purified chitosomes, 40–70 nm in diameter. B, Chitosomes that have been proteolytically activated and incubated with substrate UDP-GlcNAc and activator, GlcNAc. (S. Bartnicki-Garcia and C. E. Bracker)

synthesized by the enzyme glucan synthase, utilizing the substrate uridine diphosphate glucose:

$$\text{UDP-Glc} + (\text{Glc})_n \rightarrow (\text{Glc})_{n+1} + \text{UDP}$$

Like chitin synthase, it is a plasma membrane-bound enzyme, accepting substrate from the cytosol and feeding glucan chains into the wall. Enzyme preparations are stimulated by very low concentrations of guanosine triphosphate, GTP. The enzyme preparations can be separated into two inactive components, a membrane-bound synthase and a soluble fraction that has the properties of a GTP-binding regulatory protein. Recombining them restores activity. In *Saccharomyces cerevisiae*, the GTP-binding protein has been identified as Rho1, the protein that activates protein kinase C which has a major role in the cell by regulating the 'PKC signal transduction pathway' which controls transcription of genes involved in polarized growth, cell wall biosynthesis including $\beta(1\rightarrow3)$ glucan synthase genes, and some stress responses. Fluorescent antibody studies have shown that Rho1 is localized at sites of active growth, where it would be available for activation of β(1→3) glucan synthase. Details of the biosynthesis of β(1→6) glucan are less well understood. As mutants of several proteins active in the endoplasmic reticulum and Golgi bodies led to impaired β(1→6) glucan formation, it appeared that its synthesis, like that of mannoproteins, occurred in the normal secretory pathway. However, electron micrographs of sections of *S. cerevisiae* cells treated with gold-labelled antibody to β(1→6) glucan showed no internal labelling, but the wall was heavily labelled. Coupled with the observation that β(1→6) glucan was associated with plasma membrane-derived vesicles and not Golgi vesicles after differential centrifugation of lysed spheroplasts, this points to the plasma membrane being the site of synthesis, as for chitin and β(1→3) glucan. In some way, however, its synthesis must be regulated by products of the secretory pathway. The *N*-linked and *O*-linked mannoproteins are synthesized by progression through the secretory pathway, from initial protein synthesis by membrane-bound ribosomes into the endoplasmic reticulum, successive addition of sugar units in the endoplasmic reticulum and then in the Golgi bodies, then packaging in secretory vesicles, transport to the plasma membrane and release into the cell wall. Synthesis of the *N*-linked mannoproteins is a complex procedure, initially involving dolichol phosphate, a phosphorylated alcohol of a long-chain isoprenoid (i.e. 15 or 16 isoprene unsaturated hydrocarbon units) as a transmembrane acceptor for two *N*-acetylglucosamine units, then nine mannose units, then three glucose units, to give $Glc_3Man_9GlcNAc_2$-PP-dolichol lipid precursor. The oligosaccharide chain is transferred within the endoplasmic reticulum to an amino group in an asparagine residue in the protein, with release of dolichol pyrophosphate. The glucose units and one mannose unit are successively removed, and the nascent glycoprotein is transferred to the Golgi bodies, where large numbers of mannose units (up to 200) are added, to give a highly branched structure, with α(1→2)- and α(1→3)-linked mannoses on a backbone of up to 50 α(1→6)-linked units. Some of the mannoses may be phosphorylated, giving a net negative charge. *O*-Linked mannoproteins have the first mannose unit added to serine or threonine of the protein in the endoplasmic reticulum from dolichol-phosphate-mannose, but then the additional α(1→2) and α(1→3)-linked mannoses are added in the Golgi bodies, via GDP-mannose.

Maturation of the hyphal wall

The nascent wall at the hyphal tip is thin and plastic, the subapical mature wall is thicker and rigid (Fig. 3.8). In *Neurospora crassa* the apical wall has a thickness of about 50 nm, but by 250 μm behind the apex wall is *ca* 125 nm and comparable increases in wall thickness occur in other fungi. Three processes contribute to the rigidification. The first is the hydrogen bonding of the rigid polysaccharides, chitin and β(1→3) glucan, into microfibrils. The second is increasingly extensive covalent cross-linking between these two wall components as the wall matures. The third, which also is responsible for the increasing thickness, is the addition of the other major components, β(1→6) glucan, mannoproteins and proteins, and their cross-linking to the scaffolding of chitin-β(1→3) glucan, to eventually give a mature intermeshed multicomponent wall of uniform thickness, as described earlier (page 103; Fig. 3.5). Early stages in this process of maturation have been investigated in growing hyphae of *Schizophyllum commune*, comparing the domed apical 2 μm wall with the immediate sub-apical 2–6 μm cylindrical hyphal wall. At the apex there were no microfibrils visible in electron micrographs and the apex was removed by mechanical shear and treatment with chitinase. Pulse labelling the hyphae with tritiated glucose for 10 min gave radioactive alkali-soluble $\beta(1\rightarrow3)$ glucan at the apex; further growth in non-radioactive glucose for 30 min resulted in the glucan being mostly alkali-insoluble, indicating that it had been extensively cross-linked with $\beta(1\rightarrow6)$ glucan. This process has been termed the steady-state theory of hyphal growth, because plastic wall material is continuously synthesized and then converted into rigid lateral wall with the apex containing a steady state amount of plastic wall that can expand.

The maintenance of polarity
The maintenance of precise polarity of growth at a hyphal tip appears to be the result of polarized transport of vesicles to the apex. In addition, it grows onwards because of the progressive strengthening and thickening of the wall by cross-linking of chitin and glucan components, and the addition of further materials. The rigid tube thus created results in the internal hydrostatic pressure pushing the wall forward at the still plastic apex. Much research has been devoted to elucidating the mechanism of vesicle movement. In light of the observation that the Spitzenkörper is rich in actin and is a focus for microtubules, it seems likely that vesicles move along tracks provided by these cytoskeletal proteins. How polarity is established, and then maintained, when a spore or a yeast cell gives rise to a hypha, remains to be fully elucidated. Progress is being made in investigations of polarity in *Saccharomyces cerevisiae*, using the powerful molecular genetics that have been developed with this organism. It is clear that actin cytoskeleton plays a major role in establishing and maintaining polarity of cell growth and protein secretion in yeast. Thus mutants of *ACT1*, the actin structural gene, produce very large budded cells, indicating growth of both mother and daughter, and show delocalized deposition of chitin, which normally occurs almost only at bud necks and in bud scars. A range of molecular genetic techniques have shown that about 60 other gene products are associated with the actin cytoskeleton, including five genes for the associated cytoskeletal protein, myosin, so precise details of the interactions occurring within the cell will take a long time to elucidate. Actin patches are concentrated at the apices of growing buds and in longitudinal fibrils. Towards the end of nuclear division, actin patches appear as rings between mother cell and bud, and at the time of cytokinesis distinct patches appear in the neck regions, where the septa are to form. Thus throughout the yeast cell cycle actin patches are concentrated at regions of localized growth.

Hyphae are electrically polarized, with a circulating electric current carried by protons, usually flowing into the apical region, and out of the sub-apical trunk. It is probably the result of the proton-translocating ATPase of the plasma membrane being excluded from the apical zone. Thus the ATPase is pumping out protons sub-apically and these re-enter the cell apically for example by symport with amino acids (page 153). This discovery, made by the use of sensitive electrodes, initially led to the idea that this may be the cause of hyphal polarity, with charged protoplasmic components moving electrophoretically in response to the current. Further experiments showed instead that it is a consequence, not a cause, of the polarity. That being said, growing hyphae show directed responses, 'galvanotropism', to imposed electrical fields, and production and responses to electrical currents may be important to some hyphae in some environments.

The peripheral growth zone, protoplasmic streaming and hyphal extension
If the apical compartment in the hypha of a septate fungus is cut, the protoplasm it contains is destroyed and growth is resumed by the formation of a new hyphal tip as a branch from a sub-apical compartment. Cutting through a sub-apical compartment leads to the death of the protoplasm in the cut compartment, but because of the rapid plugging of septal pores at both ends of the compartment, other compartments including the apical compartment are commonly undamaged. Nevertheless, unless the hypha is severed at a considerable distance behind the apex, the hyphal extension rate may be greatly reduced. Systematic experiments on cutting through the hyphae of fungal colonies led to the definition of a peripheral growth zone which unless damaged contributes to the advance of the hyphae at the colony margin. The width of the peripheral growth zone varies considerably between species and varieties. It is commonly, however, many times greater than the length of the extension zone of the hypha (Table 3.4) or even of the apical compartment, so that it is likely to include a dozen or more sub-apical compartments. In *Geotrichum candidum*, however, in which the septa lack pores and have only ultramicroscopic perforations, the width of the peripheral growth zone corresponds to the length of the apical compartment. In the Zygomycetes severing hyphae causes more extensive damage than it does in septate fungi, but apical survival can occur and here too the concept of a peripheral growth zone seems applicable.

Examination of large fungal hyphae with the microscope often shows striking protoplasmic streaming occurring in the direction of the colony margin. Experiments have established that in some fungi water can be transported through hyphae for long distances. It is probable that such water movements are generated osmotically. It is postulated that the conversion of carbon sources to sugars and other soluble low-molecular-weight compounds in older parts of the mycelium leads to the osmotic influx of water and the generation of high internal hydrostatic pressures. In other areas hydrostatic pressures may be lower, and at the apices hyphal walls are elastic and extending. Mass flow of water down a hydrostatic pressure gradient towards the colony margin would hence occur. Movement of water would carry with it nutrients in solution and cytoplasmic organelles such as vesicles. Protoplasmic streaming, generated osmotically, and at a rate faster than that of the apical extension, is hence likely to act together with transport associated with cytoskeletal components, actin fibrils and microtubules, in which the peripheral growth zone supplies hyphal apices with substrates in solution and lipids and enzymes at the rates needed for hyphal extension.

Protoplasmic streaming is readily observed in fungi with large hyphae (> 10 μm diameter) such as *Neurospora* or *Rhizopus* but in fungi with small hyphae (< 5 μm diameter) such as *Aspergillus* or *Penicillium* prolonged observation is

Table 3.4 Dimensions of the extension zone and apical compartments of marginal hyphae and of the peripheral growth zone of fungal colonies

Species	Mean dimensions (μm)			Ratios	
	Extension zone (EZ)	Apical compartment (AP)	Peripheral growth zone (PGZ)	PGZ/EZ	PGZ/AC
Zygomycetes					
Rhizopus stolonifer	29	Inapplicable	8700	300	–
Actinomucor repens	14	Inapplicable	2500	180	–
Higher fungi					
Neurospora crassa	33	510	6800	210	13.3
N. crassa (another strain)	27	340	5000	190	14.7
Geotrichum candidum	2	290	420	210	1.5
Penicillium chrysogenum	?	140	500	?	3.6
Aspergillus niger	?	430	1100	?	2.6

Species with hyphae that are wide and have high extension rates also have long hyphal extension zones and wide colony peripheral growth zones. The peripheral growth zone is over 100 times the size of the extension zone, in slow as well as fast-growing fungi. It is several times that of the apical compartment in fungi with rapidly extending hyphae and striking protoplasmic streaming, but only a little larger in fungi with slowly extending hyphae in which streaming is not obvious. Data from Trinci, A. P. J. (1971). *Journal of General Microbiology* 67, 325–344; Trinci, A. P. J. (1973). *Archiv für Mikrobiologie* 91, 113–126; Collinge, A. J. & Trinci, A. P. J. (1974). *Archives of Microbiology* 99, 353–368; Steele, G. C. & Trinci, A. P. J. (1975). *New Phytologist* 75, 583–587; Trinci, A. P. J. & Collinge, A. J. (1975). *Journal of General Microbiology* 91, 355–361.

needed to detect protoplasmic movement. This slow protoplasmic streaming is probably the consequence of the resistance offered to fluid flow by narrow tubes. This has been expressed in the Hagen–Poiseuille law which states that, other factors being equal, the volume transported in unit time in tubes varies as the fourth power of the radius. Hence the rate of flow is directly proportional to the square of the radius. Narrow septal pores, however, will not appreciably retard flow, since the length of such pores is short and the law states that flow rates are inversely proportional to tube length. Slow rates of protoplasmic streaming will limit the distance over which metabolic activities can support growth at the hyphal apex. It is hence not surprising that *Aspergillus* and *Penicillium* with their narrow hyphae have much smaller peripheral growth zones than *Neurospora* and *Rhizopus*, with their wide hyphae (Table 3.4).

Although a massive protoplasmic flow towards the hyphal apex is the most obvious form of protoplasmic movement, it is not the only form that occurs. Usually it is accompanied by a less obvious reverse flow close to the hyphal wall. Nor can protoplasmic mass flow account for all movement of organelles. Thus after a nuclear division has occurred the daughter nuclei move in opposite directions. A world record, at 4 cm h^{-1}, is claimed for haploid nuclei of *Coprinus congregatus* migrating through haploid hyphae of a sexually compatible strain, to result in a dikaryotic mycelium. Also vesicles, after they approach the apex of the hypha, have to be brought into contact with the appropriate area of plasma membrane (page 112). From observations of mutants and effects of specific inhibitors of cytoskeletal proteins in hyphae of a range of species it is clear that activities of microfibrils (actin–myosin interaction) and microtubules (tubulin–dynein and tubulin–kinesin interaction) have major roles in intrahyphal transport of cell components, and thus in hyphal growth, but details of these are unclear. Actin has already been discussed in terms of hyphal polarity. Numerous experiments with microtubule inhibitors, such as the agricultural fungicide benomyl, and with mutants of tubulins and their associated motor proteins, dynein and kinesin with different fungi have given a confused picture, but the overall conclusion is a clear implication of microtubules with positioning of the Spitzenkörper, with apical growth and with organelle movements. Thus microtubule inhibitors led to the dispersal of the Spitzenkörper and reduction in linear growth rate of hyphae of *Fusarium acuminatum*, and to loss of directional growth and multiple branching in *N. crassa*. These, and many other observations, correlate with the apical localization of γ-tubulin, associated with microtubule organizing centres, and a mass of longitudinal microtubules from the sub-apical region converging on the Spitzenkörper in hyphae of *Allomyces macrogynus* described on page 111. Dynein mutants of the Ascomycete *Nectria haematococca* grew at a third of the wild-type rate, had small or ephemeral Spitzenkörpers, and showed aberrant nuclear division and limited nuclear migration, but a dynein mutant of *N. crassa* showed no loss of Spitzenkörper despite showing multiple abnormal branching and a growth rate one-fifth of wild type. In contrast, in the same study, hyphae of a kinesin mutant and the corresponding double mutant had small or ephemeral Spitzenkörpers, multiple abnormal branching and growth rates 22% and 7% of wild type, respectively. As the kinesin and dynein motors may be responsible for apical and retrograde transport of cell components along the longitudinal cytoplasmic microtubules, respectively, these results and others from other fungi are difficult to interpret at present, but do reconfirm the central role of the cytoskeleton in hyphal growth.

The duplication cycle: nuclear division and cross-wall formation
The apical compartment of filamentous fungi is often multinucleate. Nevertheless, studies of DNA synthesis, mitosis and septation have led to the recognition of a cell cycle although, because of uncertainty as to what constitutes a cell in septate fungi, the term duplication cycle is preferred.

Detailed studies have been carried out on the duplication cycle of *Aspergillus nidulans*, an Ascomycete in which the apical compartment contains about 50 nuclei, under conditions in which the duplication cycle lasts about 2 hours. The G1 phase is long and G2 phase short, but the precise lengths of both are dependent on conditions and growth rates. Between the G1 and G2 phases the period of DNA synthesis (S phase) lasts about 20 minutes or 15% of the duplication cycle. Mitosis, as in other filamentous fungi, is closed and does not involve the breakdown of the nuclear membrane. Within an apical compartment mitosis is parasynchronous, i.e. nearly but not quite synchronous. It normally occurs first with nuclei near the apex and then a wave of nuclear division proceeds through the compartment (Fig. 3.11B). If the duplication cycle is regarded as beginning when septation is completed then the first mitosis begins at 0.75 (three-quarters of the way through the duplication cycle) and all nuclei in the compartment have divided by 0.84. The M phase hence lasts about 9% of the cycle, or a little over 10 minutes. With individual nuclei, however, mitosis is completed in about 3% of the cycle (*ca* 4 minutes).

This very rapid mitosis is normal in filamentous fungi. In another Ascomycete, *Ceratocystis fagacearum*, prophase has been shown to take about 1 minute, metaphase 2 minutes, anaphase a quarter of a minute, and telophase 2 minutes. From 0.93 to 1.00 in the duplication cycle of *A. nidulans* a parasynchronous wave of cross-wall formation (septation) occurs, leading once more to an apical compartment with about 50 nuclei and a fresh set of sub-apical compartments averaging about four nuclei per compartment. Hyphal extension continues throughout the duplication cycle, with the apical compartment doubling in length in the course of a cycle and being restored to the original length by septation. Similar observations have been made on other septate filamentous fungi. Mitosis characteristically occurs about three-quarters of the way through the duplication cycle. In the Basidiomycete, *Polyporus arcularis*, with about 75 nuclei in the apical compartment, nuclear division is again followed by parasynchronous septation in which a new apical compartment of *ca* 75 nuclei and many sub-apical compartments with an average of two nuclei per compartment are produced. In monokaryons of *Schizophyllum commune* (Fig. 3.11A) nuclear division of a uninucleate apical compartment is followed by the production of a single cross-wall (septum). With the binucleate apical compartment of the dikaryon the production of a septum, clamp cell formation and nuclear movements are, as discussed earlier (page 60), coordinated to give two binucleate compartments.

Septa of most fungi are very rich in chitin, arranged as tangential microfibrils, and strongly staining with the fluorescent dye Calcofluor. The site and inwards growth of septa involves localization of a ring of actin, and septal formation and

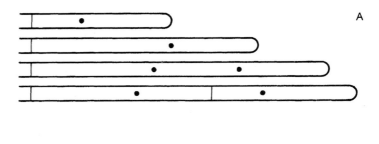

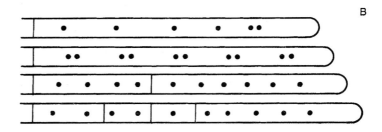

Figure 3.11 Nuclei and septa during the duplication cycle of filamentous fungi. A, A monokaryotic hypha of the Basidiomycete *Schizophyllum commune*, after septation, before mitosis, after mitosis and after further septation. Note nuclear movements. B, A hypha of the Ascomycete *Aspergillus nidulans*, the beginning of a wave of mitosis, its completion, the initiation of septation and its completion. (Diagrams based on those of Trinci in Burnett, J. H. & Trinci, A. P. J., eds. (1979). *Fungal Walls and Hyphal Growth*, pp. 319–356. Cambridge University Press, Cambridge.)

inwards growth is reversibly blocked in *Aspergillus nidulans* by the actin depolymerizing agent cytochalasin A (itself a fungal secondary metabolite) and by benomyl, a microtubule inhibitor. Thus, as for apical growth, actin filaments and microtubules are involved in the positioning of septal wall synthesis.

Branch production
The hyphae of fungi achieve exponential growth by branching. Rarely branching is dichotomous, with a hyphal apex replaced by two adjacent apices. More often, however, branching is lateral, with a new hyphal apex being formed a long way behind the existing apex. The control of branch initiation is not well understood, but it is reasonable to suppose that it occurs when the production of materials for wall extension exceeds the quantity that can be utilized at the existing apex. Vesicles containing wall-synthesizing enzymes can then be diverted to a new site, where a new Spitzenkörper is forming. In septate fungi branches often arise just behind septa. Branches usually grow forwards, but an exception is the very short branch of a clamp connection, which grows backwards and curves in to fuse with the parent hypha (page 111).

The Growth of Populations and Colonies

The form of microbial growth most readily analysed and understood is that of a homogeneously dispersed population of unicellular microorganisms. There have hence been extensive studies on the growth of bacteria and of yeasts in aerated and agitated liquid media. The principles that have emerged from such studies will be considered, and then the extent to which they are applicable to the filamentous fungi.

Exponential Growth

Specific growth rate and doubling time
A population of unicellular microorganisms provided with abundant nutrients and a suitable environment will grow. Maintenance of uniform conditions will result in balanced growth with the mean size and composition of the cells remaining constant. There will be a proportionate increase in all aspects of biomass, such as cell number, dry weight, and the nucleic acid and protein

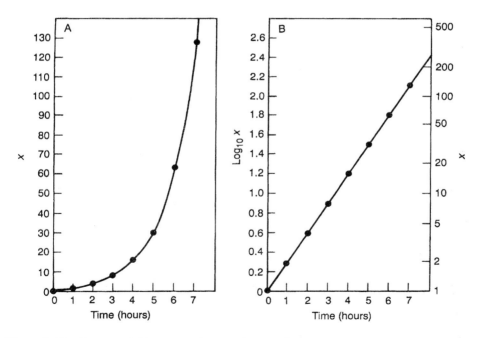

Figure 3.12   Exponential increase of a population of microorganism. The symbol x can represent cell number, or dry weight, or protein or nucleic acid content or any other convenient measure of population growth. The specific growth rate (μ) is 0.69 which gives a culture doubling time (t_d) of one hour. The value of x at 0, 1, 2, 3, 4, 5, 6 and 7 h is 1, 2, 4, 8, 16, 32, 64 and 128, respectively. A, Plot of x against time. B, Logarithmic plot of x against time. This can be made either by plotting logarithms of x on ordinary graph paper or by plotting x on semi-log paper.

content of the population. Growth may hence be monitored by measuring any of these or other convenient attributes of the population including optical density. Such measurements will show that the population grows in size more and more rapidly (Fig. 3.12A). This continual increase in the population growth rate is the inevitable consequence of a constant growth rate per cell with a continual increase in the number of cells, and is capable of simple mathematical treatment.

Let x represent the biomass of a population of microorganisms. The population growth rate at any moment can be represented by the differential coefficient dx/dt. This will be proportional to the growth rate per unit of biomass, usually represented by μ, and referred to as the **specific growth rate**, which for a given species under uniform conditions is a constant. Population growth rate will also be proportional to biomass. These relationships can be expressed by equation 1.

$$dx/dt = \mu x \tag{1}$$

Integration of equation 1 yields equation 2 in which natural logarithms (logarithms to the base e) are indicated by ln, t represents elapsed time, and x_0 the biomass when $t = 0$.

$$\ln x = \ln x_0 + \mu t \tag{2}$$

Natural logarithms can be converted into logarithms to base 10 with adequate accuracy by dividing by 2.30. Division of equation 2 by this number yields equation 3.

$$\log 10 x = \log 10 x_0 + \mu t/2.3 \tag{3}$$

Equation 2 can also be put in the form of equation 4.

$$x = x_0 \, e^{\mu t} \tag{4}$$

Since in mathematical terminology μt is here an **exponent**, and x is proportional to $e^{\mu t}$, growth according to this relationship is commonly referred to as **exponential growth**. The form of equations 2 and 3 implies that when exponential growth is occurring a plot of either ln$x$ or $\log_{10} x$ against time yields a straight line (Fig. 3.12B) instead of the parabola obtained by plotting x against time (Fig. 3.12A). Such log plots are very useful as departures from exponential growth are readily detected.

Rearrangement of equation 3 to give equation 5 enables the specific growth rate (μ) to be calculated from growth measurements.

$$\mu = 2.3 \, (\log_{10} x - \log_{10} x_0)/t \tag{5}$$

From the form of this equation specific growth rate has the dimensions of reciprocal time (l/t), and is analogous to an interest rate – thus a specific growth rate of 0.1 h^{-1} is equivalent to a 10% interest rate per hour. The 'interest', however, is not simple interest but continuous compound interest. Hence a specific growth rate of 0.1 h^{-1} leads to population doubling not in 10 hours but in a considerably shorter period (6.93 h). The relationship between specific growth rate and **population doubling time** (t_d), also known as the **culture doubling time**, can be derived by substituting t_d for t and $2x_0$ for x in equation 2 to give equation 6.

$$\ln 2x_0 = \ln x_0 + \mu t_d \tag{6}$$

Rearranging and cancelling gives equation 7.

$$t_d = (\ln 2)/\mu = 0.69/\mu \tag{7}$$

Hence a specific growth rate of 0.1 h^{-1} leads to a doubling time of 6.9 h, not 10 h as would happen with simple interest at 10% h^{-1}. The inverse relationship between μ and td means that with a value of μ twice as large ($\mu = 0.2$ h^{-1}) the doubling time would be half as long ($t_d = 3.45$ h). When balanced growth is occurring the values for specific growth rate and population doubling time are applicable to all the features of biomass.

Population doubling time and mean cell interdivision time
The increase in the number of cells in a growing population of unicellular organisms is the consequence of the division of individual cells. It is hence sometimes assumed that the doubling time (t_d) of a population is the same as the **mean interdivision time** (τ) of cells, i.e. the average time that elapses between successive divisions of individual cells. The term **generation time** is therefore often applied to both doubling time and interdivision time. The population doubling time (t_d) is in fact a little shorter than the **mean interdivision time** (τ) since the faster growing cells make a disproportionate contribution to population growth. There have been some studies in bacteria in which the relationship between τ and t_d has been examined. In *Pseudomonas aeruginosa* population doubling time is about 1% less than mean cell interdivision time, in *Bacillus megaterium* about 5% less, and in *Bacillus cereus* var. *mycoides* about 8% less. The following relationship, equation 8, in which σ is the standard deviation of the interdivision times, has been demonstrated.

$$\tau = t_d + \sigma^2/2 \tag{8}$$

It will be seen from this equation that τ and t_d can only be the same if all the cells in the population have the same interdivision time and σ hence is zero.

Batch Culture and Growth Phases

Microorganisms can be cultured either in closed systems, considered below, or in open, continuous flow systems, discussed in the next section. In a closed or **batch system** a volume of a suitable medium is inoculated and growth takes place until terminated by exhaustion of an essential nutrient or accumulation of toxic products of metabolism. A closed system can be either **homogeneous** or heterogeneous. A homogeneous distribution of organisms, nutrients and wastes is achieved in a liquid medium by some form of agitation. Agitation of culture volumes of 500 ml or less is usually carried out by means of a rotary or a reciprocating shaker. Larger culture volumes in laboratory or industrial fermenters are stirred by impellers. Heterogeneous batch cultures can be obtained using static liquid media, media gelled with agar or moist solid substrata of various kinds. When microorganisms are grown in a homogeneous batch culture, for example in liquid media in Erlenmeyer flasks on a rotary shaker, three phases of growth can be distinguished: the **lag, exponential** and **stationary** phases (Fig. 3.13). These will ultimately be followed by a phase of decline in which cell death and autolysis (cell breakdown from the release of the organism's own lytic enzymes) cause a gradual fall in biomass.

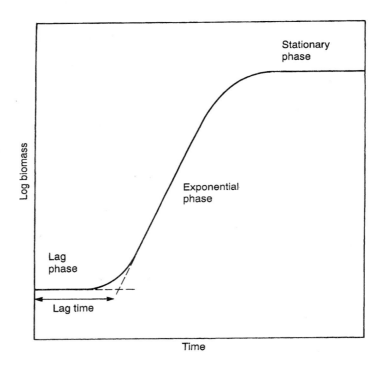

Figure 3.13 Growth phases in a batch culture of a microorganism growing in shaken liquid medium. The culture passes gradually from the lag phase into the exponential phase, but a precise lag time can be defined by extrapolation of the lag and exponential phase until they intersect. The transition between the lag and exponential phase is sometimes termed the acceleration phase but its length is difficult to determine and the term is little used. When a culture is inoculated with cells growing vigorously on an identical medium, a lag phase is absent and the culture proceeds directly into the exponential phase. Exhaustion of an essential nutrient or accumulation of a toxic product gradually brings the exponential phase to an end and leads to a stationary phase. The period of falling specific growth rate is sometimes termed the deceleration phase. Finally the stationary phase passes into a decline phase in which cell death and autolysis leads to a fall in biomass. Representations of the decline phase on the diagram would require a considerable extension of the time axis.

Lag phase

Vigorously growing organisms transferred to an identical medium at the same temperature will continue exponential growth. Under these circumstances no lag phase occurs. Organisms from a culture that is not growing, however, take some time to pass out of the resting state and to commence biosynthesis. Determination of the length of the lag phase presents problems, since the point at which perceptible increase in biomass is detected in a population of organisms that may be small is dependent on the sensitivity of the analytical method. However, a precise measure of the delay in achieving exponential growth can be obtained by proceeding as if the transition from lag phase to the exponential phase was sudden and not gradual (Fig. 3.13). This parameter of culture growth is termed the **lag time**.

Exponential phase

The exponential phase is the period during which a constant specific growth rate (μ) is maintained and biomass increase is exponential (Fig. 3.13). The specific growth rate is maximal ($\mu = \mu_{max}$) when all the nutrients required for growth are in excess and no inhibitor of growth is present. If any nutrient is present in sub-optimal amounts then the value of μ relative to μ_{max} is determined by the affinity of the organism for the limiting nutrient and by the concentration of the nutrient. The specific growth rate (μ) when plotted against nutrient concentration (s) yields a rectangular hyperbola (Fig. 3.14), just as does the rate of an enzyme reaction when plotted against substrate concentration. Such a relationship is to be expected, since enzyme reactions are involved in the uptake and utilization of the

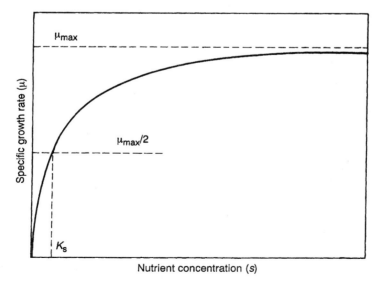

Figure 3.14 The relationship between specific growth rate (μ) and nutrient concentration (s). The maximal specific growth rate is represented by μ_{max} and half the maximum growth rate by $\mu_{max}/2$. This rate is attained with a nutrient concentration equal to the saturation constant, K_s.

nutrient. The affinity of the organism for a nutrient is expressed as the saturation constant K_s, the nutrient concentration at which μ is half μ_{max}. The equation 9 relating specific growth rate to substrate (i.e. nutrient) concentration is similar in form to the Michaelis–Menten equation for enzyme action

$$\mu = \mu_{max}.\ s/(s + K_s) \tag{9}$$

Saturation constants are very low, even for nutrients that are required in large amounts. Thus the K_s for glucose in *Saccharomyces cerevisiae* is 25 mg l^{-1} (140 µM), for *Aspergillus nidulans* 5 mg l^{-1} (28 µM) and for *Geotrichum candidum* 1 mg l^{-1} (5.6 µM). Such saturation constants can be compared with the nutrient levels usual in laboratory media, such as 10 g l^{-1} glucose – *ca* 1000 times as great. Nutrients required in smaller amounts have still lower saturation constants; thus with an arginine-requiring strain of *A. nidulans* the K_s for arginine is 0.5 mg l^{-1} (2.9 µM). Such low saturation constants (or high substrate affinities) enable rapid growth to take place even at the very low nutrient concentrations which occur in natural environments. Laboratory media will normally contain all nutrients vastly in excess of their respective saturation coefficients, so during the exponential phase of growth in batch culture the specific growth rate (μ) will normally be maximal, μ_{max}. Maximal specific growth rates have been determined for a range of fungi (Table 3.5).

Ultimately depletion of an essential nutrient will result in the lowering of μ and the end of the exponential phase of growth. As growth rate declines, reserves may be built up from a nutrient present in excess. Thus nitrogen exhaustion may be accompanied by uptake and storage of carbohydrate by the cell.

Growth during the exponential phase is balanced, with the amounts of the metabolites required for growth being finely controlled. The metabolic activity during the exponential phase is often termed **primary metabolism** in contrast with the **secondary metabolism** (see below) characteristic of the stationary phase. These terms, although useful, lack precision and have been used by different authors in somewhat different ways. Primary metabolism is generally taken to

Table 3.5   Maximum specific growth rates and doubling times for some fast-growing fungi

Species	Maximum specific growth rate (μ_{max} h^{-1})	Doubling time (min)
Oomycete		
Achlya bisexualis	0.81	51
Higher fungi		
Candida tropicalis	0.74	56
Geotrichum candidum	0.62	66
Aspergillus nidulans	0.57	73
Neurospora crassa	0.50	83
Saccharomyces cerevisiae	0.47	88

Doubling times are obtained from the relationship $t_d = 0.69/\mu$ and are usually expressed in hours but are here converted into minutes. Data from various authors.

refer to the quantitatively important metabolic pathways involved in growth, such as energy generation and the synthesis of nucleic acids, proteins, lipids and carbohydrates. Other compounds such as vitamins may be produced, but only in the small amounts essential for growth. The industrial interest of the exponential phase hence lies in the production of biomass, single cell protein, enzymes which attack nutrient macromolecules, such as amylases, cellulases and proteases, and other products characteristic of primary metabolism. Secondary metabolism, not conspicuous during the exponential phase, refers to the production of compounds not essential for growth but presumably having some other role in the life of the organism (see Chapter 8).

Stationary phase

With the supply of an essential nutrient nearing exhaustion the specific growth rate of a batch culture declines to zero and the maximum biomass is attained. This amount of biomass, termed the total yield, is determined by the quantity of limiting nutrient supplied. The relationship between the amount of the limiting nutrient and yield is normally linear over a wide concentration range. A yield constant – grams of biomass produced per gram of limiting nutrient supplied – can be calculated for each essential component of the medium. With *Penicillium chrysogenum*, for example, the yield constant for glucose is 0.43, or, if expressed in terms of carbon supplied as glucose, 1.08. Growth yields for *Candida* spp. have been determined with a range of carbon sources (Table 3.6). The yield per gram of substrate varies with the amount of energy that the carbon source can supply. The oxygen requirement is greatest (lowest Y_{oxy}) for the most reduced substrate.

When an organism passes from the exponential to the stationary phase major physiological changes will occur. The precise nature of these changes will be determined by how exponential growth is brought to an end (e.g. by carbon or nitrogen limitation) and hence can be controlled by attention to medium composition. Appropriate nutritional imbalances can result in the accumulation of metabolic intermediates, such as citric acid, or overproduction of vitamins. In addition secondary metabolism becomes important. Fungi are able to produce a wide range of secondary metabolites some of which are commercially important such as the antibiotic penicillin, the pharmacologically active ergot alkaloids, and the plant hormone gibberellic acid (see Chapter 8). Many secondary metabolites are derived from acetyl CoA and their production is favoured by excess carbon. Carbon excess also favours lipid accumulation. The possible production of lipids,

Table 3.6   Yield constants for the yeast *Candida*

Carbon source	Yield constant (g biomass g^{-1} nutrient)		
	Substrate	Substrate carbon	Oxygen
Glucose	0.51	1.28	1.30
Acetic acid	0.36	0.90	0.62
Ethanol	0.68	1.30	0.58
n-alkanes	0.81	0.96	0.35

Results for *n*-alkanes obtained with *Candida intermedia* and for other compounds with *Candida utilis*. Data from Pirt, S. J. (1975). *Principles of Microbe and Cell Cultivation*. Blackwell, Oxford.

organic acids, vitamins and other useful compounds during the stationary phase makes this stage in batch culture of great interest to fermentation technologists. Provided conditions are appropriate the stationary phase is also characterized by the onset of differentiation with structures important for survival or dispersal, such as spores or sclerotia, being produced. Differentiation may also be accompanied by commercially interesting biosynthetic activities – for example the production of some ergot alkaloids is associated with the formation of incipient sclerotia.

Decline phase

The stationary phase is followed by one of cell death, autolysis and decline in biomass. Some growth may occur in the decline phase, utilizing nutrients released by autolysis, but will be less than the biomass loss through death. Not easily detected, such growth is termed cryptic growth. The kinetics of the phase of decline have received relatively little study, and information about the phase is particularly sparse for fungi.

Growth phases in heterogeneous batch cultures

The phases of growth just described for homogeneous batch cultures can to some extent be distinguished in heterogeneous batch cultures. A lag phase can be observed and this is followed by a period in which growth is exponential. However, as metabolism proceeds there is increasing heterogeneity and some regions will be depleted in nutrients or oxygen before others. Hence growth will have come to an end in some parts of the culture while still proceeding vigorously elsewhere, and growth and stationary phases will overlap. The same will apply to the stationary and decline phases. Growth in heterogeneous batch culture will be discussed further for both yeasts (page 134) and filamentous fungi (pages 135–145).

Continuous Flow Culture

Open, continuous flow systems of culture can, like batch systems, be either homogeneous or heterogeneous. Homogeneous continuous flow systems have been employed extensively both in the laboratory and on an industrial scale and will be considered first. Precise quantitative analyses of growth in such systems have been published; here the treatment will be non-mathematical and only semi-quantitative.

Homogeneous continuous flow systems

Liquid medium in a culture vessel of appropriate design (Fig. 3.15) is inoculated. Further medium is delivered to the vessel from a reservoir at a controlled rate by a peristaltic pump or similar device. The volume of the medium in the vessel is kept constant by means of an overflow to a collecting vessel. Thus, fresh medium enters the culture vessel and partially spent medium containing microorganisms leaves it. Thorough mixing in the culture vessel is essential, and is achieved with

Impeller

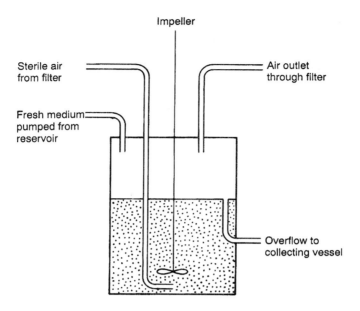

Sterile air
from filter

Air outlet
through filter

Fresh medium
pumped from
reservoir

Overflow to
collecting vessel

Figure 3.15 Homogeneous continuous flow culture – a chemostat or turbidostat. The culture is kept thoroughly stirred; sparging with sterile air contributes to mixing. Fresh medium is pumped into the culture vessel and organisms and partially spent medium leave through an overflow. In a chemostat a constant flow rate is maintained and determines the specific growth rate (μ). In a turbidostat the biomass (x) is kept constant by adjusting the flow rate through the system in response to changes in biomass. These may be sensed directly with an optical device, or indirectly by measuring a product of metabolism such as carbon dioxide.

an impeller or, in small vessels, with a magnetic stirrer. For aerobic operation, which will be usual with yeasts or filamentous fungi, sterile air must be sparged through the medium; this will contribute to mixing. When the system is fully operational, a steady state is established with a constant biomass (x) in the culture vessel growing at a constant specific growth rate (μ). This state can be maintained for weeks or even months. An important operating variable in homogeneous continuous flow culture is the **dilution rate** (D), the rate at which the medium in the vessel is replaced by fresh medium. D is determined (equation 10) by the culture volume (v) and the flow rate (f).

$$D = f/v \qquad (10)$$

Thus if the flow rate is such as to provide a volume of fresh medium equal to half the culture volume in one hour then D is 0.5 h^{-1}. During steady-state operation of the system the rate of production of microorganisms in the culture vessel (μx) will equal the rate of their departure through the overflow (Dx). Hence $\mu x = Dx$ and $\mu = D$. If D exceeds the critical value D_c at which it equals μ_{max}, then microorganisms are removed from the culture vessel faster than they can be produced and growth is terminated by 'washout'. The way in which attributes of the culture vary with D is illustrated in Fig. 3.16.

A second crucial operating variable is the concentration of the limiting nutrient in the medium supplied from the reservoir, s_{res}. As indicated earlier (page 127 and equation 9) the concentration of limiting nutrient s determines the specific growth rate μ. In the culture vessel the value of s will be influenced by the biomass (x) which in turn will be determined by the dilution rate D. The value of s will then determine the specific growth rate (μ). In steady-state operation the output of microorganisms is a product of the biomass x and the dilution rate D and is hence Dx. It increases until the dilution rate reaches D_m and then falls rapidly (Fig. 3.16). Biomass and output of organisms can be increased by increasing s_{res} but D_c remains unchanged

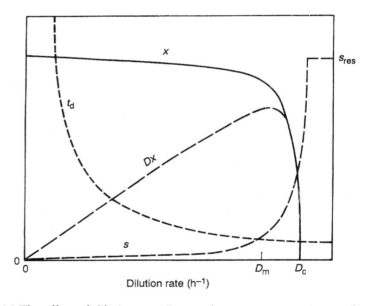

Figure 3.16 The effect of dilution rate D on a homogeneous continuous flow culture. Biomass (x) decreases slowly over a wide range of dilution rates and then more rapidly as the critical dilution rate D_c, at which complete washout occurs, is approached. The value of D_c and maximum specific growth rate (μ_{max}) are identical. The concentration of the limiting substrate (s) is determined by the number of organisms metabolizing, and hence increases as x decreases, reaching the reservoir concentration (s_{res}) at the critical dilution D_c. Since $D = \mu$, the culture doubling time (t_d) which is the reciprocal of the specific growth rate, shortens as D increases. The output of organisms (Dx) is the product of the dilution rate D and the biomass x. It increases to a maximum value at dilution rate D_m and then decreases sharply as D_c is approached. A greater supply of the limiting nutrient will lead to a higher biomass and output of organisms but the critical dilution will be unchanged since μ_{max} cannot be changed. The curves illustrated are calculated. In practice there is some deviation at high dilution rates, due to imperfect mixing, and washout being averted by growth on the walls of the culture vessel. (Based on Herbert *et al.* (1956). *Journal of General Microbiology* **14**, 601–622).

since μ_{max} cannot be altered. The specific growth rate and biomass in a homogeneous continuous flow culture can be controlled in two ways: by determining the concentration of the limiting nutrient that is supplied, as in the **chemostat**, or by adjusting the dilution rate, as in the **turbidostat**.

The chemostat

The steady state in a homogeneous continuous flow culture is stable. Any fall in biomass concentration (x) results in an increase in the level of the limiting nutrient (s), and increase in μ and hence eventually to restoration of biomass concentration. An increase in biomass concentration above the steady-state level lowers s, lowers μ and reduces x. Thus, inherent stability in the system means that it can be run at a constant specific growth rate, and biomass concentration maintained by providing the appropriate concentration of limiting nutrient from the reservoir. A homogeneous continuous flow culture controlled in this way is a chemostat.

Chemostats have been utilized in a variety of ways in the laboratory. They can provide a constantly available source of organisms in exponential growth at any desired specific growth rate. The maintenance of constant conditions in a large population over a long period permits the application of a constant selection pressure and facilitates the study of microbial mutation and evolution. Chemostats will operate effectively at very low nutrient levels and hence can simulate natural conditions for ecological studies.

The fermentation industry has in the past operated largely by batch culture and is equipped with appropriate plant. Continuously operating processes are, however, often more satisfactory for industry than batch processes, so continuous culture on the chemostat principle is likely to be increasingly employed for new fermentations. Indeed, continuous culture is the only realistic way of achieving the biomass output needed to produce useful amounts of single cell protein or lipids for food.

The turbidostat

In the chemostat the specific growth rate (μ) is controlled by means of the concentration of limiting nutrient in the medium supplied; the dilution rate is not altered. In the turbidostat the biomass concentration (x) is controlled by varying the dilution rate. The chemostat is most sensitive and reliable where biomass concentration changes only slightly with change in dilution rate. The turbidostat, however, operates effectively at dilution rates above D_m where biomass changes with change in D. A rise in the turbidity of the culture, indicative of increased biomass, is sensed by an optical device and the pump supplying medium is activated. Thus the required biomass concentration is maintained by controlling flow rate. The turbidostat has in the past been less widely employed than the chemostat, partly as a result of practical problems in optical sensing, such as growth of microorganisms on the culture vessel wall. An alternative approach which is becoming increasingly feasible with current technology is the sensing of some variable, such as carbon dioxide output or oxygen uptake which responds rapidly to any change in the output of microorganisms (Dx). As Dx varies markedly with D both above and below D_m, such an approach should render the turbidostat capable of stable operation through the whole range of biomass concentrations.

Heterogeneous continuous flow systems

The possibility of heterogeneous continuous flow or **plug-flow culture** has been investigated. Inoculum and medium are mixed and pass slowly along a tubular reactor as growth occurs. Ideally no further mixing should occur and hence the temporal sequence of growth phases of a batch culture should be replaced by a spatial sequence of growth phases along the tube. Plug-flow culture with biomass feedback is achieved on a large scale with the activated sludge process of sewage digestion. Plug-flow culture can be simulated by having a number of chemostats in series; this overcomes some of the practical difficulties of the method.

Growth of Yeast Populations and Colonies

Stirred and shaken liquid cultures

There have been many studies on the growth of yeasts in stirred or shaken liquid cultures. In batch culture the sequence of growth phases to be expected with a population of unicellular organisms occurs. Many studies have also been carried out with chemostats, in which yeasts are readily maintained for long periods in exponential growth at constant biomass.

Static liquid culture

The static liquid culture of *Saccharomyces cerevisiae* in sugary liquids is one of the oldest crafts practised by humans. Since glucose represses the oxidative metabolism of pyruvate – the Crabtree effect (page 72) – growth is fermentative and alcohol is produced. Fermentation continues even when the glucose concentration falls, because with the poor aeration of static cultures anaerobic conditions soon result, and alcohol continues to accumulate. Brewing and wine-making are discussed further in Chapter 8.

Culture on agar media

Most yeasts form small circular colonies when grown on agar media. Quantitative studies on the growth of such yeast colonies is lacking but work on bacterial colonies suggests the probable course of events (Fig. 3.17). Growth will at first be exponential until a mass of cells is produced. At this stage only the cells at the edge of the mass will have access to nutrients so a marginal growth zone will result, and exponential growth in this zone will lead to a linear increase in radius with time. Finally, increase in radius will be slowed or even halted by nutrient depletion or by the accumulation of toxic metabolites such as ethanol.

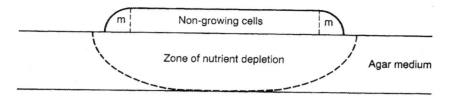

Figure 3.17 Section through a colony of unicellular organisms, such as yeast cells, on an agar medium. Growth is confined to a marginal growth zone (m) at the edge of the colony. Beneath the colony is a region in which nutrient depletion has occurred. A diffusion gradient will occur from the edge of this zone to the colony. The zone will also become one of accumulation of inhibitory metabolites. (Based on a figure by Pirt, S. J. (1975). *Principles of Microbe and Cell Cultivation*. Blackwell, Oxford.)

Growth of Filamentous Fungi in Liquid Media

Filamentous fungi when grown in static liquid culture form a mat of hyphae at the surface of the liquid. Agitation of a liquid culture by shaking, stirring or sparging results in submerged growth. The term surface culture is hence often used for static liquid culture of filamentous fungi and submerged culture for any form of agitated liquid culture. Fungi with septate hyphae grow well in submerged culture, but growth of the non-septate lower fungi is often less satisfactory. Considerable shearing forces may be generated in shaken and especially stirred culture, and the commonly superior growth of higher fungi in these conditions is probably due to the plugging of septal pores limiting any damage.

Form in submerged culture
Filamentous fungi in submerged culture may grow as a nearly homogeneous suspension of hyphae (filamentous growth) or as discrete pellets (pellet growth) (Fig. 3.18). Pellets vary in form from the extremes of loose flocculent pellets to compact, spherical ones. Sectioning a compact pellet may result in an alcoholic odour, showing that oxygen has failed to penetrate to the centre and fermentation has resulted. A pellet may autolyse in the interior while vigorous growth occurs at the margin.

One factor that can determine whether filamentous or pellet growth occurs is inoculum size. If a medium is inoculated with a very large number of propagules (spores or hyphal fragments), limited growth from each propagule can produce considerable biomass and exhaust nutrients. On the other hand if the inoculum consists of only a few propagules then the hyphae of each will have to undergo extensive growth and branching before appreciable nutrient utilization occurs. The result will be a few pellets with numbers corresponding to the number of propagules in the inoculum. Hence, some fungi (e.g. *Penicillium chrysogenum*) will yield pellet growth with a small inoculum and filamentous growth with a large inoculum. Submerged culture of other fungi (e.g. *Aspergillus nidulans*) yields exclusively pellets. This is because clumping of spores occurs so that even a massive inoculum results in relatively few points of propagation. Yet others (e.g. *Geotrichum candidum*) show exclusively filamentous growth. This is a

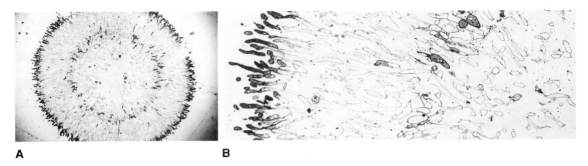

A   B

Figure 3.18   Pellet formation resulting from growth in submerged, agitated liquid culture. A, Median transverse section through a 2 mm diameter pellet of *Basidiobolus ranarum*, with autolysis at the centre. B, Enlargement showing hyphae (Keith Gull).

consequence of ready fragmentation of hyphae at septa to yield many more propagules than were initially added. Hence inoculum size, clumping and fragmentation can interact to determine whether filamentous or pellet growth occurs. However, even with those fungi that show the strongest tendency to form pellets, such as *Aspergillus nidulans*, some fragmentation must occur, or continued reduction in propagule number through dilution would render chemostat culture impossible. Growth form in submerged culture is of considerable importance in fermentation technology as it influences aeration, growth rates, power consumption for stirring and the ease with which biomass can be separated from culture medium.

Hyphal characteristics are dependent on the medium used. Thus for *Aspergillus oryzae* in a chemostat with glucose as limiting nutrient, specific growth rate increased with increased glucose concentration, from 0.025 to 0.15 h^{-1} at glucose concentrations of 1.0 and 7.0 mg l^{-1}, respectively. These hyphae had linear relationships between a number of variables and specific growth rate: minimum volume of apical compartment, from 40 to 420 μm^3; hyphal diameter, from 1.7 to 2.8 μm; and number of nuclei in apical compartment, from 3 to 10; and tip extension rate (measured separately in a flow-through chamber on a microscope stage), from 1.0 to 7.0 μm h^{-1}; while minimum length of the apical compartment increased in a non-linear fashion from 20 to 60 μm.

Growth in submerged culture

With those fungi that are capable of filamentous growth in submerged culture, the sequence of phases normal to batch culture occurs. A study of filamentous growth of *P. chrysogenum* showed a lag phase and an exponential phase with a doubling time of *ca* 14 h. On the medium employed, maximum growth was soon followed by exponential decline due to autolysis with a 'halving time' of *ca* 90 h. On a different medium a doubling time of *ca* 8.5 h during the exponential phase was obtained. If, however, conditions are such that pellet growth occurs, a logarithmic plot of biomass against time shows that exponential growth soon comes to an end and a less rapid increase in biomass follows. With pellet-forming cultures of *P. chrysogenum* this takes place when the pellets are about 0.1 mm diameter. With the less compact pellets of *Aspergillus nidulans* it does not occur until a diameter of about 2.5 mm is obtained, and with the flocculent pellets of some fungi exponential growth continues until the stationary phase approaches. With fungi forming compact pellets the exponential phase is followed by one in which a plot of the cube root of biomass against time is linear. The explanation for this relationship is that growth is limited by oxygen diffusion to an outer shell of the pellet. If this shell remains of constant width and within the shell a constant specific growth rate is maintained, a pellet will show a linear increase in radius and a cubic increase in biomass. Such a relationship has been demonstrated in several fungi, such as *Penicillium chrysogenum* and *Aspergillus nidulans*.

Growth and form in surface culture

Most fungi, if grown in static liquid culture, form a mat at the surface of the liquid. Such cultures lack homogeneity. The under surface of the mat will have access to abundant nutrients but receive little oxygen. The upper surface on the other hand will be exposed to air but may be starved of nutrients. Growth is slow compared with submerged culture, with a merging of growth phases due to lack of spatial homogeneity. An advantage of surface culture is the readiness with which sporulation occurs; sporulation is sometimes difficult to obtain in submerged culture and the production of some metabolites may be linked to sporulation. Another advantage is the simplicity of the equipment required. Growth of surface cultures of *Penicillium notatum* in a variety of bottles was the way that mass production of penicillin was pioneered in Britain in the early 1940s, but this was soon overtaken by the development of submerged fermentations.

Growth of Filamentous Fungi on Agar Media

Germ-tube growth, branching and hyphal maturation

When a spore germinates, typically it produces one or more germ tubes which elongate exponentially. This results from the autocatalytic effect of the increasing rate of uptake and metabolism of nutrients from the medium. The extension rate eventually reaches a nearly constant value, probably when transport of material from the sub-apical region becomes limiting. Exponential growth of the colony is achieved by formation of sub-apical branches, each of which becomes an apically elongating hypha showing a linear increase in length, so that the total length of the hyphal system increases exponentially due to an approximately exponential increase in the number of hyphal tips (Fig. 3.19).

Hyphae in a young colony, however, are juvenile in behaviour and undergo a poorly understood slow process of maturation. Juvenile hyphae of many species are slower growing and narrower than are mature hyphae. Thus, in experiments with *Botrytis cinerea*, the mean extension rate of leading hyphae increased from 85 μm h^{-1} at 20 h to 330 μm h^{-1} at 44 h after inoculation, and remained constant thereafter. Over the same period the mean diameter of the hyphae increased

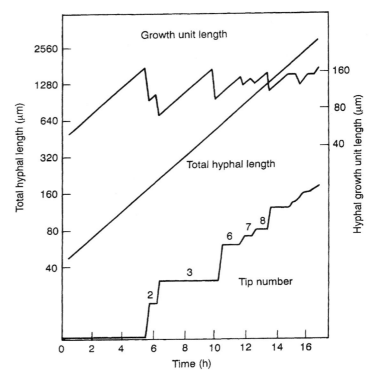

Figure 3.19   Growth of *Aspergillus nidulans* on agar medium from a single unbranched hypha *ca* 40 μm in length to an undifferentiated mycelium of *ca* 2500 μm. The length of the single hypha, and subsequently of the hyphal system, increases exponentially. The increase in tip number is at first discontinuous (e.g. 1 to 2; 2 to 3, 3 to 6) but as the number of branches become large approximates to a smooth exponential increase. As this occurs the length of the hyphal growth unit shows reduced oscillation and approaches a constant value of *ca* 100 μm. (Based on a figure in Trinci, A. P. J. (1974). *Journal of General Microbiology* **81**, 225–236.)

linearly from 5.7 to 8.7 μm. There was a direct relationship between the extension rate and the square of diameter of a hypha. In contrast, for *Mucor rouxii*, in the same study, extension rate increased dramatically from spore germination at 5.5 h to reach a value of over 300 μm h^{-1} at 10 h, after which it increased little over the next 36 h. Meanwhile, hyphal diameter increased more slowly, from 5.8 μm at 15 h to its maximum value of 8.4 μm at 26 h. There was no direct relationship between hyphal extension rate and the square of hyphal diameter. A positive relationship between hyphal diameter and extension rate for a particular species has been reported for many fungal colonies: for both variables, values for leading hyphae are greater than those for primary branches, which in turn are greater than those for secondary branches. A different style of growth is shown by hyphae of the human pathogen, *Candida albicans*. Although these hyphae still exhibit a maturation process, becoming faster growing and wider with age, differences are seen when yeast cells form germ tubes and when hyphae branch.

The germ tubes elongate at a constant rate, not in an autocatalytic fashion. This was explained by the observation that the germ tubes are formed at the expense of vacuolation of the parent yeast cells. Thus germ tube formation can be regarded as resulting from migration of the protoplasm from the yeast cell to the apex of the hypha. This phenomenon of extensive vacuolation of individual cells continues during hyphal growth, so that there is a delay, sometimes very extended, between septation (to give uninucleate hyphal cells) and production of branches behind the septa.

Juvenile hyphae, as well as being slower growing and thinner than mature ones, also differ in other ways. For example, young hyphae of *B. cinerea* and *M. rouxii* are much less susceptible than mature ones to antifungal agents, such as nikkomycin and echinocandin, inhibitors of cell wall biosynthesis. Thus leading hyphae of a mature colony of *B. cinerea* (44 h and older) burst within one minute when treated with 10 µM nikkomycin or 93.5 µM echinocandin. In contrast, juvenile hyphae (e.g. aged 16 and 24 h), treated with 100 µM nikkomycin took more that 5 min to swell and burst and those treated with 935 µM echinocandin showed little bursting, although they did stop growing.

The mechanism of maturation of a vegetative hypha is unclear, but it may prove that growth-regulating chemicals have to reach critical concentrations to allow expression of particular genes. Such a situation is seen in colonies of some bacteria, where signalling molecules such as homoserine lactone need to accumulate to allow expression of particular genes, such as those for luminescence. The observation in many cases that differentiation does not occur in submerged cultures is consistent with this idea, as gradients of putative effectors could not then accumulate. One aspect of metabolic control of hyphal development is demonstrated by experiments with *cr-1* mutants of *N. crassa*. These mutants are deficient in cyclic AMP. They form mycelia with a single size class of hyphae, 3–5 µm, in contrast to the hyphal hierarchy of their wild-type parent, which has leading hyphae of 14–20 µm, primary branches of 8–12 µm and secondary branches of 3–5 µm. When grown with 2–3 mM 8-bromocyclic AMP (an analogue of cyclic AMP), after 18–20 h some hyphal hierarchy was observed, with hyphal diameters of 3–13 µm. Thus it may be concluded that metabolism involving cAMP plays some part in the process of hyphal maturation.

Undifferentiated mycelium and the hyphal growth unit
Continued hyphal growth and branching results in the production of an **undifferentiated mycelium** (i.e. one not differentiated into concentric zones) on the surface of the agar medium. The hyphae of undifferentiated mycelia on agar are similar in width, length of extension zone, and compartment length to the hyphae formed by filamentous growth in submerged liquid culture. The branches produced by undifferentiated mycelium grow approximately at right angles to the parent hypha. The average length of hypha associated with a hyphal tip is known as the **hyphal growth unit** and is determined by dividing the total hyphal length by the number of apices. At first the hyphal growth unit fluctuates in length with branch initiation but as the increase in tip number becomes nearly continuous it approaches a constant value typical of the species (Fig. 3.19).

Differentiated mycelium and colony growth

Inoculation of an agar medium at a point with one or more spores or hyphal fragments will result in an approximately circular area of undifferentiated mycelium. Growth within this area will create a 'sink' into which nutrients will diffuse from outside and a 'source' from which the various excreted metabolites will diffuse outwards. The development of a clearly defined colony margin with outwardly growing hyphae is probably the consequence of the establishment of such gradients. The colony will then show a linear increase in radius with time as the colony margin advances at a uniform rate. Even if the colony margin was initially irregular in outline it becomes circular as growth proceeds. The colony commonly continues to increase in radius at a uniform rate until an obstacle (e.g. edge of Petri dish, another colony) is encountered. Sometimes, however, the rate of increase may gradually decline and growth may even cease without all the available medium having been exploited. This effect is known as 'staling' and is due to 'staling substances', toxic products of metabolism, such as hydrogen ions or ammonia. The occurrence of staling implies an inadequately buffered or otherwise unsuitable medium. Sometimes phases of reduced rate of spread and rapid growth alternate. The phases of rapid extension could result from the fungus having grown beyond a severely staled area or from the disappearance, by breakdown or evaporation, of a staling factor.

Fungal colonies that have acquired a clearly defined margin and are undergoing a linear increase in radius with time show spatial differentiation from margin to centre. Extending, productive, fruiting and aged zones can be recognized, although there are no sharply defined boundaries between these zones. The attributes of the four zones will now be considered.

The **extending zone** of the fungal colony, which should not be confused with the zone of extension of a single hypha, is the region in which hyphae are advancing into unexploited medium. For a given medium the number of hyphal tips per unit length of colony margin will remain constant. This means that some primary branches become converted into leading hyphae. Branch patterns are most easily studied on dilute media in which mycelial growth is relatively sparse. Whereas the branches in undifferentiated mycelium commonly grow perpendicular to the parent hypha, those of differentiated mycelium generally grow at an acute angle. This may well be a negative autotropic response (see below) to denser mycelium further back. Branching is often approximately monopodial; a leading hypha tends to produce primary branches alternately on opposite sides of the hypha, and as the primary branches mature these produce secondary branches. The mean width and extension rate of primary branches is less than that of the leading hyphae and the secondary branches are still narrower and more slowly growing. A sympodial branching pattern is sometimes seen, with a primary branch taking the place of a leading hypha that has ceased to extend. Some fungi show dichotomous branching of the apices of leading hyphae to give two leading hyphae. It has been shown that branches are arranged in a pattern that efficiently drains an area of nutrients, one analogous to that of a river system draining a valley. The production of such a pattern, and the spacing of leading hyphae at the colony margin, implies negative autotropism, with the growth of hyphae oriented so as to avoid other hyphae. The mechanism is unknown. It could be due to a positive chemotropism, with growth up a gradient of some factor depleted by metabolism, such as oxygen. Alternatively it could be a negative chemotropism to some agent produced by hyphae, such as hydrogen ions.

Behind the extending zone is the **productive zone**, the region in which the major increase in biomass occurs and differentiation can occur. Differentiation requires maturation of the hyphae, as juvenile hyphae of many fungi have a limited repertoire of differentiation. Thus those of *Mucor mucedo* do not produce the sex hormone, trisporic acid, or respond to it to produce zygophores, as do mature hyphae. Competence for sporulation of *Aspergillus nidulans* is acquired after 18–20 h growth of mycelium. Acquisition of this competence for sporulation appears to be controlled genetically rather than environmentally, as it is unaffected by continuous replacement of the medium or by concentrations of limiting nutrients, and precocious mutants have been described that can sporulate earlier.

In the **fruiting zone** net gain in biomass is ceasing, and spores are being formed, commonly on hyphae that rise above the surface of the medium. The production of aerial hyphae in many cases will be associated with the production of specific hydrophobins, which will coat the hyphal surfaces with a hydrophobic layer. The fruiting zone corresponds to the stationary phase of batch culture. Finally at the centre of the colony is the **aged zone**, corresponding to the decline phase of batch culture. Here hyphae are often highly vacuolated or even emptied of protoplasmic contents, protoplasm having been moved into spores or into younger parts of the colony. In this zone autolysis often results in the disappearance of cell walls as well as hyphal contents.

Comparison of radial extension rates of fungal colonies on agar media with specific growth rates in submerged culture have been carried out. It is found that some fungi (e.g. *Aspergillus*, *Penicillium*) that grow fast in submerged culture show slow rates of colony extension on agar, and some that grow slowly in submerged culture (e.g. *Mucor*) spread fast on agar. Others, such as *Neurospora* have both a high specific growth rate and a high rate of radial extension on agar. The fungi that spread rapidly on agar have big hyphae and wide peripheral growth zones (page 118). The rate of extension at the apex of leading hyphae depends not only on the specific growth rate but on the width of the peripheral growth zone that supplies materials to the apex. The rate of colony extension is in fact approximately the product of the specific growth rate demonstrated in submerged culture and the width of the peripheral growth zone (Table 3.7). In fungi with large hyphae it is probable that the peripheral growth zone includes most of the productive zone. In others where it is very narrow (e.g. *Geotrichum candidum*) it is approximately equivalent to the extending zone.

Growth of compact pellets in submerged culture has features in common with that of colonies on agar. After the pellet has attained a certain size diffusion of oxygen into the pellet limits growth (page 137). An extending zone is clearly present at the surface of the pellet and a productive zone in which biomass is increasing exponentially. At the centre is a non-growing zone corresponding to the aged zone. The peripheral growth zone includes part or all of the productive zone. The rate of increase in radius of pellets is rather less than that of colonies on agar media – about 60% for *Penicillium chrysogenum* and 40% for *Aspergillus nidulans*. Pellets in submerged culture usually lack a fruiting zone.

Table 3.7   The relationship between the radial growth rate of a fungal colony on agar media, the specific growth rate in submerged culture and the width of the peripheral growth zone

Species	Radial growth rate, observed (mm h^{-1})	Specific growth rate (μ h^{-1})	Growth zone width (mm)	Radial growth rate, calculated/observed
Zygomycetes				
Rhizopus stolonifer	0.97	0.14	8.7	1.26
Actinomucor repens	0.49	0.18	2.5	0.92
Mucor racemosus	0.43	0.10	3.4	0.79
Absidia glauca	0.38	0.12	2.5	0.79
Higher fungi				
Neurospora crassa[a]	4.40	0.35	10.0	0.80
Geotrichum candidum	0.20	0.35	0.4	0.70
Aspergillus niger	0.13	0.12	1.1	1.02
Penicillium chrysogenum	0.08	0.16	0.5	1.00

It is assumed that exponential growth occurs in the peripheral growth zone at the rate demonstrated in submerged culture. On this basis the rate at which a hypha spanning the growth zone should extend can be calculated by multiplying the zone width by the specific growth rate. The result obtained corresponds closely with the observed radial growth rate, as indicated in the final column. Data from Trinci, A. P. J. (1971). *Journal of General Microbiology* 67, 325–344.

[a] A different *N. crassa* strain from those in Table 3.4.

Hyphal anastomosis

Fusion between hyphae is an integral part of the sexual process in most fungi. Vegetative hyphal fusion (hyphal anastomosis) is more limited in its distribution. There are reports of its occurrence in a few Oomycetes and Zygomycetes, notably vesicular–arbuscular endophytes (page 397), but it is very common in Ascomycetes, Basidiomycetes and mitosporic fungi. As discussed earlier, vegetative hyphae commonly avoid each other and thus exploit available nutrients more effectively. Hyphal anastomosis on the other hand involves the growth of hyphae towards each other. The types of anastomosis (Fig. 3.20) traditionally recognized are hypha-to-hypha, hypha-to-peg, and peg-to-peg. Hypha-to-hypha anastomosis involves the hyphae curving towards each other. In hypha-to-peg anastomosis a hypha growing toward the sub-apical region of another hypha elicits the production of a 'peg', a short branch. Peg production need not, however, occur (see Fig. 3.20D). Peg-to-peg anastomosis occurs between parallel hyphae, short branches being produced at adjacent points on the two hyphae. In all these types of anastomosis positively autotropic behaviour results in apical contact between hyphae or pegs involved. With appropriate microscopy, the Spitzenkörper in the apex of the approaching hypha can be observed to become asymmetrically displaced just before contact so that it became re-aligned with the receptive site of fusion. Protoplasmic continuity follows the fusion. This is how the 15 hectare colony of *Armillaria bulbosa* (page 144) can be considered to be a single organism.

The circumstances under which hyphal anastomosis replaces the usual avoidance reactions between hyphae require more study. Germ-tubes commonly display negative autotropism, but those of some fungi under some conditions show positive autotropism culminating in hyphal anastomosis. In some species

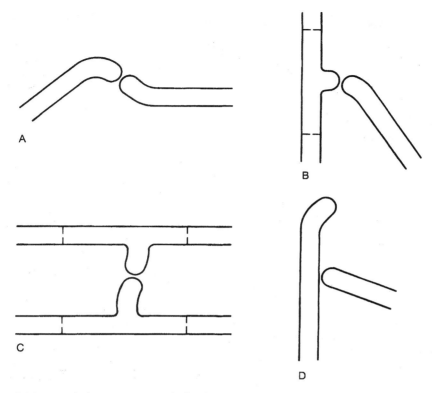

Figure 3.20   Hyphal anastomosis. The hyphae and pegs are shown growing towards each other just prior to contact and fusion. A, Hypha-to-hypha, a pair of actively growing hyphae being involved. B, Hypha-to-peg. An actively growing hypha has elicited peg formation from a compartment of an older hypha. C, Peg-to-peg, with two established hyphae about to be linked. (Based on Buller, A. H. R. (1933). *Researches on Fungi*, vol. 5. Longmans Green, London.) D, Hypha-to-side. This resembles hypha-to-peg, but a peg is not elicited as the side of a hypha, in the sub-apical region, is approached by a hyphal tip. (After Watkinson, S. C. (1978). End-to-side fusions in hyphae of *Penicillium claviforme*. *Transactions of the British Mycological Society* **70**, 451–453.)

hyphal anastomosis can be observed in young colonies derived from a single spore, and in other species in older colonies. When two mature colonies meet, the leading hyphae commonly grow past each other without fusing, but hyphal anastomosis will occur between branches. Vegetative incompatibility (page 257) will normally prevent the completion of hyphal anastomosis if the individuals involved are not genetically identical, an allelic difference at any one of a number of loci being able to terminate the process by death of the cells immediately involved in the fusion. More distantly related strains may fail to fuse.

There seems to have been a reluctance to accept the involvement of chemotropism in hyphal fusions despite the fact that action at a distance was clearly described over 100 years ago for hyphae *Botrytis cinerea*. More recent observations of germinating spores of *B. cinerea* in a flow-chamber led to the conclusion that the positive autotropism between adjacent germ tubes could only be explained by the production by cells of a macromolecular diffusible growth stimulator, with a half-life of about 10 seconds and a radius of action of about 10 μm. Although as yet we have no knowledge of specific molecules, the recent elucidation of multiple complex mating pheromone/pheromone receptor systems (discussed later) demonstrates that fungi have complex signal production, reception and transduction mechanisms. It may be that some parts of the mating signalling systems could also be used during vegetative autotropisms and fusions.

Hyphal anastomoses would seem to have several roles. They convert a radiating system of hyphae into a three-dimensional network. This will permit the transport of protoplasm to any point in the colony and allow the formation of multi-hyphal structures such as coremia, mycelial strands, sclerotia and fruit bodies. They permit damaged hyphae to be by-passed readily. By merging separate but genetically identical individuals into a single hyphal network they may eliminate wasteful competition between identical genotypes.

Substrate penetration and aerial mycelium

Not only do fungal hyphae spread rapidly over a surface, they also penetrate the substratum. Such penetration, which facilitates access to nutrients, has been relatively little studied, but occurs rapidly. On agar media, for instance, hyphae soon reach the bottom of a Petri dish, suggesting rates of hyphal advance comparable with those of surface spread.

Many species also produce aerial mycelium, vegetative hyphae rising above the substratum. In mycelium of *Schizophyllum commune*, production of a specific hydrophobin (page 102) is required for the formation of aerial hyphae. It is envisaged that hyphal tips that start to breach the liquid–air interface become coated with a layer of hydrophobin which enables them to grow up out of the aqueous film. In natural conditions, aerial hyphal formation may facilitate invasion of suitable substrates which are not in immediate contact with those already colonized. A few fungi, such as the Zygomycetes *Rhizopus* and *Absidia*, have stolons, stout, fast-growing, aerial hyphae which appear to be specialized for this role.

Senescence

Animals have a determinate life span. Growth comes to an end, and finally senescence occurs, with a decline in vigour followed by death. The growth of fungal mycelium is indeterminate, and continues for as long as nutrients are available and conditions otherwise favourable. Basidiomycete mycelium in the form of fairy rings in grassland may increase in diameter for centuries. A clone of *Armillaria bulbosa* which covers 15 hectares of forest floor in a Canadian forest has been estimated to have been spreading for 1500 years and to have achieved a weight of at least 10 000 kg. In the laboratory the mycelium of most fungi can be maintained indefinitely in the vegetative state by subculture of hyphae to fresh media. Exceptions are, however, known. In *Podospora anserina* (page 53) ageing occurs in most wild-type strains after about 25 days' vegetative growth. Hyphal extension slows down and finally ceases, hyphal tips burst, cells die, and subculture becomes impossible. Mutant strains undergoing similar senescence are known in *Neurospora crassa* and *Aspergillus amstelodami*. In *Podospora* senescence is associated with autonomously replicating plasmids derived from mitochondrial DNA. There are, however, laboratory strains with mutations to their nuclear genes which suppress the liberation or expression of senescence plasmids and in which senescence does not occur. It is perhaps significant that a coprophilous fungus, *Podospora anserina*, is the species in which senescence is most strikingly displayed in wild-type strains. In a spatially limited habitat, such as dung, only limited vegetative growth is possible. Natural selection will hence not eliminate mutations that prevent the unlimited vegetative growth that is advantageous in fungi with less restricted habitats.

The Effect of the Environment on Growth

Fungi have, in the course of evolution, diversified to exploit a wide variety of habitats. Different species hence require different conditions for optimal growth. The ways in which physical and chemical, especially nutritional, factors in the environment affect the growth of different fungi will be considered.

The Aeration Complex

Air contains about 21% oxygen. Organisms growing on a surface exposed to the air will hence experience such an oxygen level. Oxygen has, however, a low solubility in water and hence diffuses only slowly into static liquids or waterlogged materials.

Oxygen concentrations in these environments will be further lowered by the metabolic activity of any aerobic organisms present, leading to very low oxygen levels or the complete absence of oxygen. Air also contains 0.03% carbon dioxide. Under conditions of poor aeration the metabolic activity of microorganisms will result in far higher concentrations of this very soluble gas. In nature, therefore, fully anaerobic conditions imply the absence of oxygen and high carbon dioxide concentrations. As conditions become more aerobic the oxygen concentration rises, but never above 21%, and the carbon dioxide concentration falls, but never below 0.03%. Pathogenic fungi, within the bodies and/or cells of their hosts, will experience very different oxygen and carbon dioxide concentrations to those within a Petri dish. The performance of various fungi within these limits and, as can happen under some laboratory or industrial conditions, beyond them, will now be discussed.

Oxygen

Organisms can obtain energy by oxidative (respiratory) metabolism or by fermentation. The normal oxidant for respiration in fungi is oxygen itself. There

Table 3.8   Energy metabolism in relation to oxygen requirements

1. *Obligately oxidative*. Oxygen is essential for such fungi which are hence obligate aerobes. Examples: *Phycomyces, Rhodotorula*

2. *Facultatively fermentative*. Since energy can be obtained both by oxidative and fermentative processes such fungi are likely to be facultative anaerobes. Oxidative metabolism, however, provides much more energy than fermentation, so higher yields can occur under aerobic conditions. Examples: *Mucor, Saccharomyces*

3. *Obligately fermentative*. Oxygen is not needed for energy production, so there is no oxygen requirement. Since, however, oxygen may be either harmless or toxic there are two possible categories with respect to the effect of air.
   (a) *Facultative anaerobes*. Examples: *Aqualinderella, Blastocladia*
   (b) *Obligate anaerobes*. Example: *Neocallimastix*

is, however, evidence that under low oxygen tensions some fungi may make use of nitrate as an oxidant; the usefulness of this capability with non-motile organisms such as fungi may be limited by the toxicity of the nitrite formed. Respiratory metabolism results in the complete conversion of organic compounds into carbon dioxide and water with energy being generated by oxidative phosphorylation. It hence yields much more energy than fermentation in which net oxidation cannot occur and relatively little energy is made available. Details of the metabolic pathways involved in respiration and in the fermentative production of lactic acid and alcohol, all of which occur in fungi, are available in textbooks of biochemistry and microbiology. The implications for oxygen requirements of the occurrence of respiration, fermentation or both in a fungus are indicated in Table 3.8. In addition to the role of oxygen for energy generation it is also needed by fungi for metabolic steps in the biosynthesis of sterols, unsaturated fatty acids and some vitamins, and for the degradation of aromatic compounds.

Since respiration is energetically so much more rewarding than fermentation it would be expected to occur in all organisms that in their natural habitat are commonly exposed to oxygen. This applies to most fungi and few non-oxidative fungi are known. Perhaps the best-studied example of an exclusively fermentative fungus is the Oomycete *Aqualinderella fermentans* which lives on fallen fruits rich in fermentable sugars in static warm waters and obtains energy by lactic fermentation. Although exclusively fermentative in its metabolism, *Aqualinderella* can tolerate oxygen, to which it is perhaps occasionally exposed if the waters in which it lives are disturbed. Members of the Chytridiomycete genus *Blastocladia* (not to be confused with the aerobic genus *Blastocladiella*) live in similar conditions and are also obligate lactic fermenters. The above fungi are facultative anaerobes but the Chytridiomycete rumen fungi (page 37), such as *Neocallimastix*, have been regarded as obligate anaerobes. However, it has recently been shown that they show a degree of aerotolerance, and can grow when exposed to 5% O_2, but they need at least 7% CO_2. The still poorly characterized 'resting stage' of *Neocallimastix*, required for transmission from cow to calf, must be able to withstand extended exposure to oxygen.

Some fungi are exclusively oxidative in their metabolism and are hence unable to survive in completely anaerobic conditions. They are, however, often capable of growing under conditions of poor aeration, since the saturation constant for oxygen, like other nutrients, can be very low. Examples of fungi with exclusively oxidative pathways of energy metabolism are the Zygomycete *Phycomyces* and the yeast *Rhodotorula*.

Some other fungi can grow at very low oxygen tensions, because their mitochondria are able to perform bacterial-like denitrification. They can utilize nitrate or nitrite as terminal electron acceptors in anaerobic respiration, and reduce them to nitrous oxide, in a process coupled with ATP formation. The best characterized systems are those of *Fusarium oxysporum* and *Cylindrocarpon tonkinense*.

A great many fungi have both respiratory and fermentative pathways. A capacity for alcoholic fermentation is widespread in Zygomycetes, Ascomycetes and related forms and for lactic fermentation in Chytridiomycetes and Oomycetes. The most intensively studied example of a fungus with both respiratory and fermentative metabolism is the yeast *Saccharomyces cerevisiae*. Although *Saccharomyces* can achieve similar growth rates by oxidative and fermentative metabolism, in the absence of oxygen biomass yields are low (page 72). *Saccharomyces* needs sterols and unsaturated fatty acids but these can be synthesized only if oxygen is present. In its absence a suitable sterol such as ergosterol and an unsaturated fatty acid such as oleic acid or natural materials containing sterols and unsaturated fatty acids must be provided in the medium. Under anaerobic conditions mitochondria persist, but the cytochromes are produced in far smaller amounts and differ from those formed in the presence of oxygen. Another organism that can grow under both aerobic and anaerobic conditions is *Mucor rouxii*. Absence of oxygen makes the fungus incapable of synthesizing the vitamins thiamin and nicotinic acid, which hence have to be supplied, and promotes a yeast-like, instead of a filamentous, form of growth (page 40).

Organisms that are capable of both oxidative and fermentative metabolism should be capable of growth under anaerobic as well as aerobic conditions. There is, in fact, a lack of reliable information as to the capacity for anaerobic growth in the fungi. In early work elimination of oxygen was often insufficiently stringent to prove anaerobic growth and in some later work, free from this defect, a failure to detect anaerobic growth may have been a consequence of factors other than an inability to grow in the absence of oxygen. Thus flushing a culture system with pure nitrogen would eliminate carbon dioxide, commonly essential for growth (page 148), as well as oxygen, and under some conditions toxic effects of alcohol or lactic acid accumulation could terminate growth. Furthermore, a medium adequate for aerobic growth may lack the sterols, unsaturated fatty acids and vitamins required for anaerobic growth. *Saccharomyces*, *Mucor* and *Fusarium* are capable of anaerobic growth. It seems likely that although some fungi are obligate aerobes many are able to survive and perhaps show limited growth under temporarily anaerobic conditions.

Excessively high oxygen concentrations may prove toxic. Thus *Penicillium chrysogenum* suffers oxygen toxicity at an air pressure of 1.5 atmospheres. This is not of significance in nature but may be in the laboratory and industry, as fermenters are commonly operated at positive pressure to exclude contamination through minor leakage.

Carbon dioxide

Fungi, as heterotrophs, obtain energy by the degradation of organic compounds, so unlike autotrophs they cannot show net carbon dioxide fixation. Carbon dioxide is, however, an essential participant in several important metabolic reactions, and **heterotrophic carbon dioxide fixation** can be demonstrated in fungi. The tricarboxylic acid cycle intermediates α-oxoglutarate and oxaloacetate are precursors of amino acids. The intermediates are hence consumed when amino acids are synthesized from ammonia, but their levels are restored by **anaplerotic** (Greek, filling up) **reactions** which commonly involve carbon dioxide fixation. Details of these reactions in fungi are uncertain but in other organisms the carboxylation of pyruvate to yield oxaloacetate is important. Other reactions which consume carbon dioxide are steps in the synthesis of purines, pyrimidines and fatty acids.

Under natural conditions the probable universal requirement for carbon dioxide will be satisfied by the levels present in the air or by metabolically produced carbon dioxide. In the laboratory, however, some procedures, such as vigorous sparging with carbon dioxide-free gas or the elimination of oxygen (and carbon dioxide) with alkaline pyrogallol, may limit growth through carbon dioxide deficiency. There are some reports of high carbon dioxide levels being toxic; this may be due to the effect of carbon dioxide on pH in a poorly buffered medium with acid-intolerant species.

Some fungi that live in anaerobic environments and hence are normally exposed to high carbon dioxide levels have a high carbon dioxide requirement. *Aqualinderella*, for example, was only successfully cultured when provided with carbon dioxide, concentrations of 5–20% being optimal. Carbon dioxide could, however, be replaced by tricarboxylic cycle intermediates such as α-oxoglutaric acid. High carbon dioxide requirements have also been demonstrated in *Blastocladia* and in rumen chytrids. Thus *Neocallimastix* species need at least 7% CO_2 in the gas phase for growth.

In some pathogenic fungi, such as *Coccidioides immitis*, high carbon dioxide levels result in the production of yeast-like instead of filamentous growth. This also occurs in many *Mucor* species such as *M. rouxii*, but here the absence of oxygen, as well as high carbon dioxide levels, is required.

Other gases and volatile compounds

Fungi produce a variety of gaseous or volatile compounds which will be swept away by vigorous aeration but will reach relatively high levels where aeration is restricted. The metabolically active gas ethylene is produced by many soil fungi and by some plant pathogens such as *Penicillium digitatum*. This fungus attacks citrus fruits, and in stores the infection of some fruit by the fungus will result in ethylene production which will accelerate ripening of fruit rendering them more liable to attack.

Carbon Sources

Sugars

Probably all fungi can utilize glucose, a very widespread sugar occupying a central position in metabolism. Many other sugars can serve as carbon sources for fungi, with the range for any particular species likely to reflect availability in the organism's usual habitat. Transport into the cell involves transmembrane carrier proteins of relatively high specificity. Transport may take the form of facilitated diffusion, in which the attainment of equilibrium between the exterior and interior of the cell is accelerated by the availability of the carrier molecule. In the case of glucose uptake, this can be an efficient mechanism, as the cell's internal glucose concentration may be kept very low by coupling the influx with instant phosphorylation to glucose-6-phosphate by hexokinase associated with the protein carrier. Alternatively the process may be active, with the sugar entering the cell by a proton-symport mechanism. This occurs by protons being transported from inside to outside the cell membrane by the plasma membrane ATPase via hydrolysis of ATP, and re-entry of the protons driven by the **proton-motive force**, i.e. down their electrochemical gradient, via the sugar permease protein, each proton being accompanied by a sugar molecule. This results in accumulation of the sugar within the cell against its concentration gradient. *Saccharomyces* and *Neurospora* have an active transport system for glucose which is effective at low concentrations, with a K_m of *ca* 10^{-5} M. In the presence of high glucose concentrations this system is repressed and glucose enters the cell by a **constitutive** (i.e. permanently present) facilitated diffusion system with a K_m of *ca* 10^{-2} M. It is likely that most fungi have a constitutive transport system for glucose, but those for other sugars, as for galactose in *Saccharomyces*, may be **inducible**, appearing only in the presence of the substrate or certain specific inducers. Disaccharides may be transported into the cell or hydrolysed into assimilable monosaccharides by extracellular saccharases. Examples of the former are the transport via proton symport of maltose by *S. cerevisiae* or lactose by *Kluyveromyces lactis*, and of the latter the hydrolysis of sucrose to glucose and fructose by periplasmic invertase in *S. cerevisiae*.

Interconversion of hexoses within the cell permits their metabolism by the glycolytic or by the hexose monophosphate pathway. The latter path permits production of pentoses for nucleotide biosynthesis if these are not otherwise available. Evidence for the occurrence of the Entner–Doudoroff pathway via 3-keto-6-deoxyphosphogluconate, which is common in bacteria, is inconclusive for fungi. As indicated earlier (page 72) in *Saccharomyces* (and in some other yeasts) high glucose levels can suppress respiration in favour of fermentation even when oxygen is present. The utilization of carbon sources other than carbohydrates by many fungi (see below) combined with a universal need for sugars for biosynthetic purposes indicates the occurrence of gluconeogenesis (in effect a reversal of the glycolytic pathway) in these fungi.

Polysaccharides
Many fungi produce extracellular amylases permitting attack on starch. An ability to degrade cellulose is also widespread in fungi. A major source of cellulose is wood, but here the cellulose is impregnated with the highly refractory aromatic polymer lignin. Some fungi have the ability to degrade lignin, thus enabling the cellulose to be utilized. Timber decay is considered in greater detail elsewhere (pages 305–319). Chitin (poly-N-acetylglucosamine; Fig. 3.4) is a major

polysaccharide in most natural environments, as it is a component of fungal walls and exoskeletons of arthropods and many other invertebrates. Chitinolytic fungi are readily isolated from soils. Most common are Mucorales, especially *Mortierella*, and mitosporic fungi and Ascomycetes, especially the genera *Aspergillus*, *Trichoderma*, *Verticillium* and *Humicola*. Chitinases of *Trichoderma harzianum* have been studied in detail with a view to recycling chitinous waste from shellfish industries. Baiting freshwater sites with chitin yields a range of fungi, especially chytrids, such as *Chytriomyces* species and *Karlingia asterocysta* which has a nutritional requirement for chitin that can only be relieved by *N*-acetylglucosamine, i.e. it is an 'obligate chitinophile'. Fungi are rare in the sea, but the sea is rich in chitin, and a range of marine fungi degrading chitinous exoskeletons have been isolated, including the Ascomycete *Abyssomyces hydrozoicus*. Myxomycetes are a rich source of lytic enzymes, and *Physarum polycephalum* produces a complex of extracellular chitinases. Many fungal invertebrate pathogens such as *Beauveria bassiana* and mycoparasites such as *Aphanocladium album* produce chitinases during their attack on their hosts. Chitosan (polyglucosamine), a major component of cell walls of Mucorales, can be utilized by a range of soil fungi, such as species of *Rhizopus*, *Aspergillus*, *Penicillium* and *Chaetomium*.

Physarum polycephalum and *Aspergillus nidulans* can make use of some part of agarose (a major component of agar) as a carbon source, resulting in slight but definite growth in the absence of any other carbon source. There are indications that this ability is widespread in fungi, rendering agar culture unsuitable for precise studies on carbon nutrition.

Organic acids and alcohols

Since the tricarboxylic acid cycle is almost universal in fungi it might be supposed that intermediates in the cycle could act as carbon sources for most fungi. In many fungi, however, these intermediates when supplied exogenously fail to enter the cell, although in others such substances as citric acid or succinic acid can be utilized. Other organic acids, such as lactic acid, tartaric acid, acetic acid and longer chain fatty acids are assimilated by some fungi. Organic acids are supplied as salts, usually of potassium, sodium or ammonium, to avoid making the medium too acidic.

The trihydric alcohol glycerol is a good carbon source for many fungi, and some, such as *Saccharomyces* under aerobic conditions, can utilize ethanol. A few yeasts can assimilate methanol. This will be discussed below as a 'C_1' compound.

Lipids

Some fungi produce extracellular lipases and phospholipases which hydrolyse lipids and phospholipids to glycerol and fatty acids which can then be assimilated. Phospholipase activity is implicated in the pathogenesis of *Candida albicans*, by its activity in degrading the plasma membranes of host cells. In a study comparing an invasive with a non-invasive strain in a mouse model, the only difference found in putative virulence factors was that phospholipase production was four times higher in the invasive strain.

Hydrocarbons and C₁ compounds

Hydrocarbons and C$_1$ compounds

There are relatively few organisms that are able to metabolize reduced carbon compounds which contain only one carbon atom (e.g. methanol, CH_3OH) or, if they have more than one carbon atom, lack carbon–carbon bonds (e.g. dimethyl ether, CH_3OCH_3). Organisms able to utilize such 'C_1' compounds are known as methylotrophs and are of potential interest as sources of single cell protein, since methanol can be synthesized from methane, a major constituent of natural gas. Organisms capable of utilizing methane itself are known as methanotrophs. Until recently it was thought that this capability was limited to prokaryotes, but the isolation of yeasts able to grow on methane has now been reported.

Increased molecular size lowers the boiling point and melting point of hydrocarbons to give liquids in the range *ca* C_5 to C_{15} and solids for still larger molecules. Utilization of the liquid hydrocarbons is of interest in relation to fuel deterioration. The Hyphomycete *Hormoconis resinae* can utilize liquid hydrocarbons. It will grow, especially at tropical temperatures, at the interface between the water that inevitably accumulates at the bottom of fuel tanks and the kerosene above. Metal corrosion can result from the organism's metabolic activity and mycelium has grounded aircraft by clogging filters. The utilization of aliphatic hydrocarbons in the C_{16}–C_{22} range by *Candida tropicalis* and *Candida intermedia*, on the other hand, is of interest for the production of single cell protein.

Amino acids and proteins

Although amino acids and proteins are usually considered mainly as sources of nitrogen, they can also act as sole carbon sources for some fungi. The assimilation of these substances will be considered in the next section.

Nitrogen Sources

In the past a number of fungi have been claimed to fix molecular nitrogen. It is now clear that no eukaryotes have this ability, and that the earlier claims result from methodological errors combined with a remarkable capacity to scavenge traces of combined nitrogen shown by some fungi. Most fungi can, however, assimilate inorganic nitrogen as nitrate or ammonia in addition to utilizing a wide range of organic compounds.

Nitrate and ammonia

Many fungi are able to utilize nitrate, a nitrogen source that is common in soil as a result of the application of artificial fertilizers or the oxidation of ammonia by nitrifying bacteria, although some (e.g. *Saccharomyces cerevisiae*, many Basidiomycetes) cannot do so. The mechanisms for entry of nitrate remain to be clarified. At high concentrations in the medium, it could enter the cell by diffusing down a gradient created by its rapid metabolism within the cell, but at lower concentrations it is probable that active transport takes place. The enzyme nitrate reductase reduces the nitrate to nitrite, and nitrite reductase reduces the nitrite to ammonia. In *Neurospora crassa*, the nitrate reductase is a protein of molecular mass 228 000, consisting of two identical sub-units and a molybdenum-containing co-factor. The regulation of this process has been intensively studied in *Aspergillus nidulans* and *Neurospora crassa*. Expression of genes for nitrate and nitrite reductases, and for synthesizing the molybdenum cofactor, is induced only in the presence of nitrate and the absence of a favoured nitrogen source such as ammonia, glutamine or glutamate. In laboratory cultures, nitrate is commonly supplied as the sodium, potassium or (see below) ammonium salt.

In nature the decomposition of animal wastes and animal and plant remains produces large amounts of ammonia. Sometimes accumulation occurs but most will be converted into ammonium salts by interaction with acids or oxidized to nitrate by nitrifying bacteria. *Neurospora crassa* has been shown to avidly accumulate ammonium ions in response to the membrane potential. The hyphae generate a membrane potential of about −200 mV, chiefly through the outwards pumping of protons by the ATPase, and this is rapidly depolarized by addition of ammonia. This indicates that the positively charged ammonium ions are entering the hyphae via electrogenic channels. Ammonia uptake has also been studied using a radioactive analogue, [^{14}C]methylammonia. There is evidence for both low and high affinity systems, the former being responsible for bulk transport and the latter for scavenging traces of ammonia. Enzymic reactions convert the ammonia into glutamic acid or glutamine and transamination reactions give rise to other amino acids. Ammonia is the major regulator of nitrogen assimilation and in its presence the utilization of other nitrogen sources such as nitrate, amino acids and proteins is repressed. In laboratory cultures, ammonia is supplied in the form of ammonium salts. However, as the ammonium ion is taken up the ionic balance of the cell is maintained by the extrusion of hydrogen ions. This can result in the formation of highly acidic conditions in the medium, depending on the co-ionic species. Thus the provision of ammonium chloride as a nitrogen source will result in the generation of a strong acid in the medium and a pH drop; since the chloride is not assimilated hydrochloric acid is formed. With ammonium nitrate, a fall in pH as ammonium is utilized may be followed by a pH rise as nitrate utilization follows. The problems resulting from a fall in pH can be avoided to some extent by using ammonium salts with anions that are either assimilated or give weak acids, such as phosphate, citrate or carbonate, or by monitoring pH and adjusting with alkali. A few fungi are incapable of utilizing inorganic nitrogen sources and have to be supplied with amino acids, such as some members of the Saprolegniales and Blastocladiales that live on substrates rich in proteins and their breakdown products, such as dead insects.

Amino acids, amines and amides

Most fungi are able to assimilate a wide range of amino acids, amines and amides. Such substances as glutamate, glutamine, aspartate and asparagine are useful components of culture media. A common medium constituent, casein hydrolysate, an acid hydrolysate of the milk protein casein, contains all the common amino acids except tryptophan which is destroyed during hydrolysis. Uptake of amino acids is by active transport, via a range of permeases for different amino acids with overlapping specificities. Experiments with

Neurospora crassa have shown that amino acids are transported by proton-motive force, with two protons entering per amino acid. Amino acids and proteins can serve as carbon as well as nitrogen sources for many fungi.

Polypeptides and proteins

Small peptides, of up to six amino acid residues in some species such as the human pathogen *Candida albicans*, can be transported into the cell via a range of peptide permeases. They are then hydrolysed by intracellular peptidases. This renders enzymic hydrolysates from proteins such as casein or from lean meat (peptone) useful nitrogen sources for fungi. Larger peptides and proteins have to be degraded by extracellular peptidases and proteases before assimilation can occur. Extracellular protease production is a common attribute of fungi. *C. albicans* produces at least nine secreted aspartyl proteases, which are implicated in its adhesion and penetration of human tissue. A few fungi are able to attack keratin, the refractory protein of hair, feathers and skin, by producing keratinases. These fungi are hence well adapted for life on the surface of animals and have potential pathogenic capabilities. Examples include *Batrachochytrium dendrobatidis*, the recently discovered strictly keratinophilic chytrid pathogen of frogs, and the mammalian dermatophyte pathogens (see Chapter 7).

Other nitrogen sources

There are many other nitrogen-containing organic compounds that can be utilized by fungi. Purines, from turnover of nucleotides, are universally degraded and animals secrete the resultant nitrogen-rich waste products – the purine urate in humans, birds and reptiles, its product allantoin by other mammals, and its further product urea by amphibians and fish. Purine utilization has been studied in detail in *Aspergillus nidulans*. Synthesis *de novo* of a urate permease and the required catabolic enzymes is induced by urate but repressed by a favoured nitrogen source such as ammonia, glutamine or glutamate. Allantoin catabolism has been studied in detail in *Saccharomyces cerevisiae*. Eight structural genes encoding permeases and catabolic enzymes have been identified. All are repressed by the presence of a favoured nitrogen source, but only some are induced by the presence of allantoin. Some of the allantoin that has been taken up can be stored in vacuoles for later use as a nitrogen source. Allantoin is well suited as a nitrogen reservoir, as it has the highest nitrogen:carbon ratio, 3:2, than any other compound apart from urea, 2:1, but urea is too toxic to be stored in the cell. Utilization of urea, involving active uptake via urea permeases and action of urease, which splits urea into carbon dioxide and ammonia, is common in fungi. Fungi which are able to degrade exogenous chitin will, in so doing, have access to a nitrogen source as well as a carbon source.

Carbon/nitrogen (C/N) ratio

Proteins contain about 15% nitrogen. Although nitrogen is also needed for the synthesis of other cell components, such as nucleic acids and chitin, carbon is required as an energy source. A balanced medium will hence contain about ten times as much carbon as nitrogen. Hence carbon:nitrogen ratios of 10:1 or less will ensure a high protein content and a C/N ratio greatly in excess of this figure (e.g. 50:1) will favour accumulation of alcohol, acetate-derived secondary metabolites, lipids or extracellular polysaccharides. Attention to C/N ratio is hence essential in fermentation technology.

Other Major Nutrients

The elements carbon, hydrogen, oxygen, nitrogen, sulphur, phosphorus, magnesium and potassium are required by all organisms in large amounts. The first three (C, H and O) are provided in the form of organic compounds, carbon dioxide, oxygen and water, and the forms of nitrogen available to fungi have also been considered. The remaining four elements (S, P, Mg and K) can be supplied to most fungi as salts, for example magnesium sulphate and potassium phosphate (Table 6.2). An element needed only by fungi from the sea and salt lakes is sodium, which can conveniently be provided as the chloride. Non-essential elements can sometimes partially replace a requirement for an essential element; provision of sodium ions, for example, may reduce the amount of potassium needed. The requirement for sulphur and phosphorus will be considered further.

Sulphur
Most fungi are able to utilize sulphate, and this process has been extensively studied in *Neurospora crassa*. Inorganic sulphate enters the cell via two distinct sulphate permeases. After uptake, it is phosphorylated via adenosine triphosphate in two steps to give 3′-phosphoadenosine-5′-phosphosulphate, reduced to sulphite, then to sulphide, which is condensed with O-acetylserine to give cysteine, which can then be incorporated into proteins and act as intermediate in the synthesis of methionine and S-adenosylmethionine. Organic sources of sulphate can also be used, such as choline-O-sulphate, widely found in plants and many fungi as an internal sulphur store, and aromatic sulphate esters, such as tyrosine-O-sulphate. These are transported into the cell via specific permeases, which, along with sulphate permeases are strongly induced in sulphur-limited conditions, and repressed when the sulphur-containing amino acids cysteine or methionine are present in the medium. The Saprolegniales cannot utilize sulphate but need more reduced forms of sulphur such as thiosulphate or sulphide. Some members of the family as well as the Blastocladiales and the Myxomycete *Physarum* have to be provided with sulphur in the organic state, for example as methionine.

Phosphorus
Phosphate can be supplied in a culture medium as potassium dihydrogen phosphate (KH_2PO_4) or as dipotassium hydrogen phosphate (K_2HPO_4), or as the corresponding sodium salts. The dihydrogen phosphate is the more soluble and gives a more acidic medium with a pH of about 5.5 instead of near neutrality. Phosphate enters the cell by active transport as the $H_2PO_4^-$ ion, possibly via the proton-motive force together with two protons. Phosphate can be stored in the cell in the form of polyphosphate (Figure 3.3; page 98).

Trace Elements

The elements iron, copper, calcium, manganese, zinc and molybdenum are required by all, or nearly all, organisms as cofactors for enzymes and other functional proteins. The amounts needed are very low, concentrations in media in the range 10^{-6} M for iron to 10^{-9} M for molybdenum being appropriate. Some of the trace elements such as copper and zinc become toxic for some fungi at levels only a few times greater than those required for optimal growth. The reagent-grade chemicals normally used for preparing media are generally contaminated with quantities of trace elements adequate for supporting good growth, so often no deliberate addition of trace elements to media is made. Research on trace element requirements involves the rigorous purification of medium constituents.

In spite of the small quantities of trace metals required, their presence in the medium does not always ensure their availability for the fungus. This problem of non-availability is most often encountered with iron. Except under strongly acidic conditions ferrous iron in solution undergoes rapid spontaneous oxidation to the ferric state followed by precipitation as the highly insoluble ferric hydroxide. The problem is dealt with by providing appropriate concentrations of trace elements and of a chelating agent with which they form a soluble complex. Although the chemical equilibrium strongly favours complex formation, some free metal ions and chelating agent are present, and utilization of the free ion by the organisms results in further liberation of metal ions from the complex. The chelating agents act as metal buffers in a way analogous to pH buffers, keeping a constant free ion concentration available. Citrate was at one time used as a chelating agent, but now more potent chelating agents such as salts of ethylenediaminetetra-acetic acid (EDTA), which can be used in far smaller amounts, are favoured.

Many fungi, however, are themselves able to produce powerful chelating agents for iron, termed siderophores (Greek, iron-carriers). Such compounds (Fig. 3.21) are released into the medium when the iron supply is limiting growth, and chelate ferric ions. Metabolic energy is utilized in transporting the iron–siderophore complex across the plasma membrane into the cell. Siderophores are also involved in the storage of iron in the cell. In some fungi these siderophores have the same structure as those involved in transport, but in others they differ. Members of the Zygomycetes store iron in their cells by using the iron-binding protein, ferritin, as do plants and animals. Some fungi do not produce siderophores. These transport iron into the cell utilizing a ferric reductase at the cell surface. Some, when iron is limiting, release large amounts of the relatively weak chelating agent citric acid into the medium. Fungi that do not themselves produce siderophores may also utilize those produced by other fungi if these are available in the environment.

Iron, copper, manganese and zinc are supplied in culture media as cations, and molybdenum as molybdate. Molybdenum is required in very small amounts, and is essential for nitrate reduction. A requirement for molybdenum is hence most readily demonstrated when nitrate is used as the nitrogen source.

Metallic cations are strongly adsorbed by fungal cell walls, particularly those of lichen fungi, by a process of ion exchange. This bioconcentration is affected primarily by the valency of the cation, so that trivalent ions displace divalent ions

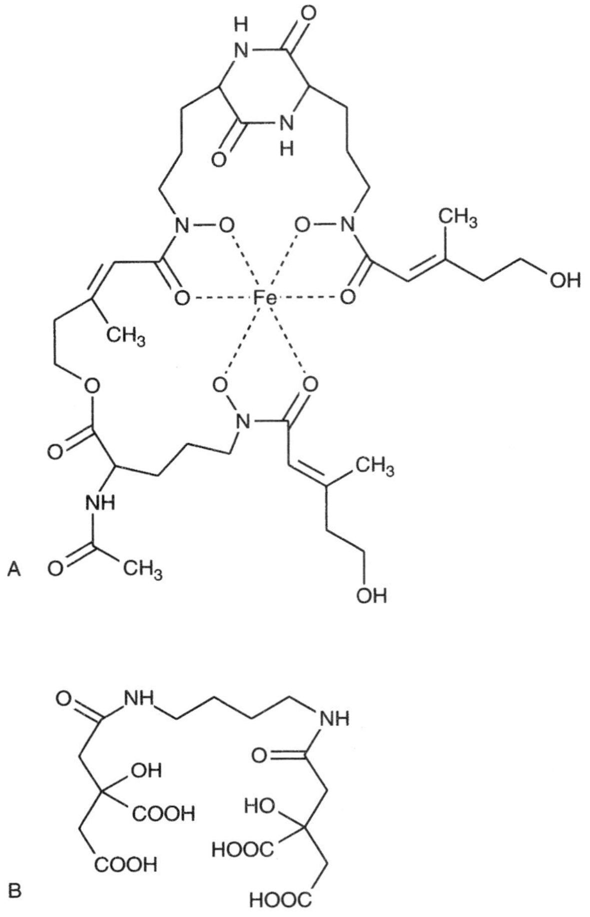

Figure 3.21 Fungal siderophores. A, Coprogen, produced by *Neurospora crassa* and some species of *Penicillium*. It is formed by the hydroxylation and acetylation of three L-ornithine molecules, and is an example of a hydroxamate siderophore. These are varied in structure, and may originate from 1, 2 or 3 L-ornithine molecules. They are widespread in Ascomycetes, mitosporic fungi and Basidiomycetes, including Basidiomycete yeasts. The coprogen–iron complex is represented. B, Rhizoferrin, produced by *Rhizopus microsporus* and other members of the Mucorales including *Mucor mucedo*. It consists of two citric acid molecules linked via amide bonds to putrescine. It is an example of a complexone or polycarboxylate siderophore, which form less stable complexes with iron than do the hydroxamate siderophores. Probably for this reason the Mucorales, unlike higher fungi but like most other eukaryotes, utilize the protein ferritin for iron storage.

which in turn displace monovalent cations. Secondarily it is affected by atomic number, so heavier cations displace lighter ones. Thus caesium displaces potassium, which in turn displaces sodium, and strontium displaces calcium. This has advantages and disadvantages in terms of heavy metal and radionuclide pollution. Thus fungi can be used for bioremediation of effluents polluted with heavy metals, and heavy metal pollution can be monitored by analysing fungi or lichens growing in particular environments. On the other hand these contaminated fungi or lichens can be, directly or indirectly, human food sources. For example the major radionuclide of long-term concern that was released to the atmosphere by the Chernobyl nuclear accident in 1986 is caesium-137, with a half-life of 30 years. Caesium is accumulated by all living organisms, because of its similarity to potassium. In fungi, however, it is strongly adsorbed, as it will displace the lighter potassium and sodium ions bound to the fungal walls. Thus in a survey in 1994–1995 of sites in Russia about 200 km north-west of Chernobyl, levels of ^{137}Cs in wild mushrooms were up to 1000-fold higher than the highest levels in agricultural products. This was reflected in body levels in the local population, which were 10-fold higher in individuals who very often ate wild mushrooms in comparison to those who never did, and in levels in the population as a whole up to 70% higher in autumn (the mushroom season) than in spring or summer. Even higher levels have been found in the reindeer-herding Samis of northern Scandinavia, where there is the very short food chain of lichen to reindeer to human (page 451). Unexpectedly, analyses in successive years show little diminution in contamination, so the problem will be with us for many years to come. In earlier decades, when there was worldwide 'fallout' of radionuclides from testing of nuclear weapons, the same population showed very high accumulations of caesium-137, strontium-90 and other radionuclides, with a seasonal variation of highest in winter and spring, when lichen is the major food source for the reindeer. Some fungi selectively concentrate particular heavy metals. An example is the common fly agaric, *Amanita muscaria*, which accumulates extraordinarily high levels of vanadium of up to 200 mg/kg.

Growth Factors and Vitamins

Some fungi, for example *Aspergillus niger* and *Penicillium chrysogenum*, require for growth the provision of only a single organic compound such as glucose. This implies the ability to synthesize from glucose and inorganic compounds the wide range of organic materials required for the structure and function of the cell. Mutations can disrupt the biosynthetic pathways involved so it is possible to obtain from *Aspergillus nidulans*, for example, a variety of strains which need to be supplied with one or more amino acids or vitamins or other such nutrients. The absence of such nutrient requirements in strains isolated from nature implies that the natural habitat of the organism is deficient in nutrients, other than the major carbon source, to an extent which would render mutant strains with additional nutrient requirements uncompetitive. Biosynthetic activities, however, require the expenditure of energy which would otherwise be available for achieving higher growth rates or yields. Hence, if a nutrient is always present in a habitat an inability to synthesize a substance will be of no disadvantage to the mutant strain but may even permit it to replace the parent strain with its unnecessary biosynthetic activity. The nutritional requirements of an organism will hence reflect the availability of nutrients in its environment.

Many fungi have rather simple nutrient requirements. Some, as indicated above, need no organic compounds other than a carbon source, and many others have only a few additional requirements for organic nutrients. It is, however, those organisms which have the most simple needs which are most readily grown in pure culture on defined media, whereas those that have elaborate requirements may not have been isolated or perhaps even detected.

Requirements for organic nutrients other than a carbon source can include amino acids (already discussed), **growth factors** (other materials required in relatively large amounts) and **vitamins** (substances needed in very small quantities). Many fungi require the vitamins thiamine (e.g. *Phycomyces*) or biotin (e.g. *Neurospora*), members of the water-soluble B group of vitamins which are of importance in animal nutrition and were originally isolated from yeast extract. Instances of requirements for most of the other B vitamins are known in the fungi but are less common. Sterols, as already mentioned (page 147) may be needed, especially under anaerobic conditions. Other growth factors which may be needed include fatty acids, purines and pyrimidines and inositol.

The slime moulds, being phagotrophic organisms, are supplied in nature with all the components of the microorganisms that they ingest. Their nutritional requirements hence may be elaborate which is probably why few have been grown in pure culture. Haem (page 24) is a requirement for *Physarum polycephalum* and perhaps all Myxomycetes. The coprophilous (dung-inhabiting) Zygomycete *Pilobolus* requires either haem or a siderophore to be supplied.

Water Availability

Living organisms consist largely of water. Hence if an organism is to grow and increase in volume it has to take up water from the environment. Growing cells and hyphae hence must have walls and plasma membranes that are permeable to water. This, however, means that water can also be lost to the environment, with the possibility that excessive loss of water could lead to desiccation and death. Whether water enters or leaves a cell depends on the difference between the water potential of the cell and that of the surrounding medium, water moving from a region of high to one of lower water potential.

Water potential is measured in units of pressure, usually either in SI metric units as megapascals (MPa) or in bar (0.987 atmospheres = 1 bar = 0.1 MPa). It is the sum of a number of components, of which the most important are osmotic potential, matric potential and turgor potential, thus (equation 11).

Water potential = osmotic potential + matric potential + turgor potential   (11)

Pure water has an **osmotic potential** of zero; this value is lowered by the presence of dissolved substances. For example, at 25°C the osmotic potential of sea water is –2.8 MPa and of saturated glucose solution –80 MPa. Osmotic potential has the same value as osmotic pressure, the pressure generated when a solution is separated from pure water by a semi-permeable membrane (one through which water can pass but the solute cannot), but is negative instead of positive. **Matric potential** results from the interaction of water with interfaces, and is of importance in determining whether water within a solid matrix, such as soil or wood, is available to a fungus. The availability of the water is determined by the radius of the pores in which it occurs (equation 12).

$$\text{Matric potential (MPa)} = -0.14/\text{radius of pore (µm)} \qquad (12)$$

Thus water in pores of radius 1 µm has a matric potential of −0.14 MPa and in pores of radius 0.01 µm, −14 MPa. This means that water present in the lumen of cells in wood is readily available to fungi, but that in the cell walls is not, and that water in a sandy or loamy soil is more accessible than that in clay. **Turgor potential** is the pressure exerted by the stretched wall of a turgid cell on the contents of the cell. Turgor potential is positive, falling to zero as a cell loses its turgor. Being positive, turgor potential increases the water potential of a cell. For example, a cell with contents having an osmotic potential of −1.5 MPa and with a wall exerting a turgor potential of 0.5 MPa has a water potential of −1.0 MPa.

The concept of water potential is widely used by plant pathologists and plant and fungal physiologists, and has a semi-official status as the correct way of expressing water availability. However, many microbiologists concerned with fungi that live in conditions of very low water availability and cause spoilage of foods and other products continue to use an alternative concept, that of water activity (a_w). This is defined in terms of the ratio between the water vapour pressure of the solution being considered (p_s) and that of pure water (p_w) (equation 13).

$$a_w = p_s/p_w \qquad (13)$$

Water activity ranges from zero (water absent) to 1.0 (pure water). The relationship between water potential and water activity is given by equation 14.

$$\text{Water potential (MPa)} = k \ln a_w \qquad (14)$$

The value of k depends on temperature and is, for example, 1.37 at 25°C and 1.35 at 20°C.

Not only is the availability of water in the surrounding liquid phase of importance to fungi, but so is the water content of the adjacent gas phase. This is because hyphae can spread over dry surfaces or rise above the substratum. In addition fungi may grow in small volumes of liquid, such as droplets on a leaf surface, that come rapidly into equilibrium with the atmosphere by gaining or losing water. The water content of the atmosphere is expressed in terms of **relative humidity**, the ratio of the water vapour pressure of the gas phase being considered, to that of a saturated atmosphere at the same temperature. It is hence the same ratio as water activity, but is usually expressed as a percentage. For example, a solution with water activity of 0.75 would be in equilibrium with a gas phase with a relative humidity of 75%.

Water potential and the fungal environment
Fungal activity, whether expressed as plant disease, mouldiness of materials, rotting of wood or the appearance of mushrooms and toadstools in fields and

woodlands, is most obvious in damp conditions. This is consistent with most fungi growing best at high water potentials, in the range 0 to −1 MPa. Commonly used media, such as those containing 2% sucrose, have a water potential in this range. At lower water potentials, rates of hyphal growth diminish, until values are reached at which growth does not occur (Table 3.9). Most wood-destroying fungi, for example, lack the ability to grow at water potentials below −4 MPa. The inability of fungi to grow at low water potentials is the basis for such traditional methods of food preservation as drying or the addition of salt or sugar. A few fungi, however, can grow at very low water potentials. Such fungi are usually termed **osmophilic** or **xerophilic**, although most are really osmotolerant, growing best at high, although able to tolerate low, water potentials. Many yeasts and members of the genera *Aspergillus*, its teleomorph *Eurotium*, and *Penicillium* are osmotolerant and of importance as agents of biodeterioration. The most osmotolerant fungi known can grow at a water potential of −69 MPa. The dry valleys of Antarctica, with a water potential of about −90 MPa, are sterile, except for transient microbial growth following snow falls and in relatively humid microenvironments such as within rocks. Some desert animals are active in still drier conditions, but their cells, provided with

Table 3.9   Water availability in different environments and approximate lower limits for growth of some fungi

Water activity	Water potential (MPa)	Examples
1.0	0	Pure water
0.996	−0.5	*Phytophthora cactorum*, lower limit
0.995	−0.7	Typical mycological media
0.98	−2.8	Sea water
0.97	−4	Most wood-destroying fungi, lower limit
0.95	−7	Bread. Leaf-litter Basidiomycetes, lower limit
0.90	−14	Ham. *Neurospora crassa*, lower limit
0.85	−22	Salami. *Saccharomyces rouxii* in NaCl solution, lower limit
0.80	−30	*Aspergillus nidulans* and *Penicillium martensii*, lower limits
0.75	−40	Saturated NaCl solution, *Aspergillus candidus*, lower limit
0.65	−60	22 molal glycerol
0.60	−69	Limit for cell growth – *Zygosaccharomyces rouxii* in sugar solutions and the mould *Monascus* (*Xeromyces*) *bisporus*
0.58	−75	Spores of some *Eurotium*, *Aspergillus* and *Penicillium* species are able to survive for several years
0.55	−80	Saturated glucose solution. DNA denatured
0.48	−90	Antarctic dry valleys

Data from various sources. Molality (molecular weight in grams per 1000 grams solute) and not molarity (MW in g per final volume of 1000 ml) is used in dealing with osmotic potentials. The lower limits of growth are those obtained at optimal temperature and nutrition; when these factors are suboptimal the limits are not so low. As indicated with *Saccharomyces rouxii*, organisms are usually more tolerant of high sugar than high salt concentrations.

metabolic water by carbohydrate metabolism and in a protected situation inside the animal, escape equilibration with the environment. The resting structures of some fungi are also able to escape equilibration with the environment through having impermeable walls, and can hence survive dry conditions until water potentials are once more adequate for growth.

Adaptation to changes in external water potentials

A growing fungus will have a water potential a little lower than that outside, allowing water to enter the cell and growth to occur. The external water potential may, however, fall, through evaporation increasing the solute concentration in the ambient medium. If the value falls below that of the fungus, then water will leave instead of entering the cell, growth will cease, the protoplast may detach from the cell wall and shrink, an effect known as plasmolysis, and desiccation and death may occur. An increase in the external water potential can occur if, for example, the ambient medium is diluted by rain or dew. This can lead to an influx of water into the cell, resulting in a high internal hydrostatic pressure with possible rupture of the cell wall and death. Fungi that live in some environments are hence exposed to the possibility of desiccation or cell rupture through fluctuations of external water potential. Such fungi can, however, cope with such fluctuations by adjusting their own water potential. This is done by raising or lowering the osmotic potential of the cell contents. Lichens are especially well adapted for life in environments with a daily pattern of alternating desiccation and hydrated state; indeed they appear to require this to maintain the balance between fungus and phototroph. For example, desert lichens are photosynthetically active only for a brief time at dawn, when they are damp with dew, but quickly become desiccated and inactive until dampened again the following morning.

One way in which the osmotic potential of a fungal cell can be lowered is by the uptake of a solute from the ambient medium. For example, *Thraustochytrium aureum*, which lives in brackish water and in the sea, can adjust its internal osmotic potential by the uptake of inorganic ions. At high concentrations, however, inorganic ions and many other solutes can change the configuration and hence the activity of enzyme molecules. Hence where very low internal osmotic potentials are required many fungi synthesize polyols, which are 'compatible solutes', having little effect on enzyme activity even at high concentrations. The polyols may be produced either from the sugars taken up from the medium or from the products of the breakdown of insoluble carbohydrate reserves. The polyol concerned in osmotic adaptation in the moderately osmotolerant yeast *Saccharomyces cerevisiae* as well as the very tolerant *Zygosaccharomyces rouxii* is glycerol. Other polyols important in osmotic adaptation in fungi are mannitol and arabitol.

When an increase in the internal osmotic potential of a fungus is needed, this can be brought about by the loss or export of solute to the environment, or by conversion of the solute to insoluble reserve materials.

There has been much work on the molecular genetics of responses of *Saccharomyces cerevisiae* to both hyper- and hypo-osmotic stress, as this has yielded important information on sensory signalling pathways in eukaryotic cells that regulate gene expression. When yeast cells are subjected to hyperosmotic stress, responses seen include temporary cessation of growth, with disaggregation of the actin cytoskeleton and loss of cell polarity, decreases in cell wall porosity and membrane permeability to glycerol, and accumulation of glycerol. When they are subjected to hypo-osmotic stress, a major response is to increase the strength of their cell walls, especially seen as an increase in chitin content, accompanied by release of glycerol to the medium which lowers the osmotic gradient across the stretched membrane. Yeast cells detect and respond to high and low extracellular osmolarity by activating two different mitogen-activated protein kinase (MAPK) signalling pathways. High osmolarity activates the HOG (high osmolarity glycerol) pathway. This pathway exhibits multiple redundancies, starting at the sensor level, where two independent branches activate the MAPK signalling cascade, a putative transmembrane osmosensor, and a protein phosphorylation relay system, which is inhibited by hyperosmotic stress. In both cases, the result is activation of the MAPK cascade, which in turn activates the *HOG1* gene. The Hog1 protein in turn induces transcription of several genes, including *GDAP1*, the structural gene for glycerol 3-phosphate dehydrogenase involved in glycerol synthesis. The cell then accumulates glycerol which raises its osmotic pressure to re-establish an osmotic gradient through its plasma membrane to allow water into the cell. Accompanying responses include a decrease in membrane permeability to glycerol. Hypo-osmotic stress activates the protein kinase C (PKC) signal transduction pathway. The PKC pathway is activated by a range of other stimuli such as nutrient stress, and it is thought that a major role for it is to maintain cell integrity by controlling cell wall assembly and perhaps membrane assembly. It has already been emphasized that the cell wall is a dynamic structure that must be maintained to resist mechanical and chemical stress. The modifications required for this entail activation of cell wall-synthesizing enzymes and vectorial transport of vesicles and wall and membrane components. The initial membrane stretch that is the consequence of hypo-osmotic stress is thought to be sensed by a mechanosensor, which activates a small GTP-binding protein, Rho1 which controls the activity of the Pkc1 protein and also of $\beta(1{\rightarrow}3)$ glucan synthase (see page 115). Pkc1 activates the MAPK cascade, and a result is the activation of genes for cell wall biosynthesis, including chitin synthase, $\beta(1{\rightarrow}3)$ and $\beta(1{\rightarrow}3)$ glucan synthases, proteins involved in mannosylation and a GPI-anchored membrane protein. Activities of these proteins lead to a strengthened wall, in particular with a higher chitin content. Chitin synthases from a range of fungi are also activated quickly following hypo-osmotic stress, probably from pre-existing zymogenic forms, giving a rapid increase in wall strength.

Cell form in relation to water stress

As indicated above, adaptation to changes in external water potential can occur through a change in the osmotic potential of the cell contents. This, however, takes a minimum of several minutes, whereas the flow of water into or out of a cell that follows a change in external water potential occurs in seconds. A fall in external water potential will hence cause plasmolysis, the protoplast with its semi-permeable membrane partly or wholly withdrawing from the cell wall until adaptation has occurred. An increase in external water potential, with a rapid flow of water into the cell, is potentially more serious. An increase in hydrostatic pressure within the cell occurs, resulting in a stretching of the cell wall and an increase in cell volume. The stretching of the wall will result in an increase in the force it exerts on the cell contents, and this increase in turgor potential will raise the water potential of the cell. If the increase in external water potential is not too great, it will be equalled by the increase in the water potential of the cell due to increased turgor, and the influx of water will cease. If, however, the increase in external water potential is large, then the increased hydrostatic pressure of the cell may lead to excessive mechanical stress in the cell wall and rupture. For example, when colonies of *Mucor rouxii* growing on agar media containing 2% glucose are flooded with distilled water, extensive bursting of hyphal tips occurs in a few seconds. Such osmotic rupture of hyphal tips after flooding with distilled water has been observed in many fungi.

It is not surprising that hyphal tips are especially liable to osmotic rupture. The strength of mature cell walls is enhanced by the cross-linking of polymers. At the hyphal apex, however, these features are lacking, since the wall has to be plastic in order to allow rapid stretching as growth proceeds. In *Neurospora crassa*, for example, the hyphal tip moves forward a distance equal to the entire length of the extension zone, 25–30 μm, in about 50 seconds, the extension rate being about 40 μm min^{-1}. Even in *Penicillium digitatum*, a fungus with the low hyphal extension rate of about 1 μm min^{-1}, the apex moves forward by the length of the extension zone in about 5 minutes. The bud of the yeast *Saccharomyces cerevisiae*, however, takes about 45 minutes to grow to a length similar to that of the mother cell, an extension rate of 0.1–0.2 μm min^{-1}. The wall of such a cell can hence have a structure in which strength is not sacrificed to a requirement for rapid extension. Yeasts are in fact typical of environments such as the surfaces of fruits and leaves, where very concentrated sugary solutions produced by exudation and evaporation may be suddenly diluted by rain. It may well be that the yeast form is an adaptation to environments liable to rapid fluctuation in water potential.

Osmoregulation in cells lacking walls

In most fungi the influx down a moderate gradient of water potential is resisted by the turgor potential exerted by cell walls. The vegetative cells of slime moulds and zoospores of Oomycetes lack cell walls but have water expulsion vesicles (contractile vacuoles) which fill with water from the adjacent cytoplasm and then discharge to the exterior. Filling normally takes a few minutes and discharge less than a second.

Hydrogen Ion Concentration

Fungi must be prepared to cope with changes in external pH to a greater extent than do animal or plant cells in their sheltered environments. For fungi, cytoplasmic pH varies very little over a wide range of external pH values. *Neurospora crassa*, for example, has cytoplasmic pH values of 7.1 and 7.4 in media of pH values 4.1 and 8.4, respectively. It is unclear exactly how cytoplasmic pH is stabilized close to neutrality. Intracellular buffering is quite inadequate to cope with acid produced during metabolism. It has been calculated that, at the rate *N. crassa* produces acid, its cytoplasmic pH ought to drop at the rate of 0.6 units per minute. That it does not must be attributed to net extrusion of protons from the cell and regulation of metabolic pathways that generate or consume protons. Addition of a permeant weak acid to the medium will cause an acidification of the cytoplasm, as it will cross the plasma membrane as free acid and ionize to its anion and a proton within the cell. This acidification could kill the cell, and is the basis of the use of food preservatives such as benzoic acid. Addition of a low concentration of butyric acid to *Neurospora crassa*, however, resulted in accelerated proton extrusion accompanied by metabolic anions, probably succinate, allowing the cytoplasmic pH to rise to maintain neutrality.

There is a lack of information on the effect of pH on fungal growth parameters, in spite of a considerable literature on growth in relation to the initial pH of media. This literature is of limited value, since fungal metabolism alters pH, either by taking up an anion or cation but not the co-ionic species, or by the production of organic acids or ammonia. Effective buffering is difficult in culture, since buffers may themselves be assimilated, or may be toxic in amounts that would be needed for efficient buffering. Hydrogen ion concentration is kept constant in industrial fermenters by the continuous monitoring of pH and the automated addition of acid or alkali. The application of such procedures with laboratory fermenters is essential for adequate study of the effect of pH on fungal growth.

Hydrogen ion concentration in a medium could affect growth either indirectly by its effect on the availability of nutrients or directly by action on the cell surfaces. It is probable that if nutrient requirements are satisfied most fungi will grow well over a broad pH range on the acid side of neutrality, e.g. pH 4–7. Some fungi, such as those which produce organic acids, are able to tolerate considerably more acid conditions.

Temperature

In natural environments, fungi can be exposed to a wide range of temperatures, with daily and seasonal variations, so active growth may not always be possible. In some seasons water may be unavailable either as a result of drought or of freezing, and at other times suitable substrates may be lacking. Hence a great many fungi have a maximum temperature for growth of 30–40°C and a minimum a few degrees above the freezing point of water. Fungi that can grow near freezing point or even a little below are termed **psychrotolerant** (from the Greek, psychros, cold), and if incapable of growth above 20°C, **psychrophilic**. A group of fungi, the snow moulds, grow on vegetation such as grass or unharvested crops when these are buried by snow. These can cause major problems if they are killing the grass or producing mycotoxins, such as *Fusarium* species producing trichothecene T-2 (Chapter 7). Many cold-tolerant yeasts are known and some of them, including some Basidiomycete yeasts, are psychrophilic. Yeasts are among the few microorganisms found in the cold dry valleys of Antarctica. Temperatures well above 40°C are often encountered in accumulations of decomposing vegetation, such as composts, and fungi such as *Mucor pusillus, Chaetomium thermophile* and *Thermoascus auranticus*, play an important role in the succession of organisms involved in successful composting. Fungi thriving in such habitats are termed **thermotolerant** if they are capable of growth at 50°C or more, and **thermophilic** if incapable of growth below 20°C. The highest temperature at which fungal growth has been recorded is *ca* 60°C. The vast majority of fungi that are neither psychrophilic nor thermophilic are sometimes termed **mesophiles,** favouring intermediate temperatures. Some mesophiles are, as indicated above, psychrotolerant or thermotolerant.

Temperature affects the growth parameters of lag time, specific growth rate and total yield differently. Consideration of the optimal temperature for growth for a species hence requires separate consideration of the way that the range of temperatures between the maximum and minimum for growth influences these parameters.

Specific growth rate

The specific growth rate of fungi, like that of other microorganisms, is very low near the minimum temperature at which growth is possible and increases steadily through most of the range at which growth can occur, until an optimum temperature is reached at which the growth rate is greatest. Through most of this range the growth rate approximately doubles for a 10°C rise in temperature, i.e. has a temperature coefficient for a 10°C increase in temperature (Q_{10}) of about 2. Such a temperature coefficient is characteristic of enzymic reactions, on which growth will depend. Above the optimal temperature the specific growth rate declines rapidly, the maximum temperature for growth being only a few degrees above the optimum. The basis for this growth failure, at what are doubtlessly unnaturally high temperatures for the species concerned, may well vary between species. Likely causes include the denaturation of one or more key enzymes and the breakdown of metabolic regulatory mechanisms.

Radial growth rate

The rate at which the hyphae at the margin of a fungal colony advance across a surface is dependent on the width of the peripheral growth zone and the specific growth rate within the zone (page 141). It has been demonstrated in a range of fungi that the width of the peripheral growth is unaffected by temperature. Hence the radial growth of a fungus will respond to temperature in a manner similar to the specific growth rate. Since radial growth rates can be determined much more rapidly than specific growth rates, probably the most detailed study of the effect of temperature on fungal growth is one for the growth of a fungal colony on an agar surface (Fig. 3.22A).

The Arrhenius equation and fungal growth

The rate (K) of a chemical reaction is related to the absolute temperature (T) by the Arrhenius equation (15), in which R is the gas constant, E the energy of activation of the reaction and A a third constant dependent upon the frequency of formation of activated complexes of the reactants.

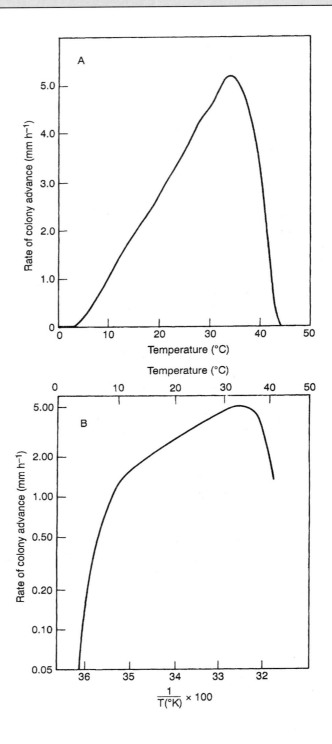

$$K = A\ e^{-E/RT} \tag{15}$$

Differentiation of this equation yields equation 16 from which equation 17 can be obtained by converting into logarithms to base 10.

$$\ln K = \ln A - E/RT \tag{16}$$
$$\log K = \log A - E/2.3RT \tag{17}$$

It will be seen that a linear relationship will result from plotting the logarithm of the reaction rate against the reciprocal of the absolute temperature.

An Arrhenius plot of the growth rate of a microorganism (Fig. 3.22B) may yield a linear relationship over a range of sub-optimal temperatures, suggesting that the growth rate in this range is governed by a single rate-limiting biochemical reaction. Inflexion points in an Arrhenius plot suggest that at the corresponding temperature a change occurs with respect to which reaction is rate limiting. It has been shown that in the bacterium *Escherichia coli* slopes and inflection points are dependent on nutritional conditions, a conclusion that also appears to be correct for *Neurospora crassa*.

Metabolism
Temperature influences primary and secondary metabolism in a variety of ways. At high temperatures an increase in nutritional requirements is sometimes observed. Thus *Saccharomyces cerevisiae* requires a supply of pantothenic acid at 38°C but not at 30°C. Temperature-sensitive mutations in which nutritional or other defects are apparent at high temperatures are well known. Such mutations are valuable for investigating many aspects of cell biology, as they can be grown at permissive temperatures and transferred to and from restrictive temperatures to observe the consequences of the defect. Metabolic regulation may break down

---

Figure 3.22 Radial growth rate of colonies of the Ascomycete *Neurospora crassa* on agar medium at a range of temperatures. The data are from the study by Ryan, F. J., Beadle, G. W. & Tatum, E. L. (1943). *American Journal of Botany* 30, 784–799. Petri dishes are inconveniently small for investigating the radial growth rate of this very rapidly growing fungus and long horizontal tubes with the lower half of the cross-section of the tube filled with agar were used. Since the width of the peripheral growth zone of fungal colonies does not vary with temperature, the figure illustrates also the way in which temperature influences specific growth rate. The study, although 60 years old, is perhaps the most detailed on the effect of temperature on the growth rate of a fungus. A, Rate of colony advance (mm h^{-1}) plotted against the temperature (°C). B, Arrhenius plot of the same data. The abscissa as before represents the rate of colony advance in mm h^{-1} but the scale is logarithmic. The lower ordinate indicates the reciprocal of the temperature (T) on the absolute scale (°K) but with the figures multiplied by 100 for convenience. The upper ordinate gives the temperature in °C; the spacing between the 10°C intervals diminishes because of the reciprocal plot. Growth rate is maximal at 33–36°C and at sub-optimal temperatures inflections in the linear relationship between temperature and growth rate occur at *ca* 12 and 5°C. The differing slopes can be interpreted as indicating ranges in which growth rate is determined by different reactions. A study of the specific growth rate of another strain of *Neurospora crassa* in liquid on a different medium demonstrates an optimum at 35–37°C and an inflection in the Arrhenius plot at 20°C.

at sub-optimal temperatures, an effect that can have useful consequences in fermentation technology. The Ascomycete *Ashbya gossypii*, for example, which will grow well at 37°C, at 28°C shows an industrially useful overproduction of riboflavin.

Total yield

The main determinant of total yield is the amount of limiting nutrient, so approximately the same total yield is to be expected throughout the temperature range in which efficient growth occurs. Maximum growth will, however, be achieved sooner at those temperatures at which the specific growth rate is high.

Lag time

High temperatures result in high metabolic rates, so the shortest lag times are to be expected in the upper part of the temperature range permitting growth.

Light

There are many reports that illumination will increase or more commonly reduce the rate at which fungi spread across an agar surface. Such effects are sometimes due to the photochemical destruction of components of the medium but in other instances a direct effect on metabolism seems likely. A common metabolic effect of light is the induction of carotenoid biosynthesis. This results in the mycelium of *Neurospora crassa*, *Fusarium aquaeductuum* and various other fungi being orange if grown in the light but colourless if cultured in darkness. The mycelium of *Phycomyces blakesleeanus* contains carotenoids even in darkness, but the amount is increased by exposure to light. It is the blue end of the visible spectrum and ultraviolet wavelengths that promote carotenoid biosynthesis, and comparison of the effectiveness of different wavelengths suggests that the photoreceptor involved is commonly a flavin. Fungal cells contain many flavoprotein enzymes which are potential photoreceptors. The biosynthesis of pigments other than carotenoids as a consequence of light action has also been demonstrated. In the mitosporic fungus *Aureobasidium pullulans*, for example, production of the black pigment melanin is stimulated by light.

Many key compounds in cell metabolism, such as porphyrins, absorb light. Light absorption by such compounds can result in the production of toxic, highly reactive forms of oxygen, which then can oxidize cell constituents. In bacteria it has been shown that carotenoids serve a protective role by inactivating the reactive oxygen species. It is probable that fungal carotenoids act in a similar way. Their frequent induction by exposure to light suggests such a possibility. Melanins, which are black or dark brown pigments, commonly occurring as wall components of spores, such as zygospores of *Mucor*, ascospores of *Neurospora* and basidiospores of *Agaricus*, have a protective role against photochemical damage and also against enzymic and chemical attack.

A number of fungi, mostly Basidiomycetes, show the property of bioluminescence. Examples include fruit bodies and mycelia of some *Panellus*,

Pleurotus, *Omphalotus* and *Mycena* species, and young rhizomorphs of *Armillaria mellea*. It has been suggested that luminescence in fruit bodies serves to attract insects to aid spore dispersal, but here and in mycelia and rhizomorphs the light may activate DNA repair, as has been shown in luminescent bacteria, or the oxidative process of generating the light may de-toxify reactive oxygen.

Vegetative Multihyphal Systems

A spreading, branching mycelium of individual hyphae is, as already discussed (page 140), highly effective in draining a substratum of nutrients. When this is completed, the fungus must reach other sources of nutrients or survive where it has grown until changed circumstances provide a further supply of nutrients. The vegetative mycelium is not well adapted for accomplishing either of these objectives. Growth through areas lacking nutrients would soon be terminated by nutrient exhaustion, and in an inactive state solitary hyphae could soon be destroyed by desiccation, insect attack or lysis by other microorganisms. Many fungi can reach fresh substrates or survive in a dormant condition by means of one or more types of spore (Chapter 4). Some, however, can advance through inhospitable regions by means of **mycelial strands** (sometimes called mycelial cords) or **rhizomorphs**, or survive in a dormant state as **sclerotia** (sing. **sclerotium**) (Table 3.10). Such multihyphal vegetative structures are readily visible to the naked eye. They are limited to the Ascomycetes, Basidiomycetes and mitosporic fungi, since septate hyphae are required for the construction of these elaborate structures. They are further limited, through the need for large amounts of nutrients for their production, to fungi with access to a massive food source, such as those that attack living plants or timber. The occurrence of mycelial strands, rhizomorphs and sclerotia in such fungi, many of which also produce spores, indicates that for these fungi such multihyphal vegetative structures fulfil some roles more effectively than do spores.

Mycelial Strands

The most intensively studied mycelial strands are those of *Serpula lacrymans* (Fig. 6.16A, B), the Basidiomycete that causes the destructive dry rot of the woodwork of buildings. The mycelium of this fungus will spread out from a food base that it has colonized, such as a block of wood, over the chippings of porous flower pots or other non-nutrient substrates. At the advancing edge of the mycelium the hyphae grow separately from each other, but a few centimetres behind the margin some main hyphae become invested by others, forming mycelial strands. Further back hyphae not associated with strands die and undergo lysis, and the nutrients released are thought to be assimilated by the strands. Such strands can continue to grow and differentiate for many months. Some of the hyphae within the strands autolyse leaving wide channels and others produce large amounts of extracellular material which ultimately makes up most of the volume of strand. In

Table 3.10 Examples of exogenous stimuli which induce initiation of multihyphal structures

Structure	Fungus	Stimulus
Mycelial strands[a]	*Serpula lacrymans*	High C/N ratio in substratum Inorganic nitrogen source Bridging between nutrient resources Presence of competing fungi Lowered water potential
Rhizomorphs	*Armillaria mellea*	Critical C/N ratio in substratum Ethanol Indoleacetic acid Aminobenzoic acid
Synnemata (coremia)[b]	*Penicillium claviforme* *Penicillium isariiforme*	Glutamic acid in substratum Light
Sclerotia[c]	*Sclerotium rolfsii*	Critical C/N ratio Threonine Lactose
	Coprinus cinereus	Ammonium
	Morchella esculenta	Bridging between poor and rich nutrient resources
Fruiting bodies[b,c]	*Coprinus cinereus*	Dark → light transfer 30 → 20°C temperature change
	Lentinus edodes	Proteinase inhibitors in substratum Dark → light transfer Temperature drop to 16°C
	Agaricus bisporus	Temperature drop Removal of carbon dioxide

[a] Also termed mycelial cords.
[b] See Chapter 4.
[c] See also Table 4.2.

the outer layers hyphae with very thick, brown walls develop to give a tough outer region. When a strand reaches a new food base a mycelium of individual hyphae is once more produced.

The strands of *Serpula lacrymans* often extend long distances through dry areas, behind plaster or over brickwork, from a food base of rotting wood to attack hitherto sound woodwork. It is probable that osmotic uptake of water at the food base generates a high hydrostatic pressure permitting mass flow of water to occur along the channels of the strand. The tough outer regions of the strand enable the hydrostatic pressure to be resisted, and may also limit water loss and protect against insect attack. The continuing growth of mycelium ahead of the strand in non-nutrient conditions implies that nutrients as well as water must be transported. Strands are initiated under conditions of nitrogen limitation and during their formation it is likely that they retrieve amino acids from the lysing non-stranded hyphae. They may thus have a role in nitrogen conservation as well as in translocation.

Many other Basidiomycetes produce mycelial strands. They are common in the soil and leaf litter of woodlands, where they are abundantly produced by many saprotrophic wood-decomposing fungi such as *Phanerochaete velutina* (pages 316–319, Fig. 6.17) In the cultivated mushroom, *Agaricus bisporus* (Fig. 1.1G) they are produced prior to fruiting. Root-infecting fungi that attack shrubs and trees often produce mycelial strands, as do many ectomycorrhiza. It is probable that numerous hyphae nourished by a mycelial strand are capable of infection when solitary, possibly ill-nourished hyphae or a germ-tube from a spore would be resisted successfully by the host. Mycelial strands vary in their mode of development and may have different roles in different species.

Rhizomorphs

The term rhizomorph means 'having the form of a root' and the rhizomorphs of the most thoroughly studied species, the honey fungus *Armillaria mellea* (synonym, *Armillariella mellea*) simulate plant roots closely. There is a 'meristem', a region of active cell division, about 25 μm behind the tip. The region inside the cylinder is known, by analogy with roots, as the medulla, and that outside as the cortex. In front of the 'meristem' is a region corresponding to a root cap protecting the meristem from damage by friction against soil particles as the rhizomorph advances. Considerable cellular differentiation occurs in both cortex and medulla as the rhizomorph matures. Further back a central lacuna develops which probably facilitates gas exchange and permits growth in regions of low oxygen tension. The rhizomorphs of *A. mellea* can advance at about five times the speed of the mycelium.

Armillaria is an important root pathogen of trees and shrubs, and rhizomorphs facilitate spread from diseased roots to those of adjacent healthy plants. Rhizomorphs, which are limited to a few Basidiomycetes, have roles similar to those of mycelial strands. They differ, however, from mycelial strands in that they advance through the soil-like roots instead of developing in the wake of an advancing mycelium of individual hyphae.

Sclerotia

Sclerotium production is widely distributed in Ascomycetes, Basidiomycetes and mitosporic fungi, and is common in species that attack herbaceous plants. Sclerotia are usually approximately spherical in form and from about a millimetre to a centimetre in diameter, depending on species. Sclerotia in the mitosporic fungal genera *Botrytis* and *Sclerotium* develop as a result of the repeated dichotomous branching of a hypha, accompanied by the production of numerous septa within the hyphae. Developmental details vary in other genera, but all involve copious hyphal branching and septation. Sclerotium production is initiated by the onset of starvation conditions or other circumstances unfavourable for continued mycelial growth. Sclerotia show cellular differentiation with a central medulla of thin-walled cells with abundant lipid and glycogen reserves, surrounded by a cortex of cells having thicker walls. Outside there may be a rind consisting of dead cells with walls impregnated with melanin.

The sclerotia of root-infecting fungi develop in the soil and those of fungi that attack the aerial parts of plants fall to the ground when mature or when the plant dies. The elongated sclerotia of *Claviceps*, called ergots (Fig. 8.17), for example, are produced in the flowers of grasses (page 54) and fall to the ground in autumn, where they survive until grasses are flowering again in late spring. Sclerotia permit prolonged survival of a plant pathogen in the soil in the absence of the host plant; strains of a fungus which lack sclerotia disappear from the soil more rapidly than do those capable of sclerotium production.

Sclerotium germination is induced in different ways in different species. Some germinate as a result of the return of appropriate conditions of light and temperature, some through exposure to nutrients, and others as a result of stimulation by substances emitted by a host plant. The sclerotia of *Sclerotium cepivorum*, which infects onion plants, are induced to germinate by organic sulphur-containing compounds (aliphatic thiols) produced by onion roots. Sclerotia of some Ascomycetes and Basidiomycetes germinate to produce the sexual stage in the fungus life cycle. In *Claviceps*, for example (Fig. 2.17), the germinating sclerotia bear structures (stromata) with fruiting bodies from which ascospores are discharged (page 54). Germination often leads to asexual sporulation, the sclerotia of *Botrytis cinerea* being capable of producing several crops of conidia. The sclerotia of root-infecting fungi such as *Sclerotium cepivorum* germinate to produce mycelium which then infects adjacent host roots.

The formation of sclerotia, resting structures which remain at the site of production, is an effective strategy for the survival of fungi infecting herbaceous plants. An area that has been suitable for a host species is likely to remain so, and may well have been sown, naturally or by humans, with seeds of the host species. Therefore the optimum tactic for once again infecting the host species may well be to stay put. Sclerotia that normally germinate with mycelium production will then infect the host species when it reappears. Those that germinate with spore production will release massive numbers of these propagules at the time of year most appropriate for infecting the host plant.

The Prevention of Fungal Growth

Fungi are important agents of biodeterioration (Chapter 6), causing the spoilage of food, timber and a wide range of other products, both in storage and in use. They are a major cause of disease in plants, causing heavy crop losses, a subject considered further in the context of plant pathology (Chapter 7). They also cause diseases of man and animals – see pages 427–444 (Chapter 7). The immense

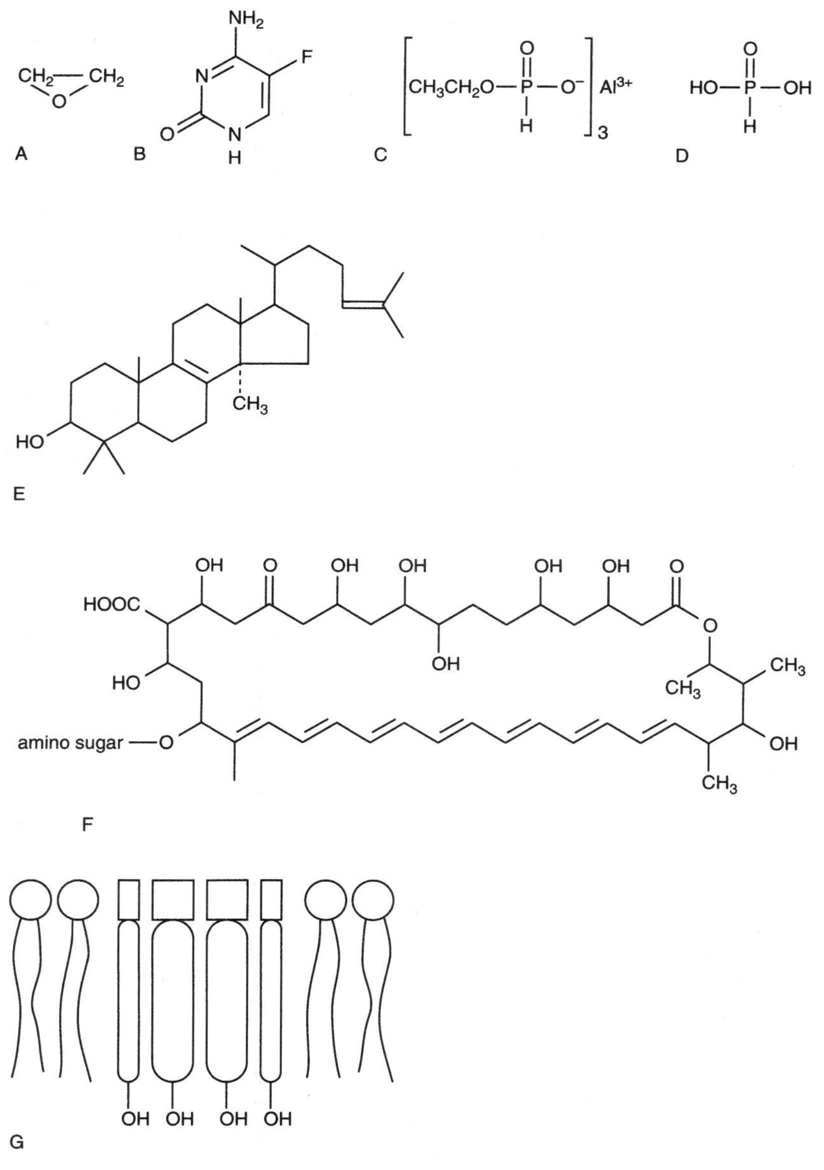

Figure 3.23 Antifungal agents and their mode of action. A, Ethylene oxide. B, 5-fluorocytosine. C, Fosetyl-Al. D, Phosphonic acid. E, Lanosterol, a precursor of ergosterol. The only methyl group indicated by lettering is that which is removed by the C-14 demethylase which is the target for imidazole antifungal agents. F, Amphotericin B. G, Suggested model for the interaction of amphotericin B molecules with the fungal plasma membrane by forming an annular pore spanning a phospholipid layer. The polar heads of the phospholipid molecules are shown as circles and the polar amino sugar groups of the polyene molecules as squares, both at the outer surface of the plasma membrane. It has been suggested that the terminal hydroxyl groups of a second inverted annulus in the inner phospholipid layer may associate with the annulus spanning the outer layer. This would give a pair of annuli and an aqueous pore spanning the entire double-layered plasma membrane.

financial losses and suffering resulting from these activities of fungi mean that the prevention of fungal growth is a problem of great practical importance. Fungal growth can be prevented by sterilizing materials and then maintaining sterility, by manipulating the environment to obtain conditions unfavourable for fungal growth, and by the use of antifungal agents. These approaches will be considered below. A further approach to restricting the growth of harmful fungi, biological control, is considered in Chapter 7.

Sterilization

Surgical equipment, pharmaceutical preparations and media for the pure culture of microorganisms must be free from all viable microorganisms. They also have to be free from any toxic chemicals, so the agents used for sterilization must have no residual activity. Hence, following sterilization access by microorganisms must be prevented.

Heat is widely used for the sterilization of surgical equipment, food in the canning industry, and culture media in both the laboratory and fermentation industry. The most heat-resistant of microbial cells, bacterial spores, have to be destroyed, and this needs a temperature of about 120°C, obtained by autoclaving. The most heat-resistant of fungal cells, such as the ascospores of *Byssochlamys fulva*, are destroyed at 90°C, so all fungi are destroyed by a heat treatment adequate to eliminate other microorganisms. Materials can also be sterilized by the more penetrating forms of **ionizing radiation**, such as **gamma radiation** from a radioactive cobalt installation, an expensive facility but appropriate for large industrial operations. Gamma radiation is used in the production of disposable plastic Petri dishes and other sterile equipment, and increasingly in the food industry. The most radiation-resistant organisms known are among the bacteria, so a dosage adequate to eliminate bacteria will also destroy fungi. Volatile or gaseous **chemical sterilants** are sometimes used for equipment or materials that would be damaged by heat. They inevitably produce some chemical changes in organic materials, since otherwise they would be unable to kill microorganisms. Ethylene oxide (Fig. 3.23A) and propylene oxide act by the alkylation of carboxyl, hydroxyl, amino and sulphydryl groups, thus altering proteins and other organic compounds. **Membrane filtration** is a sterilization method that is used in the laboratory and pharmaceutical industry with solutions that are heat sensitive and when any chemical change must be avoided. A pore size that prevents passage of bacterial cells will obviously exclude the much larger fungal cells.

Provision of Conditions Unfavourable for Growth

Many fungi grow poorly or not at all in the absence of oxygen (page 146). Hence **oxygen deprivation** can sometimes be used to prevent fungal growth. Thus if grain is stored in sealed containers with little gas space, its metabolism lowers the oxygen concentration and prevents the growth of fungi that cause spoilage. Low water activity (page 166) will prevent the growth of most fungi. Hence avoiding dampness is crucial in preventing the decay of timber and damage to other products by fungi.

Drying has long been used in food preservation, as has the provision of low water activity through the addition of sugar, as in the making of jam. Low temperature (page 164) delays fungal growth, although a few fungi will grow slowly even at very low temperatures, causing discoloration of meat in cold stores and spoilage of food in refrigerators.

Antifungal Agents

A wide range of antifungal agents are used in combating biodeterioration and in preventing or treating fungal diseases of plants. In these contexts they are commonly referred to as fungicides. Others are used for treating disease in animals and humans, and are referred to as antifungal drugs or, if produced by means of a microbial fermentation, as antifungal antibiotics. Antifungal agents differ widely in their chemical nature and in their properties and mode of action. Examples illustrating this will be considered below.

Agents of low biochemical specificity

Some antifungal agents are of low biochemical specificity, acting upon many cell constituents and processes. Since it is unlikely that any organism will lack all these constituents and processes, most species will be potentially vulnerable to such an agent. This low species specificity does not matter in some contexts, and can even be advantageous. For example, it will be beneficial if a preservative applied to wood to prevent fungal decay also prevents insect attack. Fungicides of low biochemical specificity are also dusted or sprayed on plants to prevent fungus attack, resulting in spores arriving at the plant surface being poisoned. Fungicides of low specificity include the salts and organic derivatives of various metals, inorganic forms of sulphur and some organic compounds. Some examples will be considered here.

Elemental sulphur was one of the first fungicides to be introduced, and is still used to prevent commercially important powdery mildew infections, such as that of the grape vine. The plants can be dusted with finely divided powder or sprayed with colloidal suspensions or other sulphur formulations. Sulphur can accept electrons from cytochrome b, thus interfering with electron transport and the supply of energy available to the fungal cell. In so doing the sulphur becomes reduced to the very toxic hydrogen sulphide. This can produce insoluble sulphides from the metals which act as co-factors for many enzymes, thus affecting metabolism in many different ways.

Sulphur dioxide is widely used to prevent spoilage of wines and fruit juices by yeasts and moulds. It may be added as sulphur dioxide itself, or as salts of sulphite or bisulphite. All give, in solution, a mixture of sulphur dioxide and sulphite (SO_3^{2-}) and bisulphite (HSO_3^-) ions, their ratio depending on the pH of the solution. In *Saccharomyces cerevisiae* and some other yeasts it is sulphur dioxide itself that diffuses into the cell, but in *Candida utilis* and some moulds it is sulphite that is transported into the cell by the system responsible for sulphate uptake. Sulphur dioxide, sulphite and bisulphite are all chemically reactive, and able to bring about a wide variety of chemical changes in cell components, interfering with metabolism and causing cell death. Sulphur dioxide at the low concentrations used in foods and beverages has hitherto been regarded as harmless to humans, but its ability to interact with nucleic acid components such as cytosine may lead to limitations on its use.

Bordeaux mixture is one of the earliest fungicides, but is still used to protect plants from a range of fungal diseases. It is made by mixing solutions of copper sulphate and calcium hydroxide. When sprayed on to plants it forms a deposit from which soluble fungicidal copper compounds are slowly released. These penetrate fungal spores, and bind to amino, carboxyl and thiol groups on proteins, inhibiting enzyme action. Other inorganic copper fungicides that act in a similar way are also used.

Weak acids, such as sorbic acid, benzoic acid and acetic acid, are widely used as preservatives for foods and miscellaneous domestic products. The theory is that these lipophilic, undissociated acids diffuse through the plasma membrane and dissociate in the cytoplasm, liberating substantial concentrations of protons, leading to cell death through acidification. This appears to be the mechanism for benzoic and acetic acids, but sorbic acid, an unsaturated fatty acid, with two conjugated double bonds, appears instead to act by causing membrane damage leading to leakage of potassium ions and increased permeability to protons, in a similar fashion to the polyene antifungal drugs (page 179), but lacks their specificity.

As mentioned earlier, a low level of specificity has advantages in fungicides for preventing biodeterioration. There are, however, disadvantages in low specificity in fungicides for preventing fungal diseases of plants. Such fungicides are often not very potent and large dosages are often needed. The lack of specificity may also result in some damage to plants when application is excessive or unskilled. Moreover, some fungicides of low specificity have considerable toxicity for humans and higher animals. Mercury compounds were at one time widely used for treating seeds prior to sowing to prevent attack by *Pythium* and other fungi during seed germination. However, the toxicity of mercury and its persistence in the environment has led to its disuse as a seed dressing in many countries. Similar problems exist with some other fungicides of low specificity.

Fungicides of low specificity, however, have one advantage over fungicides of high specificity. The latter often lose much of their effectiveness within a few years of being introduced whereas the low specificity fungicides do not. The effectiveness of high specificity fungicides is usually due to interaction with a single fungal enzyme, and a single mutation may confer resistance on the fungus. In the large fungal populations involved in plant disease epidemics, such a mutation and the spread of a resistant fungal strain is very probable. However, when fungicidal activity is due to interaction with many proteins, as with a fungicide of low specificity, many mutations may be needed to confer resistance. When this is so, the development of resistance in the fungus will be improbable. Elemental sulphur and Bordeaux mixture have retained their effectiveness through over 100 years of use.

Agents of high biochemical specificity

Many of the antifungal agents now available are very effective against fungi but have little adverse effect on animals and plants. Such agents are of high biochemical specificity, interacting with a single cell component or biochemical process. Their selectivity can result either from the target being present only in fungal cells, or from the agent being converted into an active form only within the fungal cell. Some selective antifungal agents have been discovered by screening microorganisms for antibiotic activity. Many others have resulted from the synthesis and testing of a large number of compounds, usually structurally related to compounds that have already shown antifungal activity. A few have been designed to take advantage of some way in which the biochemistry of fungi differs from that of other organisms. This last approach is likely to be increasingly used as knowledge of fungal biochemistry deepens. So far, however, an understanding of the mode of action of an antifungal agent has usually been obtained only after it has been in use for some years.

Fungicides of low biochemical activity have to be used as **contact fungicides** which, for as long as they persist on the plant surface, protect the plant from fungal attack. Some agents of high specificity are also used as contact fungicides, but others can be used as **systemic fungicides**. These are taken up and translocated throughout the plant, suppressing established fungal disease as well as preventing fresh infection. Clinically useful antifungal drugs also have to be of high specificity, so that they can be administered orally or by intravenous injection and reach the site of fungal infection without harming the patient. Treatment of fungal diseases is dealt with in Chapter 7, pages 438–440. The examples of antifungal agents that follow illustrate varied modes of action.

Inhibitors of mitosis Fungal mitosis and hence growth is inhibited both by the antibiotic **griseofulvin** and the unrelated synthetic **benzimidazoles**. Both act by preventing tubulin polymerization and microtubule elongation in the nucleus. Griseofulvin, given orally, is used in treating infections by the dermatophytes *Trichophyton* and *Microsporum*. The benzimidazoles, such as **benomyl**, are used for preventing or treating plant diseases caused by a wide range of fungi.

Inhibitors of nucleic acid synthesis The synthetic drug **5-fluorocytosine** (5-FC; Fig. 3.23B) is given orally to treat *Candida* and *Cryptococcus* infections. It is transported into the fungal cell by cytosine permease, and then converted into 5-fluorouracil (5-FU) by cytosine deaminase. The 5-FU is then converted by a series of enzymic steps into 5-fluorodeoxyuridylate. This inhibits the action of thymidylate synthase, a key enzyme in the synthesis of thymidine and hence of DNA. Mammalian cells lack cytosine deaminase, so it is only within the fungal cells that the 5-FC is converted into 5-FU. Hence 5-FC has very little adverse effect on the patient. Unfortunately fungal resistance to 5-FC arises frequently through mutations bringing about the loss of one or other of the sequence of enzymes that convert 5-FC into 5-fluorodeoxyuridylate.

The **acylalanines**, for example metalaxyl, are synthetic fungicides the effectiveness of which is almost wholly limited to the Oomycetes. This group of fungi, however, is responsible for some of the most important plant diseases. In Oomycetes the acylalanines inhibit RNA polymerase I, the enzyme responsible for the production of ribosomal RNA. They do not prevent sporangium germination, zoospore activity, cyst formation and cyst germination, as these activities need no additional ribosomes. The inhibition of ribosome production, however, renders these fungicides very effective against the rest of the life cycle of sensitive species. Unfortunately fungal resistance, apparently through a change in the RNA polymerase, arises frequently.

Inhibitors of protein synthesis The antibiotic **kasugamycin** is used to treat rice blast, caused by *Pyricularia oryzae*. It prevents the binding of aminoacyl t-RNA to the ribosome by itself binding to the ribosome. The recently discovered **sordarins** (Chapter 8, page 520) are showing promise for development as antifungal drugs. They specifically inhibit elongation factor 2 of the fungal protein synthetic machinery.

Inhibitors of respiration The **carboxamides**, for example carboxin, are effective against Basidiomycete plant pathogens such as rusts and smuts. They enter the mitochondria and, interacting with the succinate–ubiquinone oxidoreductase complex, inhibit electron transfer between succinate and coenzyme Q. The **strobilurins** (Chapter 8, page 524) have quickly established a major place as fungicides for control of cereal diseases. Their mode of action is by specific inhibition of fungal respiration by interacting with the cytochrome $bc1$ segment of the respiratory chain.

Inhibitor of phosphate metabolism **Fosetyl-Al**, the aluminium salt of ethyl phosphonate (Fig. 3.23C), has a fungicidal activity limited to the Oomycetes, which, however, include some of the most important plant pathogens. It penetrates plants readily, after which it is transported both upwards in the xylem and downwards in the phloem. Efficient downward transport of an applied compound is unusual, and helps make fosetyl a very effective systemic fungicide. It seems likely that fosetyl is hydrolysed within the fungus releasing phosphonic acid (Fig. 3.23D). This, it is thought, interferes with Oomycete phosphate metabolism, which has unusual features, such as the presence of unique phosphoglucans and the absence of the polyphosphate present in most other organisms.

Inhibitors of ergosterol synthesis Ergosterol is an essential component of the plasma membrane in all fungi other than Oomycetes. There are a large number of antifungal agents which act by preventing ergosterol synthesis, mostly by inhibiting demethylation at carbon atom 14 of lanosterol (Fig. 3.23E), a precursor of ergosterol. Among such antifungal agents are the **imidazoles**, for example imazalil which is used against plant pathogens, and the clinically important clotrimazole and ketoconazole and **triazoles**, for example metaconazole used against plant pathogens, and fluconazole, itraconazole and voriconazole used medically (page 439, Chapter 7). These agents interact with cytochrome P-450, an essential part of lanosterol 14-demethylase, inhibiting the enzyme. C-14 demethylation is also an essential step in the synthesis of cholesterol, the sterol of animal plasma membranes, and of the sitosterol and stigmasterol of plant plasma membranes. However, the P-450 cytochromes of the animal and plant C-14 demethylases are much less sensitive to imidazoles and to similar inhibitors. Other drugs and agricultural fungicides affecting this pathway are the **phenylmorpholines**, such as amorolfine, which inhibit Δ_{14}-reductase and Δ_7-Δ_8-isomerase steps in ergosterol biosynthesis. All of these antifungal agents lead to the accumulation of 'incorrect' precursor sterols in the fungal membranes, with disruption of growth, and for example abnormal deposition of chitin in the cell walls. Their action is mostly **fungistatic**, i.e. stopping fungi growing, as opposed to **fungicidal**, i.e. killing them. Although less desirable than fungicidal agents, fungistatic ones can still be of great therapeutic value, as the host's defence mechanisms are given a much greater chance of overcoming the infection. Other drugs affecting ergosterol biosynthesis that are of clinical use are the **allylamines**, such as terbinafine, (page 439, Chapter 7) and the **thiocarbamates**, such as tolnoftate, which inhibit the first step of the sterol pathway from squalene, squalene epoxidase, so that no sterols are produced. These are fungicidal, and are particularly active against dermatophytes.

Disruptors of plasma membrane function **Nystatin**, named from its discovery in the New York State (NY State) Public Health Laboratories, was the first **polyene macrolide** antibiotic. It is used topically against dermatophytic fungi and orally against fungal infections in the gut. It is not absorbed from the gut, and is too toxic for injection. The most valuable of the many polyene antibiotics subsequently discovered, **amphotericin B**, is given by intravenous injection to treat systemic fungal infections. It is highly toxic to a wide range of pathogenic fungi, i.e. it is a broad-spectrum antifungal agent, and it is fungicidal. Its biggest problem is its toxicity, and measures taken to combat this are discussed later (page 439, Chapter 7). The molecule of amphotericin B (Fig. 3.23F) is a rigid rod. One side of the macrolide ring is hydrophobic, with seven conjugated double bonds in the *trans* position, the other is hydrophilic with six hydroxyl groups. At one end of the molecule is an amino sugar, at the other end a hydroxyl group. The structure of nystatin is similar. The hydrophobic (polyene) portion of the molecule binds readily to ergosterol but much less readily to cholesterol. This accounts for the higher susceptibility of fungi than of mammalian cells to amphotericin B and nystatin. Some polyene antibiotics with fewer double bonds bind preferentially to mammalian cells and are of no clinical value. Amphotericin B and nystatin disrupt the plasma membrane in such a way that potassium ions are lost from the cell and protons enter, the latter causing acidification and death. It has been suggested that a side by side association of eight polyene molecules form an annulus (ring) that spans a single phospholipid layer of the plasma membrane (Fig. 3.23G), providing an aqueous pore through which ions pass, or that perhaps two such rings span the entire membrane. The **pradimicins** and related compounds are a group of antifungal antibiotics that appear to act via direct interaction with mannoproteins. For their fungicidal action, they require Ca^{2+}, apparently for binding to the cell and action on the cell membrane, with subsequent leakage of K^+, and cell death. They show promise in animal models of fungal infections.

Inhibitors of cell wall synthesis Chitin, $\beta(1\rightarrow3)$ glucan and mannoproteins are essential components of the cell walls of most fungi, and thus their syntheses are obvious targets for antifungal agents, in the way that peptidoglycan assembly is the most important target for the antibacterial antibiotics, such as the β-lactams. To date, however, this has been a disappointing area, with only one example in commercial use. The **polyoxins** and **nikkomycins** are antibiotics that inhibit chitin synthesis. They are nucleosides with a di- or tripeptide side chain and, through their configurational resemblance to UDP-N-acetylglucosamine, the substrate for chitin synthase, act as very potent and specific competitive inhibitors of the enzyme and are taken up into the cell by peptide permeases. They rapidly cause bursting of growing hyphal tips of a very wide range of fungi. Polyoxin is used in Japan to control fungal infections of plants, but resistance rapidly occurs, by acquisition of dipeptide permeases with altered specificities. The **echinocandins** and **papulocandins** and their semi-synthetic derivatives are antibiotics which specifically inhibit β(1→3) glucan synthase. They show excellent fungicidal activity, like polyoxins and nikkomycins, and cause bursting of hyphal tips of a range of fungi, but not of Zygomycetes, which lack β(1→3) glucan in their hyphal walls. Echinocandin derivatives should be available for therapeutic use in the near future, for treatment of infections by *Candida* species, *Cryptococcus neoformans* and *Pneumocystis carinii* (page 436).

Further Reading and Reference

General

Brambl, R. & Marzluf, G. A. eds. (1996). *Biochemistry and Molecular Biology*. Vol. III *The Mycota*, K. Esser & P. A. Lemke, eds. Springer-Verlag, Berlin.

Burnett, J. H. (1976). *Fundamentals of Mycology*, 2nd edn. Arnold, London.

Cooke, R. C. & Whipps, J. M. (1993). *Ecophysiology of the Fungi*. Blackwell, Oxford.

Gooday, G. W., Lloyd, D. & Trinci, A. P. J., eds. (1980). *The Eukaryotic Microbial Cell. Thirtieth Symposium of the Society for General Microbiology*. Cambridge University Press, Cambridge.

Gow, N. A. R. & Gadd, G. M., eds. (1994). *The Growing Fungus*. Chapman & Hall, London.

Gow, N. A. R., Gadd, G. M. & Robson, G. D., eds. (1999). *The Fungal Colony. Twenty-first Symposium of the British Mycological Society*. Cambridge University Press, Cambridge.

Griffin, D. H. (1994). *Fungal Physiology* 2nd edn. Wiley-Liss, New York.

Jennings, D. H. & Rayner, A. D. M., eds. (1984). *The Ecology and Physiology of the Fungal Mycelium. Eighth Symposium of the British Mycological Society*. Cambridge University Press, Cambridge.

Pringle, J.R., Roach, J.R. & Jones, E.W. (1997) *The Molecular Biology of the Yeast Saccharomyces – Cell Cycle and Cell Biology*. Cold Spring Habor Laboratory Press, New York.

Smith, J. E., ed. (1983). *Fungal Differentiation: a Contemporary Synthesis*. Marcel Dekker, New York.

Wessels, J. G. H. & Meinhardt, F., eds. (1994). *Growth, Differentiation and Sexuality*. Vol. I *The Mycota*, K. Esser & P. A. Lemke, eds. Springer-Verlag, Berlin.

Structure and Composition of Fungal Cells

Ashford, A. E. (1998). Dynamic pleiomorphic vacuole systems: are they endosomes and transport compartments in fungal cells? *Advances in Botanical Research* **28**, 120–159.

Bartnicki-Garcia, S. (1996). The hypha: unifying thread of the fungal kingdom. In B. Sutton, ed., *A Century of Mycology*, pp. 105–133. Cambridge University Press, Cambridge.

Broda, P., Oliver, S. G. & Sims, P. F. G., eds. (1993). *The Eukaryotic Microbial Genome: Organization and Regulation. Fiftieth Symposium of the Society for General Microbiology*. Cambridge University Press, Cambridge.

Buck, K. W. (1998). Molecular variability of viruses in fungi. In Bridge, P., Couteaudier, Y. & Clarkson, J., eds., *Molecular Variability of Fungal Pathogens*, pp. 53–72. CAB International, Wallingford, Oxon.

Gemmill, T. R. & Trimble, R. B. (1999). Overview of N- and O-linked oligosaccharide structures found in various yeast species. *Biochimica et Biophysica Acta* **1426**, 227–237.

Gull, K. W. & Oliver, S. G., eds. (1981). *The Fungal Nucleus. Fifth Symposium of the British Mycological Society*. Cambridge University Press, Cambridge.

Jedd, G. & Chua, N-H. (2000). A new self-assembling peroxisomal vesicle required for efficient resealing of the plasma membrane. *Nature Cell Biology* **2**, 226–231.

Kapteyn, J. C., Van Den Ende, H. & Klis, F. M. (1999). The contribution of cell wall proteins to the organization of the yeast cell wall. *Biochimica et Biophysica Acta* **1426**, 373–383.

Kershaw, M. J. & Talbot, N. J. (1998). Hydrophobins and repellents: proteins with fundamental roles in fungal morphogenesis. *Fungal Genetics and Biology* **23**, 18–33.

Kuhn, P. J., Trinci, A. P. J., Jung, M. J., Goosey, M. W. & Copping, L. G., eds. (1990). *Biochemistry of Cell Walls and Membranes in Fungi*. Springer-Verlag, Berlin.

Magliana, W., Conti, S., Gerloni, M., Bertolotti, D. & Polonelli, L. (1997). Yeast killer systems. *Clinical Microbiology Reviews* **10**, 369–400.

Markham, P. (1994). Occlusions of septal pores in filamentous fungi. *Mycological Research* **98**, 1089–1106.

McDaniel, D. P. & Roberson, R. W. (1998). γ-Tubulin is a component of the Spitzenkörper and centrosomes in hyphal-tip cells of *Allomyces macrogynus*. *Protoplasma* **203**, 118–123.

Osumi, M. (1998). The ultrastructure of yeast: cell wall structure and formation. *Micron* **29**, 207-233.

Ruiz-Herrera, J. (1992). *Fungal Cell Wall*. CRC Press, Boca Raton.

Smits, G. T., Kapteyn, J. C., Van Den Ende, H. & Klis, F. M. (1999). Cell wall dynamics in yeast. *Current Opinion in Microbiology* **2**, 348–352.

Wessels, J. G. H. (1997). Hydrophobins: proteins that change the nature of the fungal surface. *Advances in Microbial Physiology* **38**, 1–45.

Wickner, R. B. (1992). Double-stranded and single-stranded RNA viruses of *Saccharomyces cerevisiae*. *Annual Review of Microbiology* **46**, 347–375.

Growth of Cells, Populations and Colonies

Bartnicki-Garcia, S., Bartnicki, D. D. & Gierz, G. (1995) Determinants of fungal cell wall morphology: the vesicle supply centre. *Canadian Journal of Botany* **73** (Suppl. 1), S372–S378.

Burnett, J. H. & Trinci, A. P. J., eds. (1979). *Fungal Walls and Hyphal Growth. Second Symposium of the British Mycological Society*. Cambridge University Press, Cambridge.

Chant, J. (1996). Generation of cell polarity in yeast. *Current Topics in Cell Biology* **8**, 557–565.

DeMarini, D. J., Adams, A. E. M., Fares, H., De Virgilio, C., Valle, G., Chung, J. S. & Pringle, J. R. (1997). A septin-based hierarchy of proteins required for localized deposition of chitin in the *Saccharomyces cerevisiae* cell wall. *Journal of Cell Biology* **139**, 75–93.

Gooday, G. W. (1995). The dynamics of hyphal growth. *Mycological Research* **99**, 385–394.

Gooday, G. W. (1999). Hyphal interactions. In England, R., Hobbs, G., Bainton, N. & Roberts, D. M., eds., *Microbial Signalling and Communication. Fifty-seventh Symposium of the Society for General Microbiology*, pp. 307–334. Cambridge University Press, Cambridge.

Gregory, P. H. (1984). The fungal mycelium: an historical perspective. *Transactions of the British Mycological Society* **82**, 1–11.

Griffiths, A. J. F. (1992). Fungal senescence. *Annual Review of Genetics* **26**, 351–372.

Harold, F. M. (1999). In pursuit of the whole hypha. *Fungal Genetics and Biology* **27**, 128–133.

Herbert, D., Elsworth, E. & Telling, R. C. (1956). The continuous culture of bacteria: a theoretical and experimental study. *Journal of General Microbiology* **14**, 601–622.

Jennings, D. H. (1991). The spatial aspects of fungal growth. *Science Progress* **75**, 141–156.

Lew, D. J. & Reed, S. I. (1995). Cell cycle control of morphogenesis in budding yeast. *Current Topics in Genetics and Development* **5**, 17–23.

Mata, J. & Nurse, P. (1998). Discovering the poles in yeast. *Trends in Cell Biology* **8**, 163–167.

Poole, R. K., Bazin, M. J. & Keevil, W., eds. (1990). *Microbial Growth Dynamics. Special Publications of the Society for General Microbiology*, vol. 28. IRL Press, Oxford.

Wessels, J. G. H. (1999). Fungi in their own right. *Fungal Genetics and Biology* **27**, 134–145.

Xiang, X. & Morris, N. R. (1999). Hyphal tip growth and nuclear migration. *Current Opinion in Microbiology* **2**, 636–640.

Environment, Nutrients and Growth

Ayres, P. G. & Boddy, L., eds. (1986). *Water, Fungi and Plants. Eleventh Symposium of the British Mycological Society*. Cambridge University Press, Cambridge.

Banuett, F. (1998). Signalling in yeasts: an informational cascade with links to the filamentous fungi. *Microbiology and Molecular Biology Reviews* **62**, 249–274.

Blomberg, A. & Adler, L. (1992). Physiology of osmotolerance in fungi. *Advances in Microbial Physiology* **33**, 122–145.

Boddy, L., Marchant, R. & Read, D. J., eds. (1989). *Nitrogen, Phosphorus and Sulphur Utilization by Fungi. Fifteenth Symposium of the British Mycological Society*. Cambridge University Press, Cambridge.

Edwards, C., ed. (1990). *Microbiology of Extreme Environments*. Open University Press, Milton Keynes.

Gadd, G. M. (1993). Interaction of fungi with toxic metals. *New Phytologist* **124**, 25–60.

Garraway, M. O. & Evans, R. C. (1984). *Fungal Nutrition and Physiology*. Wiley, New York.

Howard, D. H. (1999). Acquisition, transport, and storage of iron by pathogenic fungi. *Clinical Microbiology Reviews* **12**, 394–404.

Jennings, D. H., ed. (1994). *Stress Tolerance of Fungi*. Marcel Dekker, New York.

Jennings, D. H. (1995). *The Physiology of Fungal Nutrition*. Cambridge University Press, Cambridge.

Marzluf, G. A. (1997). Genetic regulation of nitrogen metabolism in the fungi. *Microbiology and Molecular Biology Reviews* **61**, 17–32.

Poole, R. K. & Gadd, G. M., eds. (1989). *Metal–Microbe Interactions. Special Publications of the Society for General Microbiology*, vol. 26. IRL Press, Oxford.

Varela, J. C. S. & Mager, W. H. (1996). Response of *Saccharomyces cerevisiae* to changes in external osmolarity. *Microbiology* **142**, 721–731.

Vegetative Multihyphal Systems

Boddy, L. (1993). Saprotrophic cord-forming fungi: warfare strategies and other ecological aspects. *Mycological Research* **97**, 641–655.

Cairney, J. W. G. (1992). Translocation of solutes in ectomycorrhizal and saprotrophic rhizomorphs. *Mycological Research* **96**, 135–141.

Watkinson, S. C. (1999). Metabolism and differentiation in basidiomycete colonies. In N. A. R. Gow, G. D. Robson & G. M. Gadd, eds., *The Fungal Colony. Twenty-first Symposium of the British Mycological Society*, pp. 125–156. Cambridge University Press, Cambridge.

Willetts, H. J. & Bullock, S. (1992). Developmental biology of sclerotia. *Mycological Research* **96**, 801–816.

The Prevention of Fungal Growth

Buchenauer, H. (1990). Physiological reactions in the inhibition of plant pathogenic fungi. In G. Haug, ed., *Chemistry of Plant Protection*, vol. 6, pp. 217–292. Springer, Berlin.

Dixon, G. K., Copping, L. G.. & Hollomon, D. W., eds. (1995). *Antifungal Agents: Discovery and Mode of Action.* BIOS Scientific Publishers, Oxford.

Ghannoum, M. A. & Rice, L. B. (1999). Antifungal agents: mode of action, mechanisms of resistance, and correlation of these mechanisms with bacterial resistance. *Antimicrobial Agents and Chemotherapy* **12**, 501–517.

Kerridge, D. (1986). Mode of action of clinically useful antifungal drugs. *Advances in Microbial Physiology* **27**, 1–72.

Russell, A. D., Hugo, W. D. & Ayliffe, G. A. J., eds. (1992). *Principles and Practice of Disinfection, Preservation and Sterilization*, 2nd edn. Blackwell, Oxford.

Russell, P. E., Milling, R. J. & Wright, K. (1995). Control of fungi pathogenic to plants. In P. A. Hunter, G. K. Darby & N. J. Russell eds., *Fifty Years of Antimicrobials: Past Perspectives and Future Trends. Fifty-third Symposium of the Society for General Microbiology*, pp. 125–156. Cambridge University Press, Cambridge.

Trinci, A. P. J. & Ryley, J. F., eds. (1984). *Mode of Action of Antifungal Agents. Ninth Symposium of the British Mycological Society.* Cambridge University Press, Cambridge.

Questions

With each question one statement is correct. Answers on page 561.

3.1 Fungal viruses are transmitted by:
    a) release from infected cells by autolysis
    b) insect vectors
    c) hyphal anastomosis
    d) all of these

3.2 The chitin in the fungal cell wall is:
    a) a protein
    b) a glycoprotein
    c) a polysaccharide
    d) a lipopolysaccharide

3.3 In the cell cycle of *Saccharomyces cerevisiae*, the bud starts to be formed:
    a) before the spindle plaque has been duplicated
    b) soon after the spindle plaque has been duplicated
    c) soon after the chromosomes have been duplicated
    d) after the nucleus has divided

3.4 A fungal mycelium grows exponentially as a result of:
   a) increasing growth rates of its component hyphae
   b) branching of its component hyphae
   c) formation of aerial hyphae
   d) anastomosis of its component hyphae

3.5 The polyene antibiotics kill fungi by their interaction with:
   a) the plasma membrane
   b) the DNA
   c) protein synthesis
   d) microtubule assembly

Spores, Dormancy and Dispersal

4

The Role of Spores in Mycology and in the Life of the Organism

A suitable substratum or host plant may support the vegetative growth of a fungus for a considerable period. Ultimately, however, conditions at any site will become unsuitable for further growth. This may result from changed environmental conditions, the exhaustion of nutrients, the death of the host or alternatively, the host successfully resisting the further growth of the fungus. The fungus must then reach other sites suitable for growth, or survive in a dormant state until favourable conditions return. Fungal hyphae can advance slowly through the soil and hyphal fragments may get dispersed through the air, as may yeast cells. Furthermore, the vegetative cells of fungi are capable of surviving for a time in the absence of nutrients through a reduced metabolic rate and the utilization of reserve materials. Some fungi live in circumstances in which efficient transmission to an adjacent site can be achieved by mycelial strands or rhizomorphs, or alternatively, prolonged survival effected by sclerotia (pages 171–172). In most fungi, however, both survival in a dormant condition and dispersal are achieved through the production of spores, the process of **sporulation**.

Sporulation and Taxonomy

A great variety of forms occur among fungal spores, and the structures which bear them – **sporophores** (Greek, meaning spore carriers) – are also varied. Sporophore development is most complex when a sexual process is involved, and the resulting sporophores of higher fungi may be large, when they are commonly referred to as fruit bodies. Sporulation in fungi results in a complexity and diversity of form that contrasts with the relative simplicity and uniformity of structure found among vegetative cells. It is for this reason that taxonomists, concerned with the classification of fungi, have devoted much attention to spores and sporophores, basing fungal classification on the morphological details of sexual sporulation. When an isolate of a fungus fails to produce a sexual phase, the more limited information available from asexual sporulation permits assignment to a species within the mitosporic fungi (page 69). In the absence of sporulation, the presence of mycelial strands, rhizomorphs or sclerotia allows classification in the order Agonomycetes (page 69), but in their absence identification may be impossible, until appropriate molecular methods are developed. However, the practical importance of the yeasts, many of which do not spore, has led to yeast classification and identification based on the form of the vegetative cell and on fermentation tests as well as on the sporulation process.

Sporulation and Developmental Biology

Plants and animals have a life cycle that is obligate. Following fertilization, the zygote has to develop, through simultaneous growth and cellular differentiation, into an adult. Then, if the species is to survive, reproduction must occur to create new individuals. Although environmental conditions, such as day length, may determine the time when seeds germinate, plants flower or animals mate, the life cycle of animals and plants is largely under endogenous control and proceeds through an obligatory series of steps to a large extent independently of environmental influences. The life cycles of fungi present an interesting contrast; whether simple or complex, they are facultative, not obligate. If conditions are ideal for vegetative growth then vegetative growth continues. If a shortage of nutrients results in the growth rate falling to a low value or to zero, then a sequence of developmental steps that lead to sporulation may, if circumstances are otherwise suitable, be initiated. Hence in the laboratory, fungi may be maintained indefinitely in the vegetative state, or sporulation may be initiated when required. Not only is a low growth rate a prerequisite for sporulation, but sporulation may occur without the assimilation of exogenous reserves. The successive developmental steps in sporulation may require different environmental conditions and hence the onset of each can be controlled. The facultative morphogenesis of fungi offers interesting opportunities for developmental biology. Vegetative cells can often be produced in large amounts, even in fermenters, which facilitates biochemical study. Development can be initiated by an appropriate stimulus which may yield nearly synchronous sporulation in large amounts of material. Many fungi are suitable for genetic analysis, and mutations (including very useful temperature-sensitive mutations) affecting different developmental steps can be obtained. The study of fungal morphogenesis is proving to be relevant to understanding the complex development of animals. The study of fungal sporulation has yielded much of interest and some systems, such as the asexual sporulation of the slime mould *Dictyostelium* (pages 16–20) and the sexual sporulation of the yeast *Saccharomyces* (pages 71–73), have great promise for further fundamental developmental studies.

Sporulation and Reproduction

Sexual sporulation is sometimes referred to as 'sexual reproduction' and asexual sporulation as 'asexual reproduction'. This usage reflects the botanical origins of mycology. The production of new plants, reproduction, is by the formation of seeds through a sexual process and less commonly by the asexual budding of bulbs, corms and other such structures. The spores of fungi were hence seen as analogues to seeds in the creation of new individuals, and sporulation as 'reproduction'. This viewpoint has some validity with respect to the production of readily dispersed spores by filamentous fungi. With yeasts, however, the process by which new individuals are produced is vegetative growth and cell division, and the process of sporulation is related to survival (Ascomycete yeasts) and dispersal (Basidiomycete yeasts), rather than to increase in the number of individuals. Even in the filamentous fungi, new individuals may result from the spread of hyphae and the subsequent emptying of protoplasm from intervening mycelium. The term 'reproduction' is ambiguous in the context of microorganisms and the explicit term 'sporulation' is preferable.

Dispersal Spores and Survival Spores

Spores, as already indicated, have two major roles in the life of the fungus – dispersal to a new site or survival until favourable conditions return. Many fungi produce more than one kind of spore (Table 4.1, Fig. 4.1). It is thus possible, and common, for specialization to occur, with a single species producing some spore types that are efficiently dispersed and others that have a high capacity for dormant survival. 'Dispersal spores' separate completely from the parent mycelium and there are often launching mechanisms to get them airborne. Their

Table 4.1 The role of spores and analogous structures in the dispersal and dormant survival of representative fungi

Group and species	Role[a]			
	Dispersal, with no capacity for dormancy	Predominantly dispersal	Predominantly survival	Survival, with dispersal improbable
Slime moulds				
Dictyostelium discoideum	–	Sporangiospores	–	Macrocysts
Physarum polycephalum	Flagellates	Sporangiospores	Cysts	Sclerotia
Oomycetes				
Phytophthora infestans	Zoospores	Sporangia	Cysts	Oospores
Other lower fungi				
Allomyces macrogynus	Zoospores, zygotes	–	Meiosporangia	–
Mucor mucedo	–	Sporangioispores	–	Zygospores
Higher fungi				
Aspergillus nidulans	–	Conidia	Ascospores	Hülle cells[b]
Saccharomyces cerevisiae	–	Vegetative cells	Ascospores	–
Coprinus cinereus	–	Basidiospores, oidia	–	Chlamydospores

[a] Deduced from spore morphology, physiology and behaviour.
[b] Hülle cells are produced singly at the tips of hyphae associated with developing ascocarps. They have very thick walls and resemble chlamydospores.

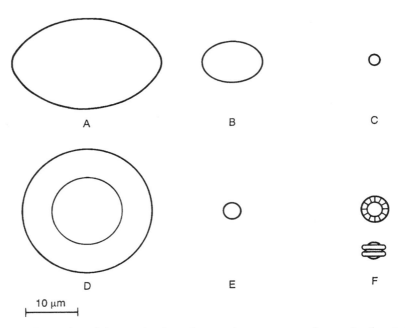

10 µm

Figure 4.1 Examples of the production of more than one type of spore by fungi. A–C, *Neurospora crassa*. A, Ascospores contain many nuclei, commonly 32 as a result of five successive mitoses. Ascospores are forcibly discharged, are capable of prolonged survival, and are responsible for the colonization of new sites (page 50). B, Macroconidia, which are produced in immense numbers, contain several nuclei. Readily released even by slight disturbance, they can be a troublesome source of contamination in the laboratory. In nature, however, they are responsible only for very local spread and for fertilizing protoperithecia (page 52). C, Microconidia are uninucleate and it is likely that their main role is in fertilization. D–F, *Aspergillus (Emericella) nidulans*. D, Hülle cells, produced on mycelium associated with ascocarps, are multinucleate and have extremely thick walls. Dispersal would seem unlikely, and it is probable that their role is prolonged survival at the site of production. E, Conidia, being produced in immense numbers and being readily released from the conidiophore, are likely to be the main agent of dispersal. F, Ascospores (polar and equatorial view) are released by rupture of ascus and ascocarp and, produced in far lower numbers than conidia, have some capacity for dispersal (indicated by capture in spore traps) and may be intermediate between Hülle cells and conidia in their capabilities.

shapes vary greatly (Fig. 4.2), depending on the requirements of their dispersal mechanisms. Since the probability of a single spore reaching a favourable new site is small, there is a tendency for them to be produced in vast numbers and hence, because of limited parental resources, to be small. Dispersal spores usually have only a moderate capacity for dormant survival, and germinate readily in the presence of nutrients. The attributes of 'survival spores' contrast with those of dispersal spores. Survival spores may not separate completely from the parent mycelium, and if they do, it is often through the lysis of the mycelium. Since they are unlikely to be dispersed, the production of vast numbers is less important than for dispersal spores. They hence tend to be large, permitting the retention of substantial nutrient reserves for when germination does occur. Lacking the need

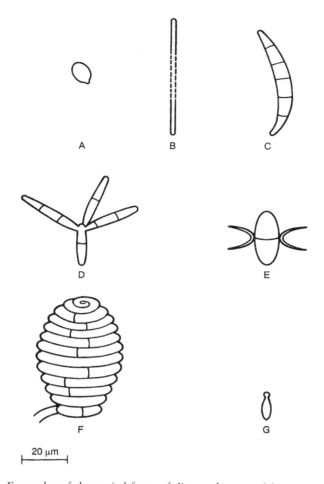

Figure 4.2 Examples of the varied form of dispersal spores. Many spores, whether they remain at the site of production or are dispersed, are approximately spherical, the form most economical in wall material. There are, however, a wide variety of other forms among dispersal forms, the form being an adaptation either for a process of active launching, for subsequent dispersal, or for retention at a suitable destination. A, Basidiospore of *Coprinus sterquilinus*, typical of actively launched basidiospores (ballistospores). B, Ascospore of *Claviceps purpurea*, very long and slender (1×100 µm). Cross-walls develop after discharge to give numerous uninucleate compartments. C, Macroconidium of *Fusarium culmorum* with several uninucleate compartments. It is a slime spore, dispersed by rain splash. D, Tetraradiate conidium of the aquatic Hyphomycete *Articulospora tetracladia*, the form probably facilitating deposition by impaction. E, Ascospore of the marine Ascomycete *Remispora maritima*. Sticky appendages probably facilitate retention on a surface after impaction. F, Helical conidium of the aquatic Hyphomycete *Helicodendron conglomeratum*. The fungus sporulates only when exposed to air, and air trapped within the helical form enables the spore to float and to be dispersed by currents at the water surface. G, Conidium of the predacious Hyphomycete *Meria coniospora*, with an adhesive knob at the end which can stick to a passing eelworm, with subsequent germination and infection. (A and E after Ingold, C. T. (1971). *Fungal Spores: their Liberation and Dispersal*. Clarendon Press, Oxford; D, F and G after Webster, J. (1980). *Introduction to Fungi*, 2nd edn. Cambridge University Press, Cambridge.)

for a form adapted for dispersal, they are often spherical, the form most economical in wall materials. Their germination normally occurs as the result of a slow decay of dormancy or in response to a specific stimulus, physical or chemical, that breaks dormancy. The concept of specialization for dormancy or for dispersal is a fruitful one in understanding the diversity of spore form and behaviour.

A very large proportion of spores that have been dispersed will fail to reach a site suitable for the growth of the fungus and will ultimately die. A few will reach a suitable site, germinate and colonize the site. Others may reach an otherwise suitable site that is already occupied by the same species. The journey of such spores is not necessarily wasted. Germination may be followed by mating with the resident, thus effecting the dissemination and propagation of parental genes. Basidiospores may often function in this way, especially in species that germinate readily on mycelium of the same species but poorly elsewhere (page 63). Often it is not clear whether the major role of a particular type of dispersal spore is to establish new colonies or to effect mating. For example, in *Neurospora crassa* (page 52; Fig. 4.1) microconidia germinate readily on the trichogynes of protoperithecia but poorly on agar media. It is hence reasonable to suppose that their role is that of mating. The macroconidia of *N. crassa* are produced in such abundance, are dispersed so readily, and germinate so easily on agar media that in the laboratory considerable care must be taken in handling the fungus so as to avoid contamination of other cultures. It would hence be reasonable to conclude that in nature macroconidia are the major vehicle for the establishment of new colonies of *N. crassa*. However, almost every new isolate from nature, even from sites that are close to each other, proves to be genetically unique. This indicates that new sites are colonized almost wholly by ascospores, the end products of mating and meiosis, even though they are produced in much smaller numbers than macroconidia. It would seem therefore that macroconidia function mainly for dispersal over short distances and mating.

Initiation of Sporulation and Sporophore Maturation

Fungi exploit the presence of abundant nutrients by vigorous vegetative growth. Some fungi can produce some types of spores even while copious vegetative growth is occurring, but many do not. The end of vegetative growth as a consequence of nutrient exhaustion, however, means that if a fungus is to survive it must be dispersed or enter a dormant state. As already indicated, for most fungi dormancy or dispersal is achieved through spore production. Hence for many fungi nutrient exhaustion leads to sporulation, whereas abundant nutrients prevent sporulation and result in abundant vegetative growth. The optimal nutritional conditions for sporulation and for vegetative growth therefore often differ. It is also advantageous to produce a type of spore when environmental conditions are such as to enable the spore type to fulfil its role. Sporulation is hence sensitive to environmental factors and may be prevented by conditions of temperature, illumination or humidity that would not adversely affect vegetative growth. Sporulation hence commonly occurs under a more restricted range of environmental conditions than does vegetative growth. For the many fungi that produce two or more types of spore with different roles, environmental conditions can determine which spore types are produced. Study of the factors influencing sporulation (Table 4.2) is not only of interest in relation to the study of developmental processes but is of considerable practical importance. It facilitates the production of sporophores in the laboratory, and thus the identification of fresh isolates from nature, and it increases our understanding of the conditions that lead to the sporulation of plant pathogens in the field and hence the spread of plant diseases. The more important factors influencing fungal sporulation will now be considered.

Table 4.2   Examples of the environmental control of developmental processes

Requirement	Process	Species
Emergence into a gas phase[a]	Conidiophore initiation	*Penicillium griseofulvum*
	Basidiocarp initiation	*Schizophyllum commune*
Depletion of exogenous assimilable nitrogen	Conidiophore initiation	*Penicillium griseofulvum*
	Basidiocarp initiation	*Schizophyllum commune*
Depletion of exogenous assimilable carbon	Asexual sporulation	*Physarum polycephalum*
	Pilus expansion	*Schizophyllum commune*
Exposure to visible light[b]	Asexual sporulation	*Physarum polycephalum*
	Basidiocarp initiation	*Schizophyllum commune*
Exposure to ultraviolet radiation[c]	Ascocarp initiation	*Pleospora herbarum*
	Conidiophore initiation	*Botrytis cinerea*
Darkness	Conidium production	*Botrytis cinerea*
	Sclerotium production	*Botrytis squamosum*

[a] Many fungi will not sporulate in submerged culture or if covered by a film of water; hyphae must merge into the gas phase before sporulation can occur.
[b] Commonly the blue end of the visible spectrum and adjacent ultraviolet wavelengths are effective.
[c] Visible light is not effective.

Culture Systems and Sporulation

A wide variety of culture systems can be utilized for obtaining fungal sporulation. The method chosen will depend upon the objectives of the investigator.

Surface culture on agar media

A great deal of attention has been given to the growth and sporulation of fungi on agar media in Petri dishes. The medium is usually centrally inoculated, and the fungus grows as a circular colony of increasing radius. Proceeding from the margin to the centre of the colony several zones can be distinguished (page 140). At the margin of the colony is an extension zone followed by a productive zone in which the major increase in biomass takes place. Then, if conditions are favourable, there is a fruiting zone in which sporulation occurs. Biosynthetic activity towards the margin of the colony utilizes the nutrients in the medium, so the fruiting zone will be one in which the nutrient levels are considerably lower than those initially provided. Hence such cultures, although often a convenient way of obtaining sporulation, fail to indicate the actual nutritional conditions required. As an alternative to central inoculation, agar media can be uniformly spread with a suspension of spores. This provides a more homogeneous surface culture, of undifferentiated mycelium (page 139), but it remains one in which the same culture has to provide conditions satisfactory for two processes, vegetative growth and sporulation, that may have differing requirements.

Agar media can be overlaid with filter paper or a permeable membrane. Fungi will grow on such a surface since nutrients will diffuse through the paper or membrane. Sporulation can be prevented by means of appropriate conditions, such as darkness in a fungus needing light for sporulation. When adequate growth has occurred the filter paper or membrane with its mycelium can be transferred to conditions suitable for sporulation. Such transfer cultures can be used to obtain sporulation in the absence of vegetative growth, to ascertain the precise conditions required for sporulation, and to determine metabolic differences between growing and sporulating cultures.

Surface culture on liquid media
Most terrestrial fungi, when inoculated into static liquid media, will produce a surface mat of mycelium. Such a mycelial mat will often sporulate on the upper surface. It is possible to remove the medium beneath a mycelial mat and to replace it with a fresh supply of the same or a different medium. Such replacement cultures, like transfer cultures with agar media, are useful for metabolic studies on sporulation.

Submerged culture
Terrestrial fungi can be grown in submerged culture by shaking the culture vessels or stirring the media. Provided that the fungus is able to grow in a filamentous rather than a pellet form (page 135), large amounts of homogeneous mycelium can be obtained. It has been found that by an appropriate change in the medium some terrestrial fungi can be induced to sporulate in submerged culture, and such cultures are valuable for physiological and biochemical investigations. It should, however, be borne in mind that such conditions are unnatural for terrestrial fungi, the developing sporophores of which are normally exposed to the atmosphere and not in direct contact with nutrient liquid.

Aquatic fungi will normally grow submerged in static liquid culture. Since sporangium formation in Oomycetes is commonly inhibited by even small amounts of the growth medium, such cultures are useful for obtaining sporulation as they can be washed repeatedly with distilled water or a dilute salts or buffer solution.

Continuous culture
Sporulation, if able to occur in submerged conditions, can be studied by means of chemostat culture (page 133). In a chemostat the growth rate is regulated by the concentration of a single growth-limiting nutrient, and at low growth rates, sporulation may occur. Conditions in chemostat culture, with one nutrient present in limiting quantities and others present in excess, may often be a good model for the natural environment of a fungus.

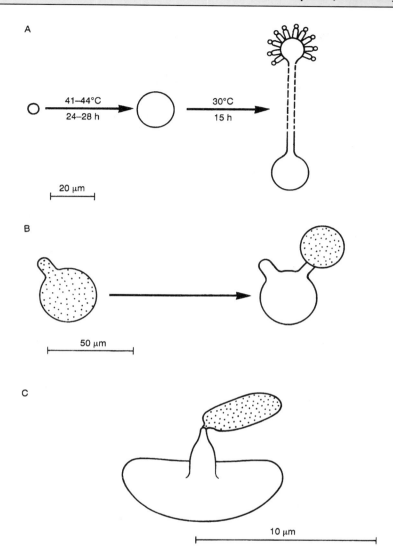

Figure 4.3 Microcycle sporulation and secondary sporulation. When, following germination, conditions prove unfavourable for growth, the life cycle may be greatly curtailed (microcycle sporulation) or effectively eliminated, with the germinated spore immediately giving rise to another spore (secondary sporulation). A, Microcycle sporulation in *Aspergillus niger*. At a temperature too high for normal germination a conidium fails to produce a germ-tube but increases in diameter. On transfer to a normal incubation temperature a hypha emerges but behaves as a conidiophore, and conidia are formed. B, Secondary conidium formation in *Conidiobolus coronatus*, a member of the Entomophthorales. On the left is the parent conidium with a papilla that had projected into the conidium, but which was everted to launch the conidium from its conidiophore. On the right the conidium, after germination on water agar, has emptied its contents into a daughter conidium borne on a short conidiophore. C, A ballistospore of the yeast *Sporobolomyces roseus* that has germinated to produce a secondary ballistospore. (A based on Anderson, J. G. & Smith, J. E. (1971). *Journal of General Microbiology* **69**, 185–197; B and C on Ingold, C. T. (1971). *Fungal Spores: their Liberation and Dispersal*. Clarendon Press, Oxford.)

Sporulation on solid substrates

Various solid substrata are useful for obtaining sporulation. A loose mixture of moist cornmeal and sand, for example, provides both nutrients and a large surface area on which sporulation can occur. The large sporophores of some Basidiomycetes are difficult to obtain in pure culture. This may be the consequence of the need for nutrient limitation to initiate sporulation along with a continued supply of nutrients for sporophore development. The provision of sterile blocks of wood provides a satisfactory substrate for the fruiting of timber-decomposing Basidiomycetes. Sporophores are initiated, presumably at sites where nutrient limitation has occurred, but continue to be supplied with the materials for further development by translocation from mycelium that is attacking the wood and assimilating nutrients at other locations.

Microcycle sporulation

Under some circumstances, spore germination may soon be followed by sporulation, the phase of vegetative growth being omitted. Such **microcycle sporulation** or secondary sporulation (Fig. 4.3) has been utilized for the study of spore production uncomplicated by vegetative growth. Microcycle sporulation can occur in nature, if spore germination is initiated but conditions then prove unsuitable for growth. Usually, however, sporulation is preceded by vegetative growth, so studies on microcycle sporulation, although providing valuable insights into developmental processes, may not be suitable for indicating the circumstances that normally initiate sporulation.

The Effect of Nutrition

The production of spores requires that nutritional conditions should be satisfactory during vegetative growth, at the time sporulation is induced and during subsequent sporophore development. The differing requirements during these three periods will now be considered.

Vegetative growth and sporulation

Some fungi are able to produce spores after only limited vegetative growth. Others, which have large fruiting bodies, must produce an extensive mycelium in order to obtain access to sufficient nutrients to form a sporophore. The differing requirements for extensive vegetative growth and fruit body initiation may be difficult to meet in the laboratory; the need to culture some wood-decomposing Basidiomycetes on blocks of timber in order to obtain fruiting bodies has already been mentioned.

Fungi are often grown on media which contain nutrient concentrations much higher than those which the organism is likely to encounter in nature. As a consequence the biomass per unit volume obtained may be greatly in excess of that which naturally occurs, and metabolism may result in the accumulation around the fungus of potentially toxic materials such as hydrogen ions or ammonia in quantities much greater than those that the organism normally encounters. Hence by the time nutrient concentrations have fallen to a level that would permit sporulation, conditions may have become unfavourable. This is the probable basis for the common observation that an agar medium giving abundant mycelial growth may fail to yield spores. Provision of lower nutrient concentrations, or of a slowly utilized carbon source such as starch, may give less abundant vegetative growth but permit sporulation to be initiated.

The nutrient balance provided during the period of vegetative growth may also be of importance in determining whether sporulation subsequently occurs. A high carbon/nitrogen ratio, for example, may facilitate the accumulation of endogenous reserve materials such as glycogen, which can then be utilized as a source of energy during sporulation. The influence of nutrient conditions during the vegetative phase on subsequent sporulation is a topic which deserves further study, utilizing methods such as continuous culture to provide precisely defined conditions during growth.

Nutrient limitation and the induction of sporulation
The exhaustion of any one of the various elements or organic compounds required by a fungus will bring growth to an end. This may occur at some distance behind a colony margin or may affect an entire mycelium. Subsequent events will be determined by which nutrient is exhausted. Depletion of assimilable nitrogen is a prerequisite for sporulation in many fungi (Table 4.2), and in such fungi the exhaustion of the carbon source while exogenous nitrogenous nutrients remain may lead to autolysis and death. It is reasonable to deduce that such fungi live in habitats in which growth is limited by the availability of nitrogen. In the Myxomycete *Physarum polycephalum*, which attacks living fungi, it is glucose exhaustion that is essential for sporulation and nitrogen exhaustion that can cause death, suggesting that in nature it is carbon limitation that normally terminates growth of this organism. In the plant pathogen *Fusarium oxysporum* a low carbon/nitrogen ratio leads to the production of survival spores (chlamydospores) whereas in other conditions only conidia are formed. Sporulation in *Saccharomyces cerevisiae* requires exhaustion of both exogenous nitrogen and assimilable sugars, but is stimulated by acetate or glyoxylate.

It is clear therefore that for many forms of sporulation the exhaustion of a key nutrient, usually assimilable nitrogen, is essential. There are, however, exceptions. If a dilute suspension of conidia of *Aspergillus nidulans* is spread on an agar medium that permits rapid growth the first conidiophores are detectable in about 24 hours. Since the time to sporulation is unchanged if membranes bearing the young colonies are transferred at hourly intervals to fresh medium, or media containing widely differing carbon and nitrogen concentrations are used, the onset of sporulation is unrelated to nutrient exhaustion. Thus although in all fungi nutrient exhaustion may promote sporulation it is clear that some forms of sporulation can accompany rapid vegetative growth. There would seem to be no point in the production of survival spores until nutrients are becoming exhausted, but it could well be of value to produce dispersal spores at any time suitable for dispersal. Differences may hence be expected with respect to the nutritional requirements for the production of different spore types.

Nutrient requirements for the completion of sporulation

Spores differ in their structure and composition from vegetative cells. Their production hence requires changed biosynthetic activities and changes in metabolic paths, and sometimes nutrients not required for vegetative growth. Thus a medium lacking sterols will support vegetative growth in *Phytophthora*, but a source of sterols must be present to permit sporulation. In other fungi sporulation has been shown to be associated with increased requirements for vitamins such as thiamine or biotin. A medium capable of supporting vegetative growth may hence be inadequate for sporulation, even if the exhaustion of a key nutrient, as discussed above, takes place.

An exogenous supply of assimilable carbon is often required for sporophore development after the process has been initiated by the depletion of the nitrogen source. In the Basidiomycete *Schizophyllum commune*, for example, initiation of fruit body primordia requires nitrogen starvation but further development of primordia needs an exogenous carbon supply. Maturation of the fruit body by pileus formation occurs when the carbon supply is exhausted (Table 4.2). Although local exhaustion of a key nutrient may be essential at an early stage in many forms of sporulation, the development of large fruiting bodies, as for example those of the cultivated mushroom, will require massive supplies of nutrients. These may be assimilated at a distance from a fruiting body and translocated through the mycelium.

The Effect of Light

Sporulation in many fungi is unaffected by light, but in others the process is influenced by light in a variety of ways.

Light requirements for sporulation

Fungal hyphae can penetrate deeply into timber, plant tissues, dung and other substrata. Efficient dispersal, however, can occur only if spores are produced at the surface and not in cavities deep inside such materials. This can be ensured by means of a light requirement for sporulation, and such a requirement for the production of dispersal spores is widespread (Table 4.2).

Some fungi have an absolute light requirement for the production of dispersal spores. These include the Myxomycete *Physarum polycephalum*, the coprophilous (dung-inhabiting) Zygomycete *Pilobolus kleinii*, the Ascomycete *Nectria haematococca* and the Basidiomycete *Schizophyllum commune*. In others, for example the mitosporic fungus *Trichoderma viride*, the production of conidia is stimulated by light but can ultimately occur in darkness. In some fungi an exposure of a few seconds to light of low intensity is sufficient to bring about sporulation whereas for others longer exposures to relatively high light intensities are needed. Light may be needed only for the initiation of sporulation, or for later stages in the process also. Even when light is not essential for sporulation the size and form of sporophores may be dependent on whether they have developed in light or in darkness.

Sporulation in the fungi mentioned above can be brought about by visible light.

There are, however, some fungi in which visible light is unable to initiate sporulation and in which it is the ultraviolet component of daylight which is effective. One example is the Ascomycete *Pleosporum herbarum* which requires exposure to ultraviolet radiation to initiate both perithecium and conidiophore formation. Other examples are the mitosporic fungi *Botrytis cinerea*, *Alternaria tomato* and *Helminthosporium oryzae* in which conidiophore production in some strains either requires or is stimulated by ultraviolet wavelengths. These and other organisms in which only ultraviolet radiation is effective are plant pathogens attacking leaves and other aerial parts of plants. Such structures are often translucent and are penetrated deeply by visible light, but probably to a much lesser extent by ultraviolet wavelengths which are strongly absorbed by many organic compounds. It is possible therefore that for plant pathogens ultraviolet radiation is a better indicator of proximity to the surface than is visible light, and sporulation induction by ultraviolet radiation advantageous.

Darkness requirements for sporulation
Many fungi that require light to initiate sporulation need a period of darkness for its successful completion (Table 4.2). In *Pilobolus kleinii* the first phase of sporulation, the production of swellings (trophocysts) in the substrate mycelium, requires illumination. Subsequent exposure to light results in the emergence of a sporangiophore from each trophocyst, but before this can occur the trophocysts need a period of darkness. The Basidiomycete *Coprinus congregatus* needs light for both sporophore initiation and maturation, but an intervening period of darkness is also required. Darkness requirements for the completion of sporulation have been studied most intensively with the plant pathogens mentioned above in which ultraviolet radiation is required for conidiophore production. In many of these fungi, for example, *Botrytis cinerea* and *Alternaria tomato*, darkness is essential in order that conidia may form on the conidiophores. Many plant pathogens show optimal sporulation when a 'warm day' (12 hours at a high temperature such as 30°C with ultraviolet irradiation) is followed by a 'cool night' (e.g. 12 hours at 20°C in darkness). Incubators capable of providing such a cycle of irradiation and darkness are often employed for obtaining sporulation of plant pathogenic fungi. The darkness requirement which often accompanies light-induced sporulation is probably a timing device to enable spore maturation to occur at the time of day or night most suitable for dispersal. In those fungi that have active mechanisms of spore discharge the launching of spores into the atmosphere is often directly controlled by the onset of light or darkness (page 221) although even in these fungi it may be helpful for spores to mature at an appropriate time. In many other fungi, such as the plant pathogens mentioned above, spore release is passive, occurring as soon as a mechanical disturbance such as wind or rain (pages 215–216) follows spore maturation. In these fungi the timing of spore maturation which permits the detachment of spore from sporophore is the only way in which the timing of release into the atmosphere can be controlled.

Most mycologists are interested in obtaining sporulation in the laboratory. Failure to sporulate in an incubator, by a fungus in which sporulation occurs in a Petri dish on the bench, often leads to the discovery of a light requirement for sporulation. The failure of such a fungus to sporulate under continuous illumination will lead to the discovery of a requirement for a period of darkness in addition to that for illumination. A requirement for darkness in the absence of any need for light is less commonly detected since such a fungus will give adequate sporulation in the normal laboratory incubator. Such a requirement may, however, be common. Thus although one strain of *Helminthosporium oryzae* mentioned above requires light for conidiophore initiation and darkness for conidium production, another strain has only the latter requirement. A requirement for darkness in the absence of any requirement for light has been found for the formation of zygospores in *Dicranophora fulva*, cleistothecia and ascospores in *Aspergillus glaucus* and sclerotia in *Botrytis squamosa*. Zygospores, sclerotia and the ascospores of *Aspergillus* all have a capacity for dormant survival, and they are probably capable of fulfilling their role whether they are produced at a surface or at a less exposed site. Their production at a surface, however, could exhaust resources at a time when they might be better utilized for the formation of dispersal spores, so a light inhibition of the formation of such structures could well be advantageous for the fungus.

Light and rhythms of sporulation

Exogenous (i.e. externally determined) rhythms of sporulation are common in fungal colonies grown with daily alternation of light and darkness on agar media (Fig. 4.4). A colony shows concentric zones of spores, each zone resulting from the light stimulation of sporulation in mycelium that was at the correct age to respond to the stimulus. The intervening non-sporing regions represent mycelium that received light only when it was too young or too old to sporulate.

Daily zones of spores may also occur on fungal colonies which have been grown in continuous darkness under constant conditions. Such an endogenous circadian ('about a day') rhythm is the expression of the operation of a 'biological clock'. There have been extensive studies of endogenous rhythms of sporulation in *Neurospora crassa*, as this has proved to be an excellent model system for studying circadian rhythms, by virtue of the availability of a wide range of mutants. These show that the rhythm of *N. crassa* has the key properties of a circadian clock: an intrinsic period of about 24 hours; responsiveness to environmental stimuli, such as light or temperature, which entrain the clock on a daily basis; a constant period under a wide range of environmental conditions such as constant high or low temperature; and it is under genetic control. Study of mutants, especially 'band' (*bd*), have shown that *Neurospora* has an intrinsic rhythm that runs in constant darkness at 21–22 hours to give bands of abundant macroconidiation separated by zones of sparse mycelium with few conidia. It can be entrained to a precise 24 hour period by exposure to a 'natural' light:dark cycle of 12 hours light:12 hours dark, and even to a very short exposure to light of only 5 minutes per 24 hours. Circadian clocks have been reported in all groups of fungi, but mostly in relation to spore discharge and this aspect will be considered further (page 221). Rhythms of sporulation which are not circadian may arise through self-sustaining metabolic cycles. An example is the formation of rings or spirals of conidia by *Nectria cinnabarina*.

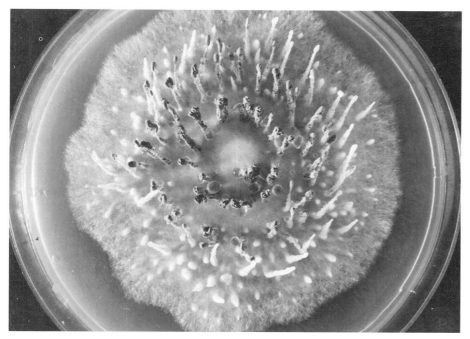

Figure 4.4   Concentric zones of synnemata (coremia) of *Penicillium claviforme*. Synnema development begins with a small bunch of branches forming at a point some way behind the colony margin. The branches aggregate into a growing apex consisting of many hyphal tips aligned parallel to each other. The tips then develop simultaneously into phialides which sporulate, apical growth coming to an end. Meanwhile the colony edge advances and new synnemata are initiated. This occurs daily in response to the alternation of light and darkness, giving the concentric zones of synnemata (John Baker).

Phototaxis and phototropism in relation to sporulation
Starved plasmodia of some Myxomycetes show positive phototaxis, migrating towards the light prior to sporulation. Positive phototaxis is also shown by the migrating grex of the cellular slime mould *Dictyostelium*. On reaching a site suitable for sporulation the grex becomes a fruiting body that will, if illumination is unilateral, be inclined towards the light instead of vertical. This is usually regarded as the result of positive phototropism, growth oriented towards the light, but since the inclined fruiting body is the consequence of the movement of amoeboid cells, its orientation is the result of a continuation of the phototactic activity already seen in the grex. The positive phototaxis which precedes spore formation in slime moulds results in spores being produced at a site optimal for dispersal, since movement towards the light means movement towards the open air.

   Positive phototropism of immature sporophores or of parts of the fruiting body is widespread in filamentous fungi. The stipes of the young fruiting bodies of many Basidiomycetes are positively phototropic. In the Ascomycetes positive phototropism can occur with the necks of perithecia (e.g. *Sordaria fimicola*), with coremia (e.g. *Penicillium claviforme*) and with conidiophores (e.g. *Aspergillus giganteus*). In Ascomycetes in which the asci line the inner side of a cup-shaped fruiting body the individual asci are phototropic. In some the whole ascus curves towards the light whereas in others curvature is limited to the tip of the ascus or may even take the form of a displacement to one side of the ascus tip of the site of the aperture through which ascospores are discharged. In the absence of such phototropic responses ascospores discharged from asci on one side of the cup would strike the opposite side of the cup; instead they are shot into the open air.

Positive phototropism of Zygomycete sporangiophores is common, and the sensory responses of the very large sporangiophores of *Phycomyces* have been intensively studied. Exposure of a sporangiophore to unilateral illumination results in the sporangiophore bending towards the source of light. This bending is the result of a greater extension on the **far** side of the growing zone than on the near side (Fig. 4.5A). The exposure of a symmetrically illuminated sporangiophore to an increased light intensity results in an increase in extension rate, known as the 'light growth response', and a decrease in light intensity to a fall in extension rate, the 'dark growth response'. Both responses are transient,

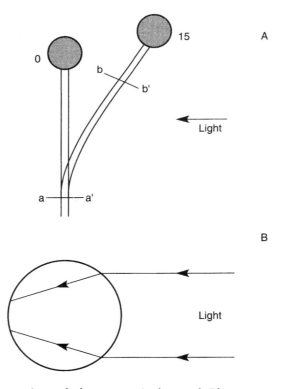

Figure 4.5 Phototropism of the sporangiophore of *Phycomyces*. A, Growth and curvature of a sporangiophore during its final phase of elongation, after sporangium formation. The position of the sporangiophore at the onset of unilateral illumination (0) and 15 minutes later (15) is shown. It can be seen that for curvature to occur wall growth has to be greater on the far side (a–b) of the sporangiophore than on the side near the light source (a'-b'). B, A diagrammatic section of the sporangiophore (as for example at a–a') showing how light is focused on the far side of the sporangiophore.

adaptation to the changed light intensity occurring within a few minutes with the restoration of extension rate to its original value. Such a system of adaptation is essential in any sensory system which is to operate over a very wide range of stimulus intensities and yet remain sensitive to small changes in the magnitude of the stimulus. The apparent paradox of a positive light growth response (stimulation of extension in a symmetrically illuminated sporangiophore) and positive phototropism (stimulation of extension on the far side of the sporangiophore) is explained through the operation of the 'lens effect'; the sporangiophore acts as a cylindrical converging lens, focusing unilateral light on the far side of the sporangiophore to produce an area of high intensity illumination (Fig. 4.5B). The interpretation of a further paradox, the coexistence of a transient light growth response with persistent phototropism, remains controversial.

Action spectra, photoreceptors and the origin of photosensitivity
The human eye can perceive light with wavelengths of about 400–750 nanometres (nm). In addition to visible light, daylight includes both shorter ultraviolet (UV) and longer infrared wavelengths. The ultraviolet component of daylight under clear conditions extends to about 290 nm. This 'near UV' radiation is of importance in nature. Radiation of shorter wavelengths ('far UV') does not occur in daylight but can be produced by various lamps. It is strongly absorbed by nucleic acids and is of interest for its consequent mutagenic and lethal effects.

Light can only produce a chemical effect if it is absorbed. Wavelengths that are strongly absorbed produce a large effect and those that are weakly absorbed a small effect. The action spectrum of a photoresponse (the effectiveness of different wavelengths in bringing it about) hence parallels the absorption spectrum of the photoreceptor compound. Action spectra can thus establish whether photoresponses have similar or different photoreceptors and in some instances may indicate the nature of the photoreceptor compound.

Very similar action spectra have been obtained for phototropism and the light growth response in *Phycomyces* (Fig. 4.6A), the suppression of a circadian rhythm of conidium production in *Neurospora crassa*, the induction of carotenoid biosynthesis in *Fusarium aqueductuum* and a wide range of other photoresponses in fungi and plants. Little response is obtained at wavelengths greater than about 525 nm, the blue end of the visible spectrum is highly effective, and effectiveness extends to UV wavelengths. It is probable that the photoreceptor is a riboflavin derivative – flavin mononucleotide (FMN) or flavin adenine dinucleotide (FAD) – forming the prosthetic group of an enzyme and that the flavin is reduced by light. For the photosuppression of the rhythm of conidial production in *N. crassa* it is probable that the photoreceptor flavoprotein is NADPH–cytochrome *c* oxidoreductase. There are, however, many flavoprotein enzymes in living organisms and it is possible that for different photoresponses different flavoproteins are involved. Furthermore, some photoresponses to the blue end of the visible spectrum have action spectra that are not readily reconciled with the absorption spectra of flavoproteins and probably have a different photoreceptor.

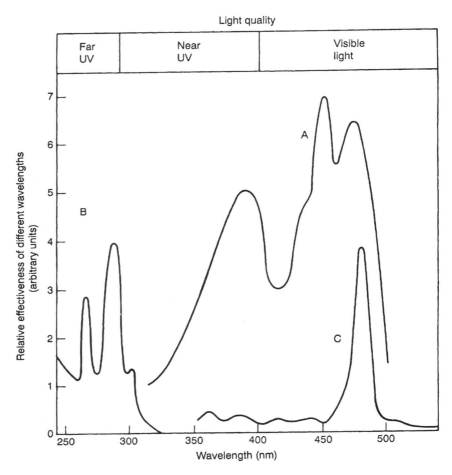

Figure 4.6   Action spectra for three fungal photoresponses. A, The light growth response of the sporangiophore of *Phycomyces*. (Based on Delbrück & Shropshire (1960). *Plant Physiology* **35**, 194–204.) In addition to the peaks in the visible and near UV regions there is a further maximum, not shown, in the far UV at 280 nm. B, Stimulation of perithecium induction in the Ascomycete *Leptosphaerulina trifolii*. (From Leach (1972). *Mycologia* **64**, 475–490.) C, Inhibition of the terminal phase of sporulation in *Stemphylium botryosum* (the anamorph of *Pleospora herbarum*). (From Leach (1968). *Mycologia* **60**, 532–546.) A peak at 280 nm is omitted. The scale on the vertical axis of the graph is to permit comparison of the effectiveness of different wavelengths for a photoresponse and not for comparing the absolute intensities required for the different responses; the threshold for the light growth response of *Phycomyces* is much lower than for the other two reactions.

A wholly different type of action spectrum is that for the promotion of sporulation in various plant pathogens, such as the stimulation of perithecium formation in the Ascomycete *Leptosphaerulina trifolii* (Fig. 4.6B). Here the maximum effect is obtained at about the lower limit of naturally occurring UV radiation at about 290 nm, and wavelengths longer than *ca* 360 nm are ineffective. The photoreceptor compound, the nature of which is unknown, has been termed mycochrome. The UV irradiation that initiates sporulation often induces the production of a class of organic compounds that have been termed mycosporines and may have an important role in sporulation.

The completion of sporulation initiated by UV radiation commonly requires darkness, and the action spectrum for the inhibitory effect of light has a maximum at 480 nm (Fig. 4.6C). One hypothesis is that the photoreceptor is a form of mycochrome with an absorption spectrum changed by the previous UV irradiation and that irradiation with blue light converts it back into the UV-absorbing form. Although it is the blue end of the visible spectrum or near UV radiation that bring about most fungal photoresponses, reactions promoted by the red end of the visible spectrum are known.

Many fungal genera have species in which sporulation is influenced by light and species in which it is not. This suggests that photosensitivity has been acquired or lost on many occasions. Both events may well have been common. Many important enzyme systems in the cell include proteins with a prosthetic group such as flavins or haem that absorb light and hence are potential photoreceptors, so it is likely that photoresponses may readily be acquired if this aids the survival of the fungus. Conversely the coupling of a response to a photoreceptor would be liable to be lost by mutation if it ceased to be of value to the fungus and thus no longer maintained by natural selection.

The Effect of other Environmental Factors

The aeration complex and exposure to a gas phase

Oxygen is required for sporulation by most fungi, even those, such as *Saccharomyces cerevisiae*, that are capable of growth under anaerobic conditions. Among the few exceptions are the rumen Chytridiomycetes and obligately fermentative aquatic fungi such as the Oomycete *Aqualinderella* and the Chytridiomycete *Blastocladia*.

Another Chytridiomycete, *Blastocladiella*, is an obligate aerobe, and will only survive anaerobic conditions in a resting state. An aquatic environment rich in organic material will, through microbial activity, become high in dissolved carbon dioxide and perhaps ultimately anaerobic. High concentrations of dissolved carbon dioxide (which in aqueous solution near neutral conditions is mostly in the form of the bicarbonate ion) result in *Blastocladiella* producing thick-walled resting sporangia with a prolonged capacity for survival, rather than the thin-walled sporangia that soon germinate. High carbon dioxide concentrations in the gaseous phase can affect sporulation in terrestrial fungi. Thus a carbon dioxide concentration of 5% will prevent fruit body initiation in *Schizophyllum commune* and pileus expansion in another Basidiomycete, *Collybia velutipes*. Perithecium formation in the Ascomycete *Chaetomium globosum*, however, is promoted by carbon dioxide concentrations of up to 10%.

One important component of the gaseous phase is water vapour. The relative humidity of the atmosphere adjacent to a sporophore will determine the rate at which water is lost to the atmosphere. In the Basidiomycete *Polyporus brumalis* it has been shown that too high a relative humidity and hence too low a rate of transpirational water loss results in fruit bodies of abnormal form, with elongated stipes and rudimentary pilei.

Many fungi, including species of *Penicillium*, are unable to sporulate when submerged but produce conidiophores and conidia readily when hyphae emerge into a gas phase (Table 4.2). It has been shown that the stimulus for sporulation is not associated with changes in the availability of oxygen, the concentration of carbon dioxide or rate of water loss, but seems to be directly associated with the change in the physical environment at the hyphal surface. *Schizophyllum commune* and some other fungi release very hydrophobic proteins, termed hydrophobins, into liquid culture media (page 102). When hyphae of such fungi emerge from a liquid into a gas phase the hydrophobins, no longer able to diffuse away, polymerize on the surface of the hyphae. The resulting hydrophobic coating would prevent further wetting of the hyphae and, depending on which hydrophobins were produced, determine whether hyphae grew singly or aggregated into structures such as synnemata or fruiting bodies.

Temperature
The interaction of light and temperature in bringing about the sporulation of plant pathogens has already been discussed. The range of temperatures at which fungi in general can sporulate tends to be narrower than that which allows growth, and the optima for sporulation are often lower than for growth.

Tropisms other than to light
The sporangiophore of *Phycomyces* shows sensory responses to a range of factors in addition to that of light. In darkness the sporangiophore shows negative geotropism, growing vertically upwards. This response is largely suppressed by light, which is a better indicator of the direction of the open air. Sporangiophores also avoid each other (negative autotropism). This response may be due to the production by sporangiophores of a growth promoting gas, so that if two sporangiophores approach each other, a higher concentration of the gas results in the space between them, growth is stimulated on the sides of the sporangiophores adjacent to the space, and bending away from each other occurs. Alternatively it may be positive aerotropism, triggered by a lowered concentration between them. This may also be the cause of the avoidance response to solid objects shown by sporangiophores and their growth into a gentle breeze (positive anemotropism). An avoidance response to a solid object is also shown by developing fruit bodies of *Coprinus cinerea*.

Geotropism is also an important factor in the development of Basidiomycete fruiting bodies. The stipes of immature fruiting bodies are commonly negatively geotropic, although this response can often be overruled by light. The gills of Agaricales are usually positively geotropic, growing downwards. This gives them the precise vertical orientation needed for efficient spore release (page 53). If, after development, the position of the fruiting body is disturbed, the gills are often capable of some geotropic reorientation to bring them once more into a vertical position.

Sexual Interaction in Sporulation

Sexual sporulation, like asexual sporulation, will occur only if nutritional and other environmental conditions are satisfactory. The sexual sporulation of self-sterile fungi has the additional requirement of cell fusion between haploid strains of different and compatible mating types. The process of cell fusion may be controlled by diffusible hormones and may involve surface interactions between the strains. Similar hormonal and surface interactions probably occur between male and female structures in self-fertile species, which are often closely related to self-sterile ones, but are less easily elucidated in such species.

Sex hormones and pheromones

The diffusible factors involved in the regulation of sexual processes in the fungi have been termed hormones and pheromones. Animal physiologists use the term hormone for a factor concerned with chemical coordination within an individual, and pheromone for a factor that is emitted and produces effects in other individuals in the same species. The distinction is an artificial one in the fungi, since it is probable that the same factors are involved in the coordination of sexual activity both in self-sterile species, in which interaction is between different individuals, and in related self-fertile species, in which interaction is between different sites in the same colony. General use, however, has settled on 'hormone' for the lower fungi, and 'pheromone' for Ascomycetes and Basidiomycetes.

The sex hormones that have been isolated from the lower fungi are isoprenoids – the sterols antheridiol and oogoniol (Fig. 2.9) from the Oomycete *Achlya*, the sesquiterpene sirenin (Fig. 2.12) from the Chytridiomycete *Allomyces*, and trisporic acid and its precursors (Figs. 2.14, 2.15) from the Zygomycetes such as *Mucor* and *Blakeslea* (Table 4.3). These three groups of hormones, although structurally very different, have some common properties: they are all specific, affecting only fungi of the same species or closely related species; they are produced in very low concentrations (apart from trisporic acids by *Blakeslea trispora*); they are active in very low concentrations; and they are unstable, being metabolized by the cells responding to them. Their production in such low concentrations has meant that very sensitive bioassays have had to be developed for their bioassay. Similar hormone systems are very likely to be involved in the mating of many other (perhaps all) lower fungi, but these are the only examples so far elucidated.

An increasing number of sex pheromone systems are now being characterized in the Ascomycetes and Basidiomycetes. These are all peptides, such as α-factor of *Saccharomyces cerevisiae*, or more usually lipopeptides, such as a-factor of *S. cerevisiae* and tremerogens *A-10* and *a-13* of *Tremella mesenterica* (Fig. 4.7). The pheromones are relatively small peptides, but are encoded by genes with much longer open reading frames, and the gene products undergo extensive post-translational modification involving successive proteolytic cleavages. The lipopeptides have a C-terminal cysteine with the sulphur atom carrying a C_{15} isoprenoid hydrocarbon side chain derived from farnesyl pyrophosphate, i.e. they are farnesylated. The terminal carboxyl group is often methylated. These modifications lead to very hydrophobic molecules. A few of the lipopeptide pheromones have been characterized 'directly', i.e. by painstaking purification and chemical analysis. An increasing number, however, have been discovered indirectly by cloning genes involved in mating. In these cases, the clue has been amino acid sequences in the open reading frames of the genes with the motif CAAX, cysteine–aliphatic amino acid–aliphatic amino acid–any other amino

Table 4.3   Isoprenoid hormones of fungi

Hormone	Molecular structure	Probable precursor	Site and specificity of synthesis	Optimal yield (M)	Sensitivity	Response of bioassay (M)
Antheridiol	Sterol ($C_{29}H_{42}O_5$)	Fucosterol	Female cells of *Achlya* species, constitutive	10^{-8}	10^{-11}	Antheridia and oogoniol by males
Oogoniol	Sterol ($C_{33}H_{54}O_6$)	Fucosterol	Male cells of *Achlya* species, induced	–	10^{-7}	Oogonia by females
Sirenin	Sesquiterpene ($C_{15}H_{24}O_2$)	Farnesyl pyrophosphate	Female gametes of *Allomyces*	10^{-6}	10^{-10}	Chemotaxis of male gametes
Parisin	?	?	Male gametes of *Allomyces*	?	?	Chemotaxis of female gametes
Trisporic acid	Apocarotenoid[a] ($C_{18}H_{26}O_4$)	Retinal	(+)/(–) Cells of species of Mucorales, in collaboration	10^{-6} (*Mucor mucedo*); 10^{-3} (*Blakeslea trispora*)	10^{-8}	Zygophores by (+) and (–) strains

[a] i.e. a degradation product of carotene, $C_{40}H_{56}$, via retinal, $C_{20}H_{28}O$.

Adapted from Gooday, G. W. (1999). Mating and sexual interaction in fungal mycelia. In N. A. R. Gow, G. D. Robson & G. M. Gadd, eds., *The Fungal Colony. Twenty first Symposium of the British Mycological Society*, pp. 261–282. Cambridge University Press, Cambridge.

acid. This is characteristic of a farnesylation site, with the cysteine sulphur atom being farnesylated followed by cleavage of the 'AAX' tripeptide, and probable methylation of the free carboxyl group.

Saccharomyces cerevisiae α-factor is produced by cells of mating type α, two unlinked genes, *MFα1* and *MFα2* being responsible for its formation. The translation product of *MFα1*, 165 amino acid residues in length, begins with a 19 amino acid signal sequence, and includes four tandem repeats of the α-factor sequence. The signal sequence is necessary for the passage of the transcript into the endoplasmic reticulum, and hence of the final product of the proteolytic cleavage of the transcript, α-factor, to the cell exterior. A knowledge of this signal sequence has been utilized in biotechnology to enable heterologous proteins synthesized by genetically manipulated *S. cerevisiae* to be released into the medium (page 538). The transcript of *MFα2* includes two copies of α-factor, one of them differing by two amino acids (legend, Fig. 4.7) from that coded by *MFα1*. α-Factor is active at 10^{-8} M, and binds to an α-factor receptor located in the cell membrane of cells of mating type **a**. The binding of α-factor to a cell results in the activation of genes that control further steps in mating. These steps include the arrest of the cell cycle in G1 phase at 'start' (page 106). The chromosomes are hence in the form of single double-stranded copies. The same effect occurs in cells of mating type α, under the influence of **a**-factor. This, during subsequent karyogamy, facilitates the attainment of diploidy rather than, for example, accidental triploidy. The cells, now gametes, elongate at one end to become 'shmoos' (after the amorphous character in Al Capp's Li'l Abner cartoon strip). The cell wall of the shmoo undergoes changes in composition: within 15 minutes, shmoos develop a cell surface mating agglutinin in the form of fibrous material at the tapered end of the shmoo, the 'conjugation tube' (Fig. 4.8) and there is a big increase in chitin content and an increase in the glucan:mannan ratio. The mating-type-specific agglutinin, interacting with the complementary agglutinin on cells of the opposite mating type following contact, results in agglutination (i.e. sticking together) of a mating pair. This occurs after the tapered end of the **a**-cell has developed into a conjugation tube growing chemotropically towards the source of α-factor – the nearest α-cell. *S. cerevisiae* α-factor also acts on cells equivalent to *S. cerevisiae* **a**-cells in *Saccharomyces kluyveri*, *Saccharomyces exiguus*, *Hansenula wingei* and *Hansenula anomala*, in all instances inducing shmoo formation.

S. cerevisiae **a**-factor is produced by **a**-cells, two unlinked genes being involved. The transcript from each includes a single copy of **a**-factor, the copies differing by a single amino acid residue, and maturation includes methylation and farnesylation. The effects of **a**-factor are similar to those of α-factor, but are exercised on α-cells.

Sex pheromones have been isolated from a number of Ascomycete yeasts, including *S. kluyveri* and *S. exiguus*, which have α-factors similar to those of *S. cerevisiae* (legend, Fig. 4.7), and *Schizosaccharomyces pombe*, in which P (for mating type Plus) cells produce a peptide pheromone, P-factor, and M (Minus) cells produce farnesylated, methylated lipopeptide, M-factor. Isolation of pheromones from mycelial Ascomycetes, in which a single mating type produces both male and female structures, is more difficult, but evidence for the occurrence of sex pheromones has been obtained from several species. For example, in

Neurospora crassa microconidia of each mating type produce a factor which attracts trichogynes from protoperithecia of the other mating type. These factors have not been isolated, nor have comparable factors in the closely related *Ascobolus*. However, sex pheromones have been characterized in the Ascomycete *Cryphonectria parasitica*, the chestnut blight fungus, in an unexpected way. Infection by the hypovirus that infects this fungus (page 95) leads to low virulence, poor asexual sporulation and female infertility. A gene, the expression of which is repressed in infected fungi, proved to encode amino acid sequences with the CAAX motifs. This led to the identification of three genes encoding multiple copies of pheromones, *Mf2/1* and *Mf2/2* expressed in a *Mat-2* mating type strain, and *Mf1/1* in a *Mat-1* strain. Evidence to support this includes: (1) although both mating types have all three genes, the genes are expressed in a mating-type-specific fashion; (2) the gene structure and post-translational processing signals are similar to those of other fungal pheromone genes; (3) a null mutant of *Mf2/2* is sexually sterile; (4) a synthetic peptide based on the *Mf1/1* predicted pheromone sequence inhibits germination of conidia (which can act as gametes) in a mating-type specific manner like the G1 arrest caused by α- and a-factors in *S. cerevisiae* (see above).

The Basidiomycete *Tremella mesenterica* includes a yeast phase in its life cycle, in which the mating process is initiated. Mating is controlled by two loci, with two alleles at one locus, and many at a second, a system known as modified tetrapolar incompatibility (page 251). The alleles at the *A* locus are termed *A* and *a*. The pheromone tremerogen *A-10* (Fig. 4.7) is produced by *A*-cells. It acts on *a*-cells, causing growth arrest in the G1 phase, cell elongation and the production of conjugation tubes. Tremerogen *a-13* (Fig. 4.7) is produced by *a*-cells, and has similar effects to tremerogen *A-10*, but on *A*-cells. Hence the pheromones of *T. mesenterica* resemble in their effects those of *S. cerevisiae*. Pheromones similar to those of *T. mesenterica* are found in a growing number of Basidiomycetes. For example, the formation of *Ustilago maydis* mating hyphae is triggered by a complementary pair of such pheromones, *a1* (13 amino acids) and *a2* (9 amino acids). The resultant mating hyphae then grow chemotropically towards the sources of the pheromone.

Mating among many Basidiomycetes is characterized by the extraordinary lengths to which they go in order to maximize their chances of outbreeding. The tetrapolar mating systems of *Schizophyllum commune* and *Coprinus cinereus* give rise at more than 20 000 and 12 000 mating types, respectively, giving outbreeding potentials better than 98%. In *S. commune*, mating type is collectively determined by two tightly linked multi-allelic loci, *a* and *b*, for *A* and two, *a* and *b*, for *B*. In the worldwide population, there are 9 and 32 specificities, respectively, for *Aa* and *Ab* and 9 each for *Ba* and *Bb*. Compatible mating requires a difference at either of the two *A* loci and either of the two *B* loci. Thus *a* and *b* of each complex are functionally redundant. The complex *A* loci code for interacting homeodomain transcription factors (see below), and the complex *B* loci code for complementary pheromones and their receptors. Two complex *B* mating type loci, *Ba1* and *Bb1*, have been cloned. Each contains a single pheromone receptor gene and three pheromone genes. The pheromone receptor proteins encoded by these genes, Bar1 and Bbr1, and a further one, Bar2, have homology with the pheromone receptors of *S. cerevisiae* (recognizing a-factor),

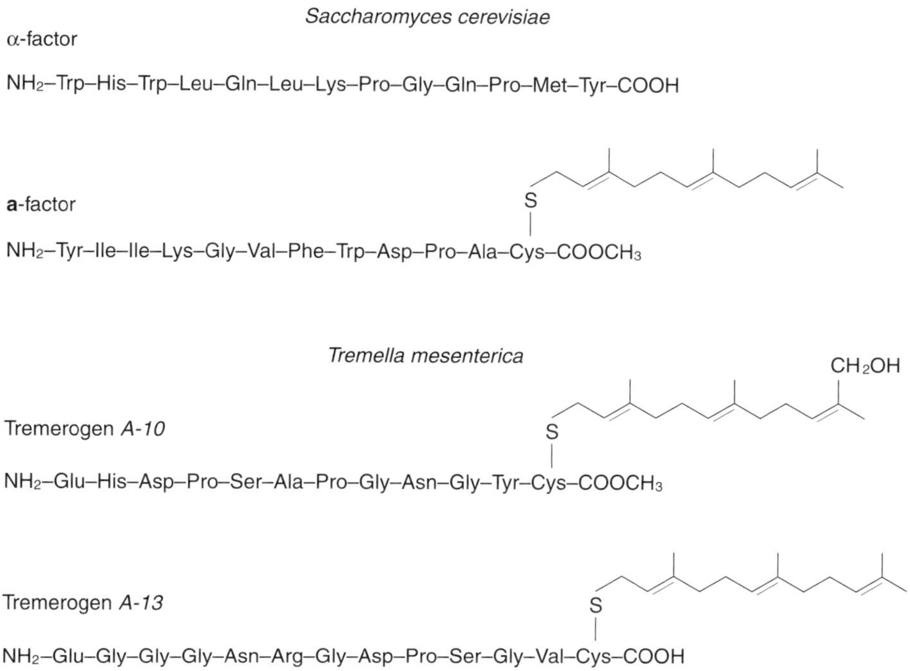

Saccharomyces cerevisiae

α-factor

NH₂–Trp–His–Trp–Leu–Gln–Leu–Lys–Pro–Gly–Gln–Pro–Met–Tyr–COOH

a-factor

NH₂–Tyr–Ile–Ile–Lys–Gly–Val–Phe–Trp–Asp–Pro–Ala–Cys–COOCH₃

Tremella mesenterica

CH₂OH

Tremerogen *A-10*

NH₂–Glu–His–Asp–Pro–Ser–Ala–Pro–Gly–Asn–Gly–Tyr–Cys–COOCH₃

Tremerogen *A-13*

NH₂–Glu–Gly–Gly–Gly–Asn–Arg–Gly–Asp–Pro–Ser–Gly–Val–Cys–COOH

Figure 4.7 The sex pheromones of the Ascomycete yeast *Saccharomyces cerevisiae* and the Basidiomycete *Tremella mesenterica*. The α-factor of *S. cerevisiae*, a tridecapeptide, is produced by cells of mating type α. A small proportion of the α-factor differs from that shown in the replacement of Gln4 by Asn and Lys by Arg. The α-factors of *Saccharomyces kluyveri* and *Saccharomyces exiguus* are closely related to that of *S. cerevisiae*, with five and four amino acid replacements respectively. *S. cerevisiae* a-factor, produced by cells of mating type **a**, occurs in two forms, one having Leu instead of Val. Methylation of the carboxyl group and farnesylation of the S atom of the cysteine render the molecule highly hydrophobic. The tremerogens are the sex pheromones of *Tremella*, 'trem' referring to the genus, 'erogen' to sexual activity, *A* and *a* to the mating type of the producer cells and the numerals to the strains used. Many sex pheromones in other Ascomycetes and Basidiomycetes share similar features to these structures.

U. maydis and *Sch. pombe*. These pheromone receptor proteins are part of the large family of receptor proteins with seven transmembrane domains, involved in signal transduction through G-proteins. The six pheromone precursors encoded by *Ba1* and *Bb1* were recognized by having the 'CAAX' box. There are no obvious differences between the *a* and *b* series, and each pheromone gene is unique in sequence. DNA sequencing of two *B* loci of *Coprinus cinereus* shows that each has six genes for pheromone precursors. These genes code for peptides with the 'CAAX' box for farnesylation at their COOH-termini, but no conservation at their NH₂-termini. Three other genes encode putative pheromone receptors with homology to those of *S. cerevisiae* and *S. commune*.

The discovery that the *B* complex encodes pheromones and pheromone receptors was a surprise, as they have no obvious role in *S. commune* or of

C. cinereus, because initial fusion between two hyphae in these species is independent of mating type. The *B* mating type genes control two distinct processes in mating. Firstly, they regulate nuclear migration following fusion of two monokaryons with different *B* genes. This process involves enzymic dissolution of the complex dolipore septa, and nuclear migration occurs at great speed, estimated at up to 3 mm h^{-1}, leading to very rapid establishment of dikaryotic mycelium. Pheromone genes are expressed during this process, suggesting a direct role in nuclear migration. Thus the pheromones may diffuse ahead of the nuclei, preparing the hyphae for their migration by triggering the activation of lytic enzymes involved in septal dissolution, and by regulating organization of cytoskeletal motor protein functions, or they may play roles in communication between the two types of nuclei. They may also play a role in hyphal behaviour during the hook-cell fusion during clamp connection formation, as this also is controlled by *B* mating type genes.

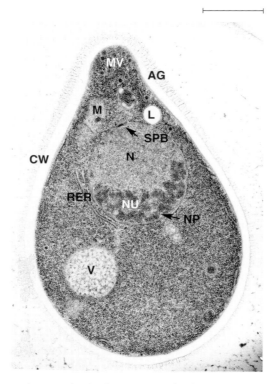

Figure 4.8 Electron micrograph of a thin section of a shmoo of *Saccharomyces cerevisiae*, fixed by freeze-substitution. The **a**-mating type cell has been treated with 3 μM α-factor for 80 min. Note the elongated shape, bounded by the cell wall (CW); an accumulation of microvesicles (MV) at the growing tip; cell surface agglutinin (AG) particularly on the surface of the elongating zone; nucleus (N), with nuclear pore (NP), spindle pole body (SPB) at its apex, and nucleolus (NU) at its base; mitochondrion (M); vacuole (V) with dispersed contents, probably polyphosphate; lipid droplet (L); and rough endoplasmic reticulum (RER) with the membranes lined with ribosomes on their outer faces; and cytoplasm packed with many ribosomes showing as tiny darkly staining dots. (From Baba, M. *et al.* (1989). *Journal of Cell Science* **94**, 207–216.) Scale bar represents 1.0 μm.

Cell surface interactions

S. cerevisiae α-factor, acting on **a**-cells, brings about the synthesis of **a**-agglutinin, and α-factor causes α-cells to produce α-agglutinin. The two agglutinins differ in size and structure, but both are glycoproteins with a high mannose content. Staining of cells with fluorescent antibodies to the agglutinins shows that the agglutinins are located at the surface of the shmoos, particularly around the elongating tips (conjugation tubes) (Fig. 4.8, Plate 7). The two agglutinins are complementary, and interacting between each other cause agglutination only between cells of opposite mating type. Fusion between the tips of conjugation tubes follows their adhesion. Complementary mating agglutinins have been found in a range of yeasts, in some species being induced by the mating pheromone but in others, for example *Hansenula wingeii*, being constitutive.

Genetic control of sexual interaction

Mating type genes are usually regulatory genes, controlling the activity of other genes. The most detailed studies on their mode of action are with *Saccharomyces cerevisiae* (Fig. 4.9). In this species whether a haploid cell is of mating type **a** or

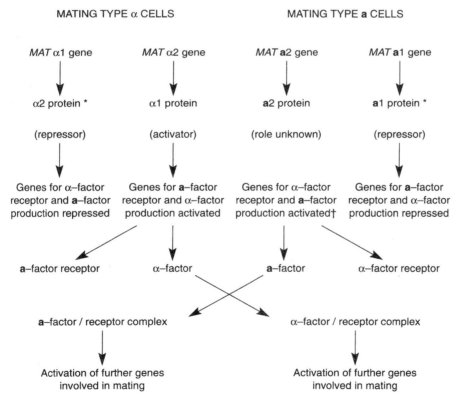

Figure 4.9   Summary of the sequence of events involved in the genetic control of mating in *Saccharomyces cerevisiae*. In mating type **a** cells the genes for the production of **a**-factor and for the α-factor receptor († above) are controlled by a general transcription activator, and, in the diploid cells resulting from mating, the α1 and **a**1 proteins (*above) cooperate to repress activities specific to haploid cells.

α is determined by genes at the *MAT* locus. *MAT*α and *MAT*a both contain two genes, α*1* and α*2*, and **a***1* and **a***2*. The genes at both loci are divergently transcribed, that is separately and in the opposite direction. The α1 protein is a transcription activator for the genes that code for the sex pheromone α-factor and the receptor for a-factor. The α2 protein is a repressor, preventing the transcription of a-specific genes in α cells. The a1 protein represses α-specific genes in a-cells. It also, in the a/α diploid resulting from mating, forms a heterodimer with the α2 protein which has the effect of repressing genes specific for the haploid phase, such as a gene that suppresses meiosis. The function of the a2 protein is unknown. In the a-cell, unlike the α-cell, a general transcription factor suffices for the activation of the genes controlling the production of the sex pheromone and the pheromone receptor. In both mating types the binding of the sex pheromone produced by the opposite mating type brings about the activation of further genes involved in mating, with the consequences already indicated above.

An increasing number of Ascomycete mating type genes are being cloned. This is establishing that the DNA sequences that determine the two mating types within a species differ greatly. In *S. cerevisiae* it has been shown that the α1 and a1 proteins are coded by totally dissimilar sequences of 747 and 642 base pairs respectively, and that the α2 sequence is partly homologous with the a2 sequence and partly dissimilar. In the fission yeast *Schizosaccharomyces pombe* the sequences of 1104 and 1128 base pairs determining the two mating types are completely dissimilar. In *Neurospora crassa mt A* has a unique sequence of 5300 and *mt a* one of 3200 base pairs. Comparable differences are being demonstrated in an increasing number of Ascomycetes. Traditionally the alternative DNA sequences that determine the two mating types in Ascomycetes have been regarded as alleles of a mating type gene. Alleles of a gene, however, usually have very similar DNA sequences, sometimes differing only by a single base pair. It has hence been suggested that the alternative sequences at Ascomycete mating type loci should be termed idiomorphs (Greek, meaning distinct forms), rather than alleles.

Cell fusion in *Ustilago maydis* is followed by the development of a dikaryotic mycelium only if the cells that fused differ with respect to the *mt b* locus. It is probable that the proteins produced by identical *mt b* loci interact to give a dimer that represses transcription of the genes involved in dikaryon growth, but that any heterodimer resulting from the interaction of differing *mt b* loci fails to do so. The *mt b* locus is hence a regulatory locus, as are the *A* and *B* loci that control mating in *Coprinus cinereus* and *Schizophyllum commune* (Chapter 2).

Spore Liberation

Survival spores are liberated by the lysis of the hyphae or fruiting bodies on which they are borne. Such lysis may take a long time, but since the role of survival spores is to persist for an extended period in the dormant state, this does not matter. Dispersal spores, however, need to be liberated as soon as possible so that dispersal can occur and new colonies be established. Dispersal is often by air currents, and for this to take place spores must not only become detached but become airborne. In order to become airborne they must traverse the layer of still air adjacent to the surface on which they were produced. This may be accomplished through a variety of launching mechanisms. Some are passive, utilizing energy from the environment, and some are active, utilizing energy generated within the fungus. Whether launching is active or passive, spores must be detached, and this can occur through enzymic mechanisms or by the rupture of the connection between spore and sporophore at the moment of launching by the forces involved in launching.

Passively launched spores have been grouped into dry spores and slime spores, although as explained below it might be better to contrast dry spores and readily wettable spores. Dry spores are those that have a hydrophobic surface and are difficult to wet, extreme examples being the conidia of *Penicillium* and *Aspergillus*. Even the most hydrophobic spores are capable of ultimately being wetted, perhaps with the aid of naturally occurring surfactants; otherwise germination could not occur. Some spores with hydrophobic surfaces are produced along with mucilage and moisture and form a slimy mass. Such spores, for example the conidia of *Fusarium*, are appropriately designated slime spores. There are other readily wettable spores such as the conidia of *Botrytis*, which are not accompanied by slime. The distinction between dry spores and slime spores is not an absolute one, there being a range from the most hydrophobic to hydrophilic spore surfaces. The distinction is, however, useful since the ways that the launching of dry spores and readily wettable spores can occur differ. Launching mechanisms, active and passive, will now be considered.

Passive Launching of Readily Wettable Spores

Some readily wettable spores are retained separately on conidiophores until they are liberated. Others form a slimy mass on the substratum or a stalked spore drop.

The stalked spore drop

The sporangia of the cellular slime mould *Dictyostelium* and of members of the Mucorales such as *Mucor* itself have a very similar appearance even though they are produced in wholly different ways. The resemblance persists when the *Dictyostelium* or *Mucor* sporangium disintegrates to form a terminal droplet. The perithecium of the Ascomycete *Ceratocystis* has a long slender neck from which ascospores exude to form a spore droplet. The conidia formed at the end of the branched conidiophores of the mitosporic fungus *Cephalosporium* also run together to form a droplet, as do those of the synnemata of another mitosporic fungus, *Graphium*. These examples from taxonomically diverse fungi are all instances of a widespread phenomenon, the **stalked spore drop**. The length of the stalk in the examples cited varies from about 50 µm to about 1 mm. Such a length makes it likely that the droplet would come into contact with the body of a passing insect, spores from the

A

B

C

D

Figure 4.10 Fruit bodies of some Basidiomycetes lacking ballistospores. A, The earth-star, *Geastrum triplex*, in which at maturity the outer layer of the fruit body splits and opens to give a stellate pattern. This exposes the papery inner layer. Raindrops striking the fruit body then result in puffs of spores emerging through the apical pore. B, Young fruit bodies of the puff-ball *Lycoperdon perlatum*. The wall of the mature fruit body is papery with an apical pore, and puffs of spores are expelled when raindrops strike. C, The bird's nest fungus, *Crucibulum laeve*. The 'nest' (splash cup) is about 5 mm in diameter and contains several 'eggs' (peridioles). When a raindrop falls into the splash cup, peridioles containing thousands of spores are hurled up to a metre and adhere to vegetation. The spores are probably dispersed further through the droppings of animals that eat the vegetation. D, The stinkhorn, *Phallus impudicus*. The 'egg' on the left will burst as rapid stipe elongation gives a mature fruit body as on the right. The dark, fertile area at the top of the fruit body is coated with a sugary, sticky, stinking solution, attracting flies which disperse the spores that it contains. (A–D, John and Irene Palmer.) Further illustrations of fruit bodies are available at http://www.wisc.edu/botany/fungi/html

droplet adhere to it, and be transported to another site. Should the spores not be carried away by insects then they may be launched by rain splash, as discussed below. Finally if neither insects nor rain arrive, then the spores may dry and become detached and launched into the atmosphere by wind or mechanical agitation. Unlike dry spores, slime spores are ill adapted to such a mode of launching and rather strong winds or vigorous agitation are required.

Rain splash, drip splash and splash cups

Rain will speedily wet vegetation, covering it with a film of water. The mucilage of slime spores will be diluted so that they will soon be suspended in the film, as will other wettable spores. A drop falling into a film of water will bring about the launching of hundreds of smaller droplets of varying sizes, with the same total volume as the original drop but composed of a mixture of water from the drop and from the film. The number of droplets produced and the height to which they are propelled will depend on the size of the original drop and its speed at impact. Raindrops vary in diameter from 0.2 to 5 mm but drips from vegetation can be larger. A large raindrop (0.5 mm diameter) striking a film of water 0.1 mm deep at its terminal velocity of about 8 m s^{-1} can launch up to 5000 droplets of diameter 5–2400 μm. If the film of water is a dense suspension of spores, then a couple of thousand of the droplets can contain spores. A shower (1 mm rainfall) on an area 1 m^2 is capable of launching about 10^9 droplets, so rain splash and drip splash can be very effective in getting wettable spores into the air. One group of fungi have evolved splash cups to utilize the kinetic energy of raindrops to launch large (a few mm diameter) spore-containing bodies (peridioles). These Basidiomycetes are called bird's nest fungi (Fig. 4.10C) as the peridioles lying in the splash cup look like eggs in a nest. The peridioles are carried to about a metre by droplets which arise from any raindrop falling into the cup, adhere to vegetation and are probably further dispersed via the gut of a herbivore.

Mist pick-up

Spores of some fungi, such as the mitosporic fungal plant pathogen *Verticillium albo-atrum*, which cannot be detached by currents of dry air, are readily removed by a fine spray from an atomizer. It is probable that in nature wind-blown mist is effective in detaching wettable spores from the conidiophores of many fungi.

Passive Launching of Dry Spores

The sporophores of dry-spored fungi can rise above the static layer of air adjacent to solid objects, and dry spores do not adhere to the surface on which they are produced. The forces needed to get them airborne once they are detached may hence be small. It is possible that gravity alone may suffice to launch spores produced on the lower surface of leaves and twigs, although this is difficult to establish, as slight air currents are almost impossible to eliminate. There are basically two ways in which dry spores may be launched, mechanical disturbance and electrostatic repulsion between spore and sporophore, and for spore launching in many fungi it is probable that both are involved.

Mechanical disturbance

Mechanical disturbance can take a wide variety of forms, and more than one form can be brought about by an event. Thus an animal moving through vegetation, or a harvesting machine moving through a crop, will shake plants and also create air currents. The air currents can shake more distant leaves and also act directly on spores. Forms of mechanical disturbance that have been investigated are as follows. **Wind** can dislodge spores, the minimum speed required ranging from 0.4 to 2 m s^{-1} depending on spore type. **Convection currents** resulting from the heating of a surface and hence adjacent air by the sun can lift spores and carry them aloft. A drop of rain hitting a dry surface will both shake the object and cause a puff of air (**rain tap and puff**). Both the shaking and the puff can dislodge spores. Rain striking the wall of a mature puff-ball or similar fungus (Fig. 4.10A, B) will operate a **bellows mechanism**, expelling large numbers of spores through an aperture. In some Myxomycetes a **censer mechanism** operates, the mature fruiting body being a basket-like dispenser on a slender stalk, readily shaken by air currents. In other Myxomycetes, and in the Gasteromycete *Podaxis*, the fruiting body contains elaters, hairs which twist in response to changes in relative humidity. These hygroscopic movements scatter the spores. Hygroscopic movements of sporophores have been observed in a wide range of other fungi, but their significance is uncertain.

Electrostatic repulsion
Leaves, and hence the fungi growing on them, carry an electric charge. The magnitude of this charge varies during the day as changes occur in relative humidity and the amount of infrared radiation. Since spores and sporophores will carry the same charge, repulsion occurs between them, and a sharp increase in charge will result in the rupture of the connection between the spore and sporophore and the launching of the spore. Electrostatic repulsion has been shown to be of importance in the launching of dry spores of fungi from taxonomically diverse groups.

Active Launching of Spores – Fungus Gunnery

The active discharge of spores is widespread in fungi. As compared with passive launching, an investment in complex mechanisms and large fruiting bodies is often required, but the abundance of actively launched spores in the air indicates that there are compensating advantages. Thus active discharge is not dependent on the fortuitous availability of rain, wind or mechanical disturbance at adequate intensity, and spores may be ejected to an optimal distance rather than one dependent on the uncontrolled magnitude of an external force. The requirements for the active discharge of spores in a way optimal for a species are responsible for much of the diversity of form that occurs in fungal fruitbodies and sporophores. The varied details of the discharge process and associated fruit body form and structure have fascinated mycologists and a great deal of information on the subject is available. There are basically, however, a limited number of mechanisms.

Bursting of turgid cells
In many Ascomycetes, such as *Pyronema*, *Sordaria*, *Neurospora* and *Claviceps*, ascospores are forcibly discharged from the ascus. Such asci are tubular, with the

ascospores linearly arranged inside the ascus. It is probable that as ascus maturity approaches, the osmotic potential of the ascus falls and hence its hydrostatic pressure increases, ending in a lid being forced open or blown off at the apex of the ascus, and the ejection of a jet of fluid and spores. The distance to which ascospores travel ranges from a few millimetres to about 30 cm and depends on the momentum and hence the size of the ascospores, the greatest distance being reached in species where the ascospores are large. Asci are commonly phototropic, ensuring that the direction of discharge is optimal for avoiding obstructions. In a few Ascomycetes 'puffing', the simultaneous discharge of thousands of asci in a fruiting body, creates turbulence which assists in the dispersal of the ascospores.

Violent discharge of spores by the bursting of a turgid cell occurs in some other fungi, such as the mitosporic fungus *Nigrospora* and the Zygomycete *Pilobolus* ('cap thrower'). In the latter genus a spore mass – the entire sporangium – is hurled over a metre, from the dung on which the fungus grows onto adjacent grass, where it sticks firmly to the leaf surface (page 41).

Cell shape changes
The spores of some fungi are launched by a sudden change in cell shape, as for example with the aeciospores of *Puccinia*. These spores form a compact mass in which individual spores are polyhedral. Beginning at the outside of the mass, the spores absorb water and suddenly become spherical. In doing so a force is exerted which detaches and launches the spore. In the mitosporic fungus *Deightoniella* the cell shape change that launches the conidium occurs in the conidiophore. Evaporation from the conidiophore causes its shrinkage and distortion. Finally the mechanical tension produced exceeds the tensile strength of the water in the cell, a bubble of water vapour appears, and the tension in the cell wall is released. The resulting reversion to the original cell shape catapults the conidium from the conidiophore. In the Gasteromycete *Sphaerobolus* a spore mass lies in a cup-shaped fruit body. At maturity a sudden release of tissue tensions leads to the eversion of the inner lining of the cup, hurling the spore mass a metre or more.

The ballistospore
In many Basidiomycetes, including the macrofungi *Coprinus*, *Schizophyllum*, the common mushroom *Agaricus bisporus* and the yeasts *Itersonilia* and *Sporobolomyces*, basidiospores are actively discharged. Such basidiospores are termed **ballistospores** (Figs. 4.11, 4.12). In the macrofungal fruit bodies the spores are shot horizontally about 0.1–0.2 mm into the free space between the gills, or for *Boletus* into the centres of the pores (Figure 2.22), allowing them to fall down into the air below the cap. In the yeasts the spores are shot about 1 mm away from the substrate. Colonies growing on agar in an inverted Petri dish will give a mirror image of spores on the underside of the lid, leading to the term 'mirror yeasts'. The following account of the mechanism of discharge of the ballistospores has been deduced by a combination of microscopic and microphysiological observations of both types of spore. The ballistospore is attached to the tip of the sterigma and water from the atmosphere condenses as a thin meniscus around the apical body of the spore, and a separate droplet, known

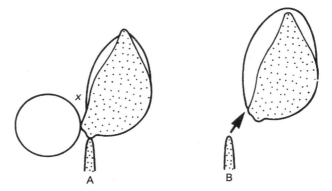

Figure 4.11   The discharge of a ballistospore. The spore and the sterigma on which it rests are stippled. A, A liquid drop is growing at the hygroscopic projection near the base of the spore. It is probable that the area immediately around the projection is hydrophobic, preventing the spreading of water from the projection and allowing the drop to grow. Other parts of the spore surface are readily wettable and may already, as indicated, carry condensation. After a few seconds growth the drop makes contact with the wet, or wettable, spore surface at x. B, The drop flows onto the spore, providing the impetus, as indicated by the arrow, for the launching of the spore (see text). (Figure based on that in Webster *et al.* (1989). *Transactions of the British Mycological Society* **91**, 199–203.)

as 'Buller's drop' after Buller's detailed descriptions of basidiospore discharge in 1922, grows from a hygroscopic projection at the base of the spore (Fig. 4.11A). Electron microscopic observations have shown that the drop is surrounded by a thin pellicle, possibly of hydrophobins, which is in continuity with a pellicle surrounding the spore. Using a micropipette, individual drops, each about 1×10^{-13} litres, were collected from spores of *Itersonilia*. Up to four drops could be collected each time in the micropipette, ejected as a droplet into mineral oil and subjected to microchemical analysis. The results showed that the drops had about 3×10^{-14} moles of mannitol and about 2×10^{-14} moles of glucose and fructose, giving a total of 5×10^{-14} moles of osmolyte. It is envisaged that mannitol and sugars are exuded onto the projection, making it the focus for hygroscopic condensation of water, and that as the drop grows the centre of mass of the spore moves from the centre of the spore to just above the sterigma (Fig. 4.11). Finally the growing drop fuses with the meniscus of the liquid over the main body of the spore, decreasing the surface free energy within the liquid and rapidly moving the centre of mass to the centre of the spore, so that the spore and liquid gain kinetic energy and momentum in the same direction as the moving centre of mass, exerting a force on the sterigma, breaking the connection between spore and sterigma, and discharging the spore (Fig. 4.11A). This mechanism has been termed a 'surface-tension catapult'. In *Itersonilia* the whole process takes 30–45 s, and calculations indicate that the spore reaches a velocity of about 1 m s^{-1} in about 1 μs, so its flight of 1 mm takes 1 ms, with an initial acceleration of about 25 000 g. As stated above, Buller's drop and the liquid on the spore body

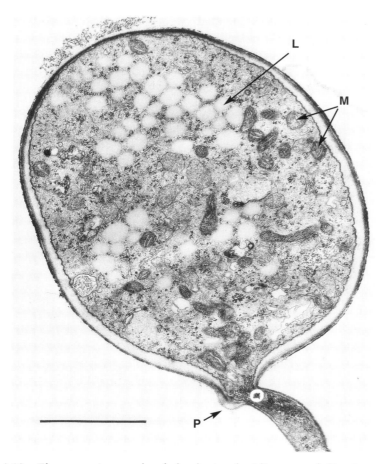

Figure 4.12 Electron micrograph of developing basidiospore of *Coprinus cinereus*, showing accumulation of lipid droplets (L); mitochondria with well-developed cristae (M); development of thicker and more complex wall than that of the sterigma of the basidium; constriction at the attachment site between spore and sterigma; and projection (P), enclosed by a pellicle, that is the site of the hygroscopic condensation of Buller's drop. (D. J. McLaughlin). Scale bar represents 2 μm.

form by condensation of moisture from the atmosphere, rather than by exudation from the cell (which was at one time the favoured theory). This hygroscopic condensation is favoured by the underside of mushrooms being cooled by evaporative cooling, to be at several degrees below air temperature in dry conditions with a light breeze.

Zoospore Release

Zoosporangium induction and maturation can be brought about in many Oomycetes and Chytridiomycetes by washing the mycelium with distilled water.

Zoospore release then requires further exposure to distilled water or to a very dilute salts solution. In nature a high ambient osmotic potential implies abundant water, either due to an aquatic environment or to recent rain or flooding. Such conditions will favour zoospore dispersal.

The mechanisms involved in zoospore release vary between groups. In *Phytophthora* and *Pythium* osmotically generated hydrostatic pressure has a role in the transfer of zoospores, or of protoplasm that then cleaves into zoospores, from the sporangium into a thin-walled vesicle that subsequently ruptures. In *Saprolegnia* zoospores seem at first to be forcibly discharged from the zoosporangium by hydrostatic pressure, after which the remaining zoospores swim out. In *Allomyces* swimming appears to be responsible for zoospore release.

Timing of Spore Liberation

A large proportion of the spores that are launched are airborne for a short period only, and if such spores reach a suitable substratum, prompt germination is desirable so as to exploit resources before the arrival of competitors. It is hence advantageous for spores to be launched when environmental conditions are favourable for both dispersal and germination. What constitutes favourable conditions will depend on the lifestyle of the species and the form of the spore. Small, thin-walled spores can be produced in very large numbers but may be liable to die through desiccation. Such spores are best launched when relative humidity is high. Larger, thick-walled spores may be nearly immune to desiccation, so can be launched at times when wind, which can be desiccating for some spores, is likely to occur and to promote dispersal. Although atmospheric conditions are continually changing, nights tend to be cool, damp and still and days relatively warm, dry and windy. Spore-trapping experiments, especially those carried out in orchards, cornfields or other sources of spores of characteristic types, show a daily fluctuation in the abundance of spores, some types being commonest at night, others in early morning, yet others at noon and some both morning and evening. This suggests, and laboratory experiments with controlled environments prove, that environmental conditions can determine the time of spore launching.

Passively launched spores

The liberation of passively launched spores will be dependent on the time of spore maturation but also on the occurrence of appropriate external conditions that provide the energy for spore liberation. Such external conditions may be periodic or they may not. Rain can occur at any time so the launching of slime spores and other readily wettable spores will not be periodic. The liberation of dry spores that depend on mechanical disturbance for launching will also tend to be irregular. A darkness requirement for spore maturation in many dry-spored plant pathogens ensures the production of an abundant supply of fresh spores overnight. It is likely that many such dry-spored plant pathogens are electrostatically launched. The change in charge that is effective in launching is brought about by exposure to infrared radiation and a fall in relative humidity, events that occur at and soon after dawn, resulting in a morning peak in spore liberation. A rise in relative humidity may also cause electrostatic spore liberation, resulting in a subsidiary evening peak in some species.

Actively launched spores

The liberation of actively discharged spores is not dependent on an external source of energy but, since it needs to occur at times optimal for dispersal and subsequent germination, it is often periodic. Examples of non-periodic liberation are provided by the plant pathogens *Gaeumannomyces graminis* and *Venturia inaequalis*, the fruiting bodies of which discharge their ascospores soon after being wetted by rain. In many fungi the timing of spore discharge is controlled by endogenous circadian rhythms, like those that have been described for sporulation (page 198). These include some species that discharge spores during the day, such as release of sporangia by the Zygomycete *Pilobolus sphaerosporus* and ascopores by *Hypoxylon coccineum*, and others by night, such as ascospores by *Daldinia concentrica* and basidiospores by *Sphaerobolus stellatus*.

Spore Dispersal

The dispersal of fungal spores is passive. Air currents disperse spores that have been actively or passively launched into the air. Water currents carry the spores of aquatic fungi, including zoospores that have swum away from the site of their production. Insects or larger animals disperse the spores of some fungi. Plant-infecting fungi may be passengers when seeds are dispersed, being carried as mycelium within or on the seed or as spores adhering to the seed surface. Dispersal through the air and by the agency of water, animals and seeds will now be considered.

Dispersal Through the Air

The air spora

Fungal spores constitute a major component of the **air spora** – the spores present in the air (Fig. 4.13). Because of the importance of airborne fungal spores in initiating plant disease and in causing allergies in humans there have been extensive studies on the air spora. A variety of methods have been devised for determining the types and numbers of spores present in the atmosphere (Fig. 4.14). Reliable quantitative data can be obtained only by means of devices that trap the spores present in known volumes of air. A device that has been utilized extensively is the Hirst spore trap. Air is sucked at a measured rate through a slit and spores in the air are impacted on a sticky slide which is moved slowly past the slit by clockwork. A single slide will carry the record of 24 hours of continuous

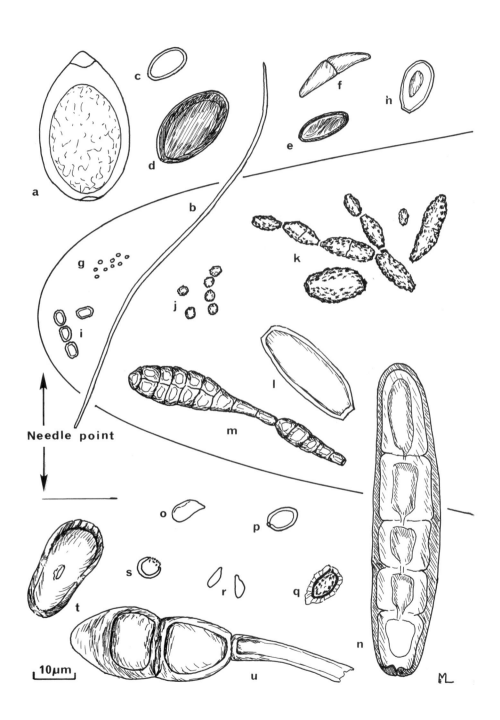

Needle point

10µm

sampling. Examination of the slide with a microscope allows a wide range of spore types and sometimes spores of individual species to be recognized, and their numbers in known volumes of air and at various times of day determined. The Burkard spore trap (Fig. 4.14A) is similar in principle. The actively discharged spores of Ascomycetes and Basidiomycetes are abundant when conditions are damp, whereas the passively launched dry spores of such mitosporic fungi as *Cladosporium* and *Alternaria* are abundant when the weather is dry. Spore trapping demonstrates daily rhythms of spore frequencies in the air, reflecting rhythms of spore liberation. The type and numbers of spores trapped also depend markedly on season and on the nature of the adjacent vegetation, including crops. Counts of microbial particles corresponding to 30 000 per cubic metre of air have been found at 1.5 metres above ground in summer in the English countryside. About a quarter of the particles were the spores of mirror yeasts, about a quarter other basidiospores and about a sixth ascospores. Spore concentrations in the air of large cities are about half those in the country. A general conclusion from studies on the air spora is that the fungal spores in the air mostly come not from the ground but from fungi on plants or from large fungal fruit bodies that rise above the soil surface. The soil itself is a sink into which most airborne spores finally disappear rather than their source.

Convection, wind and eddy diffusion

Spores liberated into the air are dispersed by air currents. The sun warms the earth's surface causing convection currents which carry spores vertically upwards. Wind will tend to move spores horizontally. Wind-borne particles such as spores do not, however, move in straight lines, since turbulence results in eddies. The way in which spores are transported from a point source under various meteorological conditions can be appreciated by the observation of smoke plumes from bonfires or chimneys. A plume moves downwind but spreads both horizontally and vertically through eddying and also becomes fainter. Such **eddy diffusion** affects spores just as it affects smoke particles, so with increasing

---

Figure 4.13 The air spora. Examples of spores found in spore traps, to show a range of sizes and shapes. The curved line is of a needle point. a, Sporangium of the Oomycete *Phytophthora infestans* (potato blight). b–f, Ascospores. b, *Claviceps purpurea* (parasite of grasses). c, *Pyronema omphalodes* (saprotroph). d, *Sordaria fimicola* (dung inhabitant): e, *Xylaria polymorpha* (wood decay). f, *Didymella* sp. (parasite of barley; spores cause late summer asthma). g, Conidium of an Actinomycete, *Thermoactinomyces vulgaris* (cause of 'farmer's lung'), to show the smaller size of prokaryote spores. h–n, Conidia of Ascomycetes and mitosporic fungi. h, *Botrytis* sp. (necrotrophic parasite and saprotroph of plants). i, *Penicillium chrysogenum* (saprotroph). j, *Aspergillus fumigatus* (a cause of asthma and aspergillosis). k, *Cladosporium* sp. (a cause of hay fever). l, *Erysiphe* sp. (powdery mildew). m, *Alternaria* sp. (sooty mould on leaf surface, can cause hay fever). n, *Helminthosporium* sp. (plant parasite). o–u, Basidiomycetes, with o–r, ballistospores. o, *Armillaria mellea* (root pathogen). p, *Serpula lacrymans* (dry rot). q, *Ganoderma applanatum* (bracket fungus). r, *Sporobolomyces* sp. (phylloplane yeast, can cause hay fever and asthma). s, Teliospore of *Ustilago avenae* (smut of oats and barley). t, Uredospore of *Puccinia graminis* (rust of grasses and cereals). u, Teleutospore of *P. graminis*. (Drawing by Maureen Lacey, Rothamsted Experimental Station.)

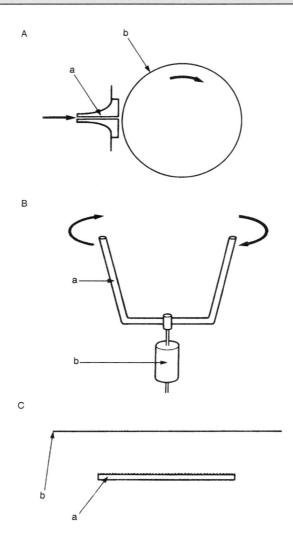

Figure 4.14   Three types of spore trap in use at Rothamsted Experimental Station. A, Burkard spore trap. Air is drawn at a rate of 10 litres per minute through a slit, opposite which a drum, covered in sticky transparent tape, is rotated slowly at a constant rate. Spores drawn in with the air stream impact on the tape, which is later removed from the drum and examined in sections under a microscope, each section corresponding to a known trapping period. A continuous trace of the concentration of different types of spores in the air for periods of up to a week can thus be obtained. a, slit; b, drum. B, Rotorod spore trap. The rotor arms are covered with a cellulose strip coated with wax. The rods are rotated at 3500 revolutions per minute for a fixed sampling period, then the spores on the tape are counted. The volume of air swept by the rods can be calculated from their dimensions and speed. The simplicity and small size of this trap compared with the Burkard trap enable it to be used to compare spore concentrations between closely spaced sites such as the upper and lower levels within a cereal crop. However, it does not provide a continuous record over long periods as can the Burkard trap. a, rotor arm, b, motor. C, Greased microscope slide (a), exposed to the air and protected by a rain shield (b). This is simple and direct but not quantitative. (From details supplied by John and Maureen Lacey, Rothamsted Experimental Station.)

distance from a source a spore cloud will get bigger but become more dilute. A fall in spore concentration from a point source to about 100 metres downwind is readily demonstrated. A small proportion of the spores that are released into the atmosphere may travel great distances. A summer day sampling flight detected a cloud of *Cladosporium* spores over England, another at about 400 km from the coast, over the North Sea and a third at about 650 km, near the Danish coast. Spores of *Cladosporium* are liberated under dry conditions during the day, and the two clouds detected over the North Sea were interpreted as spores released in England during the two previous days and carried by a west wind. Between the *Cladosporium* clouds were high concentrations of ascospores and basidiospores, such as are liberated under damp conditions in darkness.

Rain splash

A raindrop striking a film of water that contains spores will result in the launching of numerous droplets, many of which will contain spores. In this way large numbers of spores will get into the air, but the larger droplets will follow a parabolic trajectory and be deposited at distances of up to 1 metre. The dispersal of the spores that were contained in these droplets may then be continued by the action of further raindrops. Rain splash may therefore act as a mechanism for the effective dispersal of spores over short distances as well as being an agent by which spores are launched.

Dispersal of Spores by Flowing Water

Zoospores and their cysts

Zoospores are produced by many aquatic fungi and under wet conditions by some important plant pathogens such as *Phytophthora*. Zoospores are able to swim (Fig. 4.15) in films of water on plants and in soil as well as in ponds and rivers. Outbreaks of *Phytophthora* infections spread rapidly under conditions in which zoospores are produced. Although *Phytophthora* can swim at speeds of up to 160 µm s^{-1}, changes in swimming direction are frequent, so that in 10 hours of swimming through wet sand zoospores reach only about 6 cm from their source rather than the 6 m that would otherwise be expected. They can, however, be carried over the surface of soil or through soil by flowing water and in this way be transported rapidly from plant to plant. The zoospores of some aquatic fungi encyst soon after they are produced and are probably transported as cysts by water currents.

Conidia of aquatic Hyphomycetes

Leaves of deciduous trees and shrubs that fall into well-aerated streams are attacked by aquatic Hyphomycetes. These produce conidia that are transported by flowing water. In many species the conidia are tetraradiate, consisting of four arms diverging from a central point. Long slender sigmoid conidia are also common. The advantage derived from these characteristic spore forms probably concerns effective deposition on surfaces at the end of their journey.

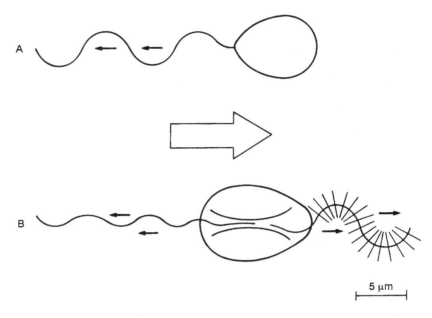

Figure 4.15   The motility of fungal zoospores. The large arrow indicates the direction in which the zoospores are swimming and the small arrows the direction that waves travel along the flagella – away from the body of the zoospore. A, Zoospore of the Chytridiomycete *Blastocladiella emersonii*. Waves travelling backward along the single smooth posterior flagellum generate a thrust that pushes the zoospore forwards. A change in the direction of movement occurs by the development of a sharp bend at the flagellum base, changing the direction in which the head points; in effect, the flagellum has acted as a rudder. B, Zoospore of the Oomycete *Phytophthora palmivora*, with two flagella emerging from a deep groove on the ventral surface of the zoospore. The smooth posterior 'whiplash' flagellum behaves in a similar way to that of *Blastocladiella*. The anterior flagellum is a 'tinsel' flagellum – it carries stiff hairs, mastigonemes. Although the waves travel forward, the resulting movement of the mastigonemes generates a thrust that contributes to the forward movement of the zoospore. (After Carlile, M. J. (1985). The zoospore and its problems. In P. G. Ayres & L. Boddy, eds., *Water, Fungi and Plants. Eleventh Symposium of the British Mycological Society*, pp. 105–118. Cambridge University Press, Cambridge.)

Animal Dispersal

Insects play a role in the dispersal of many fungi. The fungus that causes Dutch elm disease, *Ophiostoma novo-ulmi* (page 53), produces stalked spore drops in the chambers that bark beetles of the genus *Scolytus* excavate beneath the bark of elm trees. The beetles come into contact with the spore drops, and when they leave the tree in which they have developed, carry spores that infect other trees. Many fungi produce spores in sugary and odorous secretions which attract insects. The spores are then dispersed on the surface of the insect or via the insect's gut. Examples of fungi that produce such secretions are the ergot fungus *Claviceps* (page 54), stinkhorns (Fig. 4.10D) and the rust *Puccinia graminis* (page 68).

The spores of some fungi are dispersed by larger animals. The coprophilous fungi such as *Pilobolus* (page 41) discharge their spores so that they fall on adjacent vegetation. This is then consumed by herbivores such as cattle which deposit the spores in their required substrate, dung, possibly at considerable distances from their site of production. Truffles (page 50) produce an odour attractive to rodents and other woodland animals. These excavate and eat the fruit bodies. It is probable that the spores pass unharmed through the rodent gut after which they may germinate to establish a mycorrhizal association (page 401) with a hitherto uninfected tree. The spores of stinkhorns (Fig. 4.10D) are dispersed by flies.

Seed-borne Fungi

Seeds are often efficiently dispersed. Spores or mycelium on or in the seed will be distributed with equal effectiveness. This is a lucky accident for fungi with other means of dispersal, but some fungi have evolved partial or complete dependence on seeds for their distribution. Seed-borne fungi are of particular interest to plant pathologists, as outbreaks of some important plant diseases are almost always due to contaminated seed.

Seed-borne fungi occur in a very wide range of plants. Here the topic will be illustrated by reference to infections of the Gramineae (grasses and cereals), a family in which the 'seed' is strictly speaking a dry indehiscent fruit, the caryopsis. In wheat plants infected with the stinking smut *Tilletia caries*, hyphae penetrate the developing caryopsis to give an interior filled with brandspores. During harvesting the hollow infected grains are broken and the brandspores, which are resting spores, adhere to the healthy grain. Infection results when the grain germinates. The problem can be dealt with by fungicidal seed dressings. In the loose smut of wheat and barley, *Ustilago nuda*, the brandspores are dispersed by wind, and if they reach the stigma of a healthy flower germinate and infect the ovary. The resulting caryopsis contains resting mycelium of the fungus. When 'seed' is distributed and germinates, the mycelium becomes active to give infected seedlings. Hot water treatment (50°C for 15 minutes) can destroy the fungus without rendering the seed non-viable. The grass *Lolium* often has a systemic infection with an endophytic fungus. The fungus penetrates developing ovules to give infected seeds. Crossing of infected and uninfected plants indicates that transmission occurs solely via the ovule. No spores are formed and the endophyte appears to be wholly dependent on seed infection for its survival.

Spore Deposition

Deposition from Air

Observations have been carried out on the number of lesions on susceptible plants at varying distances from a strong source of spores of a plant pathogenic fungus.

The deposition of spores at varying distances from a massive artificial release of spores of a non-pathogenic fungus has also been studied. Both types of investigation show that spore deposition is greatest immediately adjacent to the site of spore liberation and that the vast majority of spores are deposited within 100 metres. Such deposition will be of major importance for the spread of a plant disease within a crop. Although only a small proportion of spores travel further, these will be of importance for the transmission of the fungus over longer distances. The processes responsible for spore deposition from the atmosphere are impaction, sedimentation, and wash-out by rain.

Impaction

An air stream is deflected as it approaches a solid object such as a stem or leaf. Particles carried in the air stream such as spores will also be deflected, but rather less because of their momentum, and may be impacted upon the object. Impaction will be most likely upon slender objects, which cause the least deflection of the air stream, and with high wind speeds and large spores, both of which give greater momentum and minimize spore deflection. The propagules of stem and leaf parasites such as the sporangia of the Peronosporales, the uredospores of rusts and the conidia of *Helminthosporium* are large enough for efficient impaction on stems and leaves. The sporangiospores of the Mucorales and the conidia of such mitosporic fungi as *Aspergillus*, *Penicillium* and *Trichoderma* are small enough to render impaction improbable: these fungi are saprotrophs and colonize dead material likely to be lying on the ground.

Sedimentation

Particles tend to fall under the influence of gravity. Their terminal velocity is reached when the acceleration due to gravity is balanced by friction between the particle and air. This, for a smooth sphere, can be calculated from Stoke's law, and will be proportional to the square of the radius. For a given mass, departures from smoothness and a spherical form will decrease the terminal velocity of a particle. Spores may be neither smooth nor spherical, but experiments on their rate of fall in still air show an approximate agreement with the predictions of Stoke's law. The smallest spores, such as the basidiospores of *Sporobolomyces*, fall at about 0.1 mm s^{-1}, and the largest, such as the uredospores of rusts, at about 10 mm s^{-1}.

Under natural conditions the air is rarely still, and the effect of sedimentation on particles the size of spores is negligible in comparison with convection, wind and turbulence. Still air does occur, however, in the laminar boundary layer which extends for about a millimetre from surfaces and sometimes, if the air is relatively still, to greater distances. Eddies may bring spores to the edge of the boundary layer and then sedimentation may occur. It is likely that much spore deposition occurs in this way.

Wash-out by rain

Raindrops descending through the air can pick up both dry spores and slime spores and deposit them on the ground and on vegetation. Large raindrops are most effective in capturing spores, and large spores – those above about 5 μm diameter – are most readily captured. A heavy fall of rain after a dry period can have a spectacular effect in removing spores from the atmosphere.

Deposition from Water

Zoospores

The way in which zoospores arrive at their destination has been studied most thoroughly with the important plant pathogen *Phytophthora*, which infects plant roots. During their growth through soil, roots slough off cells and also release lubricating macromolecular compounds. Microbial attack on both will liberate water-soluble substances of low molecular weight. Such substances are also emitted by roots and include amino acids, sugars and carboxylic acids. If the soil is waterlogged and hence poorly aerated, ethanol will be released as a result of glycolytic activity by roots. There are thus a wide variety of compounds likely to occur as diffusion gradients in the soil water adjacent to roots. Zoospores able to respond to such compounds by swimming up the gradient – positive **chemotaxis** – will reach the root surface. Positive chemotaxis to root exudates and many of their components, such as amino acids and ethanol, has been shown in zoospores of many species of *Phytophthora* and *Pythium*. There is increasing evidence, however, that another response, **electrotaxis**, is of greater importance in close proximity to the root. Plant roots generate external electric currents and fields due to the flow of protons and other ions into and out of growing and wounded regions. These vary in different conditions, but can be as much as $500\,\text{mV}\,\text{cm}^{-1}$ for a highly resistive soil water. Experiments with zoospores of *Phytophthora palmivora* and *Pythium aphanidermatum* have shown that they respond electrotactically, i.e. by swimming to the anode or cathode, respectively, in fields as low as $2–5\,\text{mV}\,\text{cm}^{-1}$, with saturation at $100\,\text{mV}\,\text{cm}^{-1}$. Exposing paralysed suspensions of zoospores to electric fields has shown that there is a charged dipole across the anterior 'tinsel' flagellum (Figs. 4.15, 4.16), and the posterior 'whiplash' flagellum. For *Ph. palmivora* and *Py. aphanidermatum*, the anterior flagella have negative and positive charges, respectively. This difference correlates with the observation that the zoospores are anodotactic and cathodotactic, respectively. It also correlates with different responses of the two zoospores to rye grass roots: *Ph. palmivora* aggregates at the anodic root tip; *Py. aphanidermatum* at the cathodic subapical region, and at wounds, which are also cathodic.

After arriving at the root surface, the zoospores become immobilized and encyst. Root exudates, amino acids, pectins and other macromolecules released by roots have been shown to cause zoospore encystment. Another response of swimming zoospores then ensues, swimming towards encysted spores, i.e. autotaxis leading to autoaggregation. This appears to be species specific, as mixtures of *Ph. palmivora* and *Py. aphanidermatum* are only attracted by aggregates of the same species. Prior to encystment, zoospores have three types of vesicles in their peripheral cytoplasm: large dorsal vesicles, and two types of smaller vesicles distributed dorsally and ventrally. These smaller vesicles contain preformed glycoproteins for the cyst primary coat and a Ca^{2+}-dependent adhesin, respectively. Within two minutes of commencement of encystment, these

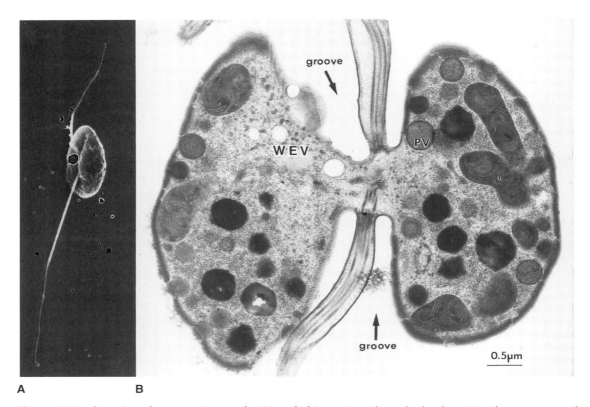

A   **B**

Figure 4.16 Scanning electron micrograph (A) and thin section through the face (B) of a zoospore of *Phytophthora palmivora*. The anterior 'tinsel' flagellum is towards the top of the micrographs, but the tinsel decorations are not visible in these preparations. Note the central groove; longitudinal sections through the two flagella showing their microtubules; the water expulsion vesicle (WEV) – also known as the contractile vacuole; and peripheral vesicles (PV) containing glycoprotein adhesins and cyst wall components. (From Cho, C. & Fuller, M. S. (1989) *Canadian Journal of Botany*, **67**, 1493–1499.)

glycoproteins are exocytosed to encapsulate the spore and to stick it to the plant surface. There is no clear role for the larger vesicles as their contents become randomly distributed. Finally, more material is released to the wall from smaller vesicles, and a range of β-glucans are synthesized to strengthen it. Within about 20–30 minutes the cyst forms a germ tube from its former ventral face, i.e. the face adhering to the plant surface, initiating invasion of the root.

Zoospores of some saprotrophic aquatic fungi are attracted by amino acids, likely to be released by suitable substrates and to form diffusion gradients in the still water close to a surface. As with the zoospores of parasitic fungi, concentrations of amino acids higher than those causing chemotaxis can bring about immobilization and encystment.

Conidia of aquatic Hyphomycetes
The conidia of aquatic Hyphomycetes are commonly either tetraradiate with four long arms or long, slender and sigmoid. Their long arms or overall length

probably increases their chances of being swept against an underwater object. Having struck such an object, a tetraradiate spore comes to rest with three arms in contact with the surface, and a sigmoid spore with contact at two points. Such multiple contact, which is followed by adhesion at the points of contact, probably diminishes the chance of detachment by water currents.

Dormancy and Survival

Between the completion of sporulation and the beginning of germination the spore is in a quiescent state. This period of dormancy may, depending on circumstances, last only a few hours or for many years. During dormancy morphological changes do not occur and the metabolic rate is much lower than in vegetative cells of the same species. It is usual to distinguish between exogenous dormancy and endogenous or constitutive dormancy. **Exogenous dormancy** is that imposed by the lack of a suitable environment for germination: the temperature, for example, may be inappropriate or essential nutrients may be lacking. It is likely that the time and place of germination of dispersal spores (page 187) is often controlled by exogenous dormancy. **Endogenous** or **constitutive dormancy** depends on structural or metabolic features of the spore, and may require precise and unusual conditions for its termination. It can prevent the germination of survival spores (page 188) under conditions that otherwise seem favourable for germination. The distinction between exogenous and endogenous dormancy is not clear cut and it is likely that in many spores both have a role.

Dormancy Prior to Spore Liberation

Spores may remain for a long time on the sporophore on which they were produced, but germination of such spores prior to their liberation is rarely seen. It is clearly desirable that spores should remain dormant until they are liberated, but there is little understanding as to how such dormancy is maintained. In some instances it is possible that lack of water or nutrients is responsible. The conidia of the powdery mildew *Erysiphe graminis*, however, can germinate on glass slides in the absence of either nutrients or liquid water, and without any spore swelling that would indicate water uptake from the atmosphere. Here, therefore, dormancy has to be attributed to some inhibitory effect at the site of sporulation which is lost when individual conidia are dispersed. Sporangiospores and conidia of a range of Zygomycetes, Ascomycetes and mitosporic fungi when placed on cellulose film laid on the mycelium of the parent species fail to germinate, but do so when the film is transferred to water agar. It appears that elusive factors from the parent mycelium are inhibiting germination, and it is possible that such inhibition is responsible for the dormancy of the spores of many species prior to liberation. The germination of the sporangiospores of a Zygomycete, *Chaetocladium jonesii*, parasitic on other fungi, is not inhibited by the mycelium of the parent or other species, so here another explanation must be sought for *in situ* dormancy. One way in which *in situ* germination of spores could be prevented is by the presence of germination inhibitors within the spores – see below. Other forms of endogenous dormancy will be effective in survival spores from the time of their formation.

Dormancy and Survival following Spore Liberation

Self-inhibition of spore germination

Very dense suspensions of spores, or large numbers close to each other on the surface of an agar medium, may fail to germinate. Such self-inhibition of spore germination has been observed in many fungi, but intensively studied in only a few. In some species several inhibitory compounds have been isolated from spores, so there is sometimes doubt as to which are active in maintaining dormancy. The spores of the bean rust (*Uromyces phaseoli*) and those of the rusts of sunflower, maize, snapdragon and peanut contain a cinnamic acid derivative (Fig. 4.17A) that inhibits germ-tube emergence at very low concentrations, 10^{-11} M being effective for the spores of peanut rust. A closely similar compound occurs in the spores of the wheat stem rust (*Puccinia graminis tritici*). Germ-tube emergence occurs through a pore in the spore wall and it appears that the inhibitor prevents the enzymic dissolution of the mannan protein complex that plugs the pore. Quiescence (dormancy) in the conidia of the tobacco downy mildew (*Peronospora tabacina*) is maintained by a β-ionone derivative (Fig. 4.17B) termed quiesone, which is thought to inhibit protein synthesis. An adenine derivative, discadenine (Fig. 2.4D), inhibits the swelling which is a stage in the germination of the spores of the cellular slime mould *Dictyostelium discoideum*. In the spores of some fungi the self-inhibitor of germination is elemental sulphur. Self-inhibitors of fungal spore germination may well have as their main role the maintenance of spore dormancy prior to liberation. However, dispersal may at

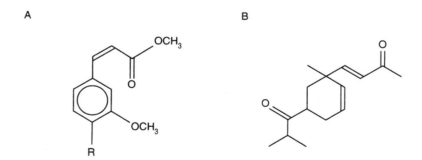

Figure 4.17 Self-inhibitors of germination, which prevent spores germinating when overcrowded. A, Cinnamic acid derivatives inhibit germ-tube emergence in rust uredospores. The inhibitor for bean rust (*Uromyces phaseoli*) is methyl *cis*-3,4-dimethyoxycinnamate (R = —OCH3) and for wheat stem rust (*Puccinia tritici graminis*) is methyl *cis*-ferulate (R = —OH). B, A β-ionone derivative, quiesone, inhibits protein synthesis and hence germination in conidia of tobacco downy mildew, *Peronospora tabacina*.

first be inefficient, and self-inhibitors may prevent germination until adequate separation has occurred. When this has been achieved, loss of inhibitors will take place more rapidly, by passage into the atmosphere if the inhibitors are volatile, or alternatively by diffusion into the aqueous phase which is usually needed for germination.

Mycostasis

Natural, non-sterile soils are able to inhibit the germination of added fungal spores. It is thought that this effect, known as **mycostasis** or **fungistasis** (see also page 337), is due to the activity of soil microorganisms, probably largely through the exhaustion of nutrients needed for germination, but perhaps also by the production of inhibitory metabolites. Susceptibility of spores to mycostasis may be advantageous, preventing germination until local microbial activity is reduced and substrates become available.

Nutrient reserves, trehalose and metabolism of dormant spores

Spores contain a variety of nutrient reserves, which may be located in membrane-bound storage vacuoles, in the cytoplasm in solution or as droplets or granules, or even in the cell wall as occurs with β-glucans in oospores. Common reserves include polyols, as in conidia of *Penicillium*, trehalose, as in sporangiospores of *Phycomyces* and *Neurospora* and, in all spores, lipids. Whereas lipid reserves are partially utilized during dormancy, metabolism of other reserves may be prevented until germination commences. Thus in dormant sporangiospores of *Phycomyces* the enzyme trehalase is in an inactive state. In dormant ascospores of *Neurospora* trehalase is active but is located outside the plasma membrane in the cell wall away from the trehalose in the cytoplasm.

As well as being a nutrient reserve, trehalose has another, perhaps more important, role as a highly efficient protecting agent, by maintaining structural integrity of the cytoplasm under conditions of environmental stress. In some spores it can be present in very high concentrations, at which it could have a powerful stabilizing effect on proteins, and give protection against desiccation. Two hypotheses have been proposed to explain its protective role. One is that trehalose replaces water molecules that are hydrogen-bonded to the surface of proteins and other macromolecules. The multiple hydrogen bonds from the hydroxyl groups of trehalose (Fig. 3.3B) would result in a stronger stabilizing effect under adverse conditions of desiccation, heat and freezing than the hydrogen bonds with water. Another is that a high concentration of trehalose has the tendency to form a glass rather than crystals upon desiccation and freezing, and that this glass would embed the macromolecules and preserve their native shapes.

Metabolic rates of dormant spores are much lower than those of active vegetative cells of the same species. Thus in dormant ascospores of *Neurospora* the rates of oxygen consumption and carbon dioxide production are 1–4% of the rates in vegetative cells. It hence seems that some metabolic activity is needed for maintaining viability. Perhaps some cell constituents are spontaneously degraded under aerobic conditions and at normal temperatures, and the limited metabolism of dormant spores permits their replacement. Dormant spores can, however, retain viability at temperatures, such as those occurring with liquid nitrogen refrigeration, where metabolism – and indeed virtually all chemical reactions – cannot occur.

A dramatic example of the resistance of some fungal spores to environmental stresses is the finding that 100% of *Sordaria macrospora* ascospores germinated after being prepared in a variety of ways and viewed in a scanning electron microscope. Up to 10% of spores even survived these treatments after chemical fixation (glutaraldehyde/formaldehyde, followed by osmium tetroxide). Thus these spores, with low water content and thick, pigmented walls, survived rapid freezing, freeze thawing, freeze drying, high vacuum conditions, irradiation in an electron beam, and chemical fixatives. Fungi have been grown from glacial ice cores from Greenland, ranging in age from several hundred to 140 000 years old. Stringent precautions were taken to avoid contamination by contemporary fungi. Morphological and molecular characteristics of eight of the isolates allowed their identification as species of *Penicillium* (3), *Cladosporium* (2), *Ulocladium* (1), a Basidiomycete, *Pleurotus*, and an unknown Ascomycete. Thus it is clear that some fungal spores have remarkable capacities for prolonged viability in the absence of metabolism.

Form and wall composition of spores with prolonged survival
Dormant spores are exposed to a hostile environment. Hazards include solar radiation, desiccation and attack by other microorganisms. A sphere is the form that has the minimum surface in relation to volume, and departures from a spherical form in spores are usually explicable in terms of the requirements of dispersal. Spores capable of prolonged survival are usually spherical or nearly so, presenting the least possible surface to the environment. Their cytoplasm often contains very high concentrations of trehalose (see above).

The composition of the wall is also important with respect to protection. Melanins, black pigments produced by the oxidative polymerization of tyrosine or phenol derivatives, occur in the walls of many spores. Examples include the sporangiospores and zygospores of *Mucor mucedo*, the conidia of *Alternaria* and *Aspergillus nidulans* and the basidiospores of the mushroom *Agaricus bisporus*. There is no evidence that melanin can be degraded by microorganisms or any enzyme, so melanin probably confers protection against microbial attack as well as against solar radiation (page 168). A chemically even more refractory compound is sporopollenin, a complex highly cross-linked polymer also found in exines of pollen grains. This is less common than melanin, but occurs in the walls of some spores noted for prolonged survival, such as the ascospores of *Neurospora* and the zygospores of *Mucor mucedo* and *Phycomyces*.

Spore Activation: the Breaking of Dormancy

Spore germination (page 237) involves the initiation of biochemical activities absent in the dormant phase, a gradual increase in metabolic rates towards the levels normal in vegetative cells and morphological changes, the most striking of which is the emergence of a germ-tube. The circumstances under which the dormant phase will end and spores become active varies between spore types within a species and between different species. This is not surprising, as different spore types have different roles, and each species has its own niche in the microbial community.

General requirements for spore activation

There are some requirements for spore germination that are very widespread or universal. Most species need liquid water for germination, although for some a high relative humidity suffices, as with the spores of some powdery mildews in which contact with liquid water causes death. Most fungi are obligate or facultative aerobes, and the presence of oxygen is usually needed for spore activation. A requirement for carbon dioxide, probably as a primer for various biochemical reactions, is common. The temperature limits for spore germination vary, but tend to be narrower than for vegetative growth.

Nutrient requirements for spore germination

Water-soluble, low-molecular-weight nutrients, such as sugars or amino acids, are of frequent occurrence in nature, although at any site their presence is likely to be brief, owing to microbial activity. They may well, however, indicate locations suitable for fungal growth, as for example when they are exuded from plant roots or diffuse from dead organic materials. Hence they could provide useful signals for spore germination in saprophytic fungi and in some parasites, such as those that infect plant roots. Nutrient requirements for spore germination are in fact widespread in fungi. Thus the macroconidia of the root-infecting fungus *Fusarium culmorum* require a carbon and a nitrogen source for germination. Nutrients are also needed for the activation of the macroconidia of the saprophyte *Neurospora*. However, some fungal spores, such as the ascospores of *Neurospora*, can germinate in pure water, although such spores may require prior activation by exposure to specific physical conditions or chemical agents.

Specific physical and chemical stimuli for spore germination

The spores of many fungi germinate readily when diluted and spread on the surface of an agar medium or added to a liquid medium. Others, however, may fail to germinate even when conditions are ideal for mycelial growth. The spores of wood-destroying Basidiomycetes are often easy to germinate, but not those of fleshy agarics such as the mushroom *Agaricus bisporus*. Germination of mushroom basidiospores occurs most readily in the presence of mycelium of the same species, and chemicals emitted by the mycelium seem to be responsible (Fig. 4.18). Similar effects have been observed in other agarics. Basidiomycete mycelium is often long-lived, and the stimulation of basidiospore germination by mycelium of the same species suggests that the role of spores in Basidiomycetes may sometimes be the transmission of genes rather than the establishment of new colonies. In plant pathogens spore germination may be brought about by chemicals emitted at low concentration by the potential host (Fig. 4.18). The germination of teliospores of the safflower rust, *Puccinia carthami*, for example, are brought about by polyacetylenes produced by the host. In other rusts a wide

A $CH_3 \cdot CH = CH \cdot (C \equiv C)_4 \cdot CH = CH_2$ B $CH_3 \cdot (CH_2)_7 \cdot CHO$

    $CH_3 \cdot (C \equiv C)_5 \cdot CH = CH_2$ $CH_3 \cdot (CH_2)_7 \cdot CH_2OH$

C $\begin{matrix} CH_3 \\ \diagdown \\ \diagup \\ CH_3 \end{matrix} CH_2 \cdot CH_2 \cdot COOH$ D $(CH_2 = CH \cdot CH_2 S -)_2$

    $\begin{matrix} CH_3 \\ \diagdown \\ \diagup \\ CH_3 \end{matrix} CH_2 \cdot CH_2 \cdot CH_2OH$

Figure 4.18 Compounds stimulating germination at low concentrations. Propagules of parasitic and mycorrhizal fungi germinate in response to metabolites released by their hosts, and those of some Basidiomycetes to products of the mycelium of the same species, which presumably facilitates dikaryon formation. The stimulants include hydrocarbons, fatty acids, aldehydes and alcohols. A, Polyacetylenes stimulate germination of uredospores of the safflower rust, *Puccinia carthamni*. B, Nonanal and nonanol induce spore germination in a wide range of rusts and smuts. C, Germination of basidiospores of *Agaricus bisporus* is induced by isovaleric acid and isoamyl alcohol. D, Organic sulphur compounds from onions and garlic, such as diallyl sulphide, stimulate germination of sclerotia of *Sclerotium cepivorum*.

range of hydrocarbons, alcohols, aldehydes, ketones and terpenes have been shown to promote spore germination, and often these are compounds likely to be emitted by host plants. The spores of some fungi require a period at a low temperature followed by return to a higher temperature before they can germinate. Such a requirement may prevent autumn germination of spores having the role of surviving over winter. Germination of some spores requires a short exposure to a high temperature. The ascospores of *Neurospora*, for example, are activated by 10–20 minutes at 50–60°C. *Neurospora* is often found on burnt vegetation in the tropics, and the heat requirement may be relevant to this habitat. Ascospore activation in *Neurospora* can also be brought about by furfural, a product of the partial combustion of the hemicellulose component of wood. Dormancy is also broken by various organic solvents at higher concentrations. Studies on the activation of *Neurospora* ascospores indicate that the effective agents bring about a change in the permeability of lipoprotein membranes and allow trehalase to act upon trehalose, a major nutrient reserve (Fig. 3.3B). Many of the chemicals that break the dormancy of fungal spores are lipophilic, and changes in the permeability of the plasma membrane or of internal membrane systems may be widely involved in the breaking of dormancy. Requirements for light, darkness, or light followed by darkness have been reported for spore germination in some plant pathogens, but detailed studies are few. Such requirements may be timing devices to ensure that germ-tube emergence occurs at the time of day or night optimal for infection. Germination of the tetraradiate spores of aquatic Hyphomycetes requires contact with a solid object. No procedure is known for stimulating the germination of the zygospores of *Mucor mucedo*; germination of members of a population is spread over a long period and it is likely that gradual endogenous changes are responsible.

Spore Germination

Morphological Changes During Germination

Germination of the cysts (encysted zoospores) of the Chytridiomycete *Blastocladiella* (page 36) has been studied intensively. Within 30 minutes of encystment the nuclear cap of the zoospore disintegrates to release ribosomes, and bulging is seen at one point on the cyst wall. Within an hour the bulge becomes a slender hypha, the rhizoid, which penetrates the substratum and subsequently branches. The cyst nucleus undergoes mitosis within 3 hours, and before a day has elapsed the cyst has swelled to become the multinucleate vegetative thallus.

In most fungi spore germination takes the form of the production of a hypha, the **germ-tube**, which by its elongation and branching establishes the mycelium (Fig. 4.19). The spores of fungi commonly swell in the course of germination. Swelling may be slight, but in some fungi a many-fold increase in spore volume occurs. Spore swelling starts within an hour of encountering conditions suitable for germination and continues for several hours. The volume increase involves water uptake, but at the same time the dry weight of the spore increases and wall growth – not merely wall stretching – takes place. It can be shown with approximately spherical spores that diameter increase is linear with time. This implies that the circumference increases linearly with time, since the circumference of a sphere varies as the radius. The linear relationship is probably the result of wall synthesis – the extension of a mature hypha, also limited by wall

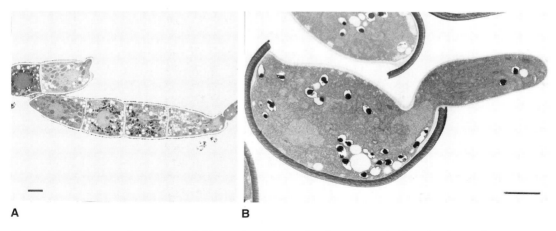

A   B

Figure 4.19 Electron micrographs of thin sections of germinating spores. A, Germinating multiseptate conidia of the mitosporic fungus *Cercosporidium personatum*, with germ-tubes emerging from ends of spores, with their walls continuous with those of the parent spores. B, Germinating ascospores of *Daldinia concentrica*, with their thin cell walls continuous with the innermost layer of the complex spore wall. The densely stained thick outer wall has been ruptured during germination. Note the nucleus migrating into the germ-tube. In both species, note the numerous translucent lipid droplets. (A from C. W. Mims; B from A. Beckett.) Scale bars represent 2 μm.

synthesis, is linear with time. Since spore swelling involves much more than water uptake the term spherical growth may be more appropriate than swelling.

Germ-tube emergence occurs some hours after germination has been initiated. When cysts of *Blastocladiella* or *Phytophthora* germinate, the germ-tube wall is continuous with the wall of the cyst, itself often recently formed since in these fungi germination usually rapidly follows encystment. Different arrangements occur in terrestrial fungal spores. In some, such as the conidia of *Botrytis cinerea* and *Cercosporidium personatum* (Fig. 4.19A), the germ-tube walls are extensions of the inner layer of the spore wall. In many spores, such as the sporangiospores of *Mucor rouxii* and ascospores of *Daldinia concentrica* (Fig. 4.19B), an early stage in germination is the synthesis of a new wall beneath the existing spore wall, and the germ-tube wall is an extension of the new wall. The old spore wall is distended and finally fractured by germ-tube emergence. In arthrospores of *Geotrichum candidum* the new wall is limited to the region where emergence occurs. In uredospores of rusts, emergence is by the enzymic dissolution of a spore plug. In most spores, however, emergence occurs not at a specific site but by the rupture of the old spore wall. Both pressure from the growing germ-tube and enzymic attack can be involved.

Metabolic Changes During Germination

DNA, RNA and protein synthesis do not occur in the dormant spore. Gradual utilization of lipid reserves and low respiration rates indicate that there is some metabolic activity, probably related to the maintenance of viability at normal temperatures. Zoospores have additional metabolic activity supporting motility and regulating entry of water and ions, but in their lack of biosynthetic activity resemble dormant spores. Some of the enzymes that will be active in the germination of dormant spores or zoospores are present in such cells. They are, however, either in an inactive form (e.g. trehalase in *Phycomyces*), or separated from their substrate (e.g. trehalase in *Neurospora*) or otherwise prevented from functioning (e.g. chitin synthase in *Blastocladiella* zoospores). Ribosomes and some classes of messenger RNA may be present but inactive. In *Blastocladiella* a translation inhibitor is present but lost at the start of germination. During germination the various forms of metabolic activity are restored to the level characteristic of the vegetative cell. The sequence of events for any spore type is far from being fully known, and as there are differences between spores with respect to the first step, spore activation, the sequence of events probably varies considerably between species. In some spores, including *Blastocladiella* zoospores, RNA synthesis starts within about 15 minutes of the induction of germination. As germination proceeds the number of free ribosomes falls and that of polysomes increases, indicating that messenger RNA is being translated. As would be expected from increased biosynthetic activity and respiration, ultrastructural changes occur in germination, including increases in amount of endoplasmic reticulum and changes in the number and form of mitochondria.

Growth and Orientation of Germ-tubes

Germ-tubes show an exponential increase in length until a maximum extension rate is reached, after which increase in length is linear with time. The same applies to branches arising from the germ-tube. The topic is considered further in Chapter 3 (pages 137–138); here factors influencing the orientation of germ-tubes will be discussed.

Autotropism

If two spores are in contact, the germ-tubes tend to emerge at the most distant points on the spore circumference and to grow away from each other. This tendency for germ-tubes to avoid each other, **negative autotropism**, is most pronounced with a clump of germinating spores. Negative autotropism presumably reduces any competition between germ-tubes for nutrients. The chemical basis for negative autotropism is uncertain. There are indications that in some spores it is due to positive chemotropism towards oxygen (**positive aerotropism**) up a gradient created through oxygen utilization by spores. In higher fungi **positive autotropism** of germ-tubes sometimes occurs, resulting in hyphal anastomosis (Chapter 3, pages 142–144). In *Botrytis cinerea* both positive and negative autotropism of germ-tubes has been observed, the latter at high carbon dioxide concentrations likely to be indicative of excessive crowding of spores.

Other chemotropic responses

Growth of germ-tubes towards hosts or substrates has often been observed, and is presumably due to chemotropism, growth up or down a chemical gradient. Little, however, is known for most fungi about the nature of the chemicals involved. Positive **chemotropism** to amino acids occurs with germ-tubes of the Oomycetes *Achlya* and *Saprolegnia* and rhizoids of the Chytridiomycete *Blastocladiella*. Germ-tubes from *Phytophthora* grow towards plant roots. The nature of the attractant is unknown, but there is evidence that these germ-tubes grow down an oxygen gradient, which could well be produced by the metabolic activity of roots. Growth up an oxygen gradient may occur with the germ-tubes of fungi that infect plants through stomata (Chapter 7, page 367), and germ-tubes from ascospores of the wood-destroying fungus *Chaetomium* show positive chemotropism to volatile factors emitted by wood.

Phototropism

Germ-tubes from conidia of *Botrytis cinerea* and the uredospores of some rusts show negative phototropism, growing away from the light. This may promote penetration of leaf tissues by these pathogens.

Thigmotropism (contact guidance)

The orientation of the germ-tubes of some plant pathogens is determined by the physical form of the plant surface, a response that may facilitate infection. Germ-tubes of rust species, such as *Puccinia hordei* on barley leaves, grow at 90° to the longitudinal arrangement of epidermal cells on their host grass leaves (Fig. 7.2). This is thought to maximize their probability of encountering stomata, their site of invasion, which are arranged in staggered longitudinal rows. In contrast, germ-tubes of some other species grow along the junctions between leaf epidermal cells. Experiments with a range of species strongly suggest that germ-tubes, and mature hyphae, have mechanoreceptors, possibly involving stretch-activated ion channels, which sense the topography of their underlying substrate.

Responses to Adverse Conditions

Spore germination in a fungus, as already discussed, is initiated by signals that are usually indicative of conditions favourable for the vegetative growth of the species. The signals, however, are not unfailingly reliable, so dormancy may be broken under conditions unsatisfactory for subsequent growth. One response, observed in many species, is to relapse into the dormant state. This will, for example, occur after heat activation, if the ascospores of *Neurospora* are placed under anaerobic conditions or *Phycomyces* sporangiospores are not provided with nutrients. Relapse into dormancy will occur only if unfavourable conditions are soon sensed. If germination has proceeded to the stage of spore swelling then resort to other options is necessary. Shortage of key nutrients, overcrowding (which may be the same in its effects) or temporary high temperatures may lead, by microcycle sporulation (page 194), to spore production rather than vegetative growth. The compartments of the macroconidia of *Fusarium*, if these spores are placed in conditions unfavourable for growth, become chlamydospores instead of forming germ-tubes. The encysted zoospores of *Phytophthora*, exposed to root exudates, form germ-tubes, but placed in water lacking nutrients a cyst will germinate to release a zoospore. Under natural conditions this response may provide a second chance of reaching a host. Alternate forms of germination also occur in *Mucor*, where absence of oxygen combined with high carbon dioxide concentrations lead, in some species, to sporangiospores budding like yeasts instead of producing germ-tubes.

Further Reading and Reference

General Works

Beckett, A., Heath, I. B. & McLaughlin, D. J. (1974). *An Atlas of Fungal Ultrastructure*. Longman, London.

Berry, D. R., ed. (1988). *Physiology of Industrial Fungi*. Blackwell, Oxford.

Buller, A. H. R. (1909–1950). *Researches on Fungi*, vols. 1–7. Longmans Green, London. Reprinted (1958) by Haffner Publishing Co., New York.

Burnett, J. H. (1976). *Fundamentals of Mycology*, 2nd edn. Arnold, London.

Chui, S.-W. & Moore, D., eds. (1996). *Patterns of Fungal Development*. Cambridge University Press, Cambridge.

Cole, G. T. & Samson, R. A. (1979). *Patterns of Development in Conidial Fungi*. Pitman, London.

Cooke, R. C. & Whipps, J. M. (1993). *Ecophysiology of the Fungi*. Blackwell, Oxford.

Griffin, D. H. (1994). *Fungal Physiology*, 2nd edn. Wiley, New York.

Madelin, M. F., ed. (1966). *The Fungus Spore*. Butterworths, London.

Moore, D. (1998). *Fungal Morphogenesis*. Cambridge University Press, Cambridge.

Read, N. D. (1994). Cellular nature and multicellular morphogenesis in higher fungi. In D. S. Ingram & A. Hudson, eds., *Shape and Form in Plants and Fungi*, pp. 251–269. Linnean Society, London.

Weber, D. J. & Hess, W. M., eds. (1976). *The Fungal Spore: Form and Function*. Wiley, New York.

Wessels, J. G. H. & Meinhardt, F., eds. (1994). *Growth, Differentiation and Sexuality*. Vol. I. *The Mycota*, K. Esser & P. A. Lemke, eds. Springer-Verlag, Berlin.

Sporulation and its Genetic and Environmental Control

Adams, T. H., Wieser, J. K. & Yu, J-H. (1998). Asexual sporulation in *Aspergillus nidulans*. *Microbiology and Molecular Biology Reviews* **62**, 35–54.

Banuett, F. (1992). *Ustilago maydis*, the delightful blight. *Trends in Genetics* **8**, 174–180.

Banuett, F. (1995). Genetics of *Ustilago maydis*, a fungal pathogen that induces tumors in maize. *Annual Review of Genetics* **29**, 179–208.

Casselton, L. A. & Olisnicky, N. S. (1998). Molecular genetics of mating recognition in basidiomycete fungi. *Microbiology and Molecular Biology Reviews* **62**, 55–70.

Corrochano, L. M. & Cerdá-Olmedo, E. (1992). Sex, light and carotenes: the development of *Phycomyces*. *Trends in Genetics* **8**, 268–274.

Davey, J., Davis, K., Hughes, M., Ladds, G. & Powner, D. (1998). The processing of yeast pheromones. *Cell and Developmental Biology* **9**, 19–30.

Dyer, P. S., Ingram, D. S. & Johnstone, K. (1992). The control of sexual sporulation in the Ascomycotina. *Biological Reviews* **67**, 421–458.

Elliott, C. (1994). *Reproduction in Fungi: Genetical and Physiological Aspects*. Chapman & Hall, London.

Gooday, G. W. (1992). The fungal surface and its role in sexual interactions. In J. A. Callow & J. R. Green, eds., *Perspectives in Plant Cell Recognition, Society for Experimental Biology Seminar Series*, vol. 48, pp. 33–58. Cambridge University Press, Cambridge.

Gooday, G. W. (1999). Mating and sexual interaction in fungal mycelia. In N. A. R. Gow, G. D. Robson, & G. M. Gadd, eds., *The Fungal Colony. Twenty-first Symposium of the British Mycological Society*, pp. 261–282. Cambridge University Press, Cambridge.

Gooday, G. W. & Adams, D. J. (1993). Sex hormones and fungi. *Advances in Microbial Physiology* 34, 69–145.

Gressel, J. & Rau, W. (1983). Photocontrol of fungal development. In W. Shropshire & H. Mohr, eds., *Photomorphogenesis. Encyclopedia of Plant Physiology*, vol. 16B, pp. 603–639. Springer-Verlag, Berlin.

Herskovitz, I. (1988). Life cycle of the budding yeast *Saccharomyces cerevisiae*. *Microbiological Reviews* **52**, 536–553.

Kües, U. (2000). Life history and developmental processes in the Basidiomycete *Coprinus cinereus*. *Microbiology and Molecular Biology Reviews* **64**, 316–353.

Lovett, J. S. (1975). Growth and differentiation in the water mold *Blastocladiella emersonii*. *Bacteriology Reviews* **39**, 345–404.

Moore, D. (1994). Tissue formation. In N. A. R. Gow & G. M. Gadd, eds., *The Growing Fungus*, pp. 423–465. Chapman & Hall, London.

Moore, D., Casselton, L. A., Wood, D. A. & Frankland, J. C., eds. (1985). *Developmental Biology of the Higher Fungi. Twelfth Symposium of the British Mycological Society*, vol. 10. Cambridge University Press, Cambridge.

Moore-Landecker, E. (1992). Physiology and biochemistry of ascocarp induction and development. *Mycological Research* **96**, 705–716.

Raju, N. B. (1992). Genetic control of the sexual cycle in *Neurospora*. *Mycological Research* **96**, 241–262.

Ramsdale, M. (1999). Circadian rhythms in filamentous fungi. In N. A. R. Gow, G. D. Robson & G. M. Gadd, eds., *The Fungal Colony. Twenty-first Symposium of the British Mycological Society*, pp. 75–107. Cambridge University Press, Cambridge.

Smith, J. E., ed. (1983). *Fungal Differentiation: a Contemporary Synthesis*. Marcel Dekker, New York.

Staben, C. (1994). Sexual reproduction in higher fungi. In N. A. R. Gow & G. M. Gadd, eds., *The Growing Fungus*, pp. 383–402. Chapman & Hall, London.

Turian, G. & Hohl, H. R., eds. (1981). *The Fungal Spore: Morphogenetic Controls*. Academic Press, London.

Watling, R. & Moore, D. (1993). Moulding moulds into mushrooms: shape and form in higher fungi. In D. S. Ingram, ed., *Shape and Form in Plants and Fungi*. Academic Press, London.

Wessels, J. G. H. (1992). Gene expression during fruiting in *Schizophyllum commune*. *Mycological Research* **96**, 609–620.

Wessels, J. G. H. (1993). Fruiting in the higher fungi. *Advances in Microbial Physiology* **34**, 147–202.

Zhang, L., Baasiri, R. A. & Van Alfen, N.K. (1998). Viral repression of fungal pheromone precursor gene expression. *Molecular and Cellular Biology* **18**, 953–959.

Spore Liberation, Dispersal, Survival and Germination

Aylor, D. E. (1990). The role of intermittent wind in the dispersal of fungal pathogens. *Annual Review of Phytopathology* **28**, 73–92.

Carlile, M. J. (1985). The zoospore and its problems. In P. G. Ayres & I. Boddy, eds., *Water, Fungi and Plants. Twelfth Symposium of the British Mycological Society*, pp. 105–118. Cambridge University Press, Cambridge.

Chambers, S. M., Hardham, A. R. & Scott, E. S. (1995) *In planta* immunolabelling of three types of peripheral vesicles in cells of *Phytophthora cinnamomi* infecting chestnut roots. *Mycological Research* **99**, 1281–1288.

French, R. C. (1992). Volatile chemical stimulators of rust and other fungal spores. *Mycologia* **84**, 277–288.

Gow, N. A. R., Campbell, T. A., Morris, B. M., Osborne, M. C., Reid, B., Shepherd, S. J. & van West, P. (1999). Signals and interactions between phytopathogenic zoospores and plant roots. In R. R. England, G. Hobbs, N. J. Bainton & D. M. Roberts, eds., *Microbial Signalling and Communication. Fifty-seventh Symposium of the Society for General Microbiology*, pp. 285–305. Cambridge University Press, Cambridge.

Gregory, P. H. (1973). *The Microbiology of the Atmosphere*, 2nd edn. Leonard Hill, Aylesbury.

Ingold, C. T. (1971). *Fungal Spores: their Liberation and Dispersal*. Clarendon Press, Oxford.

Leach, C. M., Hildebrand, P. D. & Sutton, J. C. (1982). Sporangium discharge by *Peronospora destructor*: influence of humidity, red–infrared radiation, and vibration. *Phytopathology* **72**, 1052–1056.

Money, N. P. & Webster, J. (1985). Water stress and sporangial emptying in *Achlya* (Saprolegniaceae). *Botanical Journal of the Linnean Society* **91**, 319–327.

Nagarajan, S. & Singh, D. V. (1990). Long distance dispersion of rust pathogens. *Annual Review of Phytopathology* **28**, 139–153.

Nilsson, S. (1983). *An Atlas of Airborne Fungal Spores in Europe*. Springer-Verlag, Berlin.

Sussman, A. S. & Halvorson, H. O. (1966). *Spores, their Dormancy and Germination*. Harper & Row, New York.

Thevelein, J. M. (1996). Regulation of trehalose metabolism and its relevance to cell growth and function. In R. Brambl & G. A. Marzluf, eds. *Biochemistry and Molecular Biology*. Vol. III *The Mycota*, K. Esser & P. A. Lemke, eds. pp. 395–420. Springer-Verlag, Berlin.

Questions

With each question one statement is correct. Answers on page 561.

4.1 The role of which spores is predominantly that of dispersal?
   a) chlamydospores
   b) ascospores
   c) conidia
   d) zygospores

4.2 The role of which spores is predominantly that of survival?
   a) oospores
   b) zoospores
   c) basidiospores
   d) oidia

4.3 *Mucor* species produce:
   a) zoospores
   b) sporangiospores
   c) conidia
   d) oidia

4.4 When trying to characterize an unknown sex pheromone of an ascomycetous yeast, an informed guess is that it will prove to be:
   a) a sterol
   b) an isoprenoid
   c) a lipoprotein
   d) a lipopeptide

4.5 Zoospores of *Phytophthora* have:
   a) one posterior flagellum
   b) two identical flagella
   c) two dissimilar flagella
   d) a bundle of several flagella

4.6 In which circumstances will ascospores of *Neurospora* not geminate:
   a) in water after heat treatment
   b) in nutrient medium after heat treatment
   c) in nutrient medium with added vitamins
   d) in water after treatment with organic solvents

Genetic Variation and Evolution

5

Fungal Genetics – Scope and Limitations

Fungi possess features that render many ideal for genetic research. Most fungi can be grown in pure culture under controlled conditions. This means that environmental variability which could conceal or be confused with genetic differences can be minimized. Many have predominantly haploid life cycles with most genes expressed in the haploid phase. Mutant alleles of such genes are readily detected. This contrasts with the situation in predominantly diploid organisms, where dominant alleles will be expressed in the diploid phase, with recessive alleles only detectable when homozygous. Many fungi produce uninucleate haploid cells at some stage in their life cycle, which simplifies the production and isolation of mutants. Nutrient requirements are usually simple, providing a good starting point for the production of nutritional mutants. Other factors facilitating genetic research with fungi are high growth rates, very large numbers of progeny from each cross, short life cycles, small genome size and low chromosome numbers. Some or all of the above favourable features occur in *Neurospora crassa*, *Aspergillus nidulans* and *Saccharomyces cerevisiae*. Hence these fungi are among the genetically best understood organisms, and their study has enabled fundamental biological advances to be made.

Higher animals lack most of the above attributes of fungi and other microorganisms that facilitate genetic research. They have, however, other features that enable the genetic diversity of a species in nature to be appreciated more readily than is usual for microorganisms. Individuals can be recognized with the naked eye and estimates of population size obtained. It is often clear that individuals differ from each other, and the frequency of genetic variants at different times and places can be determined. These features have led, with higher animals and plants, to the development of **population genetics**, concerned with genetic diversity and genetic changes in natural populations. The absence of the above favourable features in microorganisms has until recently delayed the development of the population genetics of fungi. However, much is known from laboratory studies about the mechanisms of variation in fungi, and molecular methods are now providing an increasing amount of information about the genetic variability of fungal species in nature. The concepts of population genetics, applied to this knowledge, can provide an understanding of how fungal populations vary and evolve. Caution is, however, needed in applying concepts that were developed with diploid organisms to the predominantly haploid fungi.

Individuals, Populations and Species

Three concepts essential for population genetics are the individual, the population and the species.

The concept of the **individual** presents no problems with yeasts, where each cell is clearly an individual. What constitutes an individual with mycelial fungi is more debatable. In reaching a conclusion it seems reasonable to take into account two factors, genetic identity and physical continuity – that of protoplasm or hyphal wall. Thus a colony that has developed from a single spore will be an individual. At one time it was thought that colonies might arise in nature by hyphal fusion of genetically different mycelia, and the extent to which a genetic mosaic should be considered an individual would be questionable. The occurrence of vegetative incompatibility (page 257), however, means that such mosaics are likely to be rare. Spores once separated from a parent colony can be regarded as individuals. When, as happens in some fungi, the colony ultimately undergoes fragmentation, the resulting fragments can be regarded as individuals. The sum of individuals that have arisen without genetic change from a common ancestor constitutes a **clone** (Fig. 5.1). Ensuring that a culture is from a single clone ('cloning') is an important step in the isolation of a fungus, necessary to ensure the genetic uniformity of an isolate.

A **population**, for the ecologist, consists of all the individuals of the same kind that occupy an area suitable for them. For the geneticist a population consists of the individuals which, as a result of proximity and genetic similarity, readily interbreed and exchange genes. Transfer of genes between populations is less common and is termed **gene flow** (Fig. 5.2). Sometimes populations may be easy to recognize, as with members of a species on a small island, but often they are less clearly delimited. Even so, the concept remains a useful one for both geneticists and ecologists.

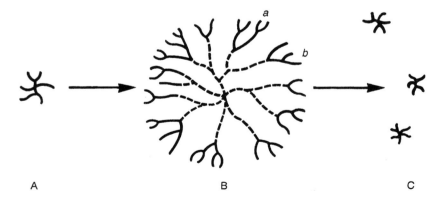

A   B   C

Figure 5.1   The physical individual and the clone. A, A young fungal colony. B, The fungus has spread and the older hyphae (---) have died, dividing the colony into several physical individuals (e.g. a,b). All these physical individuals remain, however, a single genetic individual, a clone. C, Asexual sporulatoin and dispersal has established further colonies. Since genetic change has not occurred, these too remain members of the same clone.

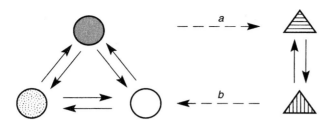

Population 1   Population 2

Figure 5.2   Populations and gene flow. A population, in the genetical sense, is the sum of individuals that can readily interbreed and exchange genes. There may be many clones within a population – three are shown for Population 1. These can readily exchange genes (→). Transfer of genes to another population, such as Population 2, is infrequent. This may be due to distance, resulting in infrequent contacts, or to genetic disharmony, with mating failures or poor fertility. If gene transfer between populations (− − →), known as gene flow, does occur, it may be in one direction (e.g. *a*) or in both (*a* and *b*).

The **species** is the fundamental unit of biological classification, but there are both theoretical and practical difficulties in defining and delimiting species. The species concept most useful to the geneticist is one that defines the normal limits of genetic exchange and has been termed the **biological species**. In a fungus with a sexual process a biological species will consist of all the populations between which successful mating and the production of viable offspring occurs. Delimiting the species requires considerable study, since isolates of a fungus from different parts of the world must be brought together and mating tests carried out. Such mating tests may yield an unambiguous demarcation. Sometimes, however, matings between different isolates may be partially successful, yielding few progeny or ones of slightly or greatly reduced vigour. Under such circumstances it may be difficult to decide whether two isolates belong to the same biological species. As an alternative to carrying out matings, molecular methods now make it possible to delimit biological species by determining the limits within which genetic recombination has occurred. This procedure has the advantage of being applicable to fungi which have no obvious mating.

An enormous variety of fungi are known, but the vast majority have not received the detailed study needed for the confident delimitation of biological species. Practical considerations, however, require that such fungi be assigned to existing or newly created species. A newly discovered fungus is only accepted as a new species after it has been described and named in accordance with the International Code of Botanical Nomenclature. A Latin binomial has to be provided, consisting of a generic followed by a specific name. Often a new species is sufficiently similar to one already known to be assigned to an existing genus, but sometimes it is necessary to create a new genus to accommodate it. A Latin description of the type specimen has to be published, and the specimen deposited in a herbarium. Thereafter all subsequent specimens that sufficiently resemble the type specimen are considered to belong to the same species. Taxonomists can differ greatly in what they regard as a sufficient resemblance. Some are 'splitters', assigning fungi with small differences to separate species, and others are 'lumpers', putting fungi with considerable differences in the same species. In the genus *Fusarium*, for example, some authorities have recognized scores of species and others less than a dozen. As knowledge increases, species may be split or combined and new specific names applied. This has to be done in accordance with the rules of the International Code for new names to be valid. Changes should be made only for very good reasons. About 70 000 fungus species are known, and a classification that was lacking in stability would lead to chaos.

The entity that carries a specific name that has been assigned in accordance with the International Code has been termed the **nomenspecies**.

Life Cycles, Mating Systems and Genetic Exchange

Life Cycles and the Sexual Process

The sexual process in fungi, as in other organisms, involves three key steps. **Cell fusion (plasmogamy)** occurs between two haploid cells, typically uninucleate and from separate organisms that differ genetically. The resulting cell hence has two types of haploid nucleus and can be termed a dikaryon. **Nuclear fusion (karyogamy)** then occurs giving a cell with a single diploid nucleus. Finally **meiosis** restores the haploid state, converting a diploid cell into four haploid cells. If there are genetic differences between homologous chromosomes, genetic recombination can occur during meiosis in two ways. First, the independent assortment of chromosomes in meiosis means that a resulting cell will receive a part of its set of chromosomes from one of the partners in the preceding sexual fusion and a part from the other. Secondly, during meiosis crossing-over occurs – homologous chromosomes break and rejoin in such a way as to exchange genetic material. Although the above three steps are universal features of the sexual process, organisms differ greatly in what happens between the steps, resulting in a diversity of life cycles. The five basic types of life cycle in fungi are described below.

Haploid   In fungi with a haploid life cycle, growth and nuclear proliferation are confined mainly to the vegetative phase, and nuclear fusion swiftly follows cell fusion. In the mycelial Ascomycetes (Fig. 2.18) meiosis comes soon after nuclear fusion. In some other fungi there may be a long delay before meiosis occurs, but with the diploid structure, such as the zygospore of Zygomycetes, dormant.

Haploid–dikaryotic This type of life cycle occurs in the Hymenomycetes (Fig. 2.20) and Gasteromycetes. The haploid mycelium undergoes extensive vegetative growth. Then, if it encounters another haploid mycelium which is compatible, they fuse to give a dikaryotic mycelium. This too can undergo extensive growth. The diploid phase is brief, nuclear fusion and meiosis both occurring in the basidia immediately prior to basidiospore formation.

Haploid–diploid With such a life cycle vegetative growth can occur in both haploid and diploid phases. In the Chytridiomycete *Allomyces* (Fig. 2.10) and some yeasts such as *Saccharomyces cerevisiae* the haploid and diploid phases are morphologically similar, but with diploid cells usually larger than haploid ones. In Myxomycetes (Fig. 2.5) the haploid phase is amoeboid and the diploid phase plasmodial.

Diploid In Oomycetes the vegetative mycelium is diploid, and meiosis, which occurs in the antheridia and oogonia, is soon followed by cell and nuclear fusion to restore the diploid state (Fig. 2.8). Nuclear fusion following soon after meiosis also occurs in some yeasts, such as *Saccharomycodes ludwigii*.

Asexual In some fungi the sexual process has not been observed. Such fungi, if related to Ascomycetes and Basidiomycetes are termed mitosporic fungi, but asexual forms have arisen in all the major groups of fungi.

The possible roles of the sexual process, and the respective advantages of different types of life cycle, will be considered later (page 262).

Mating Systems and the Promotion of Outcrossing

Many fungi have genetic systems that prevent mating between genetically identical cells. Such systems have been termed **mating systems** or **breeding systems**. Fungi possessing such systems are referred to as **self-sterile**, or since different strains have to be involved for mating to occur, **heterothallic** (Greek, *hetero* = different, *thallos* = young shoot). The controlling systems have been termed **homogenic incompatibility** (Greek, *homo* = same), since mating fails if strains are identical with respect to **mating type**. Several kinds of mating system, as described below, have been recognized in the fungi. In all of them the sexual process only occurs if two strains which differ in mating type interact.

Breeding systems with two mating types
Many fungi have only two mating types. Examples considered in Chapter 2 include the cellular slime mould *Dictyostelium discoideum*, the Zygomycete *Mucor mucedo*, the Ascomycete *Neurospora crassa*, the yeast *Saccharomyces cerevisiae*, the rust *Puccinia graminis* and the smut *Ustilago violacea*. Where detailed studies have been carried out, the two mating types have been found to differ with respect to the allele present at a single mating type locus. Since mating takes place between haploid cells or mycelium that differ at the mating type locus, the system ensures that mating does not take place between the genetically identical progeny of a single haploid cell. A diploid cell that arises from mating will carry both mating types, and when meiosis occurs the resulting haploid progeny will be half of one mating type and half of the other. The probability that an encounter between two individuals derived from a single diploid will result in mating (the **inbreeding potential**) is 50%. Since the species has only two mating types, the probability of an encounter between two unrelated individuals resulting in mating (the **outbreeding potential**) is also 50% (Table 5.1). Thus although

Table 5.1   Inbreeding and outbreeding potential of different mating systems

System	Inbreeding potential (%)[a]	Outbreeding potential (%)[b]
Two mating types	50	50
Multiple alleles at one locus – 'bipolar incompatibility'[c]	50	nearly 100
Two alleles at one locus, multiple alleles at another – 'modified tetrapolar incompatibility'[c]	25	50
Multiple alleles at two loci – 'tetrapolar incompatibility'[c]	25	nearly 100

[a] The probability that a randomly encountered sibling (product of the same diploid parent) will be compatible.
[b] The probability that a randomly encountered unrelated individual will be compatible.
[c] Terms in inverted commas used with Basidiomycetes.

selfing is prevented, the likelihood of mating between close relatives is not reduced.

Many species of *Phytophthora* (page 31) are self-sterile, with two mating types. Although the nuclei in the antheridia and oogonia are haploid, the initial sexual interaction occurs before they are formed and is between diploid mycelia that differ in mating type. The genetic basis of the difference in mating type is not fully understood. It seems likely, however, that the A1 mycelium (page 31) is homozygous for mating type (*aa*) and the A2 mycelium heterozygous with one allele dominant (*Aa*).

Unifactorial incompatibility systems with many mating types
Some Hymenomycetes, such as *Coprinus comatus* (page 63), have a single mating type locus (and are hence 'unifactorial') but a large number of mating type alleles. Mating occurs between any two haploid mycelia that differ in mating type, so the diploid basidium nucleus is heterozygous for mating type. Hence two of the haploid spores borne on a basidium will be of one mating type and two of the other. Such a system is hence sometimes termed **bipolar**. Since any diploid strain yields two mating types, the inbreeding potential, as with breeding systems having only two mating types, is 50%. Since, however, a large number of mating types can occur in a population, almost all encounters with non-sibling strains can be compatible and the outbreeding potential can approach 100% (Table 5.1). A greater **outbreeding bias** hence results. Unifactorial incompatibility systems with many mating types are common in Myxomycetes as well as Basidiomycetes.

Bifactorial incompatibility systems
In many Basidiomycetes there are mating systems with two unlinked mating type factors designated A and B. The operation of such bifactorial systems is described in Chapter 2 in the context of the Hymenomycetes *Coprinus cinereus* and *Schizophyllum commune*. Since a basidium can produce spores of four different

mating types, such systems are also termed **tetrapolar**. In a population that contains many different A and B factors the outbreeding potential will approach 100%. The inbreeding potential can, however, be as low as 25% (Tables 5.1, 5.2) giving a greater outbreeding bias than with bipolar systems. However, recombination between the subunits of the A or the B factor can occur in meiosis to yield a new mating type compatible with the parent types (page 65). The frequency of such recombination within mating factors varies greatly between strains. Where it is high, new mating types will arise with high frequency and as a result the inbreeding potential will be greatly increased. The subunit structure of the A and B factors hence may permit some evolutionary flexibility in outbreeding bias. Bifactorial incompatibility occurs in the Myxomycete *Physarum polycephalum* (page 21) as well as in Basidiomycetes.

Table 5.2   Mating among the haploid progeny from the fruiting body of a Basidiomycete with tetrapolar incompatibility. The basidium nuclei have the mating genotype A_1A_2, B_1B_2. Meiosis gives rise to haploid progeny with four possible mating type genotypes – A_1B_1, A_1B_2, A_2B_1, A_2B_2. Mating is only completed successfully if encounters are between mycelia that differ with respect to both mating type factors. The chequerboard shows that only 25% of the possible encounters satisfy this condition. The inbreeding potential is thus 25%.

	A_1B_1	A_1B_2	A_2B_1	A_2B_2
A_1B_1	−	−	−	+
A_1B_2	−	−	+	−
A_2B_1	−	+	−	−
A_2B_2	+	−	−	−

In the smut *Ustilago maydis* (page 66) and in some other Basidiomycetes that mate in a yeast phase there are many B factors but only two A factors. This in its consequences resembles typical tetrapolar incompatibility in giving an inbreeding potential of 25%, but bipolar incompatibility in having an outbreeding potential of 50% (Table 5.1). Although the system is sometimes termed **modified bifactorial incompatibility** or **modified tetrapolar incompatibility** there is no evidence that it evolved from the usual tetrapolar system; indeed from the involvement of a pair of complementary hormones (page 208; Fig. 4.7), it would seem improbable.

The effectiveness of heterothallism
It is reasonable to suppose that the elaborate breeding systems outlined above are effective in promoting genetic recombination and diversity. Now, however, advances in DNA technology have made it possible to determine not only the amount of molecular variability in natural populations, but to deduce, from the pattern of variation, if it results solely from an accumulation of mutations in clones or whether genetic recombination has contributed. Eighteen heterothallic species thoroughly studied in this way all prove to display extensive genetic recombination in natural populations. Fourteen of the species have mitospores as well as sexually produced spores (meiospores) and, with about half of these, clonal as well as recombining populations have been found.

Systems Restricting Outcrossing

Self-fertility

Self-fertile species occur in all major groups of fungi. Self-fertility does not prevent outcrossing but could reduce the likelihood of it occurring, as encounters are more likely to take place between cells of common parentage than between unrelated cells. Since in self-fertile species the sexual process can occur between genetically identical cells, self-fertility is also referred to as **homothallism** (Greek, *homo* = same). Homothallism is termed **primary** if there is no evidence of a heterothallic ancestor. If it is clear that an earlier heterothallism has been circumvented, the terms **secondary homothallism** or **pseudocompatibility** are used. Examples of such circumvention are seen in *Neurospora tetrasperma* (page 52) and *Agaricus bisporus* (page 64). Thus *N. tetraspora* (like *Neurospora crassa*) has two mating type alleles, but instead of an ascus containing eight uninucleate ascospores only four are produced, each with two nuclei, one of each mating type. The resulting vegetative mycelia and protoperithecia also contain nuclei of both mating types, so fertilization by conidia of a different strain is not required for ascus production. The cultivated mushroom, *Agaricus bisporus*, bears two binucleate spores per basidium instead of the four uninucleate spores characteristic of the basidia of most Basidiomycetes. A unifactorial incompatibility system occurs, but the two nuclei in a basidiospore are commonly of different mating type. Hence the mycelium that arises from a single basidiospore is usually able to produce fertile fruiting bodies.

Many strains of the yeast *Saccharomyces cerevisiae* are heterothallic, with two mating types, **a** and α (page 73). Some strains, however, appear to be homothallic, with mating occurring among the progeny of a single haploid cell. The apparent homothallism is the result of a switch from **a** to α or α to **a** mating type which can occur at cell division. The molecular basis of mating type

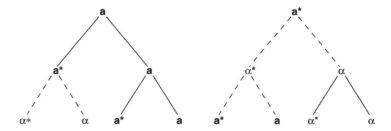

Figure 5.3   Mating type switching in the yeast *Saccharomyces cerevisiae*. Cells of the two mating types are represented by **a** and α respectively. Cells that originated as the mother cell in the previous division are marked with an asterisk, and those that originated as daughter cells (i.e. as a bud) are unmarked. Two rounds of cell division originating in a daughter cell (left) and in a mother cell (right) are shown. Cell divisions in which mating type switching occurs are represented by broken lines. It occurs when a cell that has originated as a mother divides, but not when a cell that has developed from a bud does so. The vegetative cell arising from an ascospore behaves as a daughter cell. It will be seen that the immediate progeny of a single cell, whether it was a mother or daughter in origin, will both have the same mating type. The next generation, however, will consist of two cells of one mating type and two of the other.

switching is the replacement of the genetic factor at the mating type locus by a copy of the alternative factor resident at a storage locus at which it is not expressed, there being one storage locus for each mating type. Switching is very rare in heterothallic strains, occurring about once in 10^5 divisions, but very common in homothallic strains. Homothallic and heterothallic strains differ in genotype at a locus which controls the rate of mating type switching, possessing the alleles *HO* and *ho*, respectively. The pattern of mating type switching is illustrated in Fig. 5.3. The immediate progeny of a cell consist of two cells that do not differ in mating type. This could encourage outcrossing by the immediate progeny of an ascospore or a vegetative cell that has been dispersed. In the next generation, however, cells differ in mating type and thus mating can occur even if there has been no encounter with a genetically different strain.

Mating type switching also occurs in the distantly related fission yeast *Schizosaccharomyces pombe*. It could conceivably occur, but be less easy to detect, in apparently homothallic filamentous fungi. If so, such homothallism could be reversible.

The effect of homothallism

In heterothallic species the production of meiospores requires an encounter with a complementary mating type. With homothallism this requirement is absent, which may have short-term advantages, especially if no other spore type is produced, or if the meiospores have some special role. It is not clear as to how far homothallism restricts recombination, as molecular variation in nature has been studied in a few species only. Recombination has been found in populations of the secondarily homothallic species *Agaricus bisporus* and *Neurospora crassa*, and in the latter clonal populations also.

Fertility barriers

As discussed above (pages 249–251), many fungi are self-sterile and have systems that promote outcrossing. There are, however, **fertility barriers** which limit the extent of outcrossing. The nature of such barriers is not well understood but there seem to be basically two types. One can be termed **genetic disharmony**. If, through geographical separation, there is little gene flow between two populations for a long time, genetic differences between the populations may arise. Such differences could lead to the inefficient performance or failure of some step in the complex interactions between the participants that occur in mating. As a result mating could fail completely or the progeny be few. Furthermore, efficient metabolism and normal development are dependent not only on the action of individual genes but on harmonious interaction throughout the genome. Hence even if mating is successfully completed, an unsatisfactory interaction between the genes from dissimilar parents may result in hybrids that are of poor viability or sterile, and unable to compete with either parent strain. A second type of fertility barrier is **heterogenic incompatibility** (Greek, *hetero* = different), where a failure of mating between two strains is engineered through one or a few genetic differences. Heterogenic incompatibility with respect to vegetative cell fusions has received detailed study (page 257). In the Ascomycete *Podospora anserina*, some of the interactions which prevent vegetative cell fusions also act to prevent mating. It is possible that in some fungi there are also mechanisms of heterogenic incompatibility that act solely in the sexual phase. The possible role of heterogenic incompatibility will be considered later (page 257).

Apomixis

Apomixis can be distinguished from homothallism only by detailed cytological study. All the usual morphological features of the sexual process are shown, but nuclear fusion and meiosis do not occur. Facultative haploid apomixis occurs in the Myxomycete *Physarum polycephalum* (page 22). A haploid amoeba that fails to encounter an amoeba of complementary mating type occasionally develops into a plasmodium without fusing with another amoeba or changing in ploidy. The resulting plasmodium produces spores without meiosis taking place. In some strains of *P. polycephalum* haploid apomixis occurs at high frequency, reducing the probability of outcrossing. Diploid apomixis has been found in some strains of the Myxomycete *Didymium iridis*. In these strains the amoeba nuclei are diploid and heterozygous for mating type. Such amoebae are able to develop into plasmodia without cell or nuclear fusion. Apomixis has attracted less attention in fungi but some examples are known. It is probable that if obligate apomixis persists in a strain or species for a long period the capacity for meiosis will be lost and apomixis become irreversible. This is equivalent to a loss of sexuality although the gross morphological features of sexual sporulation are retained.

Loss of sexuality and its possible consequences

There are many fungi (e.g. the mitosporic fungi) in which the patterns of sporulation associated with a sexual process – the production of a perfect stage or teleomorph – have not been observed. Sometimes this may be due to failure to isolate a complementary mating type, or to determine the correct cultural conditions for sporulation, but in most instances is likely to be due to the permanent loss of sexual sporulation. Historically there have been remarkable changes in opinion as to whether such fungi will also have lost the ability to undergo genetic recombination. In the early days of fungal genetics the prevailing view was the simple one of 'no sex, no recombination'. Then in the 1950s, with extensive laboratory work on heterokaryosis and the parasexual cycle (page 256), came the view that these would promote genetic recombination among asexual fungi in nature. Subsequently extensive vegetative incompatibility (page 257) was found in natural populations, limiting cell fusion and hence genetic exchange by parasexual processes to very closely related individuals, effectively resulting in the revival of the 'no sex, no recombination' view. Recently the molecular variability in nature of several species of mitosporic fungi has been studied. In the flax rust *Melampsora lini* and the human pathogen *Coccidioides immitis* both recombining and clonal populations have been found. Most of the other species examined have proved to be predominantly clonal but with limited recombination occurring. The methods currently available, however, do not indicate what mechanisms bring about genetic recombination in mitosporic fungi in natural populations.

Sexual Differentiation and its Relationship to Outcrossing

In many organisms **isogamy** occurs – sexual fusion is between cells which do not differ in morphology. In others there is **anisogamy** – sexual fusion is between a large (female) and a small (male) cell. The cellular slime moulds and Myxomycetes lack specialized sexual cells, and any differentiation into male and female. In Oomycetes the small male antheridia produce fertilization tubes which fuse with the large female oospheres. In Chytridiomycetes a variety of systems occurs. Thus in the *Euallomyces* subgenus of *Allomyces* fusion is between small highly motile male and larger more sluggish female gametes. In the *Cystogenes* subgenus fusion is between morphologically identical motile cells. In another Chytridiomycete, *Monoblepharis* (not mentioned in Chapter 2), motile male gametes are released from antheridia and penetrate oogonia to fertilize large static oospheres. In many Zygomycetes isogamy occurs, sexual fusion being between progametangia of equal size borne on suspensors of equal size. In some, however, such as members of the genus *Zygorhynchus*, a small progametangium borne on a small suspensor fuses with a large progametangium borne on a large suspensor. This is anisogamy, although in the Zygomycetes the term **heterogamy** is used. It is not usually regarded as being differentiation into male and female, although it appears to be an evolutionary step in that direction. In some Ascomycetes the female ascogonium is fertilized by a hypha emerging from the male antheridium. In many species, however, the ascogonium can be fertilized by conidia (e.g. *Neurospora*), arthrospores (e.g. *Ascobolus furfuraceus*) or hyphae indistinguishable from vegetative hyphae (e.g. *Ascobolus immersus*). In yeasts such as *Saccharomyces cerevisiae* mating is between morphologically indistinguishable haploid cells. In the Hymenomycetes and Gasteromycetes fusion is between vegetative hyphae but in the rusts there is some sexual differentiation, with pycniospores being transferred to receptive hyphae.

Where fungi have only two mating types there is a temptation to equate the two mating types with the two sexes. To do so is usually an error, as can be seen by considering the sexual process in a heterothallic Ascomycete. A *Neurospora crassa* colony, for example, which originates from a single haploid spore is able to produce simultaneously both female and potentially male structures – protoperithecia and conidia – but fertilization will occur only as a result of an encounter with the complementary mating type. Sexual differentiation is primarily a logistic device concerned in the efficient movement of gametes and provisioning of zygotes, whereas mating types are concerned with the promotion of outcrossing.

However, in the Oomycetes, which are only distantly related to other fungi (Chapter 2), sexual differentiation has become involved in the promotion of outcrossing. This is seen clearly in heterothallic species of *Achlya* (page 29), where self-fertility is prevented by the inability of a colony to produce antheridia and oogonia simultaneously. Sexual structures are produced only as a result of an encounter with a second colony, and which colony produces antheridia and which oogonia depends on the relative position of the two colonies in a spectrum ranging from strong female to strong male. A full understanding of the extent to which outcrossing is promoted will need the elucidation of the genetic control of such relative sexuality. A similar situation seems to occur in *Pythium sylvaticum* (page 31) and the difference between the two mating types in the heterothallic species of *Phytophthora* may be essentially a sexual one. It is important, however, when considering the sexual process in fungi other than Oomycetes, not to confuse sexual differentiation and mating systems.

Heterokaryosis, the Parasexual Cycle and Vegetative Incompatibility

In many Ascomycetes, Basidiomycetes and mitosporic fungi fusion between vegetative hyphae (hyphal anastomosis) of the same colony is common during colony development (page 142). Hyphal anastomosis is also common when two colonies of the same species come into contact, and is often seen between the hyphae emerging from germinating spores. If the hyphae that fuse carry genetically different nuclei, the colony that develops may be a **heterokaryon**, with nuclei that differ in genotype in a common cytoplasm. If they differ with respect to cytoplasmic genetic elements – mitochondrial DNA (page 89) or plasmids (page 92) or viruses (page 92) – the resulting colony may be a **heteroplasmon**, with cytoplasmic genetic factors originating from different sources.

Heterokaryons can readily be produced by bringing together different mutants that have been derived from the same parental strain. A very effective method is to place two mutants with different nutritional deficiencies on a medium that lacks either nutrient. Under such circumstances neither mutant strain will be able to thrive since each is only able to synthesize one of the missing nutrients. However, any heterokaryotic hyphae will have both types of nuclei and hence the enzymes needed for synthesizing both nutrients. They will hence proliferate and a heterokaryotic colony will soon become established.

Occasionally two haploid nuclei in a vegetative mycelium may fuse to give a **somatic diploid** nucleus. If the mycelium is heterokaryotic then the fusion may be between genetically unlike nuclei to give a heterozygous diploid nucleus. Such **diploidization** is rare, occurring perhaps once in a population of a million nuclei. In a growing colony the number of diploid nuclei (like the number of haploid nuclei) will increase by mitosis. During the division of such nuclei **mitotic crossing over** may occur. This is a rare event compared with meiotic crossing over, occurring perhaps once per five hundred mitoses. Errors in mitosis are quite common. Sometimes the mitosis of a diploid nucleus (chromosome number $2n$) may give one $2n+1$ daughter nucleus (i.e. having three copies of a chromosome) and one $2n-1$ nucleus (i.e. with only one copy of that chromosome). Such abnormalities in chromosome number (aneuploidy, see page 265) commonly lead to poor growth, and any further change in chromosome number that gives improved growth will be favoured. The loss of any one of the chromosomes present as three copies in a $2n+1$ nucleus will restore the diploid state. In a $2n-1$ nucleus the sequential loss of chromosomes present as two copies can occur to yield a haploid nucleus. Such a conversion of a diploid into a haploid nucleus (**haploidization**) may occur with about one diploid nucleus in a thousand and take several successive mitoses to accomplish. Not only mitotic crossing over but haploidization itself can result in genetic recombination, since the chromosomes lost can come from either of the homokaryons that contributed to heterokaryon formation. The inclusion of a recombinant haploid nucleus in a uninucleate spore yields a recombinant homokaryon differing from either of the original homokaryons. The sequence of events that yield such a recombinant strain has been termed the **parasexual cycle** (Fig. 5.4). The parasexual cycle has been of great value in genetic analysis. How important it is in nature is unclear, since the widespread occurrence of vegetative incompatibility, discussed below, may prevent heterokaryosis except between closely related strains.

An encounter between hyphae of different strains of the same species may result in **vegetative incompatibility**, also known as **somatic incompatibility** or **heterokaryon incompatibility**. Since vegetative incompatibility is based on genetic differences it is a form of **heterogenic incompatibility**, which can also occur in the sexual phase (page 253), although sexual heterogenic incompatibility has received less study than vegetative heterogenic incompatibility.

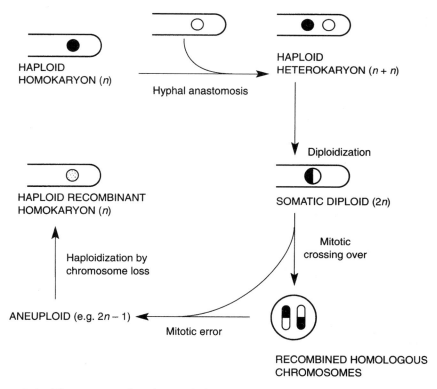

Figure 5.4 The parasexual cycle. Hyphal anastomosis between two genetically different homokaryons gives a heterokaryon. Nuclear fusion may then give a heterozygous diploid nucleus, within which recombination between homologous chromosomes (mitotic crossing over) may occur. Mitotic errors can yield aneuploids, with abnormal chromosome numbers (e.g. $2n - 1$) and commonly poor growth. Loss of chromosomes to restore the haploid state is then selectively advantageous. Since the chromosomes that are lost can have come from either of the original homokaryons, a recombinant can result, even if mitotic crossing over has not occurred.

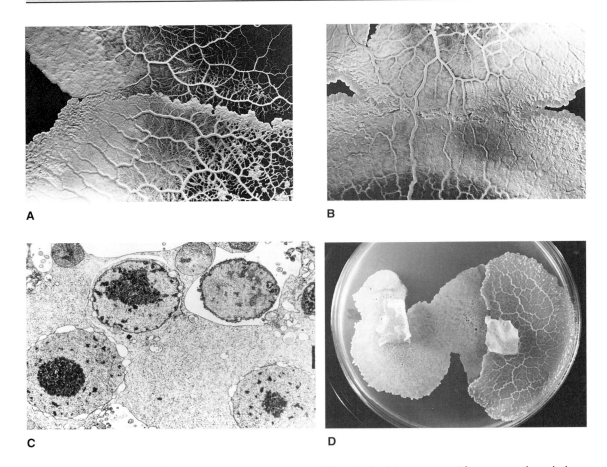

A

B

C

D

Figure 5.5   Plasmodium fusion and vegetative incompatibility in the Myxomycete *Physarum polycephalum*. A, Two plasmodia that are of different vegetative compatibility (v-c) groups with respect to fusion (i.e. differ at the *fus* locus), several hours after contact. Fusion has not occurred. B, Two plasmodia that are of the same fusion v-c group (i.e. identical at the *fus* locus) about one hour after contact. Prominent 'veins' connect the two plasmodia and extensive protoplasmic mixing is taking place. The subsequent history of the fused plasmodia will depend on whether they are of the same or different post-fusion v-c groups; if they are of the same group there will be no adverse reactions and a heterokaryon may be formed. C, Electron micrograph of an incompatibility reaction occurring after fusion of two plasmodia of different post-fusion v-c groups (i.e. different at the *let* locus). At the bottom are two nuclei of the 'killer' strain, dominant at the *let* locus, each with a large nucleolus and normal in appearance except for blebs in the nuclear membrane, usual after plasmodium fusion. At the top are two nuclei of the sensitive strain, recessive at the *let* locus. That on the left has a disintegrating nucleolus and chromatin, and is being separated from the surrounding cytoplasm by a furrow. That on the right has further disintegrated and has been expelled from the plasmodium. Mitochondria are also visible at the top of the figure. Destruction may not be limited to nuclei but can be followed by extensive cytoplasmic death – see below. Scale bar on right, 1 μm. D, Two plasmodia of different post-fusion v-c groups 17 hours after fusion. The sensitive strain on the left is dead. (A, B reprinted with permission from – Carlile, M. J. & Dee, J. (1967) The lethal interaction following plasmodial fusion between strains in a Myxomycete *Nature* **215**, 832–834. Copyright Macmillan Magazines Limited. C, reproduced by permission of the Company of Biologists Ltd from Lane, E. B. & Carlile, M. J. (1979). The lethal interaction following plasmodial fusion between two strains of the Myxomycete *Physarum polycephalum*. *Journal Cell Science* **35**, 339–354. D, reproduced with permission from Carlile, M. J. (1972). The lethal interaction following plasmodial fusion between two strains of the Myxomycete *Physarum polycephalum*. *Journal of General Microbiology* **71**, 581–590.)

Spectacular vegetative incompatibility occurs in Myxomycetes (Fig. 5.5). Plasmodia that are identical at all vegetative compatibility (v-c) loci fuse with each other to form larger plasmodia. Differences at v-c loci result in incompatibility reactions. There are two types of v-c loci, *fus* (fusion) and *let* (lethal) loci. If plasmodia differ at a *fus* locus, **fusion incompatibility** results, and the plasmodia do not fuse. If plasmodia are identical with respect to their *fus* loci, fusion occurs. Subsequently, however, **post-fusion incompatibility** can occur if there are phenotypic differences between the plasmodia with respect to one or more *let* loci. Plasmodia are diploid, and plasmodia of a strain that is homozygous and dominant at a *let* locus (e.g. *letA letA*) will fuse harmlessly with the corresponding heterozygote (*letA leta*) but can destroy the nuclei of homozygous recessive (*leta leta*) plasmodia. This destruction of nuclei will eliminate one genotype from the fused plasmodium and can also cause extensive destruction of protoplasm. If the plasmodia differ at only one incompatibility locus the reactions are unilateral, but bilateral reactions can result from plasmodia being dominant at different *let* loci. Post-fusion incompatibility reactions between two or more strains commonly result in the survival of one strain only.

There are few reports of fusion incompatibility within species of filamentous fungi. Where it does occur, as in *Thanatephorus cucumeris* (anamorph *Rhizoctonia solani*), it prevents both vegetative and sexual fusion and thus all genetic exchange. Hence in filamentous fungi it is likely that different fusion incompatibility groups constitute different biological species, and that only in rather broadly demarcated species, such as *T. cucumeris*, will fusion incompatibility be observed.

Post-fusion incompatibility is, however, widespread in filamentous fungi (Fig. 5.6). When two colonies that are in different post-fusion incompatibility groups meet, they fuse, but subsequently the protoplasm in the fused hyphal compartment, and sometimes adjacent ones, is destroyed. Such incompatibility is often visible to the naked eye as 'barrage', the formation of a demarcation zone of sparse mycelium, sometimes with black pigmentation. The post-fusion incompatibility that is usual in filamentous fungi has been termed **allelic incompatibility**, since it can occur as a result of interaction between two alleles of the same v-c locus. Incompatibility in Myxomycetes is hence also allelic incompatibility, but with the complication that heterozygosity can occur as plasmodia are diploid. In one fungus, *Podospora anserina*, not only does allelic incompatibility occur, but incompatibility can also result from interaction between alleles at different v-c loci. This has been termed **non-allelic incompatibility**. It is not known whether it occurs in other fungi; if so it is likely to be less common than allelic incompatibility. Where detailed studies have been carried out many v-c loci have been found. If a fungus with allelic incompatibility has n v-c loci with two alleles at each, then 2^n v-c groups are possible. Thus *Neurospora crassa*, on the basis of 10 known v-c loci, could have up to 2^{10} or 1024 v-c groups. It is hence not surprising that when two isolates of the same species are obtained from nature they commonly show vegetative incompatibility.

Although post-fusion incompatibility, when it occurs, normally prevents the vegetative transmission of nuclei between strains, the extent to which virus spread is affected varies. It is possible that with a slow incompatibility reaction a virus may have time to pass to adjacent and undamaged cells before cell death at the

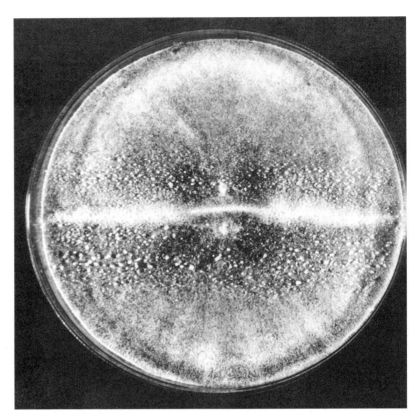

Figure 5.6 Vegetative incompatibility between strains of *Ophiostoma novo-ulmi*, the cause of Dutch elm disease. The points of inoculation of the two strains are visible just above and below the centre of the Petri dish. A 'barrage' of fuzzy white aerial mycelium, narrow between the inoculum sites and wider further out, has been produced where the strains have met. In both strains the production of synnemata (visible as white dots) has been stimulated on either side of the barrage. (C. M. Brasier, The Forest Authority Research Division.)

point of anastomosis occurs, but that with a faster reaction transmission may be prevented. Evidence for this comes from *Cryphonectria parasitica* (formerly *Endothia parasitica*), the chestnut blight fungus, in which a virus that spreads readily between strains of the same v-c group reduces the growth rate and pathogenicity of the fungus (pages 95, 208). The extent to which virus transmission takes place between strains of different v-c groups depends on the number of v-c loci at which they differ. Transmission occurs if there are differences at only a few loci, but is negligible with differences at five or more loci.

The Role and Consequences of Vegetative Incompatibility

Natural selection, as will be discussed in a later section (pages 272–274), acts mainly on individuals. Features that enable individuals to survive and to leave more progeny than other individuals will be favoured and become widespread. Thus in evaluating the **role** – rather than merely the consequences – of such a feature as vegetative incompatibility its value to individuals rather than to the species must be considered.

The fusion of small plasmodia to give a large one is a step in the Myxomycete life cycle, and hyphal anastomosis to give a network rather than merely a system of radiating and branching hyphae is a normal part of colony development in many fungi. Such cell fusions – especially in colony development – will commonly be between cells of recent common origin and hence genetically identical. Sometimes, however, fusion may be with unrelated cells. This can expose an individual and its genome to risks. Thus in *Neurospora crassa* a gene is known which enables nuclei containing it to replace other nuclei in a heterokaryon. In *Neurospora*, and in other fungi, mitochondria occur that have genomes that render them defective in their respiratory function yet capable of normal replication. Infection with such mitochondria would be harmful. *Podospora anserina* strains have a plasmid that causes senescence, and there are viruses, as indicated (page 260) that adversely affect growth rate. Vegetative incompatibility may be a defence mechanism that protects individuals against harmful nuclei, mitochondria, plasmids and viruses. There is no evidence that nucleic acid or viruses can pass through the fungal cell wall, so fusion incompatibility, where it occurs in fungi, should be a complete protection against infection with alien nucleic acid. Post-fusion incompatibility is likely to protect fully against alien nuclear and perhaps mitochondrial genomes, and against plasmids and viruses if there are differences at several v-c loci. An offensive as well as a defensive role for post-fusion incompatibility is also possible. Thus if several Myxomycete strains of the same fusion v-c group but different post-fusion v-c groups are grown together, only the strain dominant at the greatest number of differing v-c loci usually survives. Fusion incompatibility will be a complete protection against such attack; this is perhaps the reason for its widespread occurrence in Myxomycetes, where severe post-fusion reactions are common.

Although the **role** of vegetative incompatibility must be sought in its value to individuals, it will have **consequences** for the species. Where v-c groups are numerous, heterokaryon formation is likely to be rare and the parasexual cycle unimportant. Where the converse is true, heterokaryosis and parasexuality could be important. The situation in nature is unclear. Vegetative incompatibility occurs in many fungi and in the few in which there have been detailed studies many v-c groups have been found. The initial colonization of a new area or a new species, however, might sometimes be achieved by a single haploid strain. The resulting population would then at first lack vegetative incompatibility so, as mutants arose, heterokaryosis and parasexuality might bring about genetic exchange. There hence may be populations in which for a time parasexuality is important. In due course, however, genetic diversity would increase to the point where individuals would need to protect themselves from alien nucleic acid. Then mutations disrupting fusion with other strains might be favoured and vegetative incompatibility might be re-established.

The Roles and Consequences of Sex and Outcrossing

Traditionally, the role of sex has been regarded as the promotion of genetic variability through outcrossing, this variability being needed to enable the species to evolve rapidly enough to cope with competitors and environmental change. There is plenty of evidence, from the fungi and elsewhere, that outcrossing promotes variability. A species with abundant genetic variability will be more likely to produce genotypes able to cope with changed environmental conditions than will a species with little variability. A **consequence** of sex may hence be an increased capacity, through variation, to survive the challenges of a changed environment or rapidly evolving parasites or competitors. Natural selection, however, favours features of benefit to an individual and its immediate progeny and not those that might be of value to the rest of the species or remote posterity. The **roles** of sex have therefore to be sought in the value of sex to the immediate products of sexual fusion or of meiosis. The evolution of sex has been the subject of much debate in recent years, and several different roles for sex have been suggested. More than one of these suggestions could be correct, as different species have different life cycles and are exposed to different environments.

The vast majority of mutations are deleterious rather than beneficial. Different cell lineages, however, will accumulate different unfavourable mutations. Outcrossing will hence result in a dikaryon or diploid in which the effect of mutations, provided that they are recessive, will be masked by the complementary correctly functioning alleles, an effect known as **complementation**. Outcrossing may also result in **heterozygous advantage**, well known in plants and animals, in which the heterozygote performs better than either corresponding homozygote.

The recombination that occurs when the haploid state is restored by meiosis will result in some progeny with fewer deleterious alleles than either of the haploid parental strains. It has been suggested that it is this effect, counteracting a gradual tendency to accumulate unfavourable mutations, rather than the actual promotion of variability, that is the main role of sex and outcrossing. On the other hand it has been argued, with respect to plants and animals, that the role of sex is to produce new variants less susceptible than the parent strain to parasites. This concept may be applicable to fungi, in which the commonest parasites are viruses, usually transmitted by vegetative hyphal fusion. The sexual process, however, will generate, through outcrossing and recombination, progeny with new vegetative compatibility genotypes which will not fuse with and acquire the viruses of the parental v-c genotypes.

The above views on the role of sex are, however, not applicable to the many fungi that are self-fertile and in which, therefore, outcrossing will be unusual. Self-fertility may well have evolved on many occasions to guarantee the production of an essential spore type, eliminating the risk of the non-arrival of a compatible mating type. The same objective could, however, be achieved by apomixis, which seems relatively uncommon in fungi. It hence appears that sex – nuclear fusion followed by meiosis – has a role in fungi even when outbreeding will not result. One possible role is in the repair of damage affecting both strands of a DNA molecule, for which a template from a homologous chromosome is needed. Meiosis would then allow copying from the template to occur during crossing over. There is evidence for such recombinational repair in a range of organisms, including *Saccharomyces*.

The Survival of Asexual Fungi

There are many fungi in which the sexual process has not been observed, including about 14 000 mitosporic species, about 20% of all known fungi. In view of the probable benefits of genetic recombination, sex and outcrossing, how do these fungi survive? Detailed study of molecular variation in natural populations of a few mitosporic fungi has now been carried out. Although many populations prove to be clonal, in most of the species some genetic recombination was found. Molecular phylogeny (page 285) is now being carried out on an increasing number of mitosporic fungi, with the conclusion that most are closely related to species with sexual sporulation, and can be assigned to Ascomycete and Basidiomycete genera. This rather limited divergence from sexual forms indicates that they have not had a long evolutionary history. The recent discovered presence in *Candida albicans* of analogues of mating-type genes may reflect a stage in the loss of sexuality. It hence seems likely that the loss of sexuality commonly results in short-term advantages but ultimate extinction. There may be an outstanding exception to this conclusion. The sexual process appears to be lacking in a whole order of fungi, the Glomales. These fungi, the vesicular–arbuscular endophytes (page 397), as well as being abundant today, occur in fossil material of the earliest land plants, and will have diverged from other fungal groups over 400 million years ago. Molecular studies have shown that the multinucleate spores of these fungi contain a population of genetically divergent nuclei, sometimes with nucleic acid sequences so different as to be characteristic of different families. This perhaps provides an alternative to sex, with natural selection operating to promote the survival of individuals with an optimally balanced population of genetically different nuclei. Moreover the vesicular–arbuscular endophytes, as valuable partners for their plant hosts, may have been largely protected from the effects of hostile environments.

Variation, Microevolution and Speciation

Variation in the Laboratory and in Nature

Variation following isolation from nature
Some fungi are obligate parasites and can be studied only by maintaining them on the host plant or animal. Many, however, can be grown in pure culture on an agar medium. A key step in the isolation of such fungi is propagation from a single cell so as to provide genetically uniform material instead of a mixture of strains. This presents no problem with the unicellular yeasts or with filamentous fungi that produce numerous unicellular spores. Some fungi, however, sporulate poorly or not at all, and with these it is necessary to excise single hyphal tips instead. Ideally the cell type chosen should be uninucleate, or if multinucleate the nuclei should be derived from a single nucleus. It should also be haploid, so as to have only one set

of genes. With mitosporic fungi and with fungi that are self-fertile, cultures established from single haploid cells should produce all the phases in the life cycle. With self-sterile fungi, however, it is necessary to isolate two haploid strains of different mating type and to bring them together in order to obtain the sexual phase. The progeny that result from sexual sporulation may vary greatly as a consequence of genetic differences between the parental strains and of recombination during meiosis. Haploid strains that differ in mating type but are otherwise genetically identical (isogenic) or closely similar to each other can be obtained by repeated backcrossing of successive generations to one of the parental strains. Such haploid strains may, however, have a reduced ability to participate in the sexual process or may be of reduced vigour. In Oomycetes meiosis occurs in the antheridia and oogonia and is soon followed by fertilization. Here, therefore, the uninucleate cells used – commonly zoospores – are diploid.

Some fungi, such as *Fusarium*, have a reputation for being highly variable in culture, morphological changes occurring with subculture. Others, such as *Aspergillus*, are regarded as stable. It is probable that such differences between species reflect not the extent to which the fungus can vary, but how drastically conditions in culture differ from those in nature. Many *Fusarium* species, for instance, live in very dilute solutions in the water-conducting vessels of plants, whereas many *Aspergillus* species are able to attack starch or cellulose, often present in massive amounts in dead plant tissues. *Fusarium* will hence encounter in pure culture far higher nutrient concentrations than it experiences in nature, whereas for *Aspergillus*, concentrations in culture and in nature may well be similar. Hence with *Fusarium* there may be intense selection favouring any variant capable of coping with the unnatural conditions in culture more efficiently, whereas with *Aspergillus* the naturally occurring forms may already be well adapted.

Subculture itself may result in changes in the morphology of a fungus. Subculture by means of pieces of agar bearing mycelium can give a culture arising from the hyphae present rather than spores. There is then no selection against mutants defective in sporulation, so the ability to sporulate may rapidly be lost. Maintenance of sporulation may hence require subculture by means of spores, and a continued capacity for sexual sporulation may require subculture by means not merely of spores, but of sexually produced spores.

Thus both prolonged growth in artificial conditions and repeated subculture can lead to genetic change. It is hence desirable to maintain stocks of freshly isolated material, or strains with valuable attributes, under conditions that permit prolonged survival, such as liquid nitrogen refrigeration.

Production and utilization of mutants in the laboratory
A few fungi have been the subject of intensive genetical study in the laboratory. Notable examples are the filamentous Ascomycetes *Neurospora crassa*, *Podospora anserina* and *Aspergillus nidulans*, the Ascomycete yeasts *Saccharomyces cerevisiae* and *Schizosaccharomyces pombe*, and the Basidiomycetes *Coprinus cinereus*, *Schizophyllum commune* and *Ustilago maydis*. A major reason for the restriction of intensive study to a few species is that the more that is known about the genetics of a species the more useful it

becomes for further genetic research. This is because effective methods for the production and selection of mutants will have been devised, many mutants will already be available, and the location of many genes will be known. With intensively studied species most research has been carried out on strains derived by successive mutations from a single haploid strain, or by mutation and recombination from a pair of sexually compatible haploid strains. This is because the effects of a mutation are best seen by the comparison of two strains that differ only with respect to the mutation. Starting from a single wild-type strain it is possible to obtain mutants affecting a wide range of metabolic activities. The production and exploitation of such mutants in genetic mapping and the elucidation of metabolic pathways are described in genetic textbooks. Many morphological mutants are also known. Some of these differ so much from the wild-type in appearance that on morphological criteria alone they would not be recognized as belonging to the same species or even the same genus.

Variation within fungal species in nature
The extent to which a fungal species varies in nature can be determined only by studying many strains from different localities. This has so far been done with relatively few species, but with sufficient to show that with fungi, as with plants and animals including humans, there is a great deal of variation within species. Such variation can affect all aspects of the biology of a fungus.

Strains isolated from nature may differ in morphology. *Aspergillus nidulans* strains, for example, differ in colony growth form and in the amount of sexual and asexual sporulation. Strains of the same species also differ in physiology, having different growth rates and producing differing amounts of metabolic products. Many different vegetative compatibility groups occur in *A. nidulans* and in other Ascomycetes (pages 259–260) as well as in Myxomycetes such as *Didymium iridis*. Natural populations of Basidiomycetes, such as *Schizophyllum commune* (page 64) and Myxomycetes also contain many different mating type alleles. Strains of plant pathogens vary in their ability to attack different strains of their host species (pages 267–269), a topic of great practical importance in plant breeding and agriculture.

Strains may also differ in chromosome number. In some species there are strains that are polyploid, with haploid chromosome numbers that are a multiple (e.g. 2X, 4X) of the basic haploid number (X). Polyploidy has been found in the Myxomycete *Didymium iridis*, the Chytridiomycete *Allomyces*, the yeast *Saccharomyces* and in most fungal groups. Alternatively a strain may be aneuploid, with one or a few chromosomes duplicated or lost (e.g. 2X+1, 2X–1, 2X–2). Naturally occurring aneuploids have been found in *Neurospora crassa* and in *Aspergillus nidulans*. Fungal chromosomes are usually small and difficult to count. A consequence is that until recently it has been impossible to ascertain the chromosome numbers of many important fungi, except through genetic analysis and the determination of the number of linkage groups. The situation has now been transformed by the advent of **pulsed field gel electrophoresis (PFGE)**. Gentle extraction methods are used to isolate DNA without fragmentation, so that extracted DNA molecules represent entire chromosomes. The DNA is applied to an agarose gel, and electrophoresis carried out, with alternate electrical pulses from two directions. Under these circumstances an important factor in determining the speed of migration of the DNA molecules is how fast they reorient with each pulse. The longer the molecule, the greater is the resistance to reorientation, and the slower is migration. Segregation of DNA of different lengths occurs, with the DNA from the shortest chromosomes travelling furthest. Not only can the haploid chromosome number be obtained, but the size of the chromosomes can be estimated. PFGE is not only enabling the chromosome numbers of many fungi to be determined, but is revealing variation within species with respect to chromosome number, size and structure.

Progress in molecular biology, especially in DNA technology has made it possible to obtain a great deal of information on the ways in which strains differ without resort to the time-consuming procedures of genetic analysis. Studies on ribosomal DNA (pages 325–328), nuclear or mitochondrial, and on selected genes, which can be sequenced, can provide much valuable information. Methods (pages 327–328) include the analysis of Restriction Fragment Length Polymorphisms (RFLPs) and Randomly Amplified Polymorphic DNA (RAPD), and others described in reference works on molecular technology. Strains within a fungal species can also differ in the dsRNA viruses that they contain, as has been found with the Ascomycete *Cryphonectria parasitica* and the rust *Puccinia striiformis*. Proteins extracted from fungi can be subjected to gel electrophoresis and the gels stained with appropriate reagents to reveal specific enzymes. A particular enzyme activity is often controlled by genes at several different loci, with the result that it exists in several slightly different forms with differing electrophoretic mobilities. These forms, termed **isozymes**, can be distinguished by gel electrophoresis. Differences in isozyme pattern within fungal species have been demonstrated for a range of enzymes. Mutations at a locus that specifies an enzyme can result in alleles that produce altered forms of the enzyme. These forms, known as **allozymes**, can also be detected by gel electrophoresis. Mating between strains that show an allozyme difference hence results in a diploid with the allozymes of both parents. This permits the detection of heterozygotes in natural populations without recourse to breeding experiments. Differences in molecular detail between strains, especially in mitochondrial DNA and in isozyme pattern, are proving valuable in elucidating the genetic behaviour of natural populations (pages 269–278).

Genetic variation may be continuous or discontinuous. **Discontinuous variation** refers to the situation where a feature is either present or absent. An isozyme, for example, is either present or lacking, and a strain either belongs to a particular mating type or it does not. Discontinuous variation occurs where a single gene has a decisive effect. **Continuous variation** refers to the situation where a feature varies in magnitude between strains. Thus strains may show a wide range of growth rates, some differing only slightly, but with a large difference between the fastest and slowest growing strains. Continuous variation is seen where the magnitude of a feature is influenced by many genes, each with a relatively small effect. Most studies of fungal variation have been concerned with discontinuous variation. In *Aspergillus nidulans*, however, both continuous variation (e.g. in intensity of sporulation) and discontinuous variation (e.g. of vegetative compatibility type) have been studied in natural populations.

Variability resulting from the interaction of host and pathogen

Many fungi are capable of infecting plants (Chapter 7). Some such infections, as with mycorrhizal fungi, may benefit the host plant, but many fungi are plant pathogens. These cause damage to the host, such damage varying from slight to sufficiently severe to cause death. Even if the damage is slight, resources which would have been utilized in growth or reproduction are diverted to the fungus, resulting in a reduced ability of the host to compete with other members of its species. Hence a mutation that confers resistance to a pathogenic fungus will give a competitive advantage to host plants that carry it, and such plants will constitute an increasing proportion of a population at risk. When these resistant plants form a large part of the host population, any mutation in the fungus which enables it to infect these plants will be advantageous and become widespread in the fungal population.

In modern agriculture large areas are often planted with a single variety of crop plant, and a fungal strain able to attack such a variety can cause immense losses. There has therefore been extensive research on the genetic basis of the interaction between varieties of important crop plants and strains of their major fungal pathogens. Similar relationships have been found for many such interactions and have been interpreted in terms of the **gene-for-gene hypothesis**. This concept is perhaps most easily understood by considering the likely course of evolution of a host–pathogen relationship and the probable physiological basis of host resistance and pathogen virulence – its ability to infect and cause damage. Breeding experiments usually show host resistance to be dominant and susceptibility recessive. A host variety could acquire resistance to a fungal strain by the replacement of an allele for susceptibility (e.g. *r1*) by a corresponding dominant allele (*R1*) for resistance. This replacement could take place either by mutation or hybridization, and as the resistance gene is dominant it will be effective either in the homozygous (*R1R1*) or heterozygous (*R1r1*) state. It is thought that a resistance gene acts by controlling the production of protein receptor molecules at the plant cell surface capable of interacting with glycoprotein molecules located at the fungal surface. These glycoproteins are termed **elicitors** since their interaction with the plant receptor molecules elicits a defence reaction by the host (Chapter 7, pages 409–411). With the replacement of *r1* by *R1* in the host, the fungal strain will be recognizable by the host and will become **avirulent** – unable to infect and cause damage. It can become non-recognizable and hence again virulent by the loss or modification of the elicitor which is recognized by the *R1* gene product. This is achieved by the replacement of the fungal gene controlling the production of the elicitor by an allele. If the fungus is diploid, it must be homozygous for the new allele, as it is essential for virulence that no elicitor is produced. The gene for virulence is hence classifiable as recessive. Using the symbol *v* for virulence, there will have been a change in phenotype from *V1* to *v1*. The above interaction between host and pathogen phenotypes, and further possible evolution of the relationship, is illustrated in Fig. 5.7. The host can also regain resistance by a mutation (e.g. *r2* to *R2*) which allows the detection of another glycoprotein feature of the fungus, and the fungus can then regain virulence by a mutation (*V2* to *v2*) that modifies this feature. Thus genes for resistance in the host correspond to genes for virulence or avirulence in the pathogen – a gene-for-gene relationship. The degree of resistance conferred by

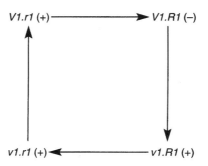

Figure 5.7 The possible evolution of a gene-for-gene relationship. A gene for host resistance is indicated by *R1* and its recessive allele giving susceptibility as *r1*. The corresponding fungal gene for avirulence, *V1*, is dominant and its allele for virulence *v1*, is recessive. The four possible combinations of host and pathogen phenotype are indicated and whether the combination results in infection (+) or failure to infect (–). The relationship can give a cyclic evolution of gene frequencies. Initially the host can acquire, through mutation or hybridization, an allele conferring resistance, resulting in the situation indicated at the top left of the figure being replaced by that shown at the top right. It is then advantageous for the fungus to replace the corresponding gene for avirulence by an allele for virulence. When this occurs the host gene for resistance loses its role. The fact that the recessive allele occurred initially indicates that it must have some useful role, and in the absence of selection for the dominant allele it could once more increase in frequency. When this happens the fungal virulence gene, *v1*, is no longer needed to secure infection and selection for it will cease. The dominant allele, the presence of which is the initial fungal population indicates that it had some useful role, will then increase in frequency completing the cycle.

a resistance gene may be partial or nearly complete – presumably clear-cut interactions between receptor and elicitor give a strong defence reaction and a less decisive interaction a weaker response.

The gene-for-gene relationship was first shown with the interaction between the flax rust, *Melampsora lini* and flax, *Linum usitatissum*. It has since been demonstrated for many other important plant diseases. The occurrence of host varieties having many different combinations of resistance genes results in fungal strains having many different combinations of virulence genes. Strains with the same combination of virulence genes constitute what has variously been termed a **pathotype**, a **biotype** or a **race**. Pathotypes can be recognized without resort to genetic analysis by testing each fungal strain against a standard set of host varieties and noting for each variety whether infection occurs and the severity of the consequences. Such tests show, for example, over 200 pathotypes of the wheat stem rust, *Puccinia graminis tritici*.

The type of resistance described above, which protects a host variety against some pathogen strains and is ineffective against others, has been termed **vertical resistance**. A further form of resistance, giving some protection against all strains of a pathogen, has been termed **horizontal resistance**. Whereas vertical resistance can be conferred by a single resistance gene, horizontal resistance is usually polygenic, resulting from the cumulative effect of many genes. Horizontal resistance is less well understood than vertical resistance. Some maintain that it is merely resistance which, as a result of the small effects of individual genes and consequent difficulties in conducting genetic studies, has not yet been shown to result from gene-for-gene interaction and to be classifiable as vertical. Vertical resistance, however, is resistance resulting from the recognition of an invader and the activation of a defence mechanism. There is no reason why there should not be host varieties that have an enhanced resistance to all members of a pathogen species – due, for example, to greater production of antifungal compounds such as phytoalexins (Chapter 7, pages 410–411) or to a lower sensitivity to enzymes or toxins that the fungus produces. It is also possible that some strains of a fungus could have enhanced pathogenicity to most varieties of a host species through, for example, a lower sensitivity to host antifungal agents or a higher production of toxins or enzymes.

Sources of Variability in a Population

Mutation

Gene mutation is the ultimate source of genetic variability. Genes vary in their mutation frequency, but one mutation per million copies of a gene per generation can be taken as an average value. The genome size for different fungi varies, but taking a genome of 10 000 genes as typical, in each generation one cell in a hundred could in some respect be a new mutant. Most of the cells in a fungal colony are capable of giving rise to further vegetative cells or to spores, so the number of cells in a fungal population that can undergo mutation and leave progeny can be very large. Furthermore, most fungi are haploid for a large part of their life cycle, so mutations will commonly be expressed, and if beneficial spread through a population by natural selection.

An example of the effectiveness of mutation in producing variation is provided by the wheat stem rust, *Puccinia graminis tritici*, in Australia where it lacks the sexual phase. A survey of 193 samples of the rust showed 16 pathotypes. Then 58 of the samples, from seven pathotypes, were examined for variation in 11 different enzyme systems. No isozyme variation was detected, indicating that the 58 samples had a common ancestor and that genetic material had not been introduced from elsewhere. The pathotype variation had hence arisen by mutation. Similar results have been obtained with several other rusts.

Recombination

Although mutation can result in new genotypes that become established in a population, such genotypes may, as indicated above, differ in rather few respects from the strains from which they were derived. On the other hand the mating of two strains that differ at many loci will result, through recombination at meiosis, in a wide variety of new genotypes. The genetic effects of the sexual process in natural populations are illustrated by studies on the amount of variability of *Puccinia graminis tritici* populations east and west of the Rocky Mountains in the USA. The sexual stage in the rust's life cycle occurs on the barberry. This was eradicated east of the Rocky Mountains in the late 1930s, preventing sexual recombination in the eastern populations since that time. West of the Rocky Mountains the barberry and hence sexual recombination in rust populations persist. Extensive surveys on the eastern rust populations have shown rather few pathotypes and isozyme patterns. Furthermore each pathotype is associated with a single isozyme pattern, different from the isozyme patterns associated with other pathotypes, showing that genetic recombination has not occurred. West of the Rocky Mountains pathotypes and isozyme variants are numerous, and the association between pathotype and isozyme pattern is random, indicating that genetic recombination has occurred. Hence the sexual process has generated virulence and isozyme diversity, and has broken any constant associations between pathotype and isozyme pattern.

Whether populations are clonal, freely recombining or with limited recombination can be established by recently developed molecular methods or by means of classical population genetics, but the latter, as applied to fungi in nature, is so laborious that relatively little information has been obtained in this way. The procedure by which the presence or absence of recombination can be inferred from data on molecular variation can, however, be illustrated by reference to genes and alleles. Consider two genes (a and b), each with two alleles (1 and 2). There are four possible genotypes, a_1b_1, a_1b_2, a_2b_1 and a_2b_2. If only two genotypes are found in a population, for example a_1b_1 and a_2b_2, then recombination between the genes is not occurring, due either to the genes being closely linked on the same chromosome or because mating and genetic recombination is absent in the population. If all four genotypes are equally frequent, then unrestricted recombination is taking place. Since there is the possibility that two genes may be linked, it is in practice necessary to consider a larger number of genes, and in a natural population most genes will have not two but many alleles. Procedures are available for deducing from the pattern of variation not only the presence or absence of recombination, but its frequency in a population. These are now applied, not to genes from various sites in the genome, but to enzymes specified by genes, or to nucleic acid variation (DNA polymorphisms) from different parts of the genome.

Such molecular studies are now providing valuable information on recombination within and between populations in nature. For example, Californian populations of the human pathogen *Coccidioides immitis* (page 435), a fungus for which a sexual stage is unknown, has been shown to have essentially unrestricted genetic recombination. However, another human pathogen that lacks a sexual morphology, *Candida albicans* (page 433), proves to be clonal with little recombination. A sexual stage, as indicated earlier in this chapter, results in recombining populations in both heterothallic and homothallic fungi, but if mitospores are also produced, clonal as well as recombining populations are likely to arise, as in *Puccinia graminis tritici* mentioned above.

Gene flow

For the geneticist a population consists of individuals that readily exchange genes. The transfer of genes between populations that are spatially separated, and are hence less likely to exchange genes, is termed **gene flow**. The molecular methods for the study of recombination discussed above also provide evidence for the presence, extent or absence of gene flow. The Californian populations of *Coccidioides immitis*, for example, prove to be almost wholly isolated from those outside California, with evidence for gene flow between them being absent. On the other hand, genetic differences between populations in the cosmopolitan Basidiomycete *Schizophyllum commune* (page 64) are mainly in allele frequences, with genetic distance correlating with geographic distance and a lack of sharp boundaries between populations. Gene flow by long-distance spore dispersal is implied, with even a limited intercontinental gene flow. As with mutation and recombination, the wheat stem rust provides an example of the potential importance of gene flow as a source of genetic novelty in a population. Major changes occurred in 1954 in the ability of *Puccinia graminis tritici* in Australia to attack different host varieties. Study of rust collections maintained from that period indicate that a change in isozyme pattern also occurred. The change was to a pattern hitherto lacking in Australia but found in African populations. The implication is that an introduction from Africa occurred, either by long-distance spore dispersal or inadvertently by humans – the latter possibility being one that is an increasing hazard as air services become swifter and more frequent.

Factors Bringing about Microevolution in a Population

Microevolution refers to small genetic changes in populations which are usually reversible and can occur with sufficient speed to be accessible to observation and experiment. Microevolution takes the form of changes in the frequency of alleles. An allele may diminish in frequency, and even disappear from a population. On the other hand it can increase in frequency and perhaps replace all alternative alleles. The factors by which a gene may thus be 'driven to fixation' are considered below.

Mutation pressure

It is theoretically possible that mutation rate may directly influence gene frequencies, an effect termed **mutation pressure**. Thus if a gene A mutates to an allele a at a finite rate, and back mutation does not occur, and there are no other factors affecting gene frequency, then A will ultimately be replaced by a in a population. If, however, back mutation occurs, neither allele will be eliminated. Instead an equilibrium will be reached with the frequencies of the two alleles depending on the rates of forward and back mutation. It can be calculated that with the relatively high mutation rate from A to a of 1 cell per 100 000 per generation, it will take 69 000 generations to halve the frequency of A in the population. Since some of the other factors affecting allele frequency can be highly effective, it is likely that mutation pressure is unimportant except where mutations are selectively neutral (page 274). Whatever the significance of mutation pressure, mutation rates may be of importance in determining rates of evolution, since in a population that lacks a beneficial allele, one that is produced at a high frequency is likely to appear sooner than one that is produced at a low frequency.

Gene flow

The frequency of alleles in a population can be affected by gene flow from other populations. The magnitude of the effect will depend on the numbers of immigrants compared with numbers in the native population, and the extent to which the immigrants differ from the natives in allele frequency. Gene flow in insects has received considerable study, which indicates major differences between species in the levels of gene flow, and also that gene flow, like the dispersal on which it depends, can be unpredictable. It is likely that allele frequencies in small populations adjacent to large ones will be influenced strongly by gene flow. Between distant populations gene flow is likely to be sporadic, but may be facilitated by intervening populations acting as stepping stones. The effect of gene flow will be to reduce genetic differences between populations, and hence may delay or prevent the evolution of populations in different geographical areas into separate species.

There is so far only limited information about the significance of gene flow in fungi, but studies on molecular variability in populations are likely to remedy this. Work on *Neurospora crassa* suggests that gene flow may be important even over long distances. Almost every naturally occurring *Neurospora* colony that is sampled proves to be genetically distinct, indicating that the colonies have arisen from single ascospores, the products of meiosis, and not from conidia. It is hence reasonable to infer that the main role of conidia, which compared with ascospores are produced in massive amounts, is to fertilize protoperithecia. Studies on the mitochondrial DNA of *N. crassa* show regional differences. Studies on the nuclear genome, however, show that regional diversity is no greater than that within populations. It seems, therefore, that gene flow resulting from the dispersal of conidia has helped maintain, with respect to the nuclear genome, genetic homogeneity throughout the species. In fertilization, however, the conidia do not contribute mitochondria, allowing geographically distinct populations of mitochondrial DNA to evolve.

Natural selection

Darwin was impressed by the success of plant and animal breeders in producing, by means of **artificial selection**, new varieties with features that were considered desirable. Artificial selection consists of selecting individuals showing the required features to a greater extent than other individuals, and breeding from them, the process being repeated over many generations. Furthermore, Darwin was aware that although organisms are capable of an exponential increase in numbers, natural populations did not usually increase in this way, their size tending to remain approximately constant. This, he realized, implied that many individuals did not survive to reproduce. He proposed that any features that favoured survival would be **naturally selected**, and tend to spread through a population, ultimately leading to evolutionary change. Natural selection of favourable variations is now generally accepted as being the major, and probably the only, basis for the acquisition of features that result in a population becoming better adapted to its environment, the process of **adaptive evolution**.

Darwin envisaged natural selection as being **individual selection**, in which the frequency of a feature increases or diminishes as a consequence of its effects on the chances of survival and reproduction of individuals displaying the feature. Subsequently the possibility that natural selection can act in other ways has been considered. For instance, a feature may become more widespread if it is slightly deleterious to an individual displaying it, but considerably enhances the survival prospects of close kin, such as the individual's progeny, which are likely to carry the gene or genes specifying the feature. The effectiveness of such **kin selection** will, however, diminish rapidly as the degree of kinship diminishes. Hence most evolutionists reject the concept of **group selection**, which envisages that features that benefit a population as a whole can result from natural selection. Competition between members of a population is intense, and a feature that confers benefits at random to competitors that may only be distantly related will not be favoured by natural selection. It is easy to slip unconsciously into group selectionist thinking. Thus features of an organism, especially in relation to sexuality (page 262), are often explained in terms of possible benefit to the species, whereas an explanation in terms of natural selection is permissible only if advantage to the individual or its close kin can be demonstrated.

Although natural selection is responsible for any adaptive genetic changes in a population, more often it is a conservative force, eliminating deviants and maintaining the *status quo*. This is because a population is normally well adapted to the environment in which it occurs, and genetic change, whether from mutation, recombination or gene flow, is more likely to result in a decrease rather than in an increase in fitness. Natural selection acting to eliminate variants that differ markedly from the average is termed **stabilizing selection**. The operation of stabilizing selection is illustrated by studies on the Basidiomycete *Schizophyllum commune*. Dikaryotic colonies isolated from fruiting bodies, and thus representing mycelium that had survived natural selection, had a limited range of radial growth rates. Dikaryons produced in the laboratory by hybridization had a much wider range of growth rates but with approximately the same mean. Hence in the natural population, strains with growth rates much lower or much higher than the mean had been eliminated by stabilizing selection. Studies on other attributes and with other fungi yield similar results. Hybridization and artificial selection can give variants differing widely from the forms occurring in nature, indicating the widespread operation of stabilizing selection.

When natural selection acts to produce change it is referred to as **directional selection**. The concept is applicable both to features showing continuous variation and those determined by single genes. Directional selection operating on a continuously varying feature will favour individuals at one extreme. Acting on features determined by single genes, it will either diminish or increase the frequency of an allele, in extreme instances bringing about the extinction of the allele or causing it to be 'driven to fixation', completely replacing alternative alleles. The appearance in the course of a few decades of several pathotypes in an Australian population of *Puccinia graminis tritici* uniform with respect to isozymes, and hence of clonal origin, is an example of the operation of directional selection. Mutation pressure could not have effected the spread of the pathotype mutants in so short a time.

Diversifying or **disruptive selection** favours individuals at the two extremes of a spectrum of variability and acts against intermediate forms. It occurs where there are two possible environments for exploitation, and success results from being highly effective in one rather than moderately effective in both. It will account for the origin within a pathogenic species of forms specialized for different hosts, such as *Puccinia graminis tritici* which attacks wheat and *P. graminis avenae* which attacks oats. The two forms can still hybridize and retain a limited ability to infect the other form's host, so a relatively recent origin by disruptive selection from a less specialized common ancestor is implied.

Natural selection may drive an allele to fixation or extinction. On the other hand it may act to maintain a balance with two or more alleles occurring at fairly high frequency. Natural selection operating in this way is referred to as balancing selection. It can occur where, in the diploid or dikaryotic phase, the heterozygote is fitter than either homozygote (heterozygous advantage – page 262). Balancing selection can also take the form of **frequency dependent selection**, in which selection increases the frequency of an allele if it is rare and decreases its frequency if it is abundant. A rare mating type allele, for example, will increase in frequency, since individuals that possess it will be able to mate with most other individuals. If, however, the allele becomes abundant, then individuals possessing it will be very likely to encounter individuals of identical mating type, with resulting failure of mating. Frequency dependent selection will also occur with the gene-for-gene relationship that determines pathogenicity (page 267). A rare virulence allele will increase in frequency since there will have been little selection for host resistance. If, however, it becomes abundant, then the corresponding gene for host resistance will also become abundant, and selection will operate against individuals carrying the now nearly useless virulence allele.

In haploid organisms a gene will be expressed either throughout or at some stage in the life cycle, and natural selection can operate to influence the frequency of its alleles. In diploid organisms a recessive allele will have no effect on phenotype in organisms heterozygous for the allele. It is hence partly sheltered from natural selection and, if deleterious, will decrease in frequency more slowly than it would if partly or completely dominant or present in a haploid organism. Its extinction is likely to be long delayed, since when low frequencies are reached it will rarely be present in the homozygous state in which it would be exposed to selection. Hence fungi that are diploid or dikaryotic through most of their life cycles are likely to carry many currently deleterious recessive alleles which, in a different genome or environment, might prove advantageous.

Many mutations are highly deleterious and some highly advantageous. Often, however, any advantage or disadvantage is slight and difficult to detect. The question has hence arisen as to whether there are **neutral mutations** which have no effect on fitness and are not subject to natural selection. Factors other than natural selection – mutation pressure, gene flow, molecular drive and random drift – would determine whether such mutations spread through a population or become extinct. Some of these factors can be of considerable magnitude. The question of whether mutations can be completely neutral is hence perhaps pointless – if selective effects are very slight allele frequency will be determined by factors other than natural selection. A question, however, that remains valid and of considerable interest is what proportion of genetic change arises from causes other than natural selection and is hence non-adaptive.

Random genetic drift and the founder effect

Theoretical studies have shown that if a population is small then chance could determine whether a neutral allele becomes extinct or increases in frequency to fixation. If a population is very small then such random genetic drift could determine the fate of an allele even in the presence of moderately strong natural selection. Random genetic drift has been demonstrated with small populations of insects in the laboratory. In nature, however, it may be unusual for a population to stay small long enough for drift to occur – the population could become extinct, grow or merge with another population. Tendencies to genetic drift will be opposed by gene flow. Hence if a fungus is abundant and widespread with copious spores capable of long distance dispersal, gene flow is likely to counteract any tendency to genetic drift. There is evidence for this in the cosmopolitan and abundant fungi *Neurospora crassa*, *Puccinia graminis tritici* and *Schizophyllum commune*.

It is generally accepted, however, that there are ways in which random events could determine the genetic structure of a population and the course of microevolution. A very small sample from a population – one or a few individuals – will not adequately represent the genetic diversity in the population. Many alleles present in the population will be absent in the sample. Such 'sampling' could occur as the result of a catastrophe almost destroying a population or by the dispersal of one or a few individuals to a new environment. The population resulting from such a **founder effect** will be genetically different from the one from which it originated. Many fungi live in environments that are highly favourable but transient, and will hence be liable to colonization from one or a few spores when they arise, and population crashes when they disappear. Founder effects are likely to occur with such fungi and, if the fungi are not highly abundant, may not subsequently be overwhelmed by gene flow.

Molecular drive

Some DNA is functional, being translated into protein or transcribed into ribosomal RNA (rRNA) or transfer RNA. Some DNA, on the other hand, specifies neither protein nor stable RNA, and appears to be functionless and without effect on phenotype. Furthermore, some DNA sequences are present as single copies and some as multiple copies, constituting 'multigene families'. Much of the single copy DNA and some multiple copy DNA, such as the rDNA which specifies rRNA, has a role, whereas some of each type apparently has not. Multiple copy DNA can occur as a series of repeated sequences, as with rDNA, or spread through the genome. Organisms with large genomes tend to have a great deal of multicopy DNA and apparently functionless DNA, but the few intensively studied fungi, with their small genomes, have rather little. The Myxomycete *Physarum polycephalum*, however, which has a high haploid DNA content, has much repetitive DNA.

Natural selection and random drift bring about the spread of a mutation through a population through the survival and reproductive success of individuals containing the mutation. However, a mutation may spread through a population as a result of its own success in survival and reproduction. The process that brings about the spread has been termed **molecular drive** and results from a variety of mechanisms of non-reciprocal DNA transfer among which gene conversion, unequal exchange at crossing over and duplicative transposition are important.

Gene conversion is most easily demonstrated with spore colour mutants in Ascomycetes (Fig. 5.8). The eight-spored asci that result from the mating of a wild-type strain and a strain with a spore colour mutation would each be expected to contain four spores having the wild-type and four the mutant colour. A small proportion of the asci, however, are found to contain wild-type and colour mutant spores in other ratios, for example 5:3 or 6:2. This results from gene conversion at meiosis. Gene conversion occurs through hybrid DNA formation within a gene, and including the site of a base sequence difference

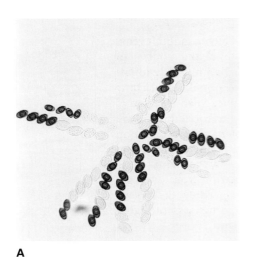

A

B

C

D

between the two alleles involved. A repair system can excise the mispaired bases from part of one DNA strand and they are replaced by bases complementary to those in the second strand. Hence following recombination there are more copies of one allele than of the other.

Gene conversion is often biased – the frequency of conversion, for example, from *A* to *a* is greater than that from *a* to *A*. Such biased gene conversion can result in the spread of an allele through a population. The frequency of gene conversion is such that its effects on the frequency of an allele will sometimes be of greater magnitude than those of natural selection and usually much greater than those of mutation pressure. Gene conversion can also occur at mitosis, not only between corresponding sites on homologous chromosomes, but between other sites, within a chromosome and between non-homologous chromosomes. This will permit a mutation to spread through a multigene family.

Misalignment of homologous chromosomes at meiosis can result in **unequal exchange at crossing over**. Following separation one chromosome has a region deleted and its partner the same region duplicated. This, where there is a series of repeated sequences, will result in one chromosome having an increase and the other a decrease in the number of repeats. Such consequences of unequal exchange during meiosis have been observed for rDNA in *Saccharomyces cerevisiae*. It can also occur in mitosis, and has been demonstrated with mitochondrial DNA in *Aspergillus nidulans*.

Duplicative transposition consists of the duplication of a DNA sequence and the insertion of one copy at a different site in the genome. A DNA sequence liable to such transposition is known as a **transposon**. The best studied fungal transposon is *Ty* ('transposon yeast'), which occurs in *Saccharomyces cerevisiae*. A terminal portion of the *Ty* sequence is known as δ, and facilitates insertion. *Ty* acts as a retrovirus. It is transcribed into RNA and encapsidated. Part of the RNA sequence encodes a reverse transcriptase which copies the RNA into a DNA

Figure 5.8 Genetic variation in Ascomycetes. Crosses were carried out between wild-type strains with pigmented ascospores and mutant strains with albino spores. As the ascus develops, its diploid nucleus undergoes meiosis to yield a tetrad, four haploid nuclei. Each nucleus then undergoes mitosis to give two genetically identical nuclei, and eight ascospores are produced. A,B, Photomicrograph and drawing of asci squeezed from a perithecium of *Sordaria fimicola*, the drawing showing ascus walls, not visible in the photograph. The narrow asci of *S. fimicola* result in ordered tetrads, in which the sequence of spore types indicates whether allele segregation occurred at the first or second division of meiotic division. Segregation at the first division gives a sequence of four wild-type and four mutant spores (a). If segregation occurs at the second division then, starting from either ascus end, two spores of one type are followed by two of the other (b). As can be seen, this can give an alternation of pairs of spore types, or four spores of one type in the middle of the ascus. C, An ascus of *Ascobolus immersus*. The width of the ascus allows spores to slip past each other to give unordered tetrads. In most asci from the type of cross carried out there are four pigmented and four albino spores. Examination of a large number of asci enables gene conversion to be detected through observation of abnormal spore ratios. In the ascus shown there has been conversion from albino to wild-type giving a ratio of 6:2 (one albino spore, at the apex, is almost entirely hidden). D, Eight ascospores discharged from an ascus. Gene conversion has here been from wild-type to albino. Any ratio from 8:0 to 0:8 is possible, depending on when and where gene conversion occurred. (C reproduced by permission of Cambridge University Press from Lamb, B. C. & Wickramaratne, M. R. T. (1973). Corresponding-site interference, synaptinemal complex structure, and 8+0 : *m* and 7+1 : *m* octads from wild-type × mutant crosses of *Ascobolus immersus*. *Genetical Research* **22**, 113–124. A,B,D, Bernard Lamb.)

sequence which then inserts into the yeast genome. *S. cerevisiae* commonly carries about 35 copies of *Ty-1*, one form of *Ty*, about 6 copies of *Ty-917*, a mutant form, and about 100 solo δ sequences.

Molecular drive will inevitably result in a process analogous to natural selection operating within the genome. Sequences that readily duplicate and insert elsewhere in the genome will tend to spread, as will mutant sequences that tend to convert rather than be converted. Molecular drive can promote the spread not only of sequences that are transcribed or translated, but also of sequences that have no other property than self-propagation. Such non-coding sequences have been referred to as **selfish DNA**. The presence of a large amount of selfish DNA will result in a larger genome and possibly increase the time needed for DNA and chromosome replication. This, and also the unnecessary consumption of materials and energy in DNA synthesis, may slightly diminish growth rate. Hence in species in which rapid growth and short generation times are important competitive features, natural selection would be expected to eliminate individuals in which the amount of non-coding DNA had become large. Those fungi that have received intensive study, mostly species with high growth rates, have only a small proportion of their genome in the form of non-coding DNA. This contrasts with many slow-growing higher organisms, in which the amount of non-coding DNA may exceed the amount that is transcribed.

Speciation

Studies on groups of organisms that have a good fossil record make it clear that in any such group species become extinct and new species appear. Where the record is exceptionally good it is possible to trace a sequence of species, each of which differs little from its predecessor, even though the difference between the first and last member of the sequence may be great. The close resemblance between successive species implies **speciation** – the origin of a new species from a pre-existing species. There are few instances where it is agreed that speciation has actually been observed. This is not surprising. First, the time needed for the amount of genetic change needed for speciation is commonly much greater than the time during which biology has been studied. Secondly, there is a lack of agreement about how different two taxa must be to qualify as being different species. There is, however, abundant evidence from the comparison of closely similar taxa, and from their attempted or successful hybridization, of speciation in progress or recently accomplished.

It is possible to envisage an entire species, as a consequence of directional selection and perhaps molecular drive and random drift, changing so much as to evolve into a different species. Such an event, however, would seem unlikely. Over the long period needed a species would be likely to break up into populations evolving at different rates and in different directions. Hence speciation will normally consist of a species splitting into two or more species through the genetic divergence of different populations.

Extensive gene flow between populations will reduce genetic divergence and hence speciation. Such gene flow is most likely between populations in the same area and in organisms in which mating is frequent and outcrossing normal. Gene flow between populations that are geographically remote will be rare, especially if dispersal is inefficient. For this reason it is probable that speciation is often **allopatric** (Greek, meaning other fatherland), with new species arising by the genetic divergence of geographically isolated populations. There are, however, circumstances, considered below, in which gene flow between populations in the same area can come to an end. Hence **sympatric** speciation – the origin of new species from populations inhabiting the same area – is possible.

Speciation in fungi with a sexual phase
Full speciation requires that two populations should diverge genetically and that gene flow between them should cease.

Genetic divergence can occur by non-selective means such as drift, or when the selective forces acting on the two populations differ. This can occur with two populations in the same area, if they occupy different habitats, as, for example, with populations of a rust species living on wheat and oats, respectively (page 273). If the populations are in different areas then environmental differences and different selective forces are also probable.

Gene flow between populations will oppose and limit genetic divergence. Such gene flow will be of little importance if the two populations are distant from each other and dispersal is inefficient. In these circumstances increasing divergence is likely to result in **genetic disharmony** (page 253) between the populations. This may affect the mechanisms involved in mating, so if the two populations once more come into contact, mating may fail, as between European and Californian strains of *Ascobolus immersus*, or be only partially successful. Such disharmony may also mean that if progeny do result, they may be few or of low viability. A complete failure of two populations to mate need not entirely prevent gene flow between them, as both may be able to mate with a third population. For example, a Californian isolate of the Basidiomycete *Merulius himantoides* cannot be crossed with a Swedish isolate, but both isolates can be crossed with an Indian isolate. Such an isolate could provide a genetic bridge between isolates that cannot mate. Where populations adapted to different habitats are in close proximity, mating between them could result in progeny ill-adapted to either environment, or, if considerable divergence had already occurred, in progeny of low viability due to genetic disharmony. Under such circumstances there will be selection for the evolution of mating barriers to prevent the production of progeny of low fitness. One or a few mutations can suffice to bring about such **heterogenic incompatibility** (page 253). It is hence likely that barriers to mating can arise both fortuitously and by natural selection. Gene flow can also be limited, and divergence promoted, by self-fertility and apomixis. In the Myxomycete *Didymium iridis* there is a high degree of isozyme uniformity within each of the three known groups of interbreeding strains, but the apomictic strains differ from each other in isozyme patterns.

Fungal populations may come to differ from each other in chromosome number, through polyploidy or aneuploidy (see page 265). Differences in chromosome number between strains will reduce the chances of hybridization and genetic exchange, because of chromosome pairing difficulties at meiosis. This will increase the likelihood of strains diverging into separate species. On the other hand, if hybridization is successful, the resulting hybrid may differ from both parental types in morphology and chromosome number and also prove competitive. Chromosome counts and hybridization experiments indicate that the Chytridiomycete *Allomyces javanicus* originated as a hybrid between two species with considerable morphological differences, *Allomyces arbuscula* and *Allomyces macrogynous*. In future, increased information on chromosomal differences within species and between closely related species, resulting from the application of pulsed field gel electrophoresis (page 265), is likely to improve understanding of the involvement of chromosomal changes in speciation.

As indicated above, some fungal species consist of several groups of strains that do not interbreed with each other. There are, however, also fungal species with a worldwide distribution such as *Neurospora crassa* and *Schizophyllum commune* which show no indication of regional differences or fertility barriers even though in any locality there is considerable genetic diversity. These fungi are abundant and produce massive numbers of readily dispersible spores, which would facilitate gene flow even between distant populations.

Speciation in asexual fungi

Molecular methods have shown extensive genetic recombination in a few apparently asexual fungi, such as *Coccidioides immitis*. Whether or not this is due to hitherto undetected mating, it makes it likely that speciation in such fungi will occur in ways similar to that in acknowledged sexual species. In a fully asexual species genetic recombination can occur *via* somatic hybridization and the parasexual cycle, but will be limited by vegetative incompatibility. Hence a mitosporic fungal species defined on traditional morphological criteria is likely to consist of a number of biological species, with some clonal and some showing genetic recombination, as has been found in the banana pathogen *Fusarium oxysporum* f. sp. (form species) *cubense*. It seems likely that in mitosporic fungi, when genetic exchange cannot occur through vegetative incompatibility, the resulting clones constitute biological species, and through subsequent evolution may develop into morphologically distinguishable species.

Classification, Identification and Evolutionary Trends

Types of Classification and their Role

Traditional classifications – natural and artificial

In dealing with diversity, classification is essential for efficient thought, and classifications of animals and plants developed long ago in many cultures. Species were recognized, and similar species grouped together. A number of species of aquatic birds, for example, were recognized as sufficiently similar to be grouped together as ducks, and other species as swans. Ducks, swans, and various other groups such as geese, pheasants and hawks were recognized as having features in common which justified their recognition as birds, a category from which groups with very different features, such as deer or apes, were excluded. The popular classification just considered is **hierarchical** – groups are recognized that are themselves members of a group more elevated in the hierarchy. It is also a **natural classification**, one based on overall similarities. Natural classifications provide a good deal of information about the entities classified. Thus knowing that an animal is a bird indicates that it is warm-blooded, feathered, egg-laying, nest-building and probably able to fly. Knowledge that it is a duck provides yet more information.

A natural classification is the usual objective of taxonomists (scientists concerned with classification), but can be achieved only when there is abundant information permitting a sound assessment of similarity. Where there is a dearth of information it may be necessary to construct a classification that is partly or wholly artificial. This may also be done to serve some special purpose. An **artificial classification** is one that is based on a few features, selected because they are readily observed or serve a special interest. Fungi, for example, can be divided into plant pathogens and those that do not damage plants, and the plant pathogens further subdivided on the basis of which plants are affected. An artificial classification, although it may be useful, provides no information other than that used to construct the classification. Two examples of artificial, but useful and widely used categories, are the yeasts (page 70) and the lichens (page 76).

The traditional approach to classification has been based on the utilization of taxonomically valuable characters, some suitable for distinguishing species and some for delimiting higher categories. This raises the question of how characters were recognized as being important. It is likely that experienced taxonomists intuitively recognized a natural group on the basis of overall similarity and took note of features widespread in that group and lacking in other groups. Then newly described organisms could be assigned or omitted from the group on the basis of such characters.

Following the publication of *The Origin of Species* in 1859 most biologists accepted that evolution had occurred – that present species had evolved from earlier species, that closely similar species had a recent common ancestor and very different species a more remote one. Hence a natural classification should reflect descent. The living species in a genus, for example, should have arisen from a common ancestral species that could also be classified within the genus. Acceptance of evolution, and the view that classification should reflect evolutionary relationships, influenced the choice of characters to be utilized in classification, as did an understanding of the basis of genetic change. It was realized that complex morphological characters that were likely to have evolved only once were suitable for delimiting higher taxonomic groups. On the other hand simple features, which could arise from one or a few mutational steps and could have come into existence in different groups, were recognized as being useful only for defining lower taxa such as species or varieties within species.

Numerical taxonomy – phenetic and cladistic
Traditional taxonomy is based on judgements, which may or may not be sound, as to which characters are of major importance, which are of lesser importance, and which should be ignored. An alternative approach, that of **numerical**

taxonomy, is to utilize as many characters as possible and to treat all characters as of equal importance. In practice numerical taxonomy is likely to yield consistent results only if at least fifty characters are utilized, and taxonomists will commonly wish to compare dozens of taxa. Very large numbers of individual comparisons are therefore involved, so numerical taxonomy developed only in the late 1950s, when adequate computers became available.

Numerical taxonomy was initially used to produce classifications that were strictly **phenetic**, based solely on similarity, and not involving any consideration as to how the sets of characters seen, for example, in two organisms might be related by both having been derived from an ancestor with an intermediate set of characters. Phenetic numerical taxonomy yields trees in which taxa are linked to those most similar, with the length of the branches indicating the extent of differences. The trees resemble that in Fig 2.1, with branch tips an equal distance from the root, but with no implications as to actual relatedness or a time scale. Phenetic numerical taxonomy was widely utilized by bacterial taxonomists, and to a more limited extent by fungal taxonomists, until the mid 1970s, when it was largely superseded by a different approach, as indicated below. The term 'numerical taxonomy' is now usually regarded as synonymous with phenetics.

A phenetic classification may well reflect ancestry but assumptions about evolution are not utilized in the construction of the classification. An alternative view is that as present taxa are the product of evolution, classifications should be based on the pattern of evolutionary divergence (phylogeny). Such classifications are termed **phylogenetic** or **cladistic**. A range of computer packages for cladistic classification is available, and different types of trees can be constructed. An 'unrooted additive tree' (Fig 1.2) indicates relatedness by its branch pattern and the amount of change by branch length, but does not locate the common ancestor. A 'rooted additive tree' also shows relatedness and amount of change, and in addition includes the starting point for the process. An 'ultrametric tree' or 'dendrogram' (Fig 2.1) gives, instead of the amount of change, the relative times of divergences. When only a few taxa are classified, there may be only a single tree that fits the available data. With a large number of taxa to be classified, there are likely to be a large number of trees that are compatible with the data, but there are computer programs for selecting the most probable tree. If, in addition, there is information as to how fast evolution has occurred or if there is adequate fossil evidence, a dendrogram may be provided with a time scale.

Phylogenetic trees and traditional classification

A phylogenetic tree is most useful to mycologists if it is converted into a classification. Traditional fungal classifications have a limited number of categories or ranks between kingdom and species, such as the five utilized in the present work, from phylum to genus (Table 2.1). The number of successive divergences in a sequence from the highest to lowest category in a phylogenetic tree is, however, likely to be greater. For example, with merely 35 fungi (Fig 2.1) there can be up to eight branch points (nodes) between them and the highest category, the kingdom. By ignoring some nodes the data could be fitted into five ranks. However, further data could favour a different classification. As more numerous and elaborate phylogenetic trees are produced, the assignment of fungi to traditional categories will become more difficult, and classifications will be liable to be unstable and short lived. This has led to suggestions that the traditional procedure of placing organisms in a hierarchy of categories of different rank should be replaced by a rankless system, in which names are given to clades, a clade being a monophyletic group, all the organisms descended from a common ancestor. International agreements on Codes of Nomenclature take a long time to be reached, but such a development may well occur within a few years.

Features used in Fungal Classification and Phylogeny

About 70 000 species of fungi are already known, presenting a massive task for those concerned with the recognition and description of new species and the grouping of taxa. Most species have hence received only limited study, so classification has had to be mainly traditional rather than numerical, and based on readily observable morphological features. There are, however, some groups of fungi which because of their practical importance have been studied more intensively. With these not only morphology but other features indicated below, have been used in classification.

Morphology
The vegetative structures of lichens are large and show a diversity of form that enables many species to be recognized with the naked eye. The fruiting bodies of many Basidiomycetes and some Ascomycetes are also large and readily identifiable by eye. Hence morphological features identifiable with the naked eye or hand lens are of importance in the classification of lichens, some other Ascomycetes, and Basidiomycetes with large fruiting bodies. Most other fungi, however, are microscopic, and study of their morphology requires the use of the light microscope. Recently some use has been made of patterns and projections on the spore surface that require scanning electron microscopy for observation. It is, however, morphology as revealed by light microscopy that is the basis for most fungal classification. There is an abundance of morphological detail associated with sexual sporulation in fungi, hence the traditional approach of giving the morphology of sexual sporulation priority in fungal classification. Where sexual sporulation is lacking, the smaller amount of morphological detail associated with asexual sporulation is employed. Where sporulation is absent, there is a dearth of morphological information to assist classification, although some use is made of sclerotium form in filamentous fungi and of cell shape and budding pattern in yeasts.

Nutrition and physiology
Yeasts are of great practical importance but have a limited range of morphological features, especially when sexual sporulation is lacking. Most yeasts, however, grow rapidly in pure culture, facilitating the use of nutritional and physiological information in their classification. Features of importance are

whether fermentation occurs or whether growth is limited to aerobic conditions, whether nitrate can be utilized as a nitrogen source and what sugars and glycosides are metabolized, and the pattern of susceptibility to a range of antifungal compounds. Only limited use has been made of nutritional and physiological criteria in the classification of other fungal groups.

Chemistry of low-molecular-weight compounds

The term **chemotaxonomy** usually refers to the utilization of compounds of low molecular weight in classification and identification, although it can equally well be applied to characterization through the analysis of the larger molecules such as nucleic acids. One approach, important in lichen taxonomy, is the identification of uncommon compounds found in a single or a few species. Lichens produce a wide variety of secondary metabolites. Some of these are pigments and present in sufficient amounts to impart colour to the lichen – a feature used taxonomically since the early days of lichenology. Other metabolites are colourless, but can give colour reactions on treatment of the lichen with various reagents, such as potassium hydroxide or potassium hypochlorite. Lichen products can also be recognized with the microscope through the crystal forms produced from extracts on a microscope slide. Thin layer chromatography is also valuable in recognizing the secondary metabolites characteristic of different species. Ascomycetes and Basidiomycetes also produce a wide range of secondary metabolites but so far they have been little utilized in taxonomy. Some use, however, has been made of the chemistry of components with an essential role in metabolism or structure and which are therefore primary rather than secondary metabolites. For example the number of isoprene units attached to the quinone nucleus in coenzyme Q (CoQ) varies, and such differences in CoQ structure have been utilized in yeast classification. An alternative to the detection of specific compounds is to obtain and analyse data on a class of compounds that are universal in their distribution but differ between species as regards which are present and their relative proportions. This approach utilizes procedures such as gas chromatography, mass spectrometry, high performance liquid chromatography (HPLC) and nuclear magnetic resonance (NMR). The data that have been obtained by such methods are subjected to numerical analysis, statistical, phenetic or cladistic depending on the nature of the data and the way they are to be employed. The fatty acids are an example of a class of compounds of universal distribution utilized in chemotaxonomy. Fatty acids vary in chain length and in the frequency and location of double bonds, the number and location of sites. Species vary in the identity of the fatty acids present and their relative proportions, and fatty acid profiles have been used in Zygomycete taxonomy to distinguish between genera and with yeasts to characterize species.

Antigenic properties

A wide range of materials will, when injected into a vertebrate such as a rabbit, induce the formation of **antibodies**. These are proteins that interact specifically with the compounds that induce their formation to give an insoluble complex. Prominent among such compounds, known as **antigens**, are proteins and carbohydrates. A fungus will contain many different proteins, and hence an extract which preserves protein structure will, on injection into a rabbit, elicit many different antibodies, each corresponding to a protein or other antigenically active compound present in adequate amounts in the fungal extract. A serum preparation from a blood sample of the rabbit will contain the antibodies formed against the injected antigens, and is hence known as an **antiserum**. Two closely related fungal strains will differ only slightly in the antigens that they contain, but with diminishing relationship the number of antigenic differences will increase. Many methods of determining antigenic similarity have been devised, one of the most effective being **immunoelectrophoresis**. An antigen preparation placed at one end of an agar gel is subjected to electrophoresis, causing the antigens present to separate by migrating through the gel at different rates. The antiserum diffuses from a well cut in the agar and antibodies interact with the antigens, each interaction producing a band of precipitated protein. Antigen preparations from closely related fungal strains will give a very similar band pattern, but more distantly related strains will have fewer bands in common. Trees indicating relatedness can then be constructed by the application of appropriate numerical methods. The procedure has been applied to some fungal genera within which species are difficult to define by traditional methods.

Carbohydrates and cell wall composition

Fungal walls have a complex structure (Chapter 3), and the major fungal groups are characterized by differences in the polysaccharides present. The walls of Oomycetes, for example, contain cellulose. The corresponding polysaccharide in most other fungi is chitin, and the amount of chitin present has proved a useful feature in yeast taxonomy. The walls of all yeasts that have been subjected to hydrolysis yield substantial amounts of glucose and mannose, but species differ in the presence or absence of smaller amounts of other sugars – fucose, galactose, rhamnose and xylose. This feature too has been utilized in classification.

Protein composition

The use of proteins in the study of natural variation within species has already been discussed (page 266). The banding patterns obtained when protein extracts or individual enzymes are subjected to gel electrophoresis are also useful in defining species and determining relationships within genera.

Nucleic acids

An organism may be of simple morphology, providing few features of value for classification, and obtaining further information by nutritional, physiological and biochemical experiments can be laborious. In every cell, however, the sequence of nucleotides in nucleic acids, nuclear or mitochondrial, represents an enormous amount of information. The number of mutational steps needed to convert one nucleic acid sequence into another is a measure of how closely two nucleic acid sequences are related. Given a set of corresponding nucleic acid sequences from a range of organisms it is possible, by means of appropriate computer programs, to construct a tree indicating how the sequences evolved from an ancestral sequence. Such **molecular phylogeny** is being used increasingly in fungal classification, with the comparison of ribosomal DNA (rDNA) sequences proving especially useful (page 325). An important development has been the **polymerase chain reaction** (page 326), which permits the production of millions of copies of a DNA molecule. This enables taxonomic work to be carried out, for example, with the DNA extracted from a single spore or from herbarium material, perhaps from a much deteriorated and irreplaceable type specimen of a species.

Fungal Identification

The correct identification of fungi is of great practical importance in plant pathology, medical and veterinary mycology, biodeterioration, biotechnology and in environmental studies. Many of the features used in classification can also be used in identification. However, whereas it may be worthwhile to spend a great deal of time and effort in establishing a sound classification, identification needs to be swift, especially where pathogens have to be recognized and action taken. This limits the possible approaches.

Morphology, which was at one time the only basis for the identification of members of most fungal groups, remains important. Many fungal species, indeed the majority, have never been the subject of physiological, biochemical or molecular study, and with these, morphological features remain the only basis for identification. Even where this is not the case, the morphology of a fungus may permit its immediate recognition, or at least its assignment to a group, allowing a decision to be made as to what other approaches are appropriate. Identifying a fungus by means of its morphology can be facilitated by consulting a key for the group. Keys are usually constructed so that a series of decisions as two which of two alternatives applies to a specimen results in identification. Yeasts have a limited range of morphological features, but great diversity with respect to their physiology. Hence for yeast identification morphology is supplemented by nutrional data and biochemical tests. Biochemical tests are also important in lichen identification, since lichens produce a wide range of metabolites that react with reagents. With a fungal species of major importance it may be worthwhile to develop a reagent reacting with that species alone, permitting its recognition, for example, in pathological material without the delay involved in obtaining pure cultures. Species-specific antibodies, including fluorescent antibodies (page 323) have proved valuable, as have probes that interact with DNA sequences unique to a strain or species (page 328).

The Course of Fungal Evolution

Types of evidence
Molecular The trees produced by molecular phenetics show whether existing organisms are closely or distantly related and the patterns of divergence that gave rise to them (Figs 1.2, 2.1). They sometimes confirm conclusions that had been reached on other grounds, for example that some of the organisms studied by mycologists are only distantly related to the mushrooms and toadstools with

which mycology began. Thus it had long been appreciated that Cellular Slime Moulds and Myxomycetes are Protozoa and that more recently that Oomycetes are Chromista that have evolved to closely mimic fungal form and lifestyle. Some of the conclusions are at first sight more surprising. For example, fungi were at one time thought of as plants, but molecular phenetics indicate that their closest relationship is with Choanoflagellates, organisms ancestral to animals. The times at which divergences between groups of organisms occurred (Fig 2.1) can be estimated by making assumptions about the rates of molecular evolution, although such estimates are tentative, and may need modification as a result of the discovery of a well preserved and accurately dated fossil (page 288). The trees produced by molecular phenetics will from now on be crucial in any consideration of the course of fungal evolution. However, in order to determine what ancestral organisms were like and the points on a tree in which important features were acquired or lost, molecular phenetics needs to be supplemented by other evidence.

Morphological and ultrastructural The evolution of a complex structure requires many steps, and it is highly improbable that it will ever be repeated. However, a structure, complex or simple, is soon eliminated by natural selection when it no longer has a useful role. Hence a complex structure which is basically similar in all organisms in which it occurs, for example the eukaryotic flagellum, is likely to have evolved only once, but may have been lost in many different lines of descent. Since some fungi, in common with many other eukaryotes, possess flagella, all present day fungi must have evolved from an aquatic flagellate ancestor. Flagella have been retained in the Chytridiomycetes but lost in the lines of descent that led to terrestrial fungi. Similarly the complex dolipore septum is likely to have evolved only once. Hence Basidiomycetes possessing this organelle, whether yeast-like or with large fruit bodies, will have had a common ancestor with such a septum. Features that are likely to have evolved independently on many occasions provide less reliable evidence of evolutionary history, but were much used in the past in constructing phylogenies. The type of reasoning employed can be illustrated by reference to the fungal fruit body, the form and structure of which is intimately related to spore dispersal, and hence subject to intense selective forces. Within a group of fungi the fruit body will occur in a variety of forms, with those of some species differing greatly from those of others. It will, however, often be possible to arrange fruit body forms in a linear or branched series so that the differences between adjacent members of the series are slight. The sequence may then indicate the way in which the most simple form gave rise to the more complex. If so, then the species ancestral to the group will have had a morphology similar to that possessed by the living species with the most simple fruit body. However, in the course of evolution structures can be simplified as well as become more complex. It is hence possible to misinterpret the direction of evolutionary change, and speculation on the basis of morphology often resulted in different phylogenies being suggested for the same group of fungi.

Nutritional and metabolic Some fungi are able to grow on a medium lacking vitamins and in which nitrogen and sulphur are supplied as inorganic salts. Other fungi have more elaborate nutritional requirements. Vitamins may be needed, nitrogen may have to be supplied as one or more amino acids and, especially in water moulds, sulphur as a sulphur-containing amino acid. Those fungi with simple nutritional requirements will have biosynthetic enzymes lacking in those with more elaborate requirements. It is difficult to evolve new enzymic activities, even under intense selection pressure, but enzymes can easily be lost when their presence is not maintained by natural selection. Hence when members of a group of fungi differ in their nutrition, it is likely that the species with the most simple needs resembles in this respect the ancestral form, and that more elaborate requirements have arisen by loss of enzymes. This reasonable hypothesis has been used in the construction of phylogenies. An enzyme, however, can be lost by a single mutation, and if it is not needed the enzyme-deficient strain will survive. The nutritional needs of members of a group are hence more likely to reflect the different environments in which they live rather than phylogeny.

Palaeontological Aquatic fungi and those associated with plants may form well-preserved fossils. Since the age of the rocks in which the fossils are found will be known from isotopic dating methods, a well-preserved fossil can provide the latest date by which a morphological feature, for example a perithecium, had evolved, and thus give a minimum age for the group of fungi with this feature. There is evidence for the occurrence of Oomycete water moulds late in the Precambrian era, and Chytridiomycetes in the Cambrian period. Fossil plants from the Devonian period, 400 million years ago, have fungi closely similar to modern vesicular–arbuscular endophytes, and perithecia with ascospores on their aerial surfaces, indicating that the Glomales and Ascomycetes had already appeared. In the next major geological period, the Carboniferous, hyphae with clamp connections have been found in Carboniferous strata, showing that Basidiomycetes had by then appeared. Work on fossil fungi has been limited compared with that on fossil animals and plants, and palaeontology has so far had a minor role in deducing the evolutionary history of fungi. Nevertheless, such evidence as there is must be taken into account.

Evolutionary trends

Habitats and ecological relationships Although the presence of zoospores in Chytridiomycetes indicates an aquatic ancestry for the fungi, the vast majority of fungi are terrestrial. Fossil evidence shows that fungi are associated with some of the earliest land plants and are hence among the earliest terrestrial organisms. So much of fungal morphology, including fundamental features such as the ascus and basidium, is associated with launching spores into the air that it is likely that most fungal diversification took place on land. Some fungi with a terrestrial ancestry, however, have returned to an aquatic environment. Examples are the marine Ascomycetes (page 350) and the aquatic Hyphomycetes (page 348).

There are saprotrophs and symbionts, either mutualistic or parasitic, in all the major fungal groups. A symbiotic relationship is a specialized one, so that it is likely that fungi were at first saprotrophs and that, in various lines of descent, parasitic or mutualistic symbionts evolved.

Plants have defence mechanisms that prevent penetration by most fungi. These defence mechanisms are, however, weak in moribund tissues, such as those that are senescent or damaged by frost in autumn. When defences are weak, fungi that are normally saprotrophs can attack plants and live as necrotrophic parasites (page 385), killing cells and deriving nourishment from their contents. If there is a sufficiently massive inoculum, such unspecialized facultative parasites may succeed in invading healthy tissues. If living plant tissues prove an important resource compared with dead plant material, then a facultative parasite will undergo selection for more effective parasitic activity. This, however, may result in greater specialization, a more limited host range, and reduced competitive saprophytic ability. Facultative parasitism may hence evolve into obligate parasitism, with the fungus unable to grow except in its away host.

A parasite that causes the death of its host may, if its competitive saprotrophic ability is poor, bring to an end its own vegetative phase, and will leave progeny only if its spores reach another individual of the host species. It may, therefore, be advantageous to limit damage to the host and to divert only a part of the host's resources to its own use, permitting the prolonged survival of the host. Many parasites are hence biotrophs (page 391), doing little damage to the host and requiring proximity to living host cells in order to survive. Natural selection does not, however, inevitably result in the gradual reduction of the pathogenicity of a parasite. A parasite strain that can invade a host and exploit its reserves rapidly is likely to produce more spores than a competing strain that attacks more slowly. There can hence be selection for increased production of toxins that destroy host resistance and enzymes that degrade tissues. There are thus two opposite strategies that can lead to success. That of biotrophy and life in equilibrium with the host is likely to be of most value for an obligate parasite with limited host range. Unrestricted attack, on the other hand, may be the best strategy for a facultative parasite with a wide host range.

Fungi early developed mutualistic relationships with vascular plants. As mentioned above, some of the earliest land plants had vesicular–arbuscular (VA) endophytes, a relationship that has persisted ever since. It is estimated that about 80% of land plants have VA mycorrhizas (VAMs), including most trees in tropical forests and herbaceous plants almost everywhere. Although associated with a very wide range of plants, VA endophytes (Glomales) show very little diversity, being classified into a few genera and a few dozen species. Trees in temperate forests, and species that grow on infertile soils in the tropics, commonly have sheathing mycorrhizas (page 401). There is evidence that these mycorrhizas evolved much later than the VA mycorrhizas. In contrast to VAMs, relatively few (although abundant) tree species are involved, but a large number of fungal species, all Basidiomycetes. The mycorrhizal relationship is clearly beneficial to the vascular plant as 'most woody species require mycorrhizas to survive, and most herbaceous plants need them to thrive'. Conversely, the mycorrhizal fungi need their vascular plant host to survive, being obligate symbionts.

Lichens (page 76) constitute another mutualistic relationship, involving a wide range of Ascomycetes and a few green or blue–green algal species. Many groups of Ascomycetes have both free-living species and species that exist only as the fungal component of a lichen. This suggests that the lichen symbiosis has evolved independently in several different groups of Ascomycetes, a conclusion now confirmed by molecular phylogeny.

Cell form and colony structure Hyphae are characteristic of true fungi, although cells that are regarded as hyphae, or resemble hyphae, occur in other groups of organisms. In higher plants the pollen tubes that penetrate the stigma and style of flowers to bring about fertilization, and the root hairs that penetrate soil to obtain water and minerals in solution, resemble hyphae in their polarized, apical growth. Hyphae occur in several algal groups, where they facilitate spread over damp soil, and in prokaryotes such as Streptomycetes (page 338), where the minute hyphae enable the penetration and degradation of refractory material to occur. In all these groups, and in the Oomycetes, an independent evolution of the hyphal form must have occurred. In the true fungi hyphae may have evolved more than once. The hyphal form is uncommon in the predominantly aquatic Chytridiomycetes, although it occurs in the intensively studied genus *Allomyces* (page 33). The essentially terrestrial Zygomycetes, Ascomycetes and Basidiomycetes are mostly hyphal. They differ, however, in their wall chemistry, and the hyphal form may have arisen independently in each of these groups. In the terrestrial fungi hyphae may initially have been useful for spreading on and penetrating damp soil, but then proved of great utility in penetrating the dead remains or living tissues of plants. It is clear that hyphae are structures of critical importance in enabling fungi to be parasitic or mutualistic symbionts of terrestrial plants from their origin to the present day.

In higher fungi – Ascomycetes, Basidiomycetes and mitosporic fungi – hyphae are normally divided into compartments by septa, although pores through the septa allow protoplasmic continuity between compartments to be maintained. The occurrence of septa allows injured areas to be swiftly sealed off with consequent efficient damage limitation (page 104). Anastomosis between vegetative hyphae, which can convert a pattern of radiating hyphae into a three-dimensional network, is also common in higher fungi (page 142). This allows protoplasm and nutrients from an extensive mycelium to be rapidly conveyed to a few points. It is a development that would seem to be a prerequisite for the evolution of large fruit bodies and sclerotia. Hyphal anastomosis, however, can permit cell fusion with hyphae from a genetically different colony, and possible infection with viruses and alien DNA. This is likely to be the reason for the evolution of vegetative incompatibility (page 257) in higher fungi.

Many Zygomycetes, Ascomycetes, Basidiomycetes and mitosporic fungi are able to occur in either a hyphal or a yeast form, and some occur exclusively as yeasts. Whereas hyphae are well adapted for penetrating plant tissues, the yeast form appears to be ideal for life on surfaces where there may be rapid fluctuations in water potential (page 163). It is likely that hyphal fungi have given rise to yeasts on many occasions. Conversely, yeasts may sometimes have given rise to hyphal descendants.

Life cycles and sexuality The fundamental features of the sexual process are cell fusion and meiosis (page 248). It is probable that the complex process of meiosis came into existence only once, early in the evolution of eukaryotes, and has persisted ever since in all major eukaryotic groups. In the fungi the sexual process has become associated with the production of spores characteristic of the class, such as zygospores, ascospores and basidiospores. Where such spores have a crucial role in the survival of a species, there will be strong selection for self-fertility, which makes the production of such spores more probable. The widespread occurrence of self-sterility and complex mating systems, however, indicates that outcrossing has important advantages, although there is a lack of agreement on the precise nature of these advantages (page 262). Mating systems have considerable flexibility and it is probable that not only have self-sterile species given rise to self-fertile ones, but that sometimes self-sterility has arisen in self-fertile lines.

In addition to spores associated with the sexual process, many fungi produce one or more spore types asexually. If the spore type associated with sexuality becomes relatively unimportant in survival or dispersal, then selection pressure for the maintenance of the sexual process will be weak, and a fungus may become wholly asexual. Such fungi can, in the short term at least, be highly successful, as is indicated by numerous mitosporic species and by the abundance of some of them, but it is doubtful whether an asexual line can persist for millions of years and give rise to diverse descendants (page 263).

Various types of life cycle occur in fungi (page 248). Many are haploid throughout their vegetative phase, and this is usually regarded as the condition from which other life cycles have evolved. In the Basidiomycetes, the group with the most extensive mycelia and the largest and most complex fruit bodies, a phase of dikaryotic vegetative growth is normal. Diploid vegetative growth occurs in the Chytridiomycete *Allomyces* (page 33) and in many Ascomycete yeasts and some Basidiomycetes. The reason for the trend to diploidy and its functional equivalent, dikaryosis, could be the gains from heterozygous advantage and the complementation of genetic defects (page 262).

Further Reading and Reference

Many of the topics considered in this chapter were first studied in organisms other than fungi, and some of the literature that is most useful in understanding basic principles is not specifically concerned with fungi. Hence in the list below the sections on fungi are preceded by ones on wider aspects of genetics and evolution.

Molecular Biology

Brown, T. A. (1999). *Genomes*. Bios, Oxford.
Lewin, B. (2000). *Genes VII*. Oxford University Press, Oxford.
Lodish, H. F., Berk, A., Zipurski, L., Matsudaira, P., Baltimore, D. & Darnell, J. (2000). *Molecular Cell Biology*, 4th edn. W. H.Freeman, Oxford.
Primrose, S. B. (1998). *Principles of Genome Analysis*, 2nd edn. Blackwell Science, Oxford.
Singer, M. & Berg, P. (1991). *Genes and Genomes*. Blackwell, Oxford.

General Genetics

Griffiths, A. J. F., Miller, J. H., Suzuki, D. T., Lewontin, R. C. & Gelbart, W. M. (2000). *An Introduction to Genetic Analysis*, 7th edn. Freeman, New York.

Hartl, D. L. & Jones, E. W. (1999). *Essential Genetics*, 2nd edn. Jones and Bartlett, Boston.
Hartl, D. L. & Jones, E. W. (2000). *Genetics: Analysis of Genes and Genomes*, 5th edn. Jones and Bartlett, Boston.

Population Genetics and Evolution

Avise, J. C. (1994). *Molecular Markers, their Natural History and Evolution*. Chapman & Hall, New York.
Barton, N. H. & Charlesworth, B. (1998). Why sex and recombination? *Science* **281**, 1986–1989.
Dawkins, R. (1982). *The Extended Phenotype: the Gene as the Unit of Selection*. Freeman, Oxford.
Dover, G. A. (1986). Molecular drive in multigene families: how biological novelties arise, spread and are assimilated. *Trends in Genetics* **2**, 159–165.
Graur, D. & Li, W. H. (1999). *Fundamentals of Molecular Evolution*. Sinauer, Sunderland, Massachusetts.
Hamilton, W. D., Axelrod, R. & Janese, R. (1990). Sexual reproduction as an adaptation to resist parasites (a review). *Proceedings of the National Academy of Sciences, Washington* **87**, 3566–3573.
Hartl, D.L. (2000). *A Primer of Population Genetics*, 4th edn. Sinauer, Sunderland, Massachusetts.
Hedrick, P. W. (2000). *Genetics of Populations*, 2nd edn. Jones and Bartlett, Boston.
Hurst, L. D. (1999). The evolution of genomic anatomy. *Trends in Ecology and Evolution* **10**, 294–299.
Hurst, L. D. & Hamilton, W. D. (1992). Cytoplasmic fusion and the nature of sexes. *Proceedings of the Royal Society, London B* **247**, 189–194.
Kimura, M. (1983). *The Neutral Theory of Molecular Evolution*. Sinauer, Sunderland, Massachusetts.
Lyons, E. J. (1997). Sex and synergism. *Nature* **390**, 19–21.
Mallett, J. (1995). A species definition for the modern synthesis. *Trends in Ecology and Evolution* **10**, 294–299.
Maynard Smith, J. (1989). *Evolutionary Genetics*. Oxford University Press, Oxford.
Maynard Smith, J. & Smith, N. H. (1998). Detecting recombination from gene trees. *Molecular Biology and Evolution* **15**, 590–599.
Michod, R. E. & Levin, B. R., eds. (1988). *The Evolution of Sex*. Sinauer, Sunderland, Massachusetts.
Stearns, S., ed. (1987). *The Evolution of Sex and its Consequences*. Birkhauser Verlag, Basel.

Genetic Reference

King, R. C. & Stansfield, W. D. (1997). *A Dictionary of Genetics*, 5th edn. Oxford University Press, New York.
O'Brien, S. J., ed. (1999). *Genetic Maps*, 6th edn. Cold Spring Harbor Laboratory Press, Cold Spring Harbor, New York.

Taxonomic and Phylogenetic Methods

Bridge, P. D., Arora, D. K., Reddy, C. A. & Elander, R. P., eds. (1998). *Applications of PCR in Mycology*. CAB International, Wallingford.

Bridge, P. D., Scott, P. R., Jeffries, P. & Morse, J, M, eds (1998) *Information Technology, Plant Pathology and Biodiversity*. CAB International, Wallingford.

Colwell, R. R. & Grigoreva, R., eds. (1987). *Current Methods for Classification and Identification of Microorganisms. Methods in Microbiology*, vol. 19. Academic Press, London.

Frisvad, J. C., Bridge, P. D. & Arora, D. P., eds. (1998). *Chemical Fungal Taxonomy*. Marcel Dekker, New York.

Hibbett, D. S. & Donoghue, M. J. (1998) Integrating phylogenetic analysis and classification in fungi. *Mycologia* **90**, 347–356.

Hillis, D. M., Huelsenbeck, J. P. & Cunningham, C. (1994). Application and accuracy of molecular phylogenies. *Science* **264**, 671–677.

Majer, D., Mithen, R., Lewis, B. G., Vos, P. & Oliver, R. P. (1996). The use of AFLP fingerprinting for the detection of genetic variation in fungi. *Mycological Research* **100**, 1107–1111.

Page, R. D. M. & Holmes, E. C. (1998). *Molecular Evolution: a Phylogenetic Approach*. Blackwell Science, Oxford.

Paterson, R. P. M. & Bridge, P. D. (1994). *Biochemical Techniques for Filamentous Fungi*. CAB International, Wallingford.

Towner, K. J. & Cockayne, A. (1993). *Molecular Methods for Microbial Identification and Typing*. Chapman & Hall, London.

Tudge, C. (2000). *The Variety of Life*. Oxford University Press, Oxford.

Weising, K. (1995). *DNA Fingerprinting in Plants and Fungi*. CRC Press, Boca Raton, Florida.

General Fungal Genetics

Bennett, J. W. & Lasure, L. L., eds. (1985). *Gene Manipulations in Fungi*. Academic Press, Orlando.

Bennett, J. W. & Lasure, L. L., eds. (1991). *More Gene Manipulations in Fungi*. Academic Press, San Diego.

Bos, C. J., ed. (1996). *Fungal Genetics: Principles and Practice*. Marcel Dekker, New York.

Peberdy, J. F., ed. (1991) *Applied Molecular Genetics of Fungi*. Cambridge University Press, Cambridge.

Sidhu, G. S., ed. (1988). *Genetics of Plant Pathogenic Fungi. Advances in Plant Pathology*, vol. 6. Academic Press, London.

Molecular Systematics, Population Genetics and Evolution of Fungi

Anderson, J. B. & Kohn, L. M. (1998). Genotyping, gene genealogies, and genomics bring fungal population biology above ground. *Trends in Ecology and Evolution* **13**, 444–449.

Barr, D. J. S. (1992). Evolution and kingdoms of organisms from the perspective of a mycologist. *Mycologia* **84**, 1–11.

Blackwell, M. (2000). Terrestrial life – fungal from the start. *Science*. **289**: 1884–1885.

Bridge, P. D., Couteaudier, Y. & Clarkson, J. M., eds (1998) *Molecular Variability of Plant Pathogens*. CAB International, Wallingford.

Burdon, J. J. (1987). *Diseases and Plant Population Biology*. Cambridge University Press, Cambridge.

Burdon, J. J. & Leather, S. R., eds. (1990). *Pests, Pathogens and Plant Communities*. Blackwell, Oxford.

Burdon, J. J. & Silk, J. (1997). Sources and patterns of diversity in plant-pathogenic fungi. *Phytopathology* **87**, 664–669.

Burdon, J. J. & Thrall, P. H. (1999). Spatial and temporal patterns in coevolving plant and pathogen associations. *American Naturalist* **153**, S15–S33.

Egger, K. N. (1992). Analysis of fungal population structure using molecular techniques. In G. C. Carrol & D. T. Wicklow, eds., *The Fungal Community: its Organization and Role in the Ecosystem*, pp. 193–208. Marcel Dekker, New York.

Gargas, A., DePriest, P. T., Grube, M. & Tehler, A. (1995). Multiple origins of lichen symbiosis suggested by SSU rDNA phylogeny. *Science* 268, 1492–1495.

Glass, N. L. & Kuldau, G. A. (1992). Mating type and vegetative incompatibility in filamentous ascomycetes. *Annual Review of Phytopathology* 30, 201–224.

Gouy, M. & Wen-Hsiung, L. (1989). Molecular phylogeny of the kingdoms animalia, plantae and fungi. *Molecular Ecology and Evolution* 6, 109–122.

Gow, N. A. R., Brown, A. J. P. & Odds, F. C. (2000). Candida's arranged marriage. *Science* 289: 256–257.

Keen, N. T. (1990). Gene-for-gene complementarity in plant–pathogen interactions. *Annual Review of Genetics* 24, 447–463.

Kohn, L. M. (1992). Developing new characters for fungal systematics: an experimental approach. *Mycologia* 84, 139–153.

Hibbett, D. S., Pine, E. M., Langer, E., Langer, G. & Donoghue, M. J. (1997). Evolution of gilled mushrooms and puffballs inferred from ribosomal DNA sequences. *Proceedings of the National Academy of Sciences, USA* 94, 12002–12006.

James, T. Y., Porter, D., Hamrick, J. L. & Vilgalys, R. (1999). Evidence for limited intercontinental gene flow in the cosmopolitan mushroom, *Schizophyllum commune*. *Evolution* 53, 1665–1677.

Metzenberg, R. L. (1991). The impact of molecular biology on mycology. *Mycological Research* 95, 9–13.

Murtagh, G. J., Dyer, P. S. & Crittenden, P. D. (2000). Sex and the single lichen. *Nature* 404, 564.

Pirozynski, K. A. & Hawksworth, D. L., eds. (1988). *Coevolution of Fungi with Plants and Animals*. Academic Press, London.

Ramsdale, M. & Rayner, A. D. M. (1997). Ecological genetics. In D. T. Wicklow & B. E. Soderstrom, eds. *Environmental and Microbial Relationships*, pp. 15–30, Vol. IV, *The Mycota* K. Esser & P. A. Lemke, eds. Springer-Verlag, Berlin.

Rayner, A. D. M. (1991). The phytopathological significance of mycelial individualism. *Annual Review of Phytopathology* 29, 305–323.

Rayner, A. D. M. (1992). Monitoring genetic interactions between fungi in terrestrial habitats. In E. M. H. Wellington, E. M. H. van Elsas & J. D. van Elsas, eds., *Genetic Interactions among Microorganisms in the Natural Environment*. Pergamon Press, New York.

Rayner, A. D. M., Brasier, C. M. & Moore, D., eds. (1987). *Evolutionary Biology of the Fungi. Twelfth Symposium of the British Mycological Society*. Cambridge University Press, Cambridge.

Reynolds, D. R. & Taylor, J. W., eds. (1993). *The Fungal Holomorph: Mitotic, Meiotic and Pleomorphic Speciation in Fungal Systematics*. CAB International, Wallingford.

Sanders, I. R. (1999) No sex please, we're fungi. *Nature* 399, 737–739.

Taylor, J. W., Geiser, D. M., Burt, A. & Koufopanou, V. (1999). The evolutionary biology and population genetics underlying fungal strain typing. *Clinical Microbiology Reviews* 12, 126–146.

Taylor, J. W., Jacobson, D. J. & Fisher, M. C. (1999). The evolution of asexual fungi: reproduction, speciation and classification. *Annual Review of Phytopathology* 37, 197–246.

Turgeon, B. G. (1998). Application of mating-type gene technology to problems in fungal biology. *Annual Review of Phytopathology* 36, 115–137.

Vilgalys, R. & Sun, B. L. (1994). Ancient and recent patterns of geographical speciation in the oyster mushroom *Pleurotus* revealed by phylogenetic analysis of ribosomal DNA sequences. *Proceedings of the National Academy of Sciences, USA* 91, 4599–4603.

Wainwright, P. O., Hinkle, G., Sogin, M. L. & Stickel, S. K. (1993). Monophyletic origins of the Metazoa: an evolutionary link with fungi. *Science* 260, 340–342.

Wolfe, M. S. & Caten, C. E., eds. (1987). *Populations of Plant Pathogens: their Dynamics and Genetics*. Blackwell, Oxford.

Worrall, J. J., ed. (1999). *Structure and Dynamics of Fungal Populations. Population and Community Biology*, vol. 25. Kluwer Academic Publishers, Amsterdam.

Zeyl, C. & Bell, G. (1997). The advantage of sex in evolving yeast populations. *Nature* 388, 465–468.

Some Journals

Annual Review of Genetics
FEMS Yeast Research
Fungal Biology and Genetics
Trends in Ecology and Evolution
Trends in Genetics

Questions

With each question one statement is correct. Answers on page 561.

5.1 In a fungus with two mating types, the probability that an encounter between two individuals derived from a single diploid will result in mating is:
   a) 100%
   b) zero
   c) 25%
   d) 50%

5.2 In a fungus with multiple alleles at two loci, A and B, the following combination results in mating:
   a) A_1B_1 with A_1B_1
   b) A_1B_2 with A_2B_1
   c) A_2B_2 with A_2B_1
   d) A_1B_1 with A_1B_2

5.3 A plant cultivar showing genetically determined vertical resistance to a fungal pathogen always has the following feature:
   a) resists all strains of the pathogen
   b) produces specific antifungal toxins
   c) resists only strains of the fungus that carry a particular avirulence gene
   d) belongs to a monocotyledonous crop species

5.4 In the parasexual cycle, the following process occurs:
   a) recombination without meiosis
   b) meiosis without recombination
   c) dikaryotization
   d) nuclear migration

5.5 The phylum of extant fungi which appears most closely to resemble the ancestral type is:
   a) Chytridiomycetes
   b) Myxomycetes
   c) Oomycetes
   d) Ascomycetes

Saprotrophs and Ecosystems

6

Features of Fungi Determining their Roles in Ecosystems

Some fungi, the parasitic and mutualistic symbionts (Chapter 7), obtain their nutrients from living organisms. Others, however, are **saprotrophs**, and obtain their nutrients from dead organisms. These latter fungi are the main group of organisms responsible for recycling the components of dead plants; the decomposition of the remains of animals and microorganisms (including fungi and bacteria) is accomplished mainly by bacteria. This activity of fungi is essential for the continuation of life on earth. The carbon cycle involves the fixation of atmospheric carbon dioxide into organic molecules by photosynthesis; the fungi play an important part in degrading these molecules and thereby replenishing the carbon dioxide of the atmosphere (Fig. 6.1). The mean value for the amount of plant remains deposited annually in the forests of temperate regions is of the order of two tons per hectare. Of this, up to a quarter is in the form of substantial woody remains which are degraded exclusively by specialized fungi (pages 305, 340).

Without degradative processes life on earth would probably come to an end after a few decades because of the accumulation of plant remains and lack of free atmospheric carbon dioxide for photosynthesis. The degradation of plant remains is also important in the cycling of other elements, particularly nitrogen (Fig. 6.2), phosphorus and potassium, which are incorporated into insoluble plant components such as cell walls. Microbes absorb these as plant material decays and release them in inorganic form into the soil whence they can be used again by plants. Fungi are better equipped for bringing about the decay of insoluble plant remains than are bacteria, both through their physical form and mode of growth and their enzymic capabilities and metabolism. Their growth form, discussed in detail in Chapter 3, enables them to penetrate plant tissues. The hyphae are of an appropriate diameter (5–20 μm) to grow inside plant cells and, particularly those specialized to decay wood, inside the tubular cells that are a major component of wood. Apical growth from a mycelium that is firmly attached to a substratum enables the hyphae to grow into tissues, generating physical pressure which together with local enzymic lysis can perforate woody cell walls. The mycelium of those fungi, mainly Ascomycetes and Basidiomycetes, that are able to degrade refractory plant remains is itself often long-lived and capable of cellular differentiation and the development of multihyphal structures such as mycelial strands and rhizomorphs (Chapter 3). The physical association between the

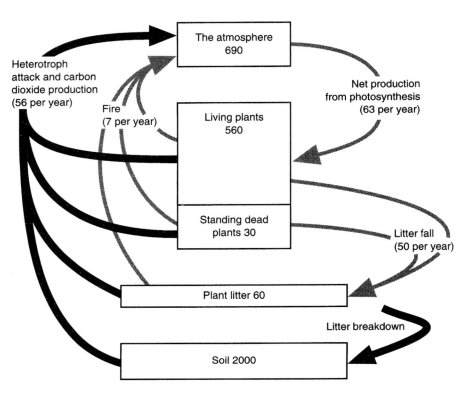

Figure 6.1 The terrestrial carbon cycle. The numbers indicate billions (10⁹) of metric tons of carbon involved in major pools and fluxes, as tentatively estimated. The atmospheric carbon is almost all in the form of carbon dioxide, and of that in plants a very large part is in the form of lignified cellulose. Slender arrows indicate physical processes or the activities of phototrophs (especially plants), and thick arrows steps in which fungi make a contribution – usually a highly significant one in view of the outstanding abilities of some fungi in attack on lignocellulose. The burning of fossil fuels, which is slowly increasing the atmospheric carbon dioxide level, is not indicated. Numerical data from Bolin *et al.* (1979); see references, page 35.

insoluble substrate and the fungal colony is therefore long-lasting, and the colony itself becomes functionally differentiated, for example by fruiting in one region and assimilating in another. Since the surface on which a colony grows may also be its food source, the term 'substrate' may be used loosely in either sense. To avoid confusion, we use the term 'substrate' in its accepted biochemical sense to mean a single chemical structure on which an enzyme acts, and 'substratum' to mean the material on or in which growth occurs. The term 'resource' will be used for a food source available to a fungal colony.

Fungal form is to a large extent determined by environment (Table 4.2, page 191). Features influenced include hyphal branching patterns, colony growth rates, sporulation, hyphal aggregation and differentiation and the adoption of a yeast form by some normally hyphal fungi. Nutrient balance, temperature, water activity, and the type, concentration and spatial distribution of resources, are all

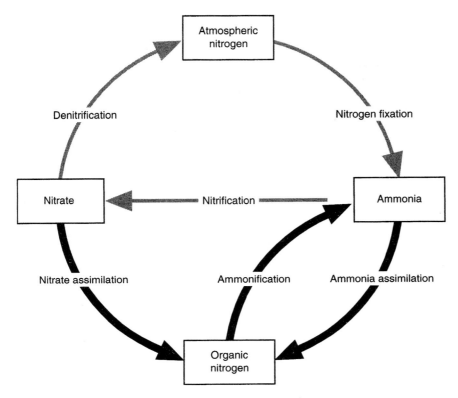

Figure 6.2 The nitrogen cycle. Thick arrows indicate processes carried out by fungi as well as many other organisms. The atmosphere contains a thousand times as much nitrogen (*ca* 10^{15} tons) as carbon, but the rate of its removal and return to the atmosphere is far less, about 250 million tons per year. About 70% of nitrogen fixation is biological, about 20% industrial and about 10% is due to other causes, such as lightning and car ignition. Only prokaryotes can carry out biological nitrogen fixation, but some fungi are indirectly involved, since some lichens have as the photosynthetic partner nitrogen-fixing cyanobacteria. Nitrification, in which nitrate oxidation yields energy, is an activity of a few prokaryotes, and denitrification, in which nitrate acts as an oxidant for respiration in the absence of oxygen, is carried out by many bacteria and a few eukaryotes. Most fungi can assimilate ammonia and many nitrate. Fungi can also participate in the attack on organic nitrogen in plant and animal remains, either assimilating amino acids, or utilizing their carbon skeletons and releasing ammonia (ammonification).

important factors (Chapter 3). These developmental changes are not usually essential for the continued existence of the organism (a mycelium may live for years without sporulating) and fungi can be regarded as having facultative morphogenesis.

Regulation of development by means of programmed responses to specific external changes (Table 3.10) is obviously advantageous to the fungi, whose immobility and heterotrophy forces upon them a life of nutritional opportunism. They must be able to produce structures such as vegetative mycelium that take maximum advantage of favourable conditions when these occur, and other structures such as spores that can survive unfavourable conditions or can be dispersed to new places. It is not surprising that alterations in form develop in response to nutrient exhaustion or imbalance. In some cases the change in form is initiated by conditions more specific than a general depletion of nutrients. The concentration of a single substance may be critical, or light of a particular wavelength may start a developmental process. This more precise response may have evolved in fungi where there is a selective advantage in ensuring that development does not occur wastefully under inappropriate conditions. An example is the light-stimulated development of spores (Chapter 4), where the response to light ensures that spores are not produced uselessly inside the substratum but only on the surface, whence they can be dispersed. Extreme chemical specificity can be shown in the development of sex organs, as in the Oomycete *Achlya ambisexualis* (page 29), where colonies of each sex produce steroid pheromones which cause development of sex organs in the colony of opposite sex. Wasteful sexual differentiation in the absence of a partner is thus avoided.

Molecular analysis shows that fungi sense signals from the environment by pathways homologous with those of plants and animals, and that the proteins concerned are highly conserved. A class of receptor common to all eukaryotic cells is the seven-helix family, in which the receptor has seven transmembrane domains which anchor it to the plasma membrane. Signal molecules such as nutrients or pheromones bind to the receptor on the outer side of the cell membrane, activating a glucose diphosphate binding protein (G protein) on the cytoplasmic side of the membrane, which initiates further processes associated with the response to the signal. Genes for the G protein, and for several downstream components of the signal transduction pathway, have been cloned from many fungi, including yeasts, filamentous saprotrophic fungi, and pathogens of plants and animals. The pathway has been found to initiate the development of specialized cells in response to mating pheromones (page 205) and nutrients, and the production of infection structures following recognition of host surfaces by plant pathogenic fungi (page 380).

Just as species differ in their morphology, so also their enzymic equipment is characteristic of particular species and groups. Although all are heterotrophic at least for carbon, they vary in their capacity for using particular substrates. Extracellular hydrolases are produced in wide variety, and almost any naturally occurring macromolecule can be degraded by some fungus (Table 6.1). Induction and repression of extracellular enzyme activity is widespread. The regulation of extracellular cellulase activity (page 314) is one of the best understood examples. The soluble products of extracellular enzyme action and other small molecules may enter hyphae by diffusion, replacing those utilized by metabolism, but more usually they are actively transported across the cell membrane by specific transport systems. It is characteristic of the fungi that some of these transport systems have a very high affinity for their transported substances, permitting rapid uptake. This also facilitates scavenging activities by fungi under starvation conditions, since nutrients can be concentrated in the mycelium from very low external concentrations. Sugars, amino acids and minerals such as phosphate are accumulated in this way. Some transport systems are constitutive, being always present. Others only become active when induced by the presence of the substrate that they transport. For example, cells of yeast, *Saccharomyces cerevisiae*, can

Plate 1 (above). Lichen photosymbiodeme
In this lichen, *Pseudocyphellaria rufovirescens*, two different photosynthetic microbes grow in symbiosis with one species of fungus (a 'photosymbiodeme'). The green lobes contain a green algal photobiont, and the dark lobes contain a cyanobacterial photobiont. Both photosynthetic members of the photosymbiodeme are also found growing separately. The photosymbiodeme is typically found at the margins of rainforests and more often in areas with higher relative humidities like stream valleys. Photo: Courtesy of T.G. Allan Green.

Plate 2 (left). A fungal endophyte
(*Biscogniauxia nummularia*) of beech (*Fagus sylvatica*). The fungus at first lives within the woody tissues of the healthy tree, causing no visible symptoms. If the tree is stressed, for example by drought, the fungus invades the bark from beneath, and erupts through it into surface lesions bearing black masses of spores. Photo: Courtesy of David Lonsdale, Forest Research.

Plate 3 (above). Ectomycorrhiza on a root of beech (*Fagus sylvatica*)
The highly branched lateral rootlets are pale tan where they are covered with a fungal mantle. Ectomycorrhizas are abundant on beech roots, and are found near the soil surface, beneath the litter layer and closely associated with decomposing plant litter like the beech nut cupule (case) on the right. They are formed by the mycelium of many common woodland fungi such as species of *Russula* and *Laccaria*.

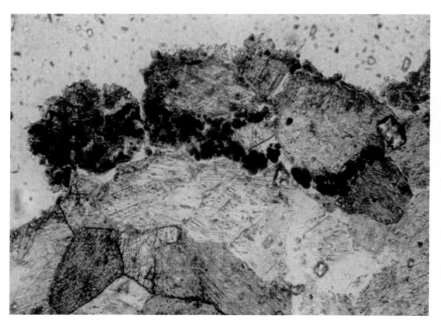

Plate 4 (above). Fungal biodeterioration of stone
A specialised rock-invading fungus, seen in a thin section of marble which the fungus has colonised. The small spherical melanised fungal cells grow inside the pores of the stone and swell, causing it to fragment.
From Gorbushina, A.A. & Krumbein, W. (1999), in J. Seckbach, ed., *Microbial Diversity*. Reproduced with kind permission of Kluwer Academic Publishers.

Plate 5 (above). An example of mycoparasitism
The bright yellow jelly fungus, *Tremella aurantica*, parasitizing fruit bodies of *Stereum hirsutum*, a bracket-forming wood-rotting fungus, on oak (*Quercus*). Hyphae of the *Stereum* host can be found within the *Tremella* fruit bodies, with the *Tremella* hyphae attached to them by haustorial cells.
Photo: Courtesy of Peter Roberts.

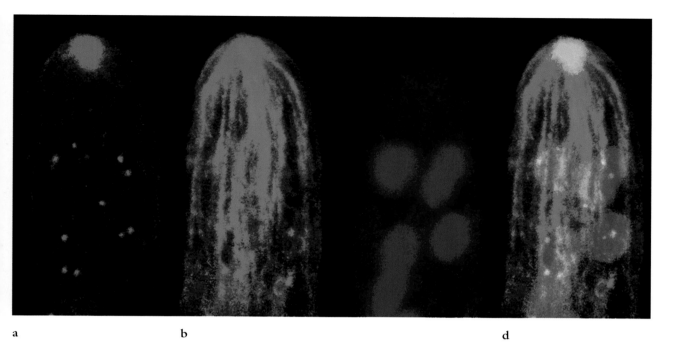

a   b   d

Plate 6 (above). Fluorescence microscopy showing structures within the hyphal apex

A single hyphal tip of *Allomyces macrogynus* triple-labelled for ɣ-tubulin, α-tubulin and DAPI, with fluorescent staining viewed by laser scanning confocal microscopy. (a) Labelling for ɣ-tubulin, with antibodies coupled with Texas red. Note red fluorescent labelling of the Spitzenkörper (the apical body) and of centrosomes. (b) Labelling for α-tubulin, with antibodies coupled with the green fluorescent dye BODIPY. Note the green fluorescent labelling of the abundant microtubules emanating from the region within the hyphal apex. (c) Labelling of nuclei with DAPI, a DNA-specific fluorescent dye. (d) Merged image of a, b and c (with the red plus green fluorescence showing orange). Note the association of the microtubules with the apical zone of ɣ-tubulin, and the centrosomes associated with sub-apical nuclei. Further studies confirmed that the Spitzenkörper is the site of apical ɣ-tubulin labelling, and that ɣ-tubulin is localized in a pair of widely separated centrosomes associated with each nucleus. The hypha is *ca.* 11 μm in diameter. Reproduced with permission from McDaniel, D. P. & Roberson, R. (1998). ɣ-Tubulin is a component of the Spitzenkörper and centrosomes of hyphal-tip cells of *Allomyces macrogynus*. *Protoplasma* **203**, 118-123.

Plate 7 (right). Fluorescence and phase-contrast microscopy showing changes in cell surface agglutinins during mating in yeast

Mating cells of *Saccharomyces cerevisiae*, with an α-shmoo and an **a**-shmoo having conjugated and produced a zygote. The cells are viewed by phase contrast microscopy (right) and fluorescence microscopy (left). The cells were treated with green fluorescent antibody (coupled with fluorescein isothiocyanate dye) specific for **a**-agglutinin, and with red fluorescent secondary antibody

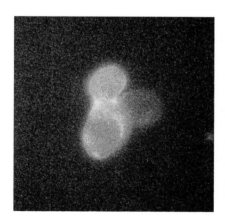 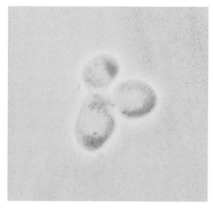

(coupled with rhodamine dye) directed against anti-α-agglutinin antibody. Note the specificity of the red and green labelling for the surfaces of the **a**-cell and α-cell, respectively. The bud representing the first diploid offspring, i.e. the zygote (at right of the three cells), is unlabelled, and contains neither **a**- nor α-agglutinin on its surface. Each shmoo is *ca.* 5 μm in diameter. Photo: Courtesy of K. Hauser and W. Tanner.

Plate 8 (top right). Dry rot
An outbreak of dry rot in a cellar. Mycelium of the fungus, *Serpula lacrymans*, feeding on a wooden floor above the cellar, has grown through the brick wall and is colonising a pile of firewood. Photo: Courtesy of John Baker.

Plate 9 (bottom right). Biological control of a forest pathogen
A fruit body of the basidiomycete *Phlebiopsis gigantea* covering the stump of a recently-felled pine tree. The stump was deliberately infected with a suspension of spores of this fungus immediately after the tree was felled during thinning of a forestry plantation. The relatively harmless parasite *Phlebiopsis* competes with, and excludes, spores of the aggressive wood rotting fungus *Heterobasidion annosum*, which can otherwise invade the trees remaining after thinning, colonising the freshly-exposed stumps. If allowed to colonise, it can then grow from tree to tree via root grafts and can cause extensive losses of timber due to butt rot. This method of biological control is approved for commercial use and applied in pine plantations throughout Britain and Poland; it is also used in many Scandinavian countries where it is applied to both pine and spruce. For further details see Pratt, J.E., Niemi, M. & Sierota, Z.H. (2000). *Biocontrol Science and Technology* **10**, 469-479. Photo: Courtesy of Joan Webber, Forest Research.

Plate 10 (above). *Entomophthora muscae*, a member of the Zygomycota that infects houseflies.
The enormous numbers of spores in a halo around this fly are discharged from sporophores growing on its body. Such enormous spore production is essential in order that a few may, when discharged, actually strike flies and continue the life cycle. Photo: Courtesy of Thomas J. Volk, University of Wisconsin-La Crosse.

Table 6.1   Some fungal enzymes which hydrolyse polymers

Substrate	Enzyme
Arabinans	Arabinofuranosidase
Callose	$1 \rightarrow 3$ Glucanases
Cellulose	Endoglucanases, cellobiohydrolases
Chitin	Chitinase
Cutin	Cutinase
DNA	Deoxyribonuclease
Hemicellulose	Hemicellulases
Lignin	Ligninases
Mannans	Mannanase
Pectic substances	Pectin methylesterase, pectate lyase, polygalacturonase
Proteins	Proteinases
RNA	Ribonuclease
Starch	Amylases
Xylans	Xylanase

take up glucose immediately it is added to their growth medium, but galactose, with an inducible uptake system, is only taken up 4 hours after it has been added. Transporting systems normally consist of single proteins, and the genes for them may be lost by mutation, producing a strain that lacks the ability to take up a particular substance.

The nutrients taken up by fungi can either be used directly in metabolism, or stored. Some fungi can carry nutrients, through the mycelium or through specialized cords (pages 341–343), to supply growth occurring at a distance from the nutrient source, a process known as translocation. For example, nitrogen compounds accumulated by mycelium ramifying through a dead tree trunk can be carried to the site where a bracket sporophore is emerging, to support growth and spore production.

The metabolism of fungi is considered in Chapter 3, and dealt with in detail in textbooks of fungal physiology. Selected metabolic activities that enable fungi to play a special role in ecosystems will be outlined here. Most natural substrates for fungi are the remains of plants, which are carbon-rich because they are mainly composed of cellulose, but may be poor in nitrogen, phosphorus and other essential nutrients. Respiration of carbon compounds provides energy in the form of ATP, which is used to drive reactions in which new materials are synthesized, and in maintaining the electrical potential difference across the cell membrane of the hyphal tip that drives the active accumulation of nutrients from the environment (page 149). Carbon compounds are also stored in the form of glycogen and lipids and are used in building cell walls. Sugar alcohols such as mannitol, arabitol and erythritol are abundant in fungi compared with plants and animals. Fungi acquiring sugars from plants in biotrophic symbioses (page 391) prevent back flow of sugar into the plant by converting glucose into mannitol, which plants are unable to utilize. Conversion of mannitol into glucose is catalysed by a dehydrogenase enzyme, producing reduced NADP which can be used in reactions such as nitrate reduction, so mannitol probably acts as a store of reducing capacity as well as a carbon reserve. Sugar alcohols are believed to be one of the forms in which sugars are translocated in mycelial cords. The disaccharide trehalose is also a common soluble sugar in fungi, both as a translocated compound and as a carbon store in many spores.

Nitrogen is the other nutrient needed in large amounts. Fungi need nitrogen not only to synthesize structural components of the cell such as proteins, nucleic acids and chitin, but also for the enzyme synthesis necessary to acquire nutrients from the environment. Fungi cannot assimilate nitrogen from the atmosphere as many prokaryotes do, but they are able to take up and incorporate nitrate and ammonium ions from their environment, as well as using many soluble organic nitrogen compounds. Fungal conversion of inorganic nitrogen compounds into amino acid and protein which animals can use is a key process in the food chain. Nitrate and nitrite are converted into ammonium by reductase enzymes, and the ammonium ion is combined with the respiratory intermediate α-ketoglutarate to produce the amino acid glutamic acid (Fig. 6.3). Addition of a further amino group to glutamic acid produces glutamine, and other amino acids are synthesized by transfer of amino groups to other intermediates. Nitrogen levels fluctuate in most natural environments and fungi can adapt their metabolism to make the most effective use of available nitrogen. Some fungi when grown on a medium rich in nitrogen, store it as protein, and as free amino acids in the cytoplasm and vacuoles, the latter containing basic amino acids such as arginine and ornithine. The stored nitrogen is used when the fungus encounters nitrogen-poor conditions. Nitrogen starvation induces metabolic processes that maintain the intracellular supply of ammonium ions, and the amino acids glutamate and glutamine, the principal nitrogen compounds needed for growth. These processes include the breakdown of internal protein, and the enhanced uptake of nitrite and nitrate ions from the environment. Many different enzymes are needed to carry

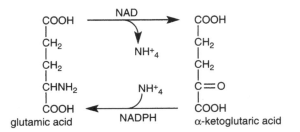

Figure 6.3 Reactions catalysed by glutamate dehydrogenase. Two forms of this enzyme are found in some fungi. Activity of NAD-dependent glutamate dehydrogenase is stimulated by high concentrations of glutamic acid. It is therefore believed to be responsible for deamination reactions, breaking down glutamic acid *in vivo*. The NADP-dependent enzyme is not activated by glutamic acid and is believed to catalyse the amination of α-ketoglutaric acid, an intermediate in the tricarboxylic acid cycle, to produce glutamic acid for use in synthesizing molecules such as proteins.

out these processes, and all of them are activated in response to nitrogen starvation. Molecular investigation using *Aspergillus* and *Neurospora* of the genetic control mechanism for sensing and responding to nitrogen starvation has shown that the system is under the control of a single gene which codes for a protein transcription factor. The level of the transcription factor depends on the nitrogen status of the mycelium, and controls the transcription of many genes encoding enzymes for nitrogen assimilation. Other controls operate to ensure that the biosynthesis of the different amino acids is kept in balance so that all are available for protein synthesis. Nitrogen metabolism is thus highly adapted to the conditions in the natural environment of fungi, in which the utilization of a relatively abundant supply of carbon is accompanied and sometimes limited by a fluctuating and less predictable nitrogen supply. By collecting soil nitrogen, degrading cellulose to soluble form, and building protein-rich biomass, the fungi convert the massive stores of carbon and energy in woody plants and their remains into forms available to animals (page 447).

Mycelium can mobilize stores of carbon and nitrogen to cope with starvation, to grow new mycelium, or develop new structures. When mycelium is transferred in laboratory cultures from a rich to a poor medium, it **autolyses**, breaking itself down. First, reserves of stored substances such as protein and glycogen are used up and then the cell walls and protoplasm are broken down, with essential molecules such as nucleic acids being the last to go. Colonies grown under natural, heterogeneous conditions use this process to mobilize and reallocate nutrients within the mycelium according to local supply and demand, removing scarce nutrients such as nitrogen and phosphorus from senescent mycelium and exhausted substrata into parts of the mycelium which are colonizing new resources and developing new structures. For example, when wood decay fungi form cords between separate wood resources, mycelium which has not made connections dies away and is broken down, while the amount of corded mycelium increases. Hyphae of brown rot fungi (page 311) are scarce within decayed wood, because older hyphae are autolysed, a process which is thought to enable the mycelium to conserve and reuse its nitrogen. Some wood decay Basidiomycetes that produce large colonies (page 342) have been found to have proteinases that are exclusively intracellular and may mobilize nitrogen for exploratory growth over nitrogen-poor areas.

Some metabolic pathways of fungi are essential at all times. Others are not, being needed for purposes such as the utilization of rarely encountered nutrients, the production of secondary metabolites required for competition with other microbes, or attack on a host plant (page 386). Studies on *Aspergillus* and some other fungi have shown that the genes encoding the enzymes of such pathways are clustered (page 513). This feature is uncommon in eukaryotes and may be an adaptation to a way of life which requires rapid responses to changes in the environment. Gene clustering may contribute to efficient regulation of pathways; a single transcription factor switches on all the clustered genes for nitrate assimilation in *Aspergillus*. Genes for the enzymes of several secondary metabolic pathways also occur in clusters, which can include up to 25 different genes occupying as much as 60 kb of DNA. Pathways for the synthesis of aflatoxin, penicillin and melanin have all been found to be encoded by gene clusters.

The scavenging activities of fungi assist them in the capture of mineral nutrients, an important part of their role in ecosystems. An example is the phosphate uptake system in mycorrhizal fungi (fungi associated with plant roots in mutualistic symbiosis, pages 394–405). These have a high affinity for phosphate. The concentration of this ion in soil is often very low, as roots and microorganisms compete for it and it also tends to form insoluble complexes with soil minerals. Mycorrhizal fungi have more efficient phosphate uptake systems than plant roots, so the association of roots with fungi can enhance phosphate uptake by plants and increase their growth.

Besides varying in their developmental potential and degradative capacity, different fungi also have characteristic nutrient requirements (Table 6.2). Some need to be supplied with a number of complex organic compounds, whereas others may be heterotrophic only for carbon. There has probably been a general evolutionary trend towards more complex nutritional needs, since synthetic activity is far more easily lost than gained.

Study of fungi reveals the close interplay that exists between the three aspects mentioned above: capacity to degrade macromolecular substrates, synthetic capacity, and variety of growth forms. It seems that as the biosphere evolved, from unicellular beginnings to complex ecosystems, the fungi must have adapted their filamentous, heterotrophic way of life to the increasingly varied niches that became available (Fig. 2.1). It is important when studying the physiology of fungi *in vitro* to remember how their specializations have been evolved in a particular ecological context. This chapter describes some natural ecosystems involving the combined activity of different saprotrophic fungal species. Fungi which are not saprotrophs, but are biotrophic – deriving their food from living plants and animals – are dealt with in Chapter 7. Their activities are also of major ecological importance.

Table 6.2   Czapek–Dox medium, widely used for the culture of fungi

Mineral base	KH_2PO_4	1 g
	$MgSO_4.7H_2O$	0.5 g
	KCl	0.5 g
	$FeSO_4.7H_2O$	0.01 g
Carbon and energy source[a]	Sucrose	30 g
Nitrogen source[b]	$NaNO_3$	2 g
Water		1 litre
If a solid medium is required	Agar	20 g

[a] An organic form of carbon is always required, as all fungi are heterotrophic. Glucose, sucrose, or starch are commonly used.

[b] Most fungi can grow with inorganic sources of nitrogen, usually provided as ammonium or nitrate salts. Some need amino acids.

The medium works well for the culture of many fungi, although it has much higher nutrient concentrations than the substrata on which most fungi grow in nature. Many fungi in addition require vitamins such as thiamine. A wide range of vitamins can be provided by including yeast extract in the medium, or alternatively pure vitamins can be added, as is essential when a fully defined medium is required. For physiological or biochemical work it may be useful to sterilize by membrane filtration (pore size *ca* 0.2 μm) rather than autoclaving, to avoid changes in chemical composition due to heat.

Substrate Groups and Nutritional Strategies

Some fungi have a rapidly completed life cycle or one that can be interrupted easily with a dormant phase, whereas others have perennial mycelium and are slow to produce spores; these kinds of behaviour are related to the substrates each species can utilize. For example, some fungi (such as Mucorales, page 38) are dependent on soluble sugars and, as these substances are not likely to remain for long in natural ecosystems, it is understandable that such fungi tend to be fast growing and quick to sporulate, enabling them to colonize rapidly and avoid competition by acting as pioneers on virgin resources – the so-called sugar fungi. The term 'strategy' is sometimes used to describe the collection of apparently adaptive responses which an organism may make to its surroundings. Strategies are believed to have evolved as a result of particular endogenous and exogenous constraints. Some of the important constraints must have been nutritional, and substrate groups such as that of sugar fungi represent the main strategies that have resulted. The course of population growth in relation to a resource which ultimately becomes limiting is described by the logistic equation

$$dx/dt = rx \, (1 - x/K)$$

in which x is population size, t is time, r is the specific growth rate and K the maximum population size that can be supported by the environment. The terms r and K are constants, r corresponding to μ in equations of exponential growth (page 128) and K to the population reached in the stationary phase of a batch culture (Fig. 3.13). The larger the value of r, the faster the potential rate of growth, as long as x, the population size, is negligible in comparison to the constant K, as it is during the phase before the resource becomes limiting. Thus organisms that, as a result of natural selection, have been endowed with the ability to increase in numbers rapidly in order to colonize a resource, as are the saprotrophic sugar fungi, are described as r-selected. The population size x stabilizes when it reaches K, the carrying capacity of the environment. Organisms which maintain a constant population size for long periods by living in equilibrium with a complex environment are termed K-selected. The slower-growing lignocellulose utilizers, with their economical use of resources, are K-selected, and likely to remain active in ecosystems that have reached the limit of their capacity to support new populations. On a virgin resource, rich in all possible nutrients for fungi, r-selected species may appear as pioneers, to be followed by a phase of crowded, vigorously competing species and ultimately by K-selected species adapted to utilizing the relatively poor residual substrates. Although theoretically possible, this idealized succession is rarely seen, since truly virgin resources are rare in nature, fungi being continuously present. Predictable successions of different fungi on particular kinds of resource do occur, as is described in the second part of this chapter, but they are more realistically seen as part of a continuous pattern of interaction between fungi and other organisms, than as the isolated event of decomposition.

Wood as a substrate for fungi
Fungal decay of wood requires the passage of enzymes out of the cell and into the insoluble substrate. Close contact between the hyphal surface and the substrate

surface is obviously necessary to minimize loss by diffusion of both the enzyme and the soluble products of its action. This close contact can be seen in electron micrographs of wood-decaying hyphae inside wood cells (Figs. 6.4, 6.5). Catabolism of this extracellular, insoluble substrate is dissimilar in several important ways from the action of hydrolytic enzymes on homogeneous, soluble substrates. The concentration of enzymes, reactants and products vary in space, and so the biochemistry of wood decay cannot be understood without consideration of its structural background. Lignin (Fig. 6.6) and cellulose (Fig. 3.4), together with hemicelluloses, make up the bulk of wood. Hemicelluloses are mixed polymers of various sugars such as xylose, mannose, arabinose and galactose, which are extractable from lignocellulosic material with mild chemical methods. They form much shorter chains than cellulose, usually under 200 sugars long. Natural or 'native' cellulose is strong, fibrous and resistant to hydrolysis because the long glucose chains (up to 7 μm) align, forming regularly packed micelles, and these are themselves aligned to form the

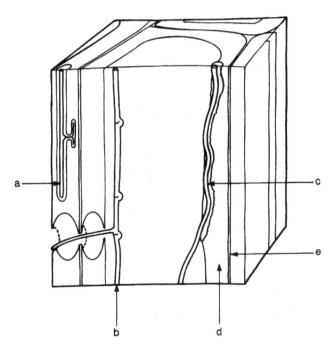

Figure 6.4 Diagram of wood cell walls showing typical modes of attack of hyphae producing different types of decay. a, Hypha of a soft rot fungus, lying within a cavity which it has dissolved in the cellulose S2 layer of a tracheid wall. b, Hypha of a brown rot fungus, lying on the inner surface of a tracheid wall without eroding a cavity, but weakening the wall structure over a relatively wide area by means of diffusible enzymes, cellulases in particular. c, Hypha of a white rot fungus eroding a channel in the wall immediately surrounding it by means of enzymes, mainly cellulase and ligninase, which are active only in the region of the hypha and do not diffuse far into the wood. d, The S2 wall of a tracheid. e, The more heavily lignified middle lamella.

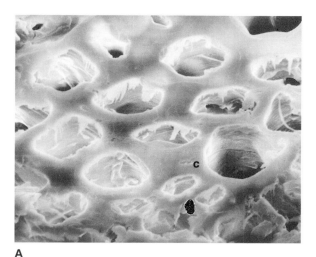

A

B

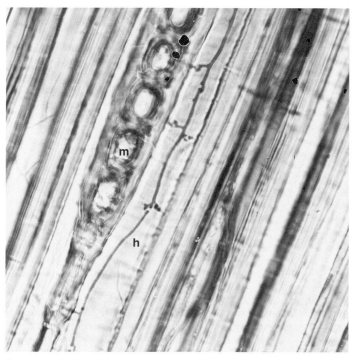

C

Figure 6.5 Wood decay; soft, white and brown rots. A, Scanning electron micrograph (SEM) of a transverse section of wood cells, showing the very thick cell walls and, deep in the walls, sectioned cavities, for example at c, caused by a soft rot fungus. B, SEM of white rot erosion troughs (t) on the inner surface of the wall of a tracheid. (A,B, M. D. Hale). C, A light micrograph showing a hypha (h) of a brown rot fungus, *Serpula lacrymans*, growing in a tracheid and branching opposite the paired pits, with the side branches entering the adjacent tracheid through the pit pair. A medullary ray (m) crosses the field (L. Jennings).

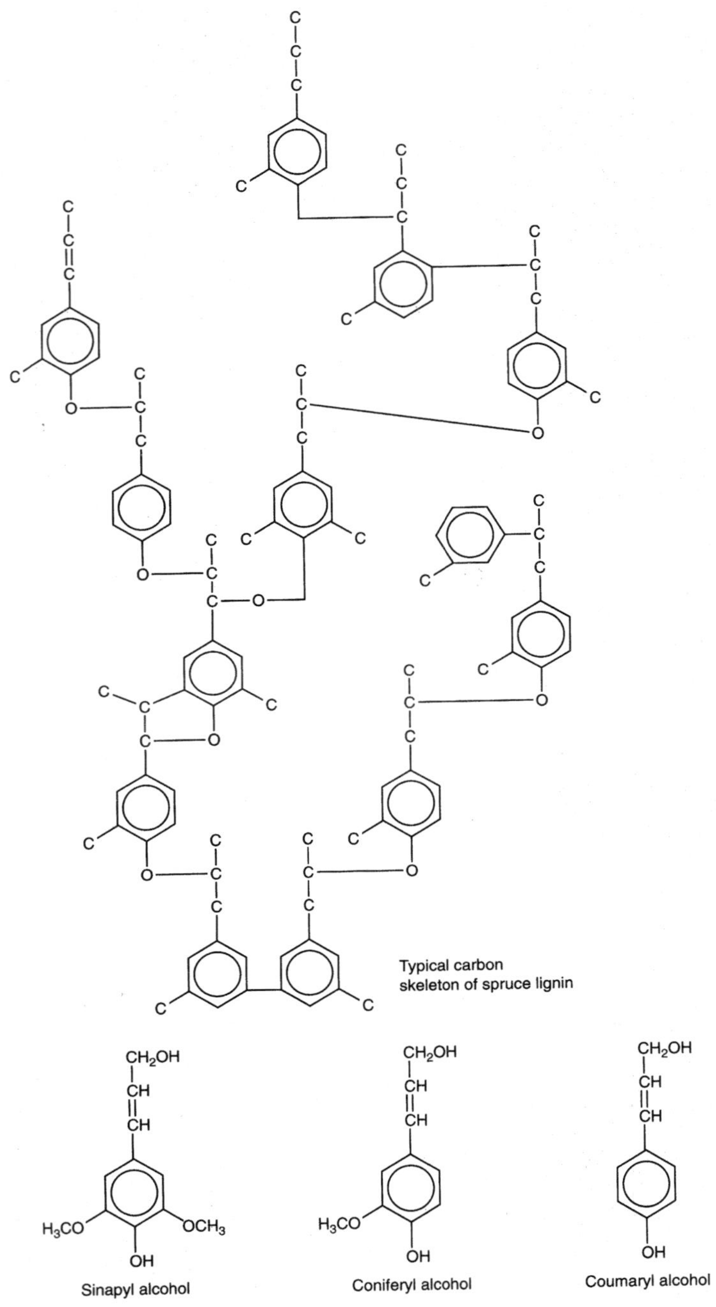

Figure 6.6 The molecular structure of lignin, a complex three-dimensional polymer formed from phenylpropanoid subunits. Coumaryl, sinapyl and coniferyl alcohols are the commonest of the phenylpropanoid subunits, and they are joined randomly in a three-dimensional, asymmetric network by various different linkages, the commonest of which are illustrated. Coniferyl alcohol is the most frequent subunit in the wood of coniferous trees.

microfibrils 4–10 nm in diameter (Fig. 6.7), which are visible in electron micrographs. Cellulose microfibrils are coated with hemicellulose and in wood tissues they are embedded in lignin, forming the material known as lignocellulose. Lignin never occurs alone, but always in this association. It is a polymer of three kinds of aromatic alcohols; coumaryl alcohol and its methoxy substituted derivatives coniferyl and sinapyl alcohols (Fig. 6.6).

Unlike cellulose, these are not linked in regular chains but are joined by numerous different bonds, aryl, ether and ester bonds being common. This gives a three-dimensional polymer of indeterminate size. Purified lignin is a dark brown, powdery material. In order to understand the structure presented to the lignocellulose-decomposing fungus, it is necessary to consider the way in which

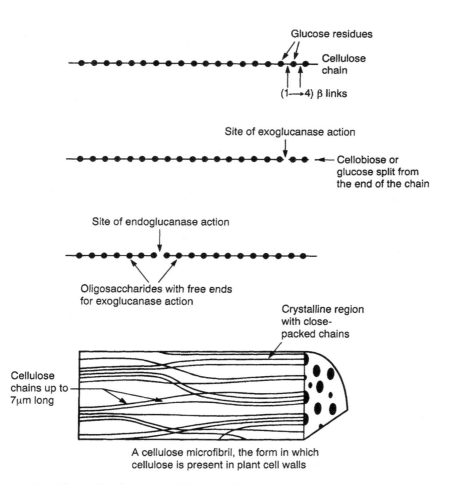

Figure 6.7 The mode of action of different cellulases on cellulose. Endoglucanases break the bonds in the middle of the chain, opening up the microfibril to further attack and increasing the numbers of exposed ends of cellulose chains on which exoglucanases can act. Exoglucanases remove monomers or dimers from the ends of the cellulose molecule.

lignin and cellulose are laid down in the plant cell, and what part and proportion of plant residues they comprise.

All the cells of plants have cellulose walls, and their contents, while they are alive, are rich in soluble nutrients for fungi. The cell walls become thickened by later deposits of cellulose and, in parts of the plant specialized for support and water transport, often lignin too. Even non-woody herbaceous plants such as cereal crops normally have a lignocellulose component, although some fibrous tissue is exclusively cellulosic, for example the fibres of cotton. In trees a woody trunk develops, and is added to year by year, by growth at its periphery. The cells destined to become part of this wood lay down their secondary wall thickening in three main stages: first a thin, lignin-rich layer (S1) is formed by addition of lignin to the existing primary cellulose wall, then a thicker, more cellulose-rich layer (S2) and finally, on the inside of the wall, a thin, heavily lignified layer (S3) (Fig. 6.8). Holes, called pits, are left in the thickened walls of water-conducting

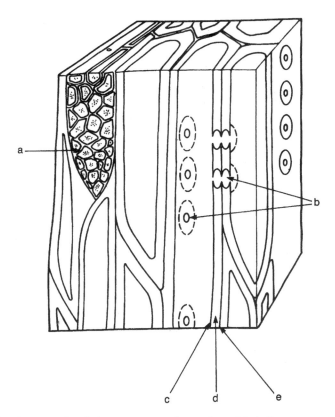

Figure 6.8 Diagram of cellular structure of wood. a, Medullary ray consisting of cells which are alive in living trees, and provide a source of readily available nutrients for fungi invading timber after the tree has been felled. b, Pits – perforations in the tracheid wall, usually coinciding in position in adjacent tracheids, and providing a pathway for hyphae to grow from one tracheid into another. c, S3 wall, with abundant lignin added to cellulose framework. d, S2 wall, mainly cellulosic and relatively less lignified. e, S1 wall, lignified.

cells. As the wall thickens the central cavity occupied by the living cell dwindles and eventually the cell dies. Thus the soluble nutrients available for fungi in wood are limited. Wood in living trees is interspersed with bands of unthickened living cells, the medullary rays, but in seasoned timber and the dead heartwood of hardwood trees even these disappear. Lignocellulose is thus a complex polymer mixture, rich in carbon but poor in all other essential nutrients for fungi, with a carbon:nitrogen ratio that may be as much as 500:1. The fungi that use wood as a sole nutrient source show considerable adaptations in their life history, morphology and biochemistry. Some are parasites that can attack living trees, and may eventually kill the whole tree, for example *Armillaria mellea* (pages 352, 371), whereas others grow in the central heartwood only, producing hollow trunks. Heartwood contains antifungal compounds such as phenols, which specialized decay fungi can degrade. Other decay fungi are exclusively saprotrophic, utilizing only dead wood.

Fungal attack on wood generally starts by hyphae penetrating through a pit into the narrow central cavity of a cell (Fig. 6.5C), after which it grows adpressed to the inner, lignified S3 layer of the wall. The way in which decay proceeds then depends on the enzymic equipment of the fungus. Those that are able to depolymerize lignin as well as cellulose produce a different effect on the cell wall from those which can only attack cellulose. This difference is reflected in the overall appearance of the rotten wood; hence the name **white rot** for rot fungi that decompose lignin, and **brown rot** for those that only decompose cellulose. **Soft rot fungi**, like brown rots, decompose only cellulose, but they attack damper wood and decay usually occurs only near the wood surface.

Brown rot fungi   Brown rot fungi are common in nature, attacking living and dead trees. Causative fungi are predominantly Basidiomycetes of the family Coniophoraceae (order Boletales), and there are fewer species of brown rot fungi than of white rot fungi. Many brown rot fungi can only bring about the decay of cellulose when it is chemically associated with lignin, and are incapable of degrading pure, non-associated cellulose. However, others are also able to attack non-woody pure cellulosic substrates. *Serpula lacrymans*, the timber dry rot fungus, belongs to this group, and will infect not only timber joists and floorboards, but also such household materials as cotton and paper. The need for lignin to be present in order for some brown rot fungi to be able to decompose cellulose is as yet unexplained.

Microscopic examination of wood after decay by a brown rot fungus shows generalized thinning of the cell walls, without any localization of decay around the hyphae, of which in any case there are very few. One explanation of this appearance is that cellulase enzymes are released from the hyphae and diffuse freely within the wood, although it has been suggested that cellulase protein molecules (page 314) are too big to diffuse through the S3 layer to their site of action in the S2 layer. The overall effect is that the wood loses its strength and shrinks, developing longitudinal and transverse cracks which eventually join, the wood breaking into dark brown cubical crumbs. Most white rot fungi (see below), and many brown rot fungi release cellulases into the culture medium. These can be assayed by a variety of techniques including measuring weight loss of pure cellulose, assay of reducing sugars, decrease in viscosity of added carboxymethyl cellulose (a pure, straight-chain soluble preparation), or the release of a dye from pure cellulose with which it has been previously combined. Cellulases are released by a large number of different fungal species, Ascomycetes and mitosporic fungi as well as Basidiomycetes. Apart from differences in the nature of the initial attack on native cellulose, the biochemistry of cellulolysis is similar in Basidiomycetes and Ascomycetes in so far as it involves the synergistic action of three kinds of enzyme – endoglucanases which can hydrolyse the b(1→4) bonds randomly in the middle of the molecule releasing various oligosaccharides, and the exoglucanases cellobiohydrolase and glucosidase which, respectively, remove cellobiose from the ends of the cellulose chains and then cleave the dimers to glucose (Table 6.3). The reason why they act synergistically is probably that endoglucanases provide an increased number of chain ends for the exoglucanase to act on (Fig. 6.8) as well as opening up the micelle structure to allow the enzyme access to the glucan chain. Molecules of the cellulose-decomposing enzymes have been photographed in the high resolution electron microscope and appear as hexagonal particles 4–7 nm in diameter, which attach to cellulose fibrils.

Table 6.3 The cellulase complex of enzymes from *Trichoderma reesei*. The enzymes that act synergistically to degrade cellulose to glucose, and the genes encoding them that have been cloned and sequenced.

Genes	Enzyme names	Activities
cbh1, cbh2	Cellobiohydrolase; 1,4-β-glucan cellobiohydrolase	Cleave cellobiose units from end of chain
egl1–egl5	Endoglucanase; 1,4-β-glucan glucanhydrolase	Hydrolyse internal bonds of cellulose
bgl1	Cellobiase; β-glucosidase	Hydrolyses cellobiose to glucose

After Kubicek, C. P. & Pentillä, M. E. (1998). Regulation of production of plant polysaccharide degrading enzymes by *Trichoderma*. In Harman, G. E. & Kubicek, C. P., eds, *Trichoderma* and *Gliocladium* Vol. 2, pp. 49–72, Taylor & Francis, London.

The nature of the initial attack by fungi on native cellulose is still uncertain. This is because the mixture of hydrolytic enzymes exuded into a culture medium, while effective in degrading soluble, straight-chain forms of cellulose such as carboxymethyl cellulose, usually do not degrade native cellulose, or do so very slowly. The highly ordered, close-packed structure of microfibrils seems difficult to break down, probably because diffusing enzymes have limited access to the fibrils. Why is a hypha which is in contact with the substrate better able to attack cellulose than is a solution of its enzymes? An hypothesis that has been put forward for the brown rot Basidiomycetes is that the hypha releases hydrogen peroxide which acts with traces of ferrous iron in wood to generate transient free radicals that cause oxidative breakage in some of the glucose pyranose rings in the cellulose chain. This would disrupt the microcrystalline structure, enabling the cellulase to act. There is much indirect evidence for this hypothesis, including the observation of decay in parts of the cell apparently inaccessible to large enzyme molecules because the pores in the S3 layer are too small for enzymes to pass through. Oxidation has also been shown to be an early step in the breakdown of both cellulose and lignin by white rot fungi as described below. No such mechanism is known to exist among the other Ascomycete, mitosporic and Chytridiomycete cellulose decomposers which have been studied, such as *Trichoderma*, *Myrothecium*, *Chaetomium* and *Neocallimastix* (page 446). However, *Trichoderma* has been found to produce an enzyme termed 'swollenin', which breaks hydrogen bonds in cellulose causing it to become hydrated and swell, presumably aiding degradation.

The structure and activity of fungal cellulases Genes for endoglucanases and cellobiohydrolases of several fungi have been cloned and sequenced (Table 6.3), and those for the enzymes of the cellulase complex of the highly cellulolytic soil fungus *Trichoderma viride* have been expressed in yeast, permitting production of pure samples of each individual enzyme. Structural investigations of enzyme molecules are helping to interpret the relationship between molecular structure and enzymic function. Many cellulases, including those with different catalytic activities, such as endoglucanases or cellobiohydrolases, share a common structural pattern. Three regions ('domains') can be distinguished in the typical cellulase molecule: a cellulose binding site, a separate active site, and a flexible connecting region that allows the enzyme to act at a distance from its point of attachment to the cellulose substrate molecule. When the binding site of a cellulase enzyme of *Trichoderma reesei* was removed by genetic engineering, the enzyme was still active in breaking down small soluble glucose polymers, but lost much of its ability to attack insoluble cellulose. Hence, whether an enzyme can act within the cellulose chain, or only at one end, can be affected by the presence of a relatively short section of the enzyme molecule unrelated to the active site. Even within the active site of the enzyme, small changes in molecular structure can also make a significant difference to enzyme activity. Differences exist in the three-dimensional structure of the active sites of two closely related cellulase enzymes, one from *T. reesei* and the other from a bacterium, that may account for their different modes of attack on cellulose. Sequence homology showed that these two enzymes, although from very different organisms, probably shared a common evolutionary origin. However, the *T. reesei* enzyme acted as a cellobiohydrolase, cleaving dimers from one end of the cellulose chain, whereas the bacterial one was able to act as an endoglucanase, breaking links internally in the cellulose chain. This major functional difference was apparently the result of a small difference in amino acid sequence producing a significant alteration in the three-dimensional structure of the enzyme protein. The effect was that the active site of the cellobiohydrolase was confined in a space within the molecule only large enough to admit a chain-end, whereas the endoglucanase had its active site exposed on the outside of the molecule (Fig. 6.9). Such differences in enzyme activity would be expected to make a significant difference to the organism's ability to utilize cellulosic resources. It seems likely that in the course of evolution particular cellulase genes have become modified as natural selection has favoured different types of cellulolytic activity in different lineages of organisms. This would explain how homologous cellulases came to be performing different functions in distantly related organisms.

Extracellular enzyme production is a potentially wasteful process for the fungus, particularly when the amounts of nutrients available to it are low. In

A

B

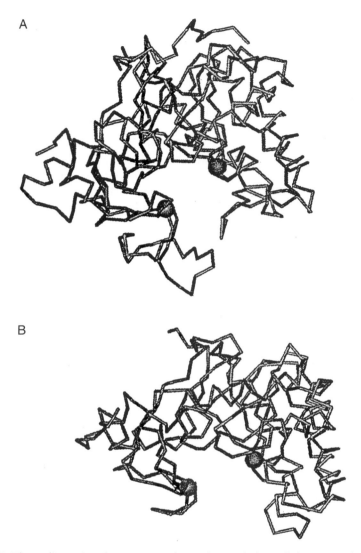

Figure 6.9 Three-dimensional structure of members of the cellulase complex. A, The *Trichoderma reesei* exoglucanase, cellobiohydrolase II (CBHII). The active site (defined by spheres) is within a tunnel. B, A bacterial endoglucanase with homology to CBHII. Here the active site is more open, enabling the enzyme to cleave a cellulose chain at any point. (After Olle Teleman, in M. Penttila & M. Saloheimo (1999); see references.)

wood, breakdown of the substrate is needed to release the nitrogen essential for growth, but as this nitrogen is also expended on synthesizing the enzymes necessary to continue breakdown, the rate of enzyme production must be controlled in order to keep the organism's nitrogen economy in balance. Positive and negative controls on cellulase synthesis are found in all cellulose-decomposing fungal groups. Positive control takes the form of induction of cellulase synthesis by the presence of cellulose. Very small quantities of cellulases are believed to be produced all the time, so that soluble products of the initial steps in breakdown may cross the cell membrane to act as inducers of increased synthesis. The disaccharide cellobiose induces cellulase synthesis in both the Basidiomycete brown rot fungus *Schizophyllum commune* and in *Trichoderma reesei*. Other disaccharides have been found to induce only one or two out of the complex of cellulases that a fungus is capable of synthesizing. Cellulase activity has been shown to be inducible in *S. commune* by thiocellobiose, which mimics the effect of cellobiose but is not metabolized. This 'gratuitous induction' could prove useful in enhancing cellulase synthesis in organisms used for commercial cellulase production. Negative control is also usual, glucose in the culture medium inhibiting cellulase synthesis even if cellulose is present. This was one of the earliest examples of the phenomenon of catabolite repression, which is widespread in fungi. Many other extracellular hydrolytic enzymes are repressed by the presence of their products of hydrolysis; for example in many fungi the numerous fungal proteases are produced only in the absence of ammonia and amino acids. This not only prevents waste but also reduces the supply of nutrients to potential competitors.

Waste of extracellular enzymes is probably minimized by the existence of a mucilaginous sheath around the growing hyphae of wood decay fungi, seen in both the brown rot and the white rot types. Enzyme molecules may be retained in this sheath and prevented from diffusing so far away into the substratum that the soluble products of enzyme action are no longer available to the hyphae. Hyphae are surprisingly sparse in wood decayed by brown rot fungi. It is thought that older parts of the mycelium are enzymically digested, allowing valuable materials such as organic nitrogen, to be reused by growing hyphae.

White rot fungi Basidiomycetes and Ascomycetes are the principal cause of white rots of wood. Species causing white rot are far more numerous than those causing brown rot, the latter being exclusively Basidiomycetes. Most white rot fungi degrade both the lignin and the cellulose, leaving a light, white, rather fibrous residue completely different from the powdery brown remains left by brown rot fungi. However, some remove lignin first, attacking the cellulose later, a process termed selective delignification. At the microscopic level the effect of the hyphae is at first much more localized than that of brown rots. Hyphae themselves are more numerous in the decayed wood and easier to find in sections prepared for microscopy. Each one can be seen to be surrounded by a local erosion zone or borehole in the early stages of attack. In the scanning electron microscope the S3, S2 and S1 layers of wood cell wall can be seen to have been eroded sequentially, with the hypha lying at the bottom of a U-shaped channel like a railway cutting (Fig. 6.5B). There is no generalized, diffuse attack on the S2 layer. White rot fungi have long been known to have biochemical features which distinguish them from brown rots. They are all capable of decomposing unassociated cellulose by means of enzymes released into the medium. In addition they almost all produce polyphenol-oxidizing enzymes. The majority of these belong to a class of polyphenol oxidases termed **laccases**, containing copper at their active site. Any culture of a white rot fungus will oxidize phenols, producing quinones. The function of phenol oxidase activity in nature is not yet fully understood. It has been suggested that the quinone can act as an oxygen donor in an enzymically catalysed oxidative attack on cellulose. Mutants of the white rot fungus *Phanerochaete chrysosporium* (ana. *Sporotrichum pulverulentum*) which are lacking in phenol oxidase activity are also impaired in their ability to degrade cellulose, and revertants to the wild type have restored cellulase activity. White rots are commoner on the hardwoods of angiosperm trees, whereas brown rots more usually occur on the softwoods of conifers. This may reflect chemical differences in the two kinds of wood. Softwood has more lignin than hardwood (25–35% compared with 18–25%), the lignin monomer is predominantly coniferyl alcohol (Fig. 6.6), and the secondary metabolites present (such as terpenoids) may differ from those of hardwoods in the extent to which they stimulate or inhibit fungal invaders.

The fact that there are two distinctively different types of attack on wood by Basidiomycetes, each with correlated biochemical and morphological features, and characteristic of different taxonomic groups of fungi, raises interesting questions about their evolution. Fossil fungi are found associated with fossil plants of the Lower Devonian period, 400 million years ago. There is no evidence about when the different kinds of wood decay evolved. The earliest land plants known from fossils contained lignin, and presumably this must have been decomposed as it does not appear to have accumulated. Specialization of fungi for the decay of different kinds of wood must have resulted from co-evolution with higher plants, with species that preferentially rot hardwoods diverging in the wake of angiosperm evolution. The enzymatic nature of lignin degradation has been harder to analyse than that of cellulolysis. As with cellulolysis one of the most difficult problems has been to establish what chemical process constitutes the initial attack on the substrate. It seemed unlikely in the case of lignin, a heteropolymer linked by a variety of chemical bonds, that a single enzyme could bring about significant depolymerization. However, a type of enzyme capable of doing this has been isolated from cultures of white rot fungi, the most intensively studied being the thermophilic white rot fungus *Phanerochaete chrysosporium*. The reaction is oxidative, requiring the presence of hydrogen peroxide, and the enzymes, lignin peroxidases and manganese peroxidases, act by releasing highly reactive, transient free radicals of oxygen which then react with parts of the lignin molecule, breaking covalent bonds and releasing a range of mainly phenolic compounds characteristic of lignin breakdown (Fig. 6.10). From the chemical nature of the fragments produced when the enzyme acts on a synthetic lignin-like dimeric molecule, it can be deduced which bonds the enzyme has broken. The enzyme does not require steric specificity – instead of a combination between the enzyme and its substrate being needed, the substrate is bombarded with free radicals generated by the enzyme. This proposed mechanism parallels that suggested for attack on cellulose by brown rots. With each, the question arises of how hydrogen peroxide is generated. Enzyme systems have been found in fungi which will do this by oxidizing glyoxylate, glucose or alcohols. This explanation for both cellulose and lignin decay is attractive because it accounts for the hitherto mysterious necessity for close contact between hypha and substrate; this is obviously essential where attack is due to a very short-lived free radical generated by the hypha. Lignin breakdown, like that of cellulose, is subject to control mechanisms. However, these are of a different kind from those involved in cellulolysis. The rate of lignin decomposition by *Phanerochaete chrysosporium*

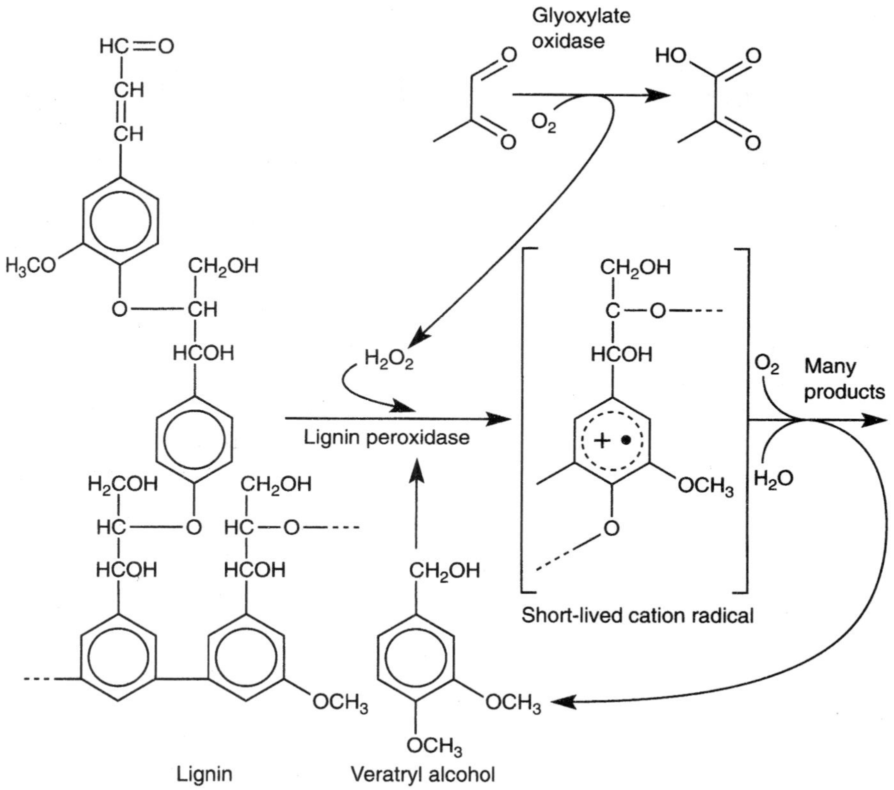

Figure 6.10 The path of lignin breakdown by *Phanerochaete chrysosporium*. Veratryl alcohol is a breakdown product which stimulates the synthesis of lignin peroxidase, and hence lignin breakdown, an example of positive feedback or metabolic amplification. (After Kirk, T. K., Burgess, R. R. & Koning, J. W. (1992). The use of fungi in pulping wood: an overview of pulping research. In G. S. Leatham, ed., *Frontiers in Industrial Mycology*, pp. 99–111. Chapman & Hall, London.)

and by the white rot fungus *Coriolus versicolor* has been measured by estimating amounts of radioactive $^{14}CO_2$ emitted by the fungus when synthetic lignin-like substances labelled with the radioactive isotope ^{14}C were added to the culture medium. The ability to release $^{14}CO_2$ developed only in older cultures because nitrogen substrates in the culture medium inhibited the onset of lignin-decomposing ability. Carbon substrates, although required for growth, had no effect, neither inducing nor repressing activity. Depletion of extracellular nitrogen not only initiated ligninolysis but also triggered several other syntheses including those of phenol oxidase, common to all white rot fungi, and an alcohol peculiar to *Phanerochaete*. This indicates that the onset of secondary metabolism in this fungus is caused by nitrogen starvation (Fig. 6.11). Other microbes, including bacteria, start secondary metabolism when nitrogen is exhausted, and the new syntheses that are involved can include those of degradative enzymes such as

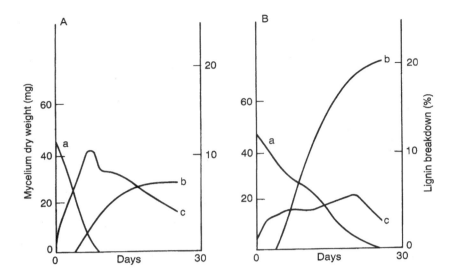

Figure 6.11 Lignin breakdown by *Phanerochaete chrysosporium* in culture. Glucose concentration, initially 56 mM (a), lignin breakdown, as shown by release of $^{14}CO_2$ from ^{14}C-labelled lignin (b) and fungal biomass (c) are shown. A, Medium with high nitrogen (24 mM ammonium tartrate). High nitrogen permits full utilization of glucose and rapid growth. Lignin is degraded slowly. B, Medium with low nitrogen (2.4 mM ammonium tartrate). The nitrogen source is limiting, so glucose utilization and growth are slow. Exhaustion of the nitrogen source, which is accompanied by a fall in the level of free intracellular amino acids, induces ligninase synthesis and a high rate of lignin breakdown. In nature this is thought to give access to nitrogen sources. (After Kirk, T. K., Schultz, E., Connors, W. J., Lorenz, L. F. & Zeikus, J. G. (1978). Influence of culture parameters on lignin metabolism by *Phanerochaete chrysosporium*. *Archives of Microbiology* **117**, 277–285, and Fenn, P. & Kirk, T. K. (1981). Relationship of nitrogen to the onset and suppression of ligninolytic activity and secondary metabolism in *Phanerochaete chrysosporium*. *Archives of Microbiology* **130**, 59–65.)

proteinases which then release more nitrogen from substrates. However, it is very unusual for an enzyme that degrades a carbon substrate like lignin to be sensitive to control by nitrogen compounds. The unexpectedness of the system prompts one to look for an advantage it might confer in nature. Such an advantage becomes obvious with consideration of the structure and chemistry of wood. Cellulose, the main energy source for wood decay fungi, is masked in wood cells by a lignin layer. The major proportion of nitrogen in the wood (90%) is covalently bound. A plausible sequence of events during decay might therefore go as follows. The fungus living on cellulose becomes limited in its advance by lack of nitrogen for protein synthesis. Nitrogen starvation triggers ligninolysis and removal of a lignin layer exposes fresh cellulose and bound nitrogen. The hyphae grow into and exploit this new source and lignin decay is repressed while the nitrogen compounds released are still present in the vicinity of the hypha. When it is all used up, lignin decay is again possible as the repression ceases to operate, and a new layer of cellulose is again exposed. This unmasking of energy and nitrogen sources is probably the main advantage conferred by the ability to decay lignin, as lignin is not in itself an adequate energy source for fungal growth. Other fungi, such as the dung fungus *Coprinus cinereus*, are able to decompose lignin in the presence of nitrogen sources.

A further observation is of interest in relation to the nitrogen economy of the whole Basidiomycete colony in its wood substrate, such as a tree stump. This is that ligninolysis is sensitive only to external nitrogen depletion and not to concentrations inside the hypha which may be high. Thus the fact that the mycelium is rich in internal nitrogen reserves does not prevent its continuing advance into the wood. Such reserves may be moved about by translocation in the mycelium, sometimes in special structures such as mycelial strands or rhizomorphs (page 342), and deployed in different parts of the colony according to need, as has been shown in some higher fungi.

White rot fungi which remove lignin from wood and lignocellulosic materials selectively, leaving a predominantly cellulose residue, are being investigated for a variety of possible uses. Wood rotted by selective delignifying fungi is used as cattle feed in southern Chile, where fungi of this type are common in the temperate rain forests. Lignocellulosic wastes such as straw, and bagasse from sugar refining processes, can be converted by suitable delignifying fungi into palatable and digestible feed for ruminant animals. In paper manufacture from wood pulp, delignifying fungi such as *Phanerochaete chrysosporium* can help to replace chemical methods of removing woody residues from cellulose fibre. The ability of this fungus to attack phenolic compounds is being exploited in **bioremediation** to remove phenolic contaminants from polluted soil and water.

Soft rot fungi   Common soft rot fungi include species of *Chaetomium*, *Fusarium* and *Paecilomyces*, which are all common soil genera, and not true wood-inhabiting fungi. Any timber which is in contact with the soil, for example fence posts or telegraph poles, is ultimately attacked by these fungi and wet wood is particularly susceptible. They are more damaging in tropical than temperate conditions, perhaps because many of the causative moulds grow best at relatively high temperatures. The fungal hyphae grow almost exclusively in the cellulose part of the wood cell wall known as the S2 layer (page 310, Figs. 6.4, 6.5). Initial colonization is by hyphae that enter the tubular wood cells and grow inside them. A lateral branch then penetrates the thin inner lignified S3 layer from within the lumen of the tube. As soon as this branch enters the cellulosic S2 layer it branches into two hyphae growing up and down the cell wall – presumably aligned with the cellulose microfibrils – giving the appearance of a T junction. The penetrating hyphae are narrow where they first advance ('proboscis hyphae') and later widen, probably because they have eroded a wider channel. Even after the fungus has died, the nature of the attack can still be diagnosed from the conically shaped cavities that remain (Fig. 6.6).

Assemblages of Fungi – Communities and Successions

So far we have considered the physiological features of individual species and groups of fungi, without reference to the ways in which they may interact within natural ecosystems. But pure cultures of single fungi rarely occur naturally, and most fungi coexist with others. A habitat usually contains many different fungal species, growing at the same time and constituting a community. Thus the examination of a forest soil by an appropriate method will yield a large number of fungal species. Re-examination of the same soil a year later will provide a similar species list. In the meantime the changing seasons, with changing soil temperature and moisture content, and the seasonal addition of substrates (e.g. leaf fall) will result in a seasonal fluctuation in the types and numbers of fungi present. The fungi present at a particular point will be influenced by the nature of the substrate at that point (e.g. fallen leaves, animal droppings). They will be affected by other microorganisms and soil animals and will change as the substrate changes through microbial attack. Such **substrate succession** has been studied in detail with a variety of substrates (e.g. Fig. 6.14). Although on a microscale substrate succession and a variety of substrates results in local heterogeneity, when a habitat as a whole is considered the vegetation present will provide the same type of substrates year after year, leading to homogeneity on a larger scale. The fungal activities in such habitats as soil are considered further (pages 330–336).

The different methods of analysis described below each give different information about the fungus flora of a sample, and an adequate description of the fungal population of any habitat requires the use of several different techniques. Even so, it is unlikely that the chosen methods will demonstrate all the fungi that are present.

Methods for the Study of Fungal Communities

Many methods have been devised for the study of the microorganisms, including fungi, present in natural habitats. **Isolation methods** can enable fungi to be grown in pure culture, identified, and lists of species compiled. **Molecular methods** (pages 323–328) are enabling mycelium and spores to be identified, increasing our knowledge of fungal population structure. Various techniques for the **microscopic examination of habitat samples** have been devised and enable the form and arrangement of fungi in the habitat to be observed. In addition the amount of fungus (**biomass**) can be measured and there are methods for the study of the **metabolic activities** of the fungi in a habitat. Many of these techniques can be applied not only to soil but also to other habitats, and to parasites and mutualistic symbionts (Chapter 7) as well as to saprotrophs.

Isolation methods

To identify a fungus by direct observation it is usually necessary to see the details of its spore-bearing structures. Since these are often absent in nature, and even if they are present are hard to see, identification of a species is usually possible only if the fungus can be induced to grow and sporulate on a culture medium.

Most natural substrata contain large numbers of microorganisms of many different kinds – fungi, bacteria, algae, protozoa and minute animals such as nematodes. Methods of isolating a single fungal species into pure culture can be divided into **direct methods**, in which observed spores or mycelium are transferred to a sterile nutrient culture medium, and **indirect methods**. In the latter a sample of substratum is used to inoculate the medium, and from the many colonies which usually develop, some are selected as the source of cells that are subsequently transferred to fresh sterile medium. The pure culture thus produced is referred to as an 'isolate'. Direct methods are very useful if the fungus produces clearly visible spores at the surface of the substratum. Fungal inhabitants of plant surfaces can often be induced to sporulate if pieces of bark, ripe seeds or other plant material are incubated under damp conditions for periods of up to several weeks. Spores of Ascomycete and mitosporic fungi and slime moulds will almost always develop and can be transferred on a sterile needle to a culture medium such as potato glucose agar. Soil fungi can be directly isolated by picking individual hyphae from a soil suspension under a microscope, cleaning and culturing them. Although tedious, this direct isolation method avoids a major problem of indirect methods in so far as any fungus isolated will have been present in the substratum as actively growing mycelium, rather than in a dormant form. **Indirect methods**, in which a sample of the substratum is incubated in sterile nutrient medium, usually produce numerous colonies of microorganisms which can be transferred into pure culture. Three widely used indirect methods for soil microorganisms are **dilution plating** and **surface plating**, both using soil suspensions, and the **Warcup plate** method using soil crumbs. In dilution plating a soil suspension is serially diluted with sterile water and a standard volume of each dilution is mixed with a known volume of melted agar at about 40°C and the mixture poured into Petri dishes and incubated. The purpose of dilution is to obtain separated colonies from which spores or mycelium can be transferred into pure culture. Dilution plating can also be used to deduce the abundance of fungal propagules – 'colony-forming units' or 'cfu' – in a sample. Surface plating, in which soil suspension is spread on the surface of an agar plate, is simpler and does not involve submerging organisms in warm agar. Its main disadvantage is that fast-growing fungi quickly overrun the plate. In Warcup's technique a very small amount of soil (5–15 mg) is dispersed in molten agar in a Petri dish without prior suspension in water. The simplicity of the technique makes it suitable for the comparison of large numbers of soil samples.

In each of the above methods the growth of a colony in the plate gives evidence of its presence in some form in the soil sample. Not all the microorganisms present in the sample will generate colonies on the plate, however. Any culture medium is unavoidably selective because of its nutrient composition and other conditions prevailing in it. Fungi of the genera *Aspergillus* and *Penicillium* usually appear abundantly since they grow well on rich nutrient media, and their antibiotics may suppress the growth of other organisms. Their frequency of isolation on soil plates probably bears little relation to the extent of growth and activity of their mycelium in the soil but is a result of their prolific sporulation, each spore head producing thousands of spores. Basidiomycetes, Chytridiomycetes, Oomycetes and slime moulds, which other methods show to be common in soils, rarely develop in soil plates. The list of fungi compiled by any one isolation method will inevitably be incomplete and the use of a variety of methods is desirable.

The selectivity of culture media becomes an advantage when the object is to isolate particular microorganisms. When isolating fungi from soil by the methods above it is usual to add antibiotics such as penicillin and streptomycin to the medium, or to acidify it to about pH 4, to make it selective for fungi and to suppress bacterial growth. Selective media have been devised for the isolation of single species. The soil-borne tree pathogen *Phaeolus schweinitzii*, for example, develops in almost pure culture when infected soil is incubated in a medium containing arsenic compounds, which inhibit most other microorganisms. Alternatively, microorganisms which have an ability to utilize an unusual substrate can be isolated from a mixed population by adding the substrate to a soil sample so that their numbers increase compared with other organisms, a procedure known as 'enrichment'. Thus keratin or cellulose decomposers can be isolated from pieces of these materials that have been buried in soil. Selective methods can also be based on features of an organism other than its nutrition. The motile spores of Oomycetes are attracted to some organic nutrients enabling this fungus to be isolated by adding small particles of a suitable bait – sterilized seeds, plant or animal material – to water samples or soil suspensions. Chytridiomycetes may be isolated with the aid of pieces of sterile 'cellophane' since alone among fungi they have motile spores and are also cellulolytic. Such baiting techniques are important when the fungal population of a habitat is being investigated because many Chytridiomycetes have little or no mycelium and do not appear in isolation procedures using agar plates.

The isolation methods described enable a list of fungi to be compiled. This is only a first step in understanding the role of fungi in a natural habitat, as such lists tell us nothing about the quantity of mycelium, its metabolic activity, the relative abundance of different species and their form and arrangement within their microhabitats. Other methods are needed to discover what the fungi are doing *in situ*, and the most obvious is to look at the material.

Direct observation

By direct observation *in situ* one may see the form and arrangement of a fungus in its habitat, but it is rarely possible to identify it, as the chances are against its sporulating at the exact time and place of observation. Both light and electron microscopes have been used. Pieces of the substratum may be fixed, embedded, sectioned and stained for light and transmission electron microscopy. This procedure has been used to show the pattern of distribution of bacteria and fungi on the roots of plants in the soil. The advantage of sectioning is that internal structures can be seen, but of course information about three-dimensional arrangement is lost. Surface observation by incident light or scanning electron microscopy at low magnification can reveal relationships between structures, such as the disposition of fungal hyphae on surfaces. For example, a scanning electron microscope picture of a leaf often shows phylloplane fungi which have traversed the surface and entered the substomatal air space via the stomata. It would be a very long and difficult task to discover this by looking at sections of a leaf. Soil surfaces have also been observed directly in this way. As an alternative to examining the substratum, it is sometimes possible to separate fungi from substrata such as soil and leaves by methods involving contact with glass or agar surfaces. Mycelium of some soil fungi will grow over, and adhere to, glass slides buried in the soil. These are usually left in place for several weeks and then removed and stained – the Rossi-Cholodny slide method. This often reveals the presence of Basidiomycete mycelium, identifiable by clamp connections (see, for example, Fig. 6.5C), which culture on agar usually fails to isolate. Other methods involve only brief contact; the presence and distribution of saprotrophic microorganisms such as yeasts on a leaf surface can be demonstrated by pressing the leaf momentarily on to an agar surface and then incubating.

Direct observation by the kind of method described above has provided much information about the place of fungi in natural habitats. Thus it is possible to categorize the growth form of soil microbes into six common types: unicellular and motile (some bacteria, zoospores, motile protists); unicellular, restricted and non-motile (some bacteria, all yeasts); restricted mycelial colonies limited to the vicinity of a colonized particle (*Penicillium*, hyphal bacteria of the genus *Streptomyces*); diffuse mycelium not associated with particular substrates (*Zygorhynchus*); mycelial strands; and plasmodia (Fig. 2.6A), which swarm over the surface of a substratum (Myxomycetes).

Molecular techniques

Fluorescent and enzyme-linked antibody techniques   The specificity with which animals make antibodies to particular substances can be used to identify fungi when other means are not available; for example, when only mycelium can be seen and there are no morphological identifying features. Antibodies to the target fungus are prepared, by standard immunological methods involving collection of serum from a rabbit injected previously with fungal antigen. The antibody produced by the rabbit can then be used to detect and identify this specific fungus, and to determine its amount, in samples which may contain other fungi. A procedure for preparing an antibody to detect the target fungus in a mixed sample is illustrated in Fig. 6.12. The washed fungal sample in which the target fungus is to be sought is exposed to the primary antibody, which binds to its spores or hyphae. Soluble fungal antigens can also be detected using the technique of enzyme-linked immunosorbent assay, described below. It is then necessary to detect the bound antibody on the fungal sample, which is done using a secondary antibody which recognizes the bound primary rabbit antibody and is also conjugated to a substance which makes it visible. This use of a secondary antibody is convenient, as anti-rabbit antibodies are commercially available, already conjugated to marker substances. It also allows a larger number of marker molecules to be attached to each primary antigen molecule because the secondary antibody binds to the primary one at more than one site per molecule. Marker molecules may be dyes such as fluorescein isothiocyanate, or enzymes which can release a dye stoichiometrically from a coloured substrate, a procedure known as **enzyme-linked immunosorbent assay (ELISA)**. The latter enables the amount of antigen in the sample to be measured, since there is a fixed ratio between the number of antibody molecules bound and the antigen. The main source of error in immunological techniques is cross reaction of the antibodies in antiserum with fungi other than that used as the antigen source. This is a consequence of many different kinds of molecules being present in the fungal material used to inoculate the animal. Only a few of these are likely to be unique to the fungal species, so some of the antigens present in the antigen preparation

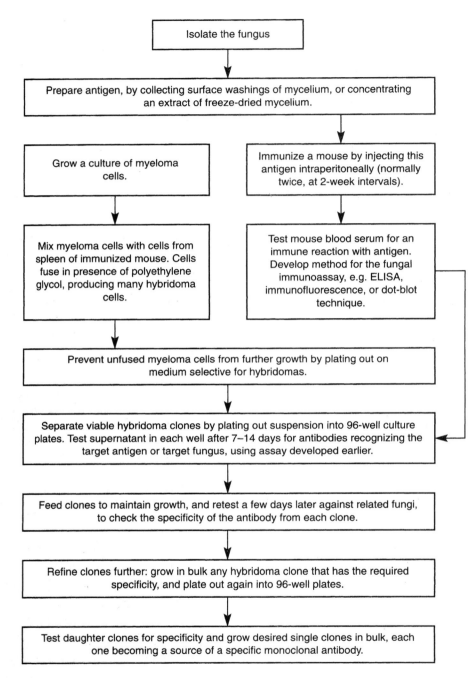

Figure 6.12 The procedure for producing a monoclonal antibody to a fungal antigen. F. M. Dewey (see Dewey *et al.*, 1993, in references, page 360.)

will be non-specific. This source of error can now be decreased by the use of monoclonal antibodies.

In the **monoclonal antibody technique** (Fig. 6.12), antibody-producing cells (lymphocytes) from an immunized animal are fused with an immortal cell line derived from tumour cells, and from the resulting cell culture single cells are used to establish lines which, being derived from single cells, produce only one specific antibody. A cell line is then selected producing an antibody specific for the fungus. This line can then be maintained as a source of large amounts of the specific antibody. The method has been used successfully, with detection of antibody binding by the ELISA method, to identify and assay *Humicola lanuginosa* mycelium, which causes spoilage of stored rice, in samples of grain, and also to detect and assay the mycelium of *Botrytis cinerea* in juice expressed from infected grapes.

DNA technology   Most fungal species have been recognized and classified on the basis of the form of their sporophores, and until recently most fungi could be identified only if sporophores were present. This has been a great obstacle to progress in fungal ecology, since fungi spend much of their life cycle as mycelia, foraging, assimilating nutrients, growing and accumulating reserves. For example, a Basidiomycete may produce a spectacular display of fruiting bodies in a wood and then, after a few weeks, not be seen again at the site for many years. However, throughout this period its mycelium may have been abundant and active in the soil, perhaps playing a vital role in plant nutrition as the fungal partner in a mycorrhizal association (page 394). Direct observation (page 322) will show mycelium in soil, but usually will not permit the recognition of individual species. An inability to identify species during their vegetative phase has been a major reason for fungal ecology being less advanced than plant or animal ecology – little progress would have been made in these sciences if plants and animals could be identified accurately only if observed reproducing! However, DNA technology is now enabling a fungal species to be identified in the absence of sporophores and even to be detected without mycelium being observed.

DNA technology permits the direct study of an organism's genome. Two individuals will be genetically identical if they have originated from a common ancestor without either recombination or mutation. All other individuals will differ genetically. Differences between the genomes of two strains that have adapted to different environments will be more marked, and those between two closely related species still greater. There are major differences in the extent to which different parts of a genome change in evolution. Some genes are so highly conserved that parts of their sequences may be nearly identical throughout a kingdom, whereas at the other extreme there are regions which can differ between individuals. Hence by examining the appropriate parts of a genome it should be possible to recognize any desired taxonomic unit, for example phylum, species or clone.

A part of the nuclear genome that has been extensively utilized in phylogeny, classification and identification, both in fungi and in other organisms is ribosomal DNA (rDNA), the region that codes for the RNA component of the ribosomes (rRNA). Unlike most other genes, rDNA genes are arranged in a repeating array of copies along a chromosome with up to 100 copies per array.

A single copy (Fig 6.13 A) has the genes for the three RNA components of the ribosome, 28S, 18S and 5.8S RNA. The 18S gene has varied enough in the course of evolution, to be utilized in assessing relationships within and between Zygomycetes, Ascomycetes and Basidiomycetes, and discriminating between the three domains of organisms, Archaea, Bacteria and Eukaryotes (Fig 1.2), and between kingdoms. The 28S gene is more variable. It has been extensively used in classification at levels from genus to phylum. The 5.8S gene is too small to contain much information, but has been used to identify fungi as Ascomycetes, Basidiomycetes and Zygomycetes. Between the 28S and 18S genes is a region in which the 5.8S gene is embedded, called the internally transcribed spacer (ITS). This, and the externally transcribed spacer (ETS) are transcribed from DNA into RNA, but then eliminated in the process that yields rRNA. The ITS is less conserved than coding regions and generally varies sufficiently between species to be utilized, as described below, in their recognition. Finally, there is a non-transcribed intergenic spacer (IGS) between rDNA copies that is still more variable, and has been used at the sub-specific level to recognize races and populations. Although, as indicated above, nuclear rDNA has regions appropriate for detecting, identifying and classifying taxa at all levels, other parts of the genome can also be employed, including mitochondrial rDNA.

An example of the way in which DNA technology is employed in species recognition is provided by the use of the ITS region in identifying the fungal partner in mycorrhizal roots. DNA is extracted from a mycorrhizal root tip. This DNA will be part plant and part fungal, with fungal rDNA being a very small part of the latter. The **polymerase chain reaction (PCR)** is then utilized. This procedure

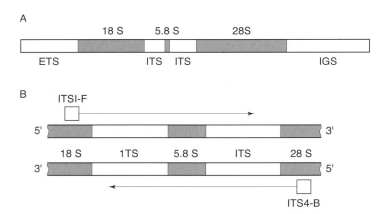

Figure 6.13 The utilization of ribosomal DNA in fungal identification. A, Single copy of ribosomal DNA (rDNA), with coding and non-coding regions. An internally transcribed spacer (ITS) lies between the genes coding for 18S and 28S RNA, with that coding for 5.8S RNA embedded within it. Beyond the RNA genes are the externally transcribed spacer (ETS) and the intergenic spacer (IGS). B, The complementary strands in the ITS region of DNA, with adjacent coding regions, showing the points at which the two primers, ITS1-F and ITS4-B, bind, and the direction of DNA replication (→), towards the 3′ end of each strand.

has revolutionized DNA technology, since it enables a selected part of a genome to be replicated repeatedly or 'amplified' until it constitutes almost all of the DNA present. Requirements for this amplification are an appropriate buffer solution, the substrate for DNA synthesis (a mixture of the four essential nucleotide triphosphates), a heat-stable DNA polymerase, a magnesium salt as co-enzyme of the polymerase, two primers (single-stranded DNA sequences of about 20 base pairs) for initiating replication and defining the template (DNA region to be copied), and a thermocycling apparatus that enables the reaction mixture to be subjected to successive temperatures for precise periods. The mixture is first exposed to 94°C for 85 s (seconds). This results in the DNA separating into the single-stranded state or 'denaturing'. The temperature is then lowered to 55°C so that the primers can bind ('anneal') to the strands. One of the primers, known as ITS1-F, is complementary to a sequence in one strand of the 18S gene of Ascomycetes and Basidiomycetes. The second primer, known as ITS4-B, is complementary to a sequence in the other strand of the 28S gene of Basidiomycetes. Neither primer will bind to plant DNA. After 55 s the temperature is raised to 72°C for 45 s. At this temperature DNA synthesis occurs, beginning at the primers and proceeding towards the 3' end of each strand (Fig. 6.13 B). With further cycles of denaturation, annealing and synthesis the proportion of such ITS DNA increases, and after 35 cycles constitutes nearly 100% of the DNA present.

The DNA that has been obtained can then be compared with ITS DNA from known fungal species to determine the species from which the sample came. Two methods of comparing DNA samples, including ITS DNA, are widely used. One is to sequence the DNA and submit the data to an appropriate DNA database; there are now several that specialize in sequence data from mycorrhizal fungi. Alternatively, use may be made of **RFLPs (Restriction Fragment Length Polymorphisms)**, pronounced 'Rif-lips'. The term refers to the effect of treating homologous samples of DNA from different strains or species with restriction endonucleases. Each restriction enzyme will recognize a short DNA sequence (usually a set of four to six base pairs) and cut the molecule at that point. A DNA molecule in which the recognized sequence occurs, for example, twice, will be cut into three fragments ('restriction fragments') of different length. A DNA molecule from a different source may have the recognition sequences differing in frequency or location, and so will yield restriction fragments differing in length or number. Most of the differences in ITS sequences found using RFLPs are the result of insertion/deletion differences (IDELS). IDELS differ among closely related taxa, producing easily recognizable differences in the lengths of restriction fragments. If these differences occur within a species, as occurs with IGS DNA, then the species is said to be polymorphic with respect to restriction fragment length. With ITS DNA such differences occur more frequently between species, so RFLP matching of sample DNA against that from known species is an effective method of determining the source of the sample. The procedure is to treat the ITS DNA with a restriction enzyme to give restriction fragments which, being of different lengths, will separate on gel electrophoresis to give a pattern. Repetition of the procedure with a second restriction enzyme which recognizes different sequences will give fragments of a different length and a different gel pattern. Two species are unlikely to be identical with respect to the RFLP pattern for more than one enzyme, so comparison of two enzymic digests of the sample with appropriate standards provides a reasonably good criterion for species identification in many genera.

The procedures for the examination of rDNA outlined above require two primers defining the region to be amplified and, for such primers to be produced, a knowledge of the DNA sequence in the appropriate region is necessary. A procedure without such limitations is known as **RAPD (Randomly Amplified Polymorphic DNAs)**, and pronounced 'rapid'. The polymerase chain reaction is employed, but with a single primer which, having a short and arbitrary sequence, binds to many sites on a genome. The result is a **fragment pattern** characteristic of a strain. The method has been used for identifying strains important in medicine, plant pathology or industry, and for studying the origin and dispersal of populations.

If a sequence at a site in a genome is known, it is possible to synthesize a complementary sequence that will bind to or 'hybridize' with that site. Such a sequence can be used, as described above, as a primer for initiating the PCR reaction and generating RFLPs. If a DNA sequence characteristic of a species (or other taxon) is known, then DNA hybridization can be used in identification. The appropriate complementary sequence, about the same size as a primer and labelled with a radioisotope or a colour reagent, is known as an **oligonucleotide probe**. In addition to allowing species recognition, oligonucleotide probes can be used to establish the nature of an organism of unknown taxonomic affinity.

DNA methodology is becoming more and more useful to fungal ecologists as the supporting technology and sequence data become more widely available. Databases of DNA sequences are growing as fast as new data can be incorporated. PCR primers of specified sequence can be synthesized cheaply and supplied quickly by commercial firms, and DNA sequencing is now routine and mechanical. Spores and mycelium, previously hard to trace and identify, have thus become much more easily identifiable, providing new tools for exploring the life histories, geographical extent and ecosystem functions of fungi.

Estimates of biomass

The biomass of mycelial fungi cannot be estimated by counting individuals because of their indeterminate growth and the difficulty of establishing the boundaries of a colony. Instead some measure is made of the amount of mycelium. The only exception is where the fungus is entirely in the form of unicells or spores, for example spores in the air (Fig. 4.13).

Methods of estimating fungal biomass as mycelium usually involve calculations of the amount of fungal material observable in a sample of known weight or volume. The initial sampling needs care, particularly if the substratum is solid, to avoid bias in the results due to the heterogeneity of distribution of mycelium within it.

Quantification may be by direct observation. Scanning electron microscopy, for example, can give an estimate of the frequency of microbes, including fungi, in soil. It has the advantage over indirect methods in that the spatial relationship between the fungi and the substratum is seen. Alternatively samples may be treated so as to free fungi from the substratum before attempting to estimate their biomass. Thus soil can be ground in water and a known volume of the suspension thoroughly mixed into a larger known volume of molten agar. Samples of this mixture are placed in the chamber of a haemocytometer slide, where the agar is allowed to set. This makes an agar film of known volume which can be removed from the haemocytometer, stained if desired, and mounted on a microscope slide. The total length of hyphae present is then recorded, using the low power objective of a light microscope. To confirm that hyphae seen in a sample of mixed fragments, such as leaf litter, are living and viable, there are several stains that are visible only after enzymic action on a precursor molecule. Thus fluorescein diacetate is taken up by mycelium but becomes fluorescent only after being acted upon by the esterases inside living hyphae. Other fluorescent dyes are available which are taken up into specific compartments of living cells, such as vacuoles and mitochondria (Plate 6).

Fungal biomass can be estimated indirectly by assaying substances characteristic of fungal cells, such as chitin or ergosterol, in the sample. Hydrolysis of chitin yields N-acetylglucosamine (Fig. 3.4), which can be estimated colorimetrically. Fungi are the only microorganisms with chitin-containing walls. However, the amount of chitin does not bear a simple relationship to biomass, as the amount and composition of the cell wall material of fungi varies with factors such as age and species. Other possible sources of error include the absence of chitin from Oomycetes and its presence in insects and mites. Chitin assay is, however, a good method for comparing the biomass of fungi in samples where these sources of error are unimportant, as in assessing the extent of growth of chitin-walled fungi within the tissues of a plant. Ergosterol is found in Basidiomycetes, Ascomycetes and mitosporic fungi, but not in plants, so it is a useful indicator of the amount of fungal material in a predominantly plant sample. A possible source of error in mixed samples are some algae and protozoa that contain ergosterol. The advantage of ergosterol over chitin as a measure of mycelial biomass is that it is a component of the cell membrane and so is only found in living mycelium. The amount of ergosterol in a sample is estimated by ultraviolet spectrophotometry of purified extracts.

Fungal biomass is much more easily measured in fluid samples such as air and water. Several kinds of spore trap (Fig. 4.14) exist for measuring the number of spores per unit volume of air drawn through the trap. In liquid samples, zoospores can be concentrated by centrifugation to a density suitable for direct counting. Alternatively, water containing the zoospores to be counted can be stabilized in a gel in which a colony develops from each zoospore, enabling the number of colony-forming spores or hyphal fragments to be estimated.

Estimates of fungal activity

Methods such as those described above are designed to estimate how much fungal material there is in a sample. A further question one may wish to ask is, how active is it, both in terms of growth and metabolic processes? Many techniques have been devised for measuring the total microbial metabolic activity in samples, particularly of soil. Rates of evolution of carbon dioxide, uptake of oxygen and breakdown of added substrates have been used to give a measure of the total activity of all heterotrophic microbes present. It is not easy, however, to distinguish fungal activity from that of bacteria and protozoa. One way is to make use of their spreading, filamentous growth form. For example a suitable sterile culture medium contained in a vessel with small holes in it may be buried in soil. Fungal hyphae, unlike bacterial colonies, can grow across an air gap through the holes to colonize the medium, and the frequency with which this occurs may give an indirect measure of the amount of growing mycelium. Unfortunately, as with any method involving growth in a culture medium, a degree of selectiveness is inescapable. Also, the presence of the sampling device itself may stimulate growth in formerly dormant mycelium, or germination of propagules such as spores or sclerotia.

Microcosms

A great deal that is relevant to the life of fungi in nature can be learnt by pure culture studies on selected species in the laboratory. For example, for a chosen species the following can be determined: the range of environmental conditions tolerated, the nutrients required, the extracellular enzymes that are produced, the maximum specific growth rate, and the factors that initiate sporulation. Pure culture studies alone, however, cannot tell us whether a species, faced with numerous competitors, will survive in a particular environment and which of its potentialities will be realized. Observations on a natural habitat, utilizing the various methods described above, can tell us whether the species is present and how abundant it is through the seasons. From a knowledge of the behaviour of the fungus in pure culture and observations on its occurrence in natural habitats it is possible to formulate hypotheses as to the significance of the activities of the fungus in an ecosystem. However, usually neither pure culture studies with competition lacking, nor observations on a natural system liable to variations outside the control of the experimenter, are adequate to confirm or refute such hypotheses. If this is so, success may be achieved with microcosms (literally, little universes), systems having some of the complexity of the natural environment but under more fully controlled conditions. For example, a microcosm for the study of soil fungi will resemble the natural environment in having unsterile soil as the substratum for growth, but with temperature, water potential and nutrient addition controlled in a way that would not be possible in nature. The use of a microcosm for the study of nutrient translocation by ectomycorrhizal fungi is illustrated in Fig 7.14A, and another for foraging by saprotrophic Basidiomycetes in Fig. 6.17A–D.

Substrate Successions

Leaves

The fungal decomposition of leaves is an essential part of the carbon cycle (page 298). Leaf decomposition has been extensively studied and is probably the best understood of natural substrate successions, even though much of our knowledge still consists of lists of fungal species rather than an understanding of the processes taking place. The sequence of fungi observed will depend on the plant species but the various leaf successions studied all have features in common. There is a remarkable consistency in the pattern of succession observed for a specific leaf type from one year to another (Fig. 6.14). Of the many fungi whose spores or other propagules are present, only some species are able to colonize the leaf at any one stage of its existence.

Typical stages in succession on the leaves of deciduous trees in temperate climates proceed as follows. Fungal colonization of the leaf surface begins as soon as the leaf emerges from the bud, and throughout the summer it provides a habitat, the **phylloplane**, for a specialized group of fungi. Hyphae traversing the leaf surface can be seen by scanning electron microscopy, in some cases entering stomata and growing in air spaces within the leaf. Dilution plating of leaf washings enables many species of saprotrophic fungi from the leaf surface to be identified. However, such washings contain not only the spores of true leaf surface inhabitants and the cells of phylloplane yeasts, but also other spores from the air. True leaf inhabitants which have some of their mycelium within the leaf tissues can be distinguished from these by sterilizing the leaf surface by brief submersion in a mild sterilant such as sodium hypochlorite solution, after which fungi present within the leaf tissue and protected by it from the sterilant will grow out on to a suitable medium. The spore fall method allows true inhabitants of the external leaf surface to be distinguished from those accidentally present. A leaf is suspended above nutrient agar in a covered dish; the spores of the actively

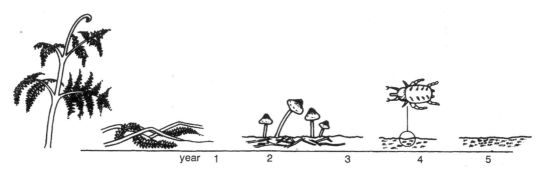

| | | year 1 | 2 | 3 | 4 | 5 |

Standing, but senescent fern is colonized in late summer by weak, host-specific pathogens such as *Rhopographus*, and by cosmopolitan conidial fungi such as *Aureobasidium*

After leaf fall, *Aureobasidium* persists, and is joined by many primary saprotrophic fungi, which break down non-lignified epidermis, cortex and phloem tissues

Basidiomycetes appear and begin to attack lignified tissues; predaceous species and Mucorales typical of soil appear as well. By 3 years half the cellulose has disappeared. Soil animals such as mites assist breakdown

At 4 years 50% of the lignin has disappeared, and by 5 years 80% of the dry weight. Fungal activity ceases by the end of the fifth or sixth year, depending on soil type; bacteria replace fungi in rotten tissues, pH rises to about 6, C/N ratio falls to about 30

Figure 6.14 The succession of fungi during the decomposition of the fern *Pteridium aquilinum* (bracken). (After Frankland J. C. (1992). Mechanisms in fungal successions. In D. T. Wicklow & G. C. Carroll, eds., *The Fungal Community: its Organisation and Role in the Ecosystem*, 2nd edn, pp. 383–401. Marcel Dekker, New York.)

sporulating species are most likely to fall from the leaf surface, initiating colonies on the nutrient agar surface. Phylloplane inhabitants include both yeasts and filamentous forms. Common yeasts on leaves belong to the genera *Candida*, *Cryptococcus*, *Rhodotorula*, *Sporobolomyces*, *Tilletiopsis* and *Torulopsis*. Some leaf fungi are dimorphic, capable of growing in both yeast and filamentous forms. *Sporobolomyces* is unusual among yeasts in producing forcibly discharged 'ballistospores' (page 217). The common name 'shadow yeasts' applied to this group comes from the 'shadow' of yeast colonies produced when a colonized leaf is placed above a nutrient agar surface on to which spores fall. It is notable that, among yeasts, phylloplane forms alone liberate spores by active discharge. Abundant production of airborne spores is typical of most leaf surface inhabitants. It may be an adaptation to a need to colonize a habitat consisting, as does the canopy of a tree, of numerous separated units. Filamentous fungi growing in the phylloplane are mainly mitosporic fungi and Ascomycetes. Often the latter produce only conidia on the living leaf and so the anamorph name (e.g. *Cladosporium herbarum*) is used, but after leaf fall perithecia are formed and the teleomorph name (*Mycosphaerella tassiana*) is applied. The anamorphic phase of this fungus is very abundant on leaves in summer and early autumn, and the conidia are a major cause of seasonal respiratory allergies.

Studies on the composition of the phylloplane mycoflora have been made on the leaves of many different kinds of plant, both temperate and tropical. A few species, for example *Aureobasidium pullulans* and *Cladosporium herbarum*, are common, being found on the leaves of geographically widely separated and unrelated plants, but many other phylloplane inhabitants occur only on particular kinds of leaf. Sometimes this can be explained by a parasitic relationship between plant and fungus, but in others there is no obvious explanation of this specificity. A study of the mycoflora of leaves of the tropical plant *Metrosideros* revealed 118 different fungi, of which 69 were absent from two other plant species in the same forest.

Nutrients for phylloplane fungi must be provided by the leaf or from the atmosphere. Substances leached from leaves by rain include sugars, pectic substances, sugar alcohols, amino acids, plant hormones, vitamins, and salts including potassium and phosphate. The amounts can be considerable. It has been shown, for example, that rainwater running off trees in a hectare of oakwood would contain about 86 kg of dissolved substances. Some leaf-dwelling insects such as aphids tap the sugary contents of cells within the leaf leading to the exudation of 'honeydew', which enhances the growth of fungi on the leaf surface. Another source of nutrients important for some fungi is pollen, which may get trapped on leaves and provides an abundant though transient food supply. A fungal genus, *Retiarius*, has been described which is highly adapted to the utilization of pollen. Its mycelial network on the leaf develops short vertical hyphae which trap wind-dispersed pollen, then other hyphae grow towards the pollen grain and invade it, and the resulting packaged mycelium acts as a rain splash dispersed propagule for the fungus. A few phylloplane inhabitants (e.g. *Aureobasidium pullulans*) are not confined to the leaf surface and air chambers, but also grow between cells. This is clear from their emergence from the internal tissues of surface-sterilized leaves incubated in moist conditions without added nutrients. Presumably they live on soluble materials available in intercellular spaces of leaves. Some of the fungi isolated by this method may be symbiotic endophytes (page 405). Some leaves exude gallic acid which is inhibitory to some fungi. Colonization of the phylloplane by new species of fungi may also be limited by the established microbial population. There is evidence that fungistatic substances are released by some phylloplane fungi, and competition for scarce nutrients probably limits spore germination and growth of hyphae, producing inhibition in the same way as occurs in soil mycostasis (page 337).

Although the leaf surface has attractive features as a fungal habitat, it can also be an extreme environment. Fungi growing on the leaf surface may be in direct sunlight, with resulting exposure to ultraviolet radiation and high intensity visible light, both potentially lethal. It is for this reason that the spores, and sometimes the hyphae, of many phylloplane fungi are dark-coloured, due to the presence of the protective compound melanin. The way in which melanin acts is not entirely clear, but its protective effect has been demonstrated. Mutants of *Gliocladium*, for example, that lack melanin are more susceptible to damage by ultraviolet radiation than is the normal dark-spored form. Phylloplane fungi producing large quantities of black spores and darkening the leaf surface are often called 'sooty moulds', of which common examples are *Aureobasidium pullulans* and *Cladosporium herbarum*. Fungi that live on leaf surfaces or in leaf litter may also be exposed to low water potentials, and to very rapid changes in water potential (pages 158–163). This is because leaf litter may dry out, and fungi on the surface of living leaves may be surrounded by a dry atmosphere or an aqueous phase in which evaporation has resulted in very high concentrations of sugars and other leaf exudates. Many species of *Aspergillus* and *Penicillium* occurring in leaf litter are xerotolerant, able to endure low water potentials in an aqueous phase or low relative humidities in the gas phase. A low water potential, however, may be succeeded by a very high one, since heavy rain may wash away a sugary solution on a leaf surface replacing it with nearly pure water. This change can cause an influx of water into the cell, resulting in a very high internal hydrostatic pressure and potential cell rupture (page 161). The yeast form is well adapted to resist such mechanical stress (page 163), and hence yeasts are common on plant surfaces.

The species composition and abundance of the phylloplane community changes as the leaf matures, senesces, dies and falls. Each stage brings a change in the phylloplane environment, as the living processes of the leaf cells decline and there is an increase in substrates for saprotrophs, and physical conditions also change. In late summer the proportion of filamentous fungi to yeasts increases. Then, as leaves die and fall, the initial surface colonizers, some of which are capable of saprotrophic growth on dead leaves, are joined and gradually replaced by exclusively saprotrophic fungi which decompose the waxes, pectins and ultimately cellulose and lignocellulose of the leaf. Genera found on deciduous leaves include *Leptosphaeria*, *Pleospora* and *Phoma*. Parasitic fungi which grew on the leaf during the summer, such as the tar spot fungus of sycamore, *Rhytisma acerinum*, may become more conspicuous at this time through sporulation. Powdery mildews (Erysiphales), whose surface mycelium was conspicuous in late summer, also produce fruiting bodies (cleistothecia) in autumn.

Fungi growing on leaf surfaces can often be identified by direct observation. Analysis of the fungal succession becomes much more difficult after the leaves have fallen and have become fragmented by animals and colonized by a large number of different fungi. The course of the succession after leaf fall depends on the abundance of decomposing organisms and animals eating plant litter, such as earthworms, mites and springtails (collembolans). The activities of these organisms in turn depend on soil and type of vegetation, as well as climatic factors. Moist leaf litter overlying a fertile, deciduous forest soil is quickly fragmented. Within the year it is incorporated into the surface layers of the soil as worm casts, faecal pellets and small pieces of debris. Experiments on the decomposition of forest litter in the presence and absence of animals have shown that weight loss of litter was increased by one-third if animals were present, fragmentation releasing soluble substances and enabling the smaller pieces to be easily attacked. The lignocellulose component of leaves is present in the form of comminuted vegetation in the surface layer, and probably provides a substrate for Basidiomycetes, for example those belonging to the genera *Pluteus* and *Lepiota*, whose fruit bodies emerge from this layer. The surface material soon becomes incorporated in lower soil layers where most of the remaining solid organic residues are degraded as described below. Burrowing soil animals, particularly earthworms, carry plant debris into the soil. On infertile soils, and on types of vegetation unfavourable for litter-eating animals and decomposing microbes, the newly fallen leaves lie in accumulating layers of litter, covering the soil surface all the year round. The layers are distinct enough to constitute separate habitats, each with its characteristic set of fungal and animal species. There is a dry, surface or 'L' layer, of intact dead leaves which is largely colonized by the mitosporic fungi and Ascomycetes which followed the phylloplane inhabitants described above. Below is a so-called 'fermentation' (F) layer which remains damp, and in which Basidiomycete mycelium grows actively, often visible to the naked eye because it is aggregated into mycelial strands. These structures are sometimes protected from attack and lysis by soil organisms by an indigestible, melanized rind. Moisture in this layer enables large populations of mites and springtails to thrive. Sometimes leaf litter colonized by Basidiomycetes is bleached, an observation which accords with the hypothesis that oxidation may be the first step in their cellulose and lignin breakdown (pages 312–316).

Accumulated leaf litter and other plant material may become damp or even water-logged. Under these conditions heat generated by fungi and other microorganisms is not readily dissipated, and high temperatures may result. This also occurs in stored plant materials, such as hay, grain or groundnuts, if these are not kept dry. The high temperatures result in the development of thermotolerant or thermophilic fungi (page 165), many of which are Ascomycetes or mitosporic fungi, include species of *Aspergillus* and *Penicillium*. Such fungi can spoil stored products, and produce dangerous levels of toxins (page 440, Table 7.5).

The Basidiomycetes present in leaf litter vary with the type of litter. For example, *Marasmius androsaceus* is common on conifer needles, whereas *Collybia dryophila* is mainly found on leaf litter from angiosperm trees. Such habitat preferences are not fully understood, but there is evidence that chemicals in leaves increase the rate of growth of some fungi but not others. For example, litter-decomposing fungi in the genera *Clavaria*, *Collybia*, *Marasmius* and *Mycena* are stimulated to grow faster by adding to the nutrient medium very small (5 ppm) amounts of substance extracted from pine needles. The substances have been identified as flavonoids, the most effective being taxifolin glycoside.

The acceleration of growth could give these species a competitive advantage and account for their success on conifer litter. This hypothesis is supported by the finding that fungi in the same habitat but with different modes of nutrition – pathogens and mycorrhiza – are not affected by taxifolin glycoside, nor are fungi from other types of leaf litter. It hence seems that specialized inhabitants of pine needle litter have evolved a growth response to their substrate. Such growth responses, involving inhibition or stimulation by substances at very low concentrations, are very common in fungi.

The activities of fungi in the F layer render the litter more accessible to animals. Fungi break down protein–phenol complexes, releasing protein, and detoxify other phenolic substances. They also assimilate nitrate and ammonium ions so that litter permeated by mycelium is enriched by organic nitrogen assimilable by animals. This phase of the succession ends as the litter fragments are consumed and incorporated into the soil.

Herbivore dung

Herbivore dung is another common natural material which is decomposed by the combined activities of a specialized group of fungi. Most dung fungi have a life cycle beginning with the passage of their spores through the gut of an animal. Hence the pattern of succession is different from that on leaves because most of the fungi are present from the moment the dung is produced. There is, however, a striking successional pattern in their fruiting, and also some of the fungal species disappear from the habitat before others. Herbivore dung is potentially a good substrate for fungal growth, containing high concentrations of soluble carbohydrates, nitrogen compounds, vitamins and minerals as well as cellulose and lignocellulose, and having a suitable moisture content. It is, however, less acidic than are the habitats of most fungi. Horse dung, for example, has a pH of about 6.5. It is widely available, and species that live only on dung occur in several classes, such as the Mucorales, Pezizales, Sordariales and Agaricales.

Specialization for growth on dung includes several physiological features. Spores may require digestion by animals to break dormancy, even by particular animal species. The Zygomycete *Pilobolus* has sporangiospores which germinate when treated with a digestive enzyme, alkaline pancreatin. Its mycelial growth in pure culture is stimulated by addition of vitamins and organic iron compounds present in dung. Dung fungi have a variety of methods of spore dispersal and many involve forcible discharge of sticky spores towards a source of light. The fact that dung fungi from widely different taxonomic groups all have spore discharge mechanisms of this kind is a striking example of convergent evolution. The spores may land on vegetation, be eaten by grazing animals, pass through the alimentary tract and be voided in the dung. They germinate to initiate a new life cycle as soon as the dung is deposited.

The fruiting bodies of dung fungi appear in a consistent sequence with the small sporangiophores of Mucorales (*Mucor*, *Phycomyces*, *Pilaira*, *Pilobolus*) being followed by the perithecia and apothecia of Ascomycetes (*Ascobolus*, *Chaetomium*, *Sordaria*) and finally by the sporophores of Basidiomycetes (*Bolbitius*, *Coprinus*). It was at one time believed that the succession of fruiting bodies resulted from the successive colonization of the dung by different species as conditions in the dung changed. It is now thought more likely that the mycelia of the different species develop simultaneously, and the sequential appearance of fruiting bodies is determined by the time that it takes each type to be formed. This explains the times of appearance of fruiting bodies, but not the failure of species that produce fruiting bodies early on to continue fruiting while soluble nutrients remain. It seems likely that hyphal interference by Basidiomycetes is responsible. Hyphae of *Coprinus heptemerus*, which is the last fungus to fruit on rabbit dung, suppress fruiting by the earlier-appearing *Pilobolus* and *Ascobolus* in culture. Contact or close approach between hyphae of *Coprinus heptemerus* and *Pilobolus crystallinus* is followed by cessation of growth of *Pilobolus* hyphae and vacuolation and granulation of the cytoplasm, attributed to alteration in the permeability of the cell membrane. If rabbit dung is inoculated with *Coprinus* spores, formation of *Pilobolus* sporangiophores is suppressed. It is likely that the Ascomycetes and Basidiomycetes survive as dung inhabitants because they can degrade cellulose and lignin, not available to most of the Mucorales. The Mucorales probably survive because, with their big hyphae and rapid hyphal extension, they have achieved sufficient growth for sporulation before their activities are terminated by hyphal interference from Basidiomycetes. The different effects on lignocellulose of Ascomycetes and Basidiomycetes isolated from dung have been shown by the use of lignocellulose labelled with ^{14}C in either the lignin or the cellulose components. Only Basidiomycetes could release $^{14}CO_2$ from both lignin and cellulose; the Ascomycetes were exclusively cellulolytic. Lignin decay by dung Basidiomycetes is probably not nitrogen-repressed as it is in the wood-inhabiting Basidiomycete *Phanerochaete chrysosporium* (page 318).

Some Important Habitats and their Fungal Communities

The substrate successions on leaves and dung described above illustrate how fungal communities may change when seen on a small scale. Overall, however, as a particular environment provides the same substrates (resources) for fungi year after year, it will acquire a constant population of fungal species. Some examples of habitats with predictable associated fungi are: soil, water (fresh and salt), surfaces of plants and animals, habitats provided by animals including humans, and environments with extreme conditions that only a few species can tolerate. Borders between habitats, such as the interface between a root and the soil, also provide a further environment to which some fungi seem adapted.

Soil
The soil consists of mineral components ultimately derived from parent rock, and chemically complex organic components derived from living organisms. Mineral particles form aggregates with colloidal clay and with organic material which may be sequestered in pores too narrow for bacteria and fungal hyphae to enter. Between these crumbs, which are of the order of 0.1 mm diameter, are spaces containing a mixture of gases, mainly oxygen, nitrogen and carbon dioxide but also others such as ethylene produced by the activities of soil organisms. Soil water is present, absorbed by organic matter, as films around mineral particles, and as water vapour. Fungi grow in the spaces between soil crumbs where

particles of organic matter or plant roots provide substrates for them. The nature of these microhabitats depends on the interaction of physical, chemical and biotic factors in the soil. Thus an increase in moisture content will reduce the proportion of the soil spaces available to the soil atmosphere, so that the gaseous exchange necessary for aerobic respiration is impeded. Further changes may then result from the increased activity of microbes with fermentative metabolism.

A great variety of soil types exists, their nature determined by climate, geology and vegetational history. Some fungal genera have representative species in most soils. The mitosporic genera *Aspergillus*, *Penicillium*, *Epicoccum*, *Fusarium*, *Trichoderma*, *Gliomastix*, *Memnoniella* and *Stachybotrys* occur frequently, as do the Zygomycete genera *Absidia*, *Mortierella*, *Mucor*, *Rhizopus* and *Zygorhynchus*. Common Ascomycetes are *Chaetomium* and *Gymnoascus*. Basidiomycete mycelium is common but is usually hard to identify since isolates rarely sporulate in culture. Myxomycetes, Chytridiomycetes and Oomycetes can usually be found in soils. There are also fungi which produce only mycelium and never sporulate such as *Rhizoctonia*.

As well as being the primary habitat of soil fungi, soil is also a repository of dormant fungal propagules including both those of true soil fungi and of fungi whose mycelial growth occurs elsewhere. Not only are spores abundant, but also sclerotia (page 171). Microsclerotia enter soil from leaf litter, where they are formed in the late phases of growth of phylloplane fungi. Some parasitic fungi are soil-borne (page 365), and for these the soil provides a reservoir of infection in which their propagules survive the absence of a host.

The nutrition of soil fungi depends ultimately on an input of organic material from plants. Estimates of the ratio between input of available carbon substrates and turnover of microbial biomass or microbial respiration in soil show that nutrients are quickly removed by microbial (fungal or other) activity and that soil is a nutrient-poor medium in which competition for limited carbon substrates is normal. A probable consequence is that soil fungi spend most of their time in a dormant state and develop into mycelial colonies only when nutrients occasionally become available.

Scarcity of nutrients is probably a major cause of the inhibitory effect of soil on germination of fungal spores, a phenomenon known as **mycostasis**. Sterilized soils do not have this effect, nor do soils to which nutrients have been added. However, mycostasis can be restored to a sterilized soil by reinoculating it with fungi. Mycostasis is probably due to the rapid removal of nutrients by soil microbes. When spores begin to germinate, they usually exude soluble nutrient substances. Washing with water to remove these causes faster loss by increasing diffusion, and stops germination. Soil microbes probably have the same effect, since they remove sugars quickly from the vicinity of spores, as is shown by the rapid evolution of $^{14}CO_2$ from soil when spores labelled with ^{14}C, or ^{14}C-labelled glucose, are added to non-sterile soil. A number of experiments with various microorganisms have confirmed that mycostasis can be mimicked by the removal of nutrients. It is possible that mycostasis can also be caused by soil microorganisms generating inhibitory compounds such as antibiotics, carbon dioxide, ethylene and ammonia.

A special type of inhibition of fungal growth can occur in some soils as a result of growing one crop plant over long periods. 'Take-all decline' is found where fields have been sown with wheat for many years. After an initial increase in disease incidence the soil becomes inhibitory to the growth of the causal fungus, *Gaeumannomyces graminis*, on wheat roots, and the disease declines. Such soils are said to be 'suppressive' to the pathogenic fungus.

Many soil Ascomycetes and mitosporic fungi produce antibiotics, of which a large number are known, although most are too toxic to humans to be of therapeutic value. Production of a specific antibiotic tends to be limited to one or a few species and, as with other secondary metabolites (page 513), they tend to be produced when carbon sources are abundant but growth is limited by lack of other essential nutrients. It is uncertain what role these substances have in the life of the fungi. The generally low concentration of nutrients in soil, and a difficulty in detecting naturally produced antibiotics, have been put forward in support of the view that antibiotics may not be of importance in nature. However, when a fungus reaches the occasional nutrient rich particle, an ability to suppress the growth of competitors by localized antibiotic production, which would not be easy to detect, would be valuable. An ability to tolerate the antibiotics produced by other species would also be of value, and a species that produces antibiotics is often resistant to antibiotics produced by other species.

The characteristics of soil Ascomycetes and mitosporic fungi are paralleled in an interesting way by those of Streptomycetes. These bacteria grow in the form of hyphae, although both hyphae and colonies are smaller than those of fungi. They sporulate freely, produce and tolerate antibiotics, and have many different kinds of extracellular enzymes that they produce abundantly. Streptomycetes probably have a similar ecology to soil Ascomycetes and mitosporic fungi, while their comparatively small size enables them to grow on substrata too small to be useful to fungi.

Agricultural soils have been more studied by mycologists than soils of natural ecosystems. They commonly differ from natural soils in containing artificially high levels of phosphate, nitrogen and potassium added as fertilizer, and lime added to raise pH. They are also subject to monoculture and disturbance by cultivation. About 25 000 different fungal species have been isolated from agricultural soils. Because soil fungi interact with the plant community above them, it is to be expected that farming and forestry will change the soil fungus flora. The nature of such changes is important in relation to plant productivity and its sustainability.

When the fungal populations of cultivated and uncultivated ecosystems are compared it is found that the species composition and physical structure of the soil fungi are altered by cultivation. Frequently disturbed arable soils tend to have fungal hyphae of shorter lengths. The absence of the surface layers of dead plant litter that build up at the soil surface in most natural ecosystems, means that the litter-inhabiting fungi and animals that live in such layers are absent. This is likely to affect nutrient cycling, since it is established that mycelium in litter layers can capture, transport and redistribute nutrients such as nitrogen, phosphorus and carbon compounds. For example, fungal hyphae have been shown to transport soil nitrogen up into the cellulose rich litter layer, promoting decomposition. Removing the litter layer may therefore remove some of the ability of the ecosystem to capture and retain soluble nutrients as they are released by decomposition, leaching and animal excretion (Fig. 6.15). Even without ploughing, chemical changes due to fertilizers and lime can alter species

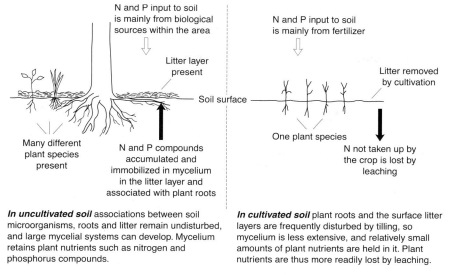

Figure 6.15 Comparison of plant nutrient cycling in cultivated and uncultivated soils. In cultivation, plant litter is removed by ploughing, and fertilizers are added to the soil. Although crop growth is enhanced, there is a loss of organisms in the litter layer, mainly fungi and insects. As a result, nitrogen compounds, being less immobilized in fungi and other soil organisms, are more easily washed out of the soil by rain. (After Swift M. J. & Anderson J. M. (1996) and Dighton, J. (1997); see references.)

composition. A comparison of limed and fertilized sheep-grazed upland soils with similar, acid soils left untreated and lightly grazed found that the more intensive farming regime produced a more fertile soil. In this soil microbes were more abundant, the proportion of bacteria to fungi was higher, and no litter layer of dead plant material developed. Without fertilizer, liming and intensive grazing, the acid soil had a higher proportion of fungi and slower decomposition, with dead, inactive fungal hyphae as well as living ones. The composition of the fungal population was also affected, with different species predominating under the different treatments.

Cultivation has been shown to reduce overall diversity of the fungi in soil. The diversity of VAM, mycorrhizal fungi (vesicular–arbuscular mycorrhizal fungi, p. 397) in cultivated soil under monoculture and woodland soil in seminatural ecosystems was compared. A hundred samples of VAM from arable soils yielded 99 fungi belonging to the genus *Glomus*, and one from another genus. By contrast, 154 fungal samples from the seminatural ecosystem yielded VAM fungi from three genera, with types much more evenly distributed between genera. This apparent decline in mycorrhizal diversity with cultivation could be one reason why it is difficult to re-establish diverse plant communities on formerly cultivated land. A direct effect of endomycorrhizal species diversity on the diversity and productivity of the plant community was found when soils were inoculated with from one to fourteen different VAM fungi and the plots sown with a mixture of seeds from a grassland plant community. Positive correlations were subsequently found between VAM diversity and the diversity, shoot, and root biomass of the plants growing on each plot. A rise in plant phosphorus and decrease in soil phosphorus suggested that plants acquired nutrients more efficiently as mycorrhizal diversity increased (Fig 7.15).

Wood

Wood which is still active in a living plant has defences against fungal attack (pages 408–416), and its functioning xylem (page 371), with cells filled with water, is probably too wet to support fungal growth. Some wood-inhabiting fungi can invade living trees, overcoming or avoiding their defences, and are thus parasites. Other fungi can only colonize dead wood. *Armillaria mellea* is an example of a parasite which can invade and kill a woody host and then live for many years as a saprotroph, utilizing the dead wood as a food source. Different wood-inhabiting fungi are specialized to occupy a wide variety of niches, using different tissues, cells or chemical components of wood as food, tolerating different antifungal substances, living on the wood of different types of tree, and preferring different combinations of physical variables. For example, Ascomycetes such as *Ceratocystis* species, the 'blue stain' fungi, utilize soluble nutrients, but do not decay wood; some of these are the ambrosia fungi that provide food for bark beetles (page 444). Others, including the Ascomycetes and Basidiomycetes that cause white and brown rots, grow inside the wood, their hyphae extending into medullary rays, entering xylem vessels (page 371), decaying the wood and forming large colonies lasting for decades. These colonies may originate from spores or hyphae already present in living wood as harmless endophytes (page 405) and only start to degrade the wood after it is dead. Some endophytic wood-decaying fungi are specific to particular tree species, for example *Biscogniauxia* (*Hypoxylon*) *nummularia* is a specific endophyte of beech, *Fagus sylvatica* (Plate 2) but others have a broad range of hosts. They are responsible for decay of wood in the forest canopy and the phenomenon of 'self-pruning' where trees shed whole branches which have died and decayed *in situ*.

Wood is a very good source of carbon compounds but other nutrients such as nitrogen and phosphorus are in relatively short supply, leading fungi which live on dead wood to evolve specialized means of acquiring them. Examples of such presumed adaptations include the nematode trapping found in members of the Basidiomycete wood-decaying genus *Pleurotus* (Fig. 7.23F), mycoparasitism (page 448; Fig. 7.25, Plate 5), formation of mycorrhiza (page 400), the ability to locate and invade colonies of saprotrophic bacteria, and formation of associations with nitrogen-fixing bacteria. As the fungal colony grows larger, phosphorus and nitrogen compounds are transported from one part of it to another. Colonies living on long-lasting woody resources such as fallen logs and tree stumps can continue to grow and develop for decades, forming clones of interconnected hyphae. Some such colonies remain within the resource in which they are initiated, but others have the ability to grow over the surrounding non-nutrient area and to locate and colonize fresh resources. This usually involves the development of mycelial cords (strands or rhizomorphs, page 342), which develop in mycelium extending from the old resource, and become consolidated by localized growth of new hyphae (Fig. 6.16) after the mycelium has begun to

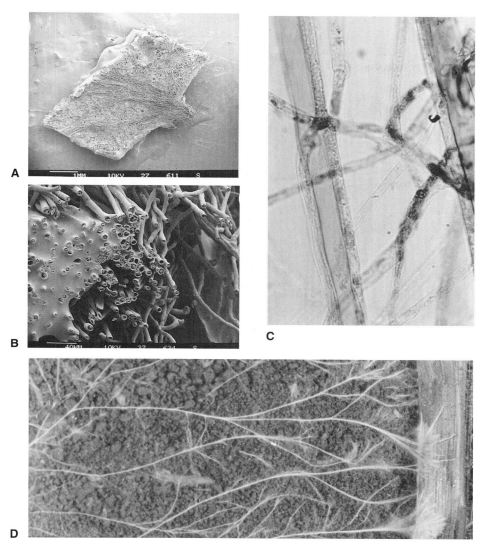

Figure 6.16 A, A scanning electron micrograph of mycelium of *Serpula lacrymans* in which a cord is forming by aggregation of hyphae behind the margin of the mycelium. B, A higher magnification view of the same cord after it had been frozen and fractured transversely. Longitudinal hyphae can be seen, some of which are empty 'vessel' hyphae, others narrow thick-walled fibres, and yet others with dense cytoplasmic contents. There are also large spaces between the hyphae and some of the hyphae are embedded in extracellular matrix material. C, Mycelial cords beginning to form in mycelium growing on the surface of a glass microscope slide. Two differentiated types of hypha can be seen. Staining with Nile blue reveals the presence of purple-staining lipid material in the wall of the wide hypha. A younger hypha with more abundant cytoplasmic contents has begun to adhere to the wider one. Further accretion of hyphae, accompanied by differentiation of very thick-walled fibres and the deposition of extracellular material, leads to the production of cords of the type seen in A. D, Cords of *Agrocybe gibberosa*, giving rise to colonizing fans of mycelium when a resource, a straw, is reached. (D, reproduced with permission from Robinson, C. H., Dighton, J. & Frankland, J. C. (1993). Resource capture by interacting fungal colonies on straw. *Mycological Research* **97**, 547–558.)

colonize the new resource. A bridge of mycelium develops which may enable scarce nutrients such as nitrogen and phosphorus to be relocated from decayed to newly colonized pieces of wood (Fig. 6.17). The colony can thus progress indefinitely from one resource to another, for example when colonizing fallen wood on the forest floor. Clones of *Armillaria* growing in forests have been found to extend for hundreds of metres.

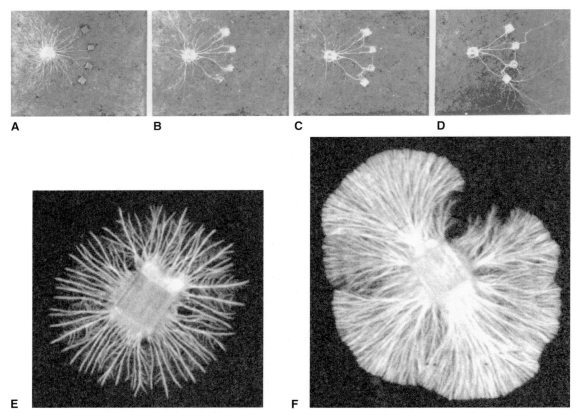

Figure 6.17   Foraging development of the mycelium of saprotrophic woodland fungi growing on the surface of compacted, but not sterilized, soil. Small blocks of wood colonized by mycelium, and 'baits' consisting of similar uncolonized blocks, were placed on the soil to mimic the effect of scattered pieces of fallen wood on the forest floor. The nutrients for growth of the fungus are derived from the wood block and carried along inside the cords to the growing margin. The mycelium was photographed at intervals. The fungus *Phanerochaete velutina*, which grows on large dead sticks and branches, is shown in A–D, which were taken over a period of six weeks after the experiment was set up. Initially the mycelium extended equally in all directions, but when a new resource was encountered cords developed in the mycelium, followed by regression of the mycelium which had not located a new resource, with further exploratory growth occurring only from the newly colonized bait. E and F show the different mycelial growth patterns of two woodland fungi, *Phallus impudicus*, the stinkhorn, (E) which grows mainly as aggregated cords, and *Hypholoma fasciculare*, the sulphur tuft, (F) with a diffuse margin of densely branched hyphae. The use of a standardized microcosm such as these trays of unsterilized soil, allows such differences in form, induced in realistic conditions by natural stimuli, to be quantified by computerized image analysis. The degree to which the mycelium is diffuse or aggregated can be quantified in terms of its fractal dimension. (Reproduced with permission from: A–D, Boddy, L. (1993). *Mycological Research* **97**, 641–645; E–F, Boddy, L., Wells, J.M., Culshaw, C. & Donnelly, D. (1999). *Geoderma* **88**, 301–328.)

The formation of cords by woodland Basidiomycetes is an example of how mycelium can develop so as to exploit an initially distant resource, fallen wood. This behaviour has been termed 'foraging' by analogy with the pursuit of food by animals. It will be subject to the same selection pressure to become more efficient, the process known as 'optimization of foraging' in animal ecology. Although some fungal structures such as germ tubes are capable of directed growth – chemotropism – in response to external stimuli, such responses have rarely been demonstrated in the hyphae of mature vegetative mycelium. However, when a mature mycelial system of a woodland cord former is examined it appears to connect separate pieces of wood by thick cords, with very little visible mycelium apart from these. This localization of mycelium is achieved as a result of preferential cord development around hyphae which have encountered resource units by chance in the course of extending across soil from a previously colonized unit. The process has been reproduced in microcosms with many different species of woodland cord formers, leading to the discovery that different species have different foraging strategies. Some species build cords in response to encounters with quite small resources, whereas others appear to ignore small resources and continue extension until they find a larger unit (Fig. 6.17). Contact with a fresh food source stimulates cord development all along the newly connected hypha. The first response is branching of the hypha making contact, quickly followed by changes in the polarity of hyphal extension, and a colony-wide coordinated response in which unconnected mycelium regresses and its resources are presumably redirected into the part of the colony which has located a food source.

There is thus a reallocation of the total biomass of the system. The spatial redistribution of mycelium and the extent to which it is aggregated into cords or spread out as diffusely branching hyphae can be quantitatively demonstrated by fractal analysis, digitized images being analysed for their fractal dimension, a measure of the degree of space-filling. It appears that a signal initiating cord formation travels through the connected mycelium from the newly encountered food source. Its nature is obscure, but there is a possibility that it could be electrical. Action potentials have been reported in cords and rhizomorphs, increasing in frequency when fresh nutrients in the form of wood or nutrient agar blocks – but not inert blocks of perspex – were placed on the mycelium. Cord development can be induced by experimental manipulations which make existing hyphae the main source of nitrogen for new growth. This suggests that a raised level of nitrogen inside the hyphae could have a role in cord initiation. Since nutrients are transported within cords, nutrient status could contribute to the signal which passes along connecting hyphae from freshly encountered food resources.

The morphology of all the cords and rhizomorphs which have been studied is similar (Figs 6.16B and 6.18). All have wide, apparently empty hyphae surrounded by extracellular matrix in which hyphae with denser cytoplasm and peripheral longitudinal fibres are embedded. They thus appear to be adapted for the rapid transport of nutrients and water. However the mechanism of transport is still obscure. Different substances can travel simultaneously in opposite directions along the same cord, indicating at least two separate channels. Mass flow of water and solutions occurs, presumably in the wide aseptate 'vessel'

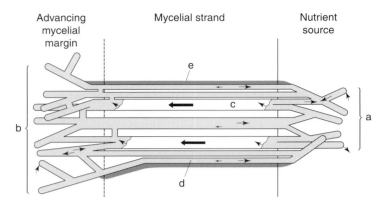

Figure 6.18 Diagram showing the relationship thought to exist in mycelial cords between the sites of nutrient uptake from wood (a), the pathways of transport within the cord, and the growing hyphae at the mycelial margin (b) where the transported nutrients are utilized for growth. The observation that different substances can move simultaneously in opposite directions in the same cord suggests the existence of separate channels. Vessel hyphae (c) are believed to be channels for pressure-driven bulk flow, with cytoplasmic transport occurring in narrower cytoplasm-rich hyphae (d). Outer layers of the cord (e) consist of fibres embedded in extracellular matrix which insulate the cord from the soil over which it grows. (After Cairney (1992). *Mycological Research* **96**, 135–141. By kind permission of Cambridge University Press.)

hyphae that run through the centre of cords. Interconnecting systems of cytoplasmic vacuoles and tubules have recently been found in hyphae of many different fungi. The tubules show peristaltic movements which carry their contents along the hyphae, and could therefore provide a cytoplasmic pathway for transport in cords. In natural environments many different wood-inhabiting fungi are likely to be present at the same time in a single woody resource (Fig. 6.19A). They remain as separate competing individuals. Even colonies of the same species arising from different spores usually remain separate, because of vegetative incompatibility (page 256), and are often separated by striking zones of melanized mycelium. 'Spalted' beech wood in which early stages of fungal colonization have produced a tracery of black melanized lines is prized by wood turners for its decorative effect.

Rhizosphere and rhizoplane
Organic nutrients for the soil mycoflora come not only from the decomposition of plant litter deposited on the soil surface and from dead roots but also from the living roots that are present throughout the soil. The advancing tip of a plant root can range from *ca* 40 μm up to *ca* 1 cm in diameter and it moves through the soil by increasing in length just behind the apex at rates of up to *ca* 1 mm a day. Covering the apex are the root cap cells, which are damaged and shed into the soil as the root advances, and extending into the soil are unicellular, delicate extensions, the root hairs, which are also continuously replaced by new growth (Fig. 7.1C). Roots also release polysaccharides and mucopolysaccharides to give a coating, sometimes termed 'mucigel', which probably lubricates their passage

Figure 6.19 Fruit bodies of wood decay fungi. A–C, Basidiomycetes. A, Stump with abundant *Coriolus versicolor*, especially emerging through the bark, the large fruit bodies of *Pleurotus ostreatus* (the gastronomically attractive oyster mushroom), *Trametes gibbosa* and *Bjerkandera adusta*. B, Massive fruit body of *Ganoderma adspersum* at base of trunk broken in storm. C, *Fomes fomentarius* on fallen trunk. D, The Ascomycete *Xylaria hypoxylon* on fallen twig. The whiteness of parts of the stromata is due to conidia. Later the conidia will be replaced by perithecia. (A–D, John and Irene Palmer.) Further illustrations of fruit bodies of wood decay fungi are available at http://www.wisc.edu/botany/fungi/html

through the soil. Other substances released by roots include a wide variety of high- and low-molecular-weight compounds – proteins, amino acids, organic acids and sugars. Such losses may be substantial. Wheat roots can lose up to 0.15% of their total assimilated carbon per week. The presence of soil microorganisms increases this loss; roots of wheat, clover and rye grass release about twice as much exudate into normal non-sterile soil as they do into sterilized soil. The water content of the soil also affects loss of substances from roots, the rate being greatly increased under dry conditions.

The advance of roots through the soil results in continuous changes in the spatial distribution of nutrients in the soil, hence the fungi, bacteria and animals utilizing nutrients derived from roots will vary in abundance and activity. Some are closely associated with the root, for example the bacteria embedded in the mucigel, symbiotic mycorrhizal fungi and the pathogens including nematodes and fungi which invade root tissues from the soil. Other organisms are not in physical contact with the root but are influenced by substances from it, some of which inhibit the growth of some fungi and stimulate that of others. Germination of some fungal propagules is affected by root exudates. Thus the sclerotia of *Sclerotinia cepivora*, a fungus which invades onion roots, causing white rot of onions, are dormant until exposed to root exudates of plants belonging to the onion family. The active components of the exudates are unsaturated thiols of low molecular weight. Plant root exudates also include siderophores (page 156) which can limit the growth of microorganisms in the vicinity by reducing the availability of iron. The region near the root in which other organisms are influenced by its activities is known as the **rhizosphere**. The habitat provided by the root surface is called the **rhizoplane**. The activities of mycorrhizal and plant disease fungi in the rhizoplane and rhizosphere are of great importance to the growth of plants, and these habitats have been the subject of many investigations (see Chapter 7).

Water

There are some fungi which spend their whole life cycle under water, and have clear adaptations to an aquatic way of life. Thus the Chytridiomycetes, with their swimming zoospores, appear to have spent their whole evolutionary history as inhabitants of water. Other fungi found in water may spend only part of their life cycle in it, and be so clearly related to groups recognized as terrestrial as to suggest that their aquatic way of life was adopted late in their evolution. There are far fewer species of aquatic fungi than terrestrial ones. However, fungi can be found in lakes, rivers, ponds, puddles, and in water-filled interstices of soil, and also in salt water – saline lakes, the open sea, the ocean depths, estuaries and the intertidal zone of the seashore. Some are saprotrophic but others live on plants or animals or other fungi. They are all microscopic or very small, the largest structure produced by any fungus in water probably being the ascocarps, measuring a few millimetres across, of marine Ascomycetes. Which species are found in a particular habitat depends on the type of substrate generated by the primary producers of that habitat, by the osmotic potential, redox potential, ionic content of the water and physical conditions such as temperature and whether the water is flowing, turbulent or stagnant. It is not surprising therefore that there is a seasonal fluctuation in the abundance of aquatic fungi, best documented for those of temperate fresh water.

Zoospore-producing fresh water fungi   Many Oomycetes and Chytridiomycetes are inhabitants of fresh water. The methods used for investigating such fungi differ from those used to isolate and identify the fungi of soils. This is because the aquatic fungi are more sparsely distributed, their mycelium is ephemeral and limited in extent, and direct observation of water samples shows only the occasional zoospore. However, although the zoospores are sparsely distributed their overall contribution to the population of microorganisms in water can be considerable. For example, the number of zoospores in a temperate lake in summer has been found to be as much as 5000 per litre of water, equivalent to an annual wet volume of 0.3 cm^3 m^{-3}.

Placing samples on to a nutrient culture medium is not useful for isolating aquatic fungi, as they are slow growing in comparison with the mycelium which develops from the spores of terrestrial fungi that have fallen into water. Several methods have been used to count zoospores. A simple, effective way of isolating fungi with swimming zoospores is baiting with particles of plant or animal material. Many Chytridiomycetes are able to attack refractory carbon sources, with species varying in which ones they are able to utilize. This enabled frequency of isolation by baiting to be used in a study of Chytridiomycete distribution. Three different kinds of substrate were used as bait: cellulose (cellophane), chitin (insect exoskeletons) and keratin (cast snakeskin). Samples of water were taken either from open water or from the muddy shore of a lake. The baits were colonized after a few days by chytrids which could then be identified under the microscope. Each kind of bait was colonized by characteristic species, probably reflecting differences in the capacity of different species to produce the appropriate enzyme to lyse the insoluble substrate. The depth of water was also important in determining which species was isolated, some being found only at the shore and others only in deep water. Also, the overall frequency of isolation was greater in water samples from the shore where the abundance of plant and animal fragments probably supports a large Chytridiomycete population and gives rise to numerous zoospores in the water and mud of the lake margin. In large lakes the number of zoospores at the margin has been found to be about ten times that at the centre. The fact that the bait was colonized within a few days and that the water sample contained no visible fungal material suggests that it was zoospores rather than resting spores which were the main colonizing agents. Chytrids such as *Olpidium* which colonize pollen grains increase in numbers when pollen is deposited on the surface of a lake, as happens in spring in lakes in coniferous forests. Other studies show that numbers of zoospores fluctuate with the seasons. For example, soil samples were baited monthly over the course of a year, to measure the frequency of isolation of different members of the Saprolegniales, and it was found that for some species this was greatest in spring and autumn. The cause of this is not known, although it is probably due to seasonal changes in climate and substrate availability. Fungi with specific substrate preferences will vary in abundance with their substrate and may become locally very numerous as a result. This is obviously so with parasitic fungi. A well-studied example is the pattern of fluctuation in numbers of the chytrids which are obligate endoparasites of diatoms (unicellular photosynthetic algae).

The diatom *Asterionella formosa* is common in lakes of the English Lake District where its population size was measured at monthly intervals over a period of several years. During the spring and summer months there were two or three peaks – 'blooms' – in diatom numbers, and each time the subsequent decline was accompanied by an increase in the proportion of diatom cells invaded by a chytrid parasite. It seems likely that there is a causal relationship between infection of the diatoms by the chytrid and their decline. Such a pattern could be the result of increase in diatoms in the summer sunlight followed by attack by chytrid zoospores. A small inoculum, maybe coming from resting spores, infects a few diatoms at first and then as each established parasite produces more zoospores an epidemic develops which eventually overwhelms the diatom population. This study showed how aquatic fungi, although so inconspicuous, can be an important element of an aquatic ecosystem.

The success of a baiting technique itself constitutes an example of the increase in numbers of a fungus which can be brought about by the presence of a suitable substrate, and similar enrichments are well known to occur in nature. However, fungi may be localized in a habitat not because of the specially suitable conditions it provides for growth but because it imposes constraints that only a few species can tolerate. Some aquatic fungi are found only in habitats where there is little oxygen. Such fungi may be obligately fermentative (Table 3.8, page 146). An example is the tropical Oomycete *Aqualinderella fermentans* which has homolactic fermentation without an oxidative pathway. Other fungi are found in habitats with very low oxygen tension although they grow better in culture if supplied with air. The Oomycete genus *Leptomitus* can tolerate a lower oxygen tension than most fungi and is unique among known fungi in being unable to utilize glucose as a carbon source and having a requirement for fatty acids. These peculiarities probably explain why *Leptomitus* is often isolated from water where pollution with sewage both provides fatty acids and results in a low oxygen tension as a result of the activity of abundant heterotrophic bacteria.

Higher fungi of fresh water   There are a considerable number of species of higher fungi which grow and sporulate in water, although their ability to do so involves the secondary adaptation of structures which originally evolved in a terrestrial environment, such as ascocarps with forcibly discharged spores, basidiospores and conidia. These aquatic fungi include a few Basidiomycetes, some Ascomycetes and many mitosporic fungi. The latter are classified as Hyphomycetes (page 69) and referred to as aquatic Hyphomycetes. Some aquatic Hyphomycetes grow on blackened leaves in stagnant water. These have conidia of a coiled (Fig. 4.2F) or basket-like shape which trap gas bubbles and enable spores to float. Many others are found in streams, and these commonly have tetraradiate spores with four equally spaced arms. Such tetraradiate spores (Fig. 4.2D) can be found easily by microscopic examination of the foam which collects at the downstream end of rapids and waterfalls. Their shape leads them to collect in the foam, between the bubbles. They form the main solid component of stream foam, and indeed the largest fungal component of fast-flowing water. The other consequence of their somewhat anchor-like shape is that they impact effectively against immovable objects in the stream. This is probably the functional advantage which has led to the evolution of this striking spore shape in different taxonomic groups. Experiments to compare efficiency of impaction of tetraradiate spores and smoothly shaped spores demonstrated this advantage and also showed that another possible effect, that of slowing sedimentation and thereby allowing a longer time for dispersal, was negligible in fast-flowing water. Tetraradiate spores of marine fungi are also found in sea foam, and spores of this shape are not confined to fungi but produced by some aquatic plants which grow and disperse their spores in fast-moving water. The tetraradiate spores of stream fungi have the capacity to sense when they impact on a surface, and germinate immediately. If a sample of foam is collected in a glass vessel the spores will develop germ-tubes as soon as they touch its sides, so presumably this is a response to contact rather than to a chemical substance. The numbers of tetraradiate spores in streams vary with the seasons and in temperate regions are highest in autumn and winter, when the autumn leaf fall supplies these fungi with their substrate, submerged leaves. Leaves which fall into streams lose their terrestrial phylloplane inhabitants such as *Aureobasidium pullulans* within a few weeks and become colonized by aquatic Hyphomycetes. During the winter these fungi play a large part in decomposing the submerged leaves. Studies in culture show that they can utilize simple sugars, cellobiose and starch, and some can attack cellulose. An important determinant of their place in the ecosystem is that, unlike Oomycete fungi, they do not require nitrogen in the form of organic compounds such as amino acids, but can utilize nitrate, present in streams from soil leaching. As a result of this, colonization of dead leaves improves their content of nitrogenous nutrients in a form which can be used as food by the aquatic fauna such as water shrimps. The fungi and animals together remove the leaf remains from the water in a way comparable to the joint activities of fungi and soil animals in the terrestrial forest ecosystem described above (page 336). About a hundred species of aquatic conidial fungi have been identified by their conidia. Some teleomorphs (sexual stages) have been discovered and usually found to be those of Ascomycetes, producing ascocarps with spores discharged into the air. A few types of tetraradiate conidia whose teleomorphic stages are not known are nevertheless suspected of having a Basidiomycete affinity because they have septa with dolipores (page 58) or what appear to be clamp connections.

The discovery of species with an aquatic anamorphic stage and a terrestrial teleomorph suggests that members of this taxonomically diverse group of fungi have evolved an amphibious mode of life suited to environments where large amounts of plant material are regularly shed into rivers and streams. The ability to produce airborne spores as well as waterborne ones would explain the otherwise puzzling observation that species identifiable by their tetraradiate spores in water can extend their area of geographical distribution upstream. Also, preinoculation of fallen leaves with aerial spores of aquatic fungi before they enter the water could account for the very rapid build-up of numbers of aquatic spores soon after leaf fall. Aquatic Hyphomycetes are widely distributed. Different species occur in acid and approximately neutral waters, and there are differences in the aquatic Hyphomycete flora of temperate and tropical areas.

Marine fungi   The largest aquatic habitat on earth is the sea. However, only about 500 species of fungi – less than 1% of the species known – have been found in the various habitats provided by the sea and seashore. There are about a hundred species of marine zoospore-forming fungi, most belonging to the Pythiales or to the exclusively marine genus *Thraustochytrium*, a member of the Labyrinthomycota (page 553). There are also higher fungi. A recent monograph listed 149 species of Ascomycetes, 56 mitosporic fungi, 4 Basidiomycetes and 177 yeasts. About 90% of the fungi found in the sea require sea water for growth and sporulation, and so are obligately marine. The vestigial nature of such land-adapted features as ascospores and basidiospores suggests that marine fungi, like those of fresh water, are secondarily adapted to life in the sea. An obvious problem in such adaptation must be the physiological and structural changes necessary for a fungus to grow in salt water, and especially in estuarine areas of fluctuating salinity. Fresh water has a low ionic content, and is high in concentrations of inorganic nitrogen and fixed carbon. Sea water, on the other hand, is low in inorganic nitrogen and fixed carbon, but has a high ionic content. It typically contains 35 g l^{-1} total salt, and of this 99.5% is in the form of a few ions present in remarkably constant proportions: Cl^-, SO_4^{2-}, HCO_3^-, Na^+, Mg^{2+}, Ca^{2+}, K^+ and Br^-. Thus in adapting to grow in sea water fungi have to overcome two problems. First, they have to maintain the internal turgor which they need in order to maintain growth at the hyphal tips, so that as the external water potential decreases with increasing salinity the fungal cells have to compensate by generating a correspondingly high intracellular concentration of osmotically active small molecules or ions. It has been shown, for example, that the marine Ascomycete *Dendryphiella salina* increases its internal glucose concentration, together with potassium ions, as the concentration of sea water in the medium is increased. Fungal life at low water potentials is discussed earlier (pages 158–164). Secondly, metabolic enzymes have to function in a cellular environment with ionic concentrations very different from those that occur in the cells of freshwater fungi. Fungi seem to have solved these problems with varying degrees of success, since species show a zonation in places such as estuaries where there is a gradient from sea to fresh water. Thus a study of fungi in the Weser estuary showed that they could be sorted into four categories depending on the saltiness of the water in which they were found. Some were found in water in the range 0.5–30% of the salinity of sea water, others throughout the brackish zone of the estuary into sea water, others could only tolerate a salinity up to and including 20% sea water, and a fourth group were found only at a fixed high salinity and did not appear to be able to tolerate any changes in salinity.

Apart from salinity, the distribution of marine fungi, like that of all aquatic fungi, is largely determined by the whereabouts of substrates. These are mainly plants or plant-derived material, although any organic material or organism is likely to serve as food for some species. Ascocarps of Ascomycetes may be found on wood which is submerged by sea water some of the time, like the timber of breakwaters. Others grow on permanently submerged wood. There are even a few species such as *Periconia abyssa* which grow at depths of up to 4000 metres, tolerating high pressure, low temperature and darkness. Wood is fairly common in the deep sea, where it collects in isolated piles. A few Basidiomycete fungi also grow on submerged wood. Some of the wood-inhabiting marine fungi degrade the cellulose of the wood cell walls causing a soft rot (see page 319). The fungi of wholly submerged substrates release their spores under water. Asci commonly release spores by dissolution of the ascus wall rather than by explosive discharge, and the ascospores usually have spiny or hair-like appendages which help in sticking the spores to the surface of a substrate (Fig. 4.2E). There is enough variation in appearance of these spiny spores for them to be useful in distinguishing between species. Some marine fungi, as mentioned above, grow on substrates which are only intermittently submerged, for example in the intertidal zone on the shore. Some of these discharge spores into the air when they are exposed by the retreating tide. There are many small organisms including microfungi which live in the interstices of sand grains on the beach, and among these are fungi which produce tetraradiate conidia like those described above for freshwater fungi of streams.

Almost a third of all known marine fungi are associated with seaweeds or plankton. Symbiotic associations include ones in which a fungus normally found growing in association with a red or brown seaweed can be isolated into pure culture, and others which are apparently obligate. Such associations are termed **mycophycobioses**. An example is the association of knotted wrack, *Ascophyllum nodosum*, with the Ascomycete *Mycosphaerella ascophylli*. The only other seaweed in which this endosymbiotic fungus is found is *Pelvetia canaliculata*. There are no intracellular invasions but the mycelium forms a network throughout the cortex and medulla, eventually producing ascocarps and spermogonia in the receptacles where the gametes of the seaweed are formed and released. It has been suggested that mycophycobioses are mutualistic. The fungus has the obvious benefit of a nutrient supply but the way the plant host benefits is less clear. One possible advantage might be that the fungal mycelium provides an internal long distance transport system which the seaweed otherwise lacks.

A more familiar symbiosis between photosynthetic cells and fungi is that of lichens (page 449, Plate 1) in which a fungus, usually an Ascomycete, and a microscopic unicellular photosynthetic organism are so closely interrelated both structurally and physiologically that the resulting combined organism resembles neither of the partners. Some lichens are marine and are found only near the sea, growing on rocks and other solid objects near the shore. There are other relationships between seaweeds and fungi that resemble that seen in lichens. For example, an association which has been found all over the world involves a fungus, *Chadefaudia corallinum*, growing intercellularly with red algae from the genera *Dermatolithon* or *Epilithon*. These algae grow as a crust over the surface of larger seaweeds which provide a firm surface although the larger seaweeds do not appear to be involved in the symbiosis, as the fungus does not penetrate them. There is little or no experimental evidence on the nature of the advantage to the red alga. It is possible that the ability of the fungus to scavenge nutrients from oligotrophic environments may play a part, as it does in terrestrial mycorrhizal associations.

Fungi in Biodeterioration

Biodeterioration is the destruction or spoilage of useful materials or artefacts by living organisms. It is a topic of great economic importance, and the subject of books, journals and international conferences. Fungi are very important agents of biodeterioration. In nature, as discussed above, fungi are active in the breakdown of plant materials, especially cellulose and lignin. The attributes that enable them to do this, such as penetrating hyphae and powerful extracellular enzymes, also enable them to attack wooden structures and stored timber (Table 6.4), cotton textiles, paper, and stored food, provided that there is sufficient air and water to permit fungal growth.

Table 6.4   Some economically important timber-decay and timber spoilage fungi

Organisms	Mode of attack	Economic importance
Basidiomycetes		
Serpula lacrymans	Brown rot	Causes 'dry rot' in interior timbers in temperate climes
Armillaria mellea	White rot; tree pathogen	Common tree pathogen and can decay damp felled timber. Difficult to control effectively, spreads by rhizomorphs
Coriolus versicolor	White rot	Causes serious decay of felled hardwoods. Detoxifies pentachlorophenol preservative
Pleurotus ostreatus	White rot	Causes decay of stored wood and is a pathogen of deciduous trees
Lentinus lepideus	Brown rot	Decays wood in contact with the ground, e.g. utility poles. Tolerant to tar oil
Phlebiopsis gigantea	Brown rot	Decays felled pine logs
Coniophora puteana	Brown rot	Decays building timbers, and wood in ground contact. Resistant to zinc and cadmium compounds in preservatives
Stereum sanguinolentum	Brown rot; tree pathogen	Decays dead stumps, logs of conifers and is also a tree pathogen
Gloeophyllum trabeum	Brown rot	Prevalent in exterior woodwork. Tolerant to copper/chrome/arsenic and pentachlorophenol preservatives
Ascomycetes and mitosporic fungi		
Phialophora fastigiata	Soft rot	Decays wood chips and also stains wood
Chaetomium globosum *Paecilomyces variotii*	Soft rots	Decay wood in ground contact
Trichoderma viride	Surface mould	Ubiquitous mould; early colonizer of timber
Cladosporium spp. *Penicillium* spp.	Surface moulds	Ubiquitous moulds; early colonizers of timber
Aureobasidium pullulans	Sap stain	Common sap stain fungus. Detoxifies tetrabutyl tin oxide
Ceratocystis pilifera	Blue stain	Common blue stain fungus and also causes soft rot in wood chip piles

Fungi are the main decomposers of timber in natural environments and will rot damp wood whenever it becomes available to them. Wood decay fungi are thus serious agents of biodeterioration of timber, in the standing tree, post-harvest and in service for various purposes including building components and transmission posts. Most fungi will only attack wood that is not too dry (below 20% of its moisture holding capacity) or too wet (above 80%) and therefore anaerobic, so fungal decay can be controlled if the wood can be kept dry. The dry rot fungus of timber, *Serpula lacrymans* (Plate 8), is unusual in being able to spread in buildings which have a source of damp in one place, by transporting water in its cords to wood in other parts of the building. It causes damage worth hundreds of millions of pounds in Britain alone. Wood from different tree species varies in its susceptibility to fungal attack, the timber of some having a high degree of natural durability due to the presence of antifungal compounds such as phenols laid down by the tree in its heartwood. Other species have wood of low durability. Wood can be rendered inaccessible to fungal decay by preservative treatment. Mixtures of copper, chromium and arsenic salts, and creosote, make wood virtually immune to fungal attack even in permanently wet situations but obviously pose a toxic threat to the environment and to those responsible for handling them. Newer, less toxic preservatives are being sought. Efforts are also being made to limit use of the most toxic preservatives to timber intended for use under the conditions most likely to lead to decay, for example prolonged contact with damp soil which both wets the wood and provides a source of fungal inoculum.

Biodeterioration occurs everywhere, but is especially important in the humid tropics where warmth and dampness encourage fungal growth. Sometimes biodeterioration is simply the consequence of the extensive biodegradation of a material, as when the dry rot fungus *Serpula lacrymans* attacks the cellulose of timber, or *Trichoderma* that of textiles. Serious biodeterioration, however, can occur even when biodegradation is limited. Food, for example, can be rendered useless by even limited growth of fungi that produce poisonous secondary metabolites, **mycotoxins** (page 440, Table 6.4). Thus grain can be made toxic by the production of ergot alkaloids by *Claviceps purpurea* (pages 521–524), vomitoxin by *Fusarium graminearum* and rubritoxin by *Penicillium rubrum*, and groundnuts by aflatoxin from *Aspergillus flavus*. Other metabolites although not toxic may impart a taint or unattractive appearance. Fungal pigments, coloured hyphae or spores may spoil the appearance of materials, as occurs with sap stains of wood. These, though not affecting the strength of timber, render it less saleable. It has recently been found that fungi can cause biodeterioration of the surface of some kinds of building stone including marble. The masonry develops a film of microscopic, slow-growing fungi, which cause the surface to crumble by infiltrating their hyphae into the pores of the stone, mechanically forcing open microscopic cracks in the structure (Plate 4). Species which attack stone surfaces in this way are highly distinctive, with thick, melanized walls and frequent septa. They seem adapted to their exposed life, with frequent fluctuations in water potential and nutrient supply. Under damp conditions fungi can ruin valuable books and works of art. In the tropics fungi have grounded aircraft. Water inevitably accumulates below the kerosene in storage tanks for aviation fuel. *Hormoconis resinae* (formerly *Cladosporium resinae*) can grow at the oil/water interface. The resulting metabolic products can corrode aeroengines, and the mycelium block fuel lines. Optical glass, such as the lenses of cameras, binoculars and microscopes, can also be ruined in the humid tropics. Sparse fungal hyphae, utilizing nutrients in dust, grease or optical cement, can spread on to and across lenses, etching the surface with acidic metabolites. The above examples are only a selection of the varied destructive activities of fungi.

Fungal biodeterioration can be prevented by rendering conditions unfavourable for fungal growth, either by controlling the water and oxygen content of the environment, or by the use of fungicides as preservatives. Many chemicals traditionally used as fungicides, such as arsenic used for dry rot treatment, are powerful general biocides. The use of such toxic materials as preservatives is undesirable unless there is no alternative. The prevention of fungal growth has been discussed earlier (pages 172–180).

Environmental Change and Fungal Conservation

A major issue in recent years has been the loss of biodiversity resulting from environmental changes brought about by human activity. Publicity has centred on the decline in numbers and threatened extinction of such spectacular wildlife as the larger mammals and birds. The problem, however, is not limited to such conspicuous fauna, but extends to all of the major groups of organisms, including the fungi.

Evidence for Fungal Decline

Evidence for fungal decline is limited to those species that can readily be observed – higher fungi with large fruit bodies, and lichens – and to areas where extensive records of the location of 'finds' have been maintained for many years. The disappearance of lichens from areas of urbanization and industrialization has been widely recognized for approaching half a century. Since then it has become clear, first in Europe but more recently in the USA, that many macrofungi have become less abundant. In several countries there has been a decline in the number of species collected on fungal forays, or observed on replicate plots in woodland, ectomycorrhizal species being most severely affected (Table 6.5). The decline is further illustrated by records of the amounts of locally collected wild mushrooms supplied to the markets of some European towns. In Saarbrucken in Germany, for example, 6–8 tonnes of fruit bodies of a mycorrhizal fungus, the chanterelle (*Cantherellus cibarius*), were sold annually in the mid 1950s, but by the mid 1970s supplies were down to about 2% of the former amount. However, annual sales of the honey fungus (*Armillaria mellea*), a tree parasite, remained at about 4 tonnes throughout this period.

Possible Causes of Fungal Decline

Fungi are dependent, directly or indirectly, on higher plants for their survival. Habitats with many different plants will provide niches for far more fungal

Table 6.5 Evidence for the decline of ectomycorrhizal fungi. Expeditions to collect and record sporophores have been made annually in the same site in the Netherlands for many years. The resulting data on the abundance and diversity of sporophores of fungi known to be mycorrhizal symbionts, parasites or saprotrophs, indicate that mycorrhizal fungi have declined.

	Average no. species per plot			Average no. sporophores per 1000 m^2		
	1972–73	1976–79	1988–89	1972–73	1976–79	1988–89
Total mycorrhizal fungi	37	32	12	4720	1110	370
Saprotrophs on soil	21	33	22	2430	920	350
Saprotrophs and parasites on wood	15	15	26	180	330	680
Total fungi with large sporophores	73	81	60	7340	2350	1400

Data from Arnolds, E. (1991), In Hawksworth, D. L., ed. *Frontiers in Mycology*, pp. 243–264. Cambridge University Press, Cambridge.

species than will those with few. In recent decades tropical forests have given way to cattle ranches, temperate woodlands have been replaced with plantations and semi-natural grassland has been ploughed and reseeded with rye grass. Such habitat loss will inevitably have reduced the abundance of many fungal species and rendered some extinct.

A decline in fungal biodiversity has occurred in many areas even where habitats have been preserved. The extent of this decline parallels the level of atmospheric pollution. Fuel combustion by power stations, cars and aircraft releases sulphur dioxide and nitrogen oxides. The extinction of lichens downwind from sources of pollution appears to result mainly from the toxic effect of sulphur dioxide. The decline in mycorrhizal fungi, on the other hand, is likely to be an indirect effect of pollution. Trees on soils poor in nitrogen form mycorrhizal associations readily, and depend on the scavenging activities of the mycorrhizal fungi for their supply of nitrogen. In heavily polluted areas soils may receive 60 kg of nitrogen per hectare from rain, in addition to that which they may receive from spray drift or run-off from heavily fertilized fields. Under such circumstances the mycorrhizal partner is not needed and associations are less readily formed.

The collecting of wild mushrooms for the market has been traditional in Central and Eastern Europe with, for example, 400 tons being sold in Munich in 1905. However, as indicated above, in some of these countries the local supply of wild mushrooms has diminished, largely due to pollution, leading to a dependence on supplies from abroad. In other countries there has in the past been only limited picking of wild mushrooms, mainly by enthusiasts for their own consumption, but in some of these countries wild mushrooms have become very popular. The diminution of local supplies in some areas and increase in popularity in others has led to an enormous increase in commercial picking, often for export, in areas where wild mushrooms are still abundant. In 1992, for example, in the forests of Northwest USA and British Columbia about 10 000 pickers were involved in the collection of about 2000 tonnes of wild mushrooms, especially morels, chanterelles, *Tricholoma* spp. and boletes, and in 1996 over 5000 tonnes of wild mushrooms were exported from Serbia. The environmental consequences of large-scale commercial picking are controversial. There is evidence that the picking of fungal fruit bodies does not in itself diminish the ability of the mycelium to produce fruit bodies in subsequent years. However, trampling, and some collection methods, such as the use of rakes to harvest *Tricholoma*, may severely damage mycelium. Furthermore, diminished availability of fruit bodies may prejudice the survival of animals that are dependent on them for food – some insects are confined to a single fungal species.

Conservation Measures

Conservation can be considered in relation to the three actual or potential causes of fungal decline mentioned above. Habitat loss is of concern to all conservationists, and any measures that lead to the preservation of areas with a diverse flora will benefit fungi. However, 'hot spots' for the diversity of one group of organisms are not necessarily 'hot spots' for a different group, so information is needed on fungal distribution, especially of fungi on national or international Red Data Lists of threatened organisms. Storing and locating such information was at one time a problem, but an increasing number of organizations are now maintaining fungal record databases, which will also be important for monitoring the effectiveness of conservation measures. Pollution abatement will also have a favourable effect on fungal survival, and lichens have returned to some areas where pollution has been reduced. The impact of fungal collection can be diminished if damaging practices are avoided; to achieve this commercial pickers should be licensed and undertake to follow a code of practice. Some countries already have such licensing, and codes of practice for commercial and recreational collectors have been introduced by mycological societies and forest authorities. Commercial collection of fungi may even have positive implications for conservation, since profits from sustained yields may be greater, especially on poor soils, than those obtainable from clear felling and alternative use.

Further Reading and Reference

General Works on the Ecology of Fungi and other Microorganisms

Andrews, J. H. (1991). *Comparative Ecology of Microorganisms and Macroorganisms.* Springer Verlag, Berlin.
Atlas, R. M. & Bartha, R. (1998). *Microbial Ecology: Fundamentals and Applications*, 3rd edn. Benjamin/Cummins. Menlo Park, California.

Bolin, B., Degens, E. T., Kempe, S. & Ketner, P., eds. (1979). *The Global Carbon Cycle. Scientific Committee on Problems of the Environment, of the International Council of Scientific Unions, Vol. 13.* Wiley, Chichester.

Carroll, G. C. & Wicklow, D. T., eds. (1992). *The Fungal Community: its Organisation and Role in the Ecosystem*, 2nd edn. Marcel Dekker, New York.

Cooke, R. C. & Whipps, J. H. (1993). *Ecophysiology of the Fungi.* Blackwell, Oxford.

Copley, J. (2000). Ecology goes underground. *Nature* **406**, 452–454.

Dix, N. J. & Webster, J. (1995). *Fungal Ecology.* Chapman & Hall, London.

Lynch, J. M. & Hobbie, J. E., eds. (1988). *Microorganisms in Action: Concepts and Applications in Microbial Ecology*, 2nd edn. Blackwell, Oxford.

Stolp, H. (1988). *Microbial Ecology: Organisms, Habitats, Activities.* Cambridge University Press, Cambridge.

Wicklow, D. T. & Soderstrom, B., eds. (1997). *Environmental and Microbial Relationships.* Vol IV, *The Mycota*, eds. K. Esser & P. A. Lemke. Springer-Verlag, Berlin.

Roles of Saprotrophic Fungi in Ecosystems

General

Andrews, J. H. & Kinkel, L. L. (1986). Colonisation dynamics: The Island Theory. In N. J. Fokkema & J. van den Heuvel, eds., *Microbiology of the Phyllosphere*, pp. 63–76. Cambridge University Press, Cambridge.

Boddy, L. (2000). Interspecific combative interactions between wood-decaying Basidiomycetes. *FEMS Microbial Ecology* **31**, 185–194.

Boddy, L., Wells, J. M., Culshaw, C. & Donnelly, D. P. (1999). Fractal analysis in studies of mycelium in soil. *Geoderma* **88**, 301–328.

Carlile, M. J. (1994). The success of the hypha and mycelium. In N. A. R. Gow & G. M. Gadd, eds., *The Growing Fungus,* pp. 3–19. Chapman & Hall, London.

Dighton, J. (1997). Nutrient cycling by saprotrophic fungi in terrestrial habitats. In: Wicklow, D. T. & Soderstrom, B., eds., *Environmental and Microbial Relationships*, Vol IV, pp. 271–279, *The Mycota*, eds. K. Esser & P. A. Lemke. Springer-Verlag, Berlin.

Frankland, J. C. (1998). Fungal succession – unravelling the unpredictable. *Mycological Research* **102**, 1–15.

Lewis, N. G. & Yamamoto, E. (1990). Lignin: occurrence, biogenesis and biodegradation. *Annual Review of Plant Physiology and Plant Molecular Biology* **41**, 455–496.

Rayner, A. D. M. & Boddy, L. (1988). *Fungal Decomposition of Wood: its Biology and Ecology.* Wiley, Chichester.

Rayner, A. D. M., Watkins, Z. R. & Beeching, J. R (1999). Self-integration – an emerging concept from the fungal mycelium. In N. A. R. Gow, G. D. Robson & G. M. Gadd, eds., *The Fungal Colony. Twenty-first Symposium of the British Mycological Society*, pp. 1–24. Cambridge University Press, Cambridge.

Ritz, K. & Crawford, J. W. (1999). Colony development in nutritionally heterogeneous environments. In N. A. R. Gow, G. D. Robson & G. M. Gadd, eds., *The Fungal Colony. Twenty-first Symposium of the British Mycological Society*, pp 49–74. Cambridge University Press, Cambridge.

Watkinson, S. C. (1999). Metabolism and differentiation in basidiomycete colonies. In N. A. R. Gow, G. D. Robson & G. M. Gadd, eds., *The Fungal Colony. Twenty-first Symposium of the British Mycological Society*, pp. 126–156. Cambridge University Press, Cambridge.

Metabolic and developmental adaptations of fungi to habitats

Arst, H. N., Jr. (1996). Regulation of gene expression by pH. In R. Brambl & G. A. Marzluf, eds., *Biochemistry and Molecular Biology*, Vol. III, pp. 235–240. *The Mycota*, eds. K. Esser & P. A. Lemke. Springer-Verlag, Berlin.

Ashford, A. E. (1998). Dynamic pleiomorphic vacuole systems: are they endosomes and transport compartments in fungal hyphae? *Advances in Botanical Research* **28**, 119–159.

Béguin, P. (1990). Molecular biology of cellulose degradation. *Annual Review of Microbiology* **44**, 219–248.

Blanchette, R. A. (1991). Delignification by wood decay fungi. *Annual Reviews of Phytopathology* **29**, 381–398.

Cullen, D. & Kersten, P. J. (1996). Enzymology and molecular biology of lignin degradation. In R. Brambl & G. A. Marzluf, eds., *Biochemistry and Molecular Biology*, Vol. III, pp. 295–312; *The Mycota*, eds. K. Esser & P. A. Lemke. Springer-Verlag, Berlin.

Felenbok, B. & Kelly, J. M. (1996). Regulation of carbon metabolism in mycelial fungi. In R. Brambl & G. A. Marzluf, eds., *Biochemistry and Molecular Biology*, Vol. III, pp. 369–380, *The Mycota*, eds. K. Esser & P.A. Lemke. Springer-Verlag, Berlin.

Gold, M. H. & Alic, M. (1993). Molecular biology of the lignin degrading basidiomycete *Phanerochaete chrysosporium*. *Microbiological Reviews* **57**, 605–622.

Henriksson, G., Johansson, G. & Petterson, G. (2000). A critical review of cellobiose dehydrogenases. *Journal of Biotechnology* **78**, 93–113.

Holzbauer, E. L. F., Andrawis, A. & Tien, M. (1991). Molecular biology of lignin peroxidases from *Phanerochaete chrysosporium*. In S. A. Leong & R. M. Berka, eds., *Molecular Industrial Mycology*, pp. 197–223. Marcel Dekker, New York.

Keller, N. P. & Hohn, T. M. (1997). Metabolic pathway gene clusters in filamentous fungi. *Fungal Genetics and Biology* **21**, 17–29.

Leonowicz, A., Matzuszewska, A., Luturek, J., Ziegenhagen, D., Wojtaswasilewska, M., Cho, N. S., Hofrichter, M. & Rogalski, J. (1999). Biodegradation of lignin by white rot fungi. *Fungal Genetics and Biology* **27**, 175–185.

Marzluf, G. A. (1993). Regulation of sulphur and nitrogen metabolism in fungi. *Annual Review of Microbiology* **47**, 31–55.

Marzluf, G. A. (1996). Regulation of nitrogen metabolism by mycelial fungi. In R. Brambl & G. A. Marzluf, eds., *Biochemistry and Molecular Biology*, Vol. III, pp. 357–368. *The Mycota*, eds. K. Esser & P. A. Lemke. Springer-Verlag, Berlin

Olsson, S. (1999). Nutrient translocation and electrical signalling in mycelia. In N. A. R. Gow, G. D. Robson & G. M. Gadd, eds., *The Fungal Colony. Twenty-first Symposium of the British Mycological Society*, pp. 25–48. Cambridge University Press, Cambridge.

Penttila, M. & Saloheimo, M. (1999). Lignocellulose breakdown and utilization by fungi. In R.P. Oliver & M. Schweizer, eds., *Molecular Fungal Biology*, pp. 272–293. Cambridge University Press, Cambridge.

Radford, A., Stone, P. J. & Taleb, F. (1996). Cellulase and amylase complexes. In R. Brambl & G. A. Marzluf, eds., *Biochemistry and Molecular Biology*, Vol. III, pp. 269–275. *The Mycota*, eds. K. Esser & P. A. Lemke. Springer-Verlag, Berlin.

Worrall, J. J., Anagnost, S. E. & Zabel, R. A. (1997). Comparison of wood decay among diverse lignicolous fungi. *Mycologia* **89**, 199–219.

Fungi in soil

Anderson, J. B. & Kohn, L. M. (1998). Genotyping, gene genealogies and genomics bring fungal population genetics above ground. *Trends in Ecology and Evolution* **13**, 444–449.

Boddy, L. (1993). Saprotrophic cord-forming fungi: warfare strategies and other ecological aspects. *Mycological Research* **97**, 641–655.

Bossier, P., Hofte, M. & Verstraete, W. (1988). Ecological significance of siderophores in soil. *Advances in Microbial Ecology* **10**, 203–220.

Bruns, T. D. & Gardes, M. (1996). Community structure of ectomycorrhizal fungi in a *Pinus muricata* forest: above and below ground views. *Canadian Journal of Botany* **74**, 1572–1583.

Cairney, J. W. G. (1992). Translocation of solutes in ectomycorrhizal and saprotrophic rhizomorphs. *Mycological Research* **96**, 135–141.

Egger, K. N. (1995). Molecular analysis of ectomycorrhizal fungal communities. *Canadian Journal of Botany* **73**, S1415–1422.

Miller, R. M. & Lodge, D. J. (1997). Fungal responses to disturbance: agriculture and forestry. In Wicklow, D. T. & Soderstrom, B., eds., *Environmental and Microbial Relationships*. Vol IV, pp. 65–84, *The Mycota*, eds. K. Esser & P. A. Lemke. Springer-Verlag, Berlin.

Rayner, A. D. M., Powell, D. A., Thompson, W. & Jennings, D. H. (1985). Morphogenesis of vegetative organs. In D. M. Moore, L. A. Casselton, D. A. Wood & J. C. Frankland, eds., *Developmental Biology of Higher Fungi, Eighth Symposium of the British Mycological Society*, pp. 249–279. Cambridge University Press, Cambridge.

Siqueira, J. O., Nair, M. G., Hammerschmidt, R. & Safir, C. R. (1991). The significance of phenolic compounds in plant–soil–microbe systems. *Critical Reviews in Plant Sciences* **10**, 63–121.

Swift, M. J. & Anderson, J. M. (1996). Biodiversity and ecology of fungi in agricultural systems. In E.-D. Schulze & H. A. Mooney, eds., *Biodiversity and Ecosystem Function*. pp. 15–41. Springer-Verlag, Berlin.

Van der Heijden, M. G. A., Klironomos, J. N., Ursic, M., Moutojis, P., Streitwolf-Engel, R., Boller, T., Wiemken, A. & Sanders, I. R. (1998). Mycorrhizal fungal diversity determines plant biodiversity, ecosystem variability and productivity. *Nature* **396**, 69–72.

Van Elsas, J. D., Trevors, J. T. & Wellington, E. M. H. (1998). *Modern Soil Microbiology*. Marcel Dekker, New York.

Wainwright, M. & Gadd, G. M. (1997). Fungi and industrial pollutants. In D. T. Wicklow & B. Soderstrom, eds., *Environmental and Microbial Relationships*. Vol IV, pp. 85–97, *The Mycota*, eds. K. Esser & P. A. Lemke. Springer-Verlag, Berlin.

Widden, P. (1997). Competition and the fungal community. In D. T Wicklow & B. Soderstrom, eds., *Environmental and Microbial Relationships*. Vol IV, pp. 135–147, *The Mycota*, eds. K. Esser & P. A. Lemke. Springer-Verlag, Berlin.

Wood, M. (1989). *Soil Biology*. Chapman & Hall, London.

Fungi on and in Plants

Andrews, J. H. & Hirano, S. S., eds. (1991). *Microbial Ecology of Leaves*. Springer-Verlag, New York.

Arora, D. K., Rai, B., Mukerji, K. G. & Knudsen, G. R., eds. (1991). *Soil and Plants*. *Handbook of Applied Mycology*, Vol. 1. Marcel Dekker, New York.

Boddy, L. (1992). Fungal communities in wood decomposition. In G. C. Carroll & D. T. Wicklow, eds., *The Fungal Community: its Organisation and Role in the Ecosystem*, 2nd edn, pp. 749–782. Marcel Dekker, New York.

Keister, D. L. & Cregan, P. B., eds. (1991). *The Rhizosphere and Plant Growth. Beltsville Symposia in Agricultural Research No. 14*. Academic Press, London.

Lynch, J. M., ed. (1990). *The Rhizosphere*. Wiley, Chichester.

Morris, C. E., Nicot, P. C. & Nguyen-The, C. (1996). *Aerial Plant Surface Microbiology*. Plenum, New York.

Waisel, Y., Eshel, A. & Kafkaki, U. (1991). *Plant Roots: the Hidden Half*. Marcel Dekker, New York.

Fungi in Aquatic Environments

Bermingham, S. (1996). Effects of pollutants on aquatic hyphomycetes colonizing leaf material in freshwaters. In J. C. Frankland, N. Magan & G. M. Gadd, eds., *Fungi and Environmental Change, Twentieth Symposium of the British Mycological Society*, pp. 201–216. Cambridge University Press, Cambridge.

Gessner, M. O., Suberkropp, K. & Chauvet, E. (1997). Decomposition of plant litter by fungi in marine and freshwater ecosystems. In Wicklow, D. T. & Soderstrom, B., eds., *Environmental and Microbial Relationships*. Vol. IV, pp. 303–332, *The Mycota*, eds. K. Esser & P. A. Lemke. Springer-Verlag, Berlin.

Goh, T. K. & Hyde, K. D. (1996). Biodiversity of freshwater fungi. *Journal of Industrial Microbiology and Biotechnology* **17**, 328–345.

Hyde, K. D., Jones, E. B. G., Leano, E., Pointing, S. B., Poonyth, A. D. & Vrijonoed, L. L. P. (1998). Role of fungi in marine ecosystems. *Biodiversity and Conservation* **7**, 1147–1161.

Moss, S. T., ed. (1986). *The Biology of Marine Fungi*. Cambridge University Press, Cambridge.

Methods for Identification and Enumeration of Fungi

Andrews, J. H. (1986). How to track a microbe. In N. J. Fokkema & J. van den Heuvel, eds., *Microbiology of the Phyllosphere*, pp. 14–34. Cambridge University Press, Cambridge.

Bridge, P. D., Arora, D. K., Reddy, C. A. & Elander, R. P., eds. (1998). *Applications of PCR in Mycology*. CAB International, Wallingford.

Brundrett, M., Bougher, N., Dell, B., Grove, T. & Malajczuk, N. (1996). *Working with Mycorrhizas in Forestry and Agriculture*. Australian Centre for International Agricultural Research, Canberra.

Bruns, T. D., Szaro, T. M., Gardes, M., Cullings, K. W., Pan, J. J., Taylor, D. L., Horton, T. R., Kretzer, A., Garbelotto, M. & Liss, Y. (1998). A sequence database for the identification of ectomycorrhizal basidiomycetes by phylogenetic analysis. *Molecular Ecology* 7, 257–272.

Cannon, P. (1996). Filamentous Fungi. In G. S. Hall, ed., *Methods for Examination of Organismal Biodiversity in Soils and Sediments*. CABI Bioscience, Wallingford.

Dewey, F. M. (1992). Detection of plant invading fungi. In J. M. Duncan & L. Torrance, eds., *Techniques for the Rapid Detection of Plant Pathogens*, pp. 47–62. Blackwell, Oxford.

Drakos, D. J. (1991). Methods for the detection, identification and enumeration of microbes. In J. H. Andrews & S. S. Hirano, eds., *Microbial Ecology of Leaves*, pp. 14–34. Springer Verlag, New York.

Egger, K. N. (1992). Analysis of fungal population structure using molecular techniques. In G. C. Carroll & D. T. Wicklow, eds., *The Fungal Community: its Organisation and Role in the Ecosystem*. 2nd edn, pp. 193–208. Marcel Dekker, New York.

Foster, L. B., Kozak, K. R., Loftus, M. G., Stevens, J. J. & Ross, I. K. (1993). The polymerase chain reaction and its application to filamentous fungi. *Mycological Research* 97, 769–781.

Fouly, H. M. & Wilkinson, H. T. (2000). Detection of *Gaeumannomyces graminis* varieties using the polymerase chain reaction with variety-specific primers. *Plant Disease* 84, 947–951.

Frisvad, J. C., Bridge, P. D. & Arora, D. K., eds. (1998) *Chemical Fungal Taxonomy*. Marcel Dekker, New York.

Gardes, M. & Bruns, T. D. (1993). ITS primers with enhanced specificity for basidiomycetes – application to the identification of mycorrhizae and rusts. *Molecular Ecology* 2, 113–118.

Gardes, M. & Bruns, T. D. (1995). ITS-RFLP matching for identification in fungi. In J. P. Clapp, ed., *Species Diagnostic Protocols: PCR and Other Nucleic Acid Methods*. *Methods in Molecular Biology* 50, 172–186. Humana Press, Ottowa.

Grigorova, R. & Jarvis, J. R., eds. (1990). *Techniques in Microbial Ecology, Methods in Microbiology*, vol. 22. Academic Press, London.

Hall, G. S. (1996). *Methods for the examination of organismal diversity in soils and sediments*. CAB International, Wallingford.

Lacey, J. & Venette, J. (1995). Outdoor air sampling technique. In Cox, C. S. & Wathes, C. M. eds., *Bioaerosols Handbook*, pp. 407–471. CRC Lewis Publishers, Boca Raton, FL.

Mitchell, J. I., Roberts, P. J. & Moss, S. T. (1995). Sequence or structure? A short review on the application of nucleic acid sequence information to fungal taxonomy. *The Mycologist* 9, 67–75.

Norris, J. R., Read, D. J. & Varma, A. K., eds. (1991–92). *Techniques for the Study of Mycorrhiza. Methods in Microbiology*, Vols 23, 24. Academic Press, London.

Biodeterioration

Allsopp, D. & Seal, K. J. (1986). *Introduction to Biodeterioration*. Arnold, London.

Arora, D. K., Mukerji, K. G. & Marth, E. H., eds. (1991). *Foods and Feeds. Handbook of Applied Mycology*, Vol. 3. Marcel Dekker, New York.

Bhatnagar, D., Lillehoj, E. B. & Arora, D. K., eds. (1991). *Mycotoxins in Ecological Systems. Handbook of Applied Mycology*, Vol. 5. Marcel Dekker, New York.

Eaton, R. A. & Hale, M. D. C. (1993). *Wood: Decay, Pests and Protection*. Chapman and Hall, London.

Gorbashina, A. A., Krumbein, W. E., Hamman, C. H., Penina, L., Soukharjevski, S. & Wollenzien, U. (1993). The role of black fungi in colour change and biodeterioration of antique marbles. *Geomicrobiology* **11**, 205–221.

Hennebert, G. L., Boulenger, P. & Balon, F. (1990). *La Mérule: Science, Technique et Droit*. Editions Ciao, Brussels.

Jennings, D. H. & Bravery, A. F. (1991). *Serpula lacrymans: Fundamental Biology and Control Strategies*. Wiley, Chichester.

Krogh, P., ed. (1987). *Mycotoxins in Food*. Academic Press, London.

Lee, L. S., Bayman, P. & Bennett, J. W. (1992). Mycotoxins. In Finkelstein, D. B. & Ball, C., eds., *Biotechnology of Filamentous Fungi: Technology and Products*, pp. 463–503. Butterworth-Heinemann, Boston.

Pitt, J. I. & Hocking, A. D. (1997). *Fungi and Food Spoilage*, 2nd edn. Blackie Academic and Professional, London.

Sinha, K. K. & Bhatnagar, D. (1998). *Mycotoxins in Agriculture and Food Safety*. Marcel Dekker, New York.

Smith, J. E. & Moss, M. O. (1985). *Mycotoxins: Formation, Analysis and Significance*. Wiley, Chichester.

Zabel, R. A. & Morrell, J. J. (1992). *Wood Microbiology: Decay and its Prevention*. Academic Press, San Diego.

Environmental Change and Fungal Conservation

Arnolds, E. (1995). Conservation and management of natural populations of edible fungi. *Canadian Journal of Botany (Supplement 1)* S987–998.

Arnolds, E. J. M. (1997). Biogeography and conservation. In Wicklow, D. T. & Soderstrom, B., eds., *Environmental and Microbial Relationships*. Vol. IV, pp. 115–131, *The Mycota*, eds Esser, K. A & Lemke, P. A. Springer-Verlag, Berlin.

Cannon, P. F. (1997). Strategies for the rapid assessment of fungal diversity. *Biodiversity and Fungal Conservation* **6**, 669–680.

Frankland, J. C., Magan, N. & Gadd, G. M., eds. (1996). *Fungi and Environmental Change. Twentieth Symposium of the British Mycological Society*. Cambridge University Press, Cambridge.

Hawksworth, D. L. (1997). Fungi and international biodiversity initiatives. *Biodiversity and Fungal Conservation* **6**, 661–668.

Janos, D. P. (1996). Mycorrhizas, succession and the rehabilitation of deforested lands in the humid tropics. In: Frankland, J. C., Magan, N. & Gadd, G. M., eds., *Fungi and Environmental Change. Twentieth Symposium of the British Mycological Society*. Cambridge University Press, Cambridge.

Pegler, N., Boddy, L., Ing, B. & Kink, P. M. (1993). *Fungi of Europe: Investigation, Mapping and Recording*. Royal Botanic Gardens, Kew.

Pilz, D. & Molina, R., eds. (1996). *Managing Forest Ecosystems to Conserve Fungus Diversity and to Sustain Wild Mushroom Harvests*. US Department of Agriculture Forest Service (Pacific Northwest Research Station), Portland, Oregon.

Rowe, R. F. (1988). The commercial harvesting of wild edible mushrooms in the Pacific Northwest of the United States. *Mycologist* **11**, 10–14.

With each question one statement is correct. Answers on page 561.

6.1 The most important role of fungi in the global carbon cycle is:
   a) weathering of limestone rocks
   b) carbon dioxide fixation
   c) degrading wood and other plant remains to carbon dioxide
   d) immobilization of carbon compounds in mycelium

6.2 It is essential to include in a growth medium for any fungus:
   a) an organic nitrogen source
   b) an organic sulphur source
   c) an organic phosphorus source
   d) an organic carbon source

6.3 The mitosporic genera *Trichoderma, Penicillium, Aspergillus, Fusarium* and *Epicoccum* are an assemblage of fungi characteristic of:
   a) the herbivore gut
   b) the sea
   c) leaf surfaces
   d) the soil

6.4 Initial attack on the lignin molecule by white rot fungi involves the production of:
   a) oxidizing free radicals
   b) cellobiohydrolase
   c) endoglucanase
   d) siderophores

6.5 The identity of a fungal hypha can be established in the absence of any fruiting structures, using:
   a) a specially raised monoclonal antibody
   b) absorption spectroscopy
   c) scanning electron microscopy
   d) culture on a variety of test media

Parasites and Mutualistic Symbionts

7

Parasitism, Mutualism and Symbiosis

Some organisms live in close contact with organisms of another species and obtain nutrients from them or benefit in other ways. De Bary, a nineteenth century mycologist and plant pathologist, described such a close relationship as a **symbiosis**, a common life. Where one of the species is larger than the other it is referred to as the **host**: De Bary noted that a symbiotic relationship might be **parasitic**, in which only one species benefits, or **mutualistic**, with both partners benefiting. A parasite that causes perceptible damage to its host is termed a **pathogen**.

The term symbiosis has sometimes been restricted to mutualistic symbioses, leaving no term that covers both parasitism and mutualism. This is unfortunate. Determining whether a relationship is parasitic or mutualistic may require detailed studies on the transfer of nutrients between the symbionts. Moreover, whether an interaction is parasitic or mutualistic can depend on the stage that it has reached and on environmental conditions. On the other hand, simple morphological investigations can establish the occurrence of close contact between the species. These considerations have led to the readoption of De Bary's terminology by some biologists, especially those with ecological interests. On the other hand, most research on the interaction of fungi and plants has been motivated by the crop losses caused by parasitic or pathogenic fungi, and most of the information available concerns such fungi. Hence discussion of the infection of plants by fungi commonly refers to the latter as the 'pathogen' or 'parasite', a practice followed in the present chapter.

Fungi and Plants

Fungal hyphae are well suited for penetrating plant tissues, as has already been discussed with reference to the utilization of dead plant material by free-living fungi (page 297). This ability is widely exploited for the invasion of living plants and the establishment of a symbiotic relationship. Some fungi are facultative symbionts, found both as free-living saprotrophs and within plants. Others are obligate symbionts, growing only in association with living plants. The hyphae of some symbiotic fungi penetrate plant cells, whereas those of others grow through tissues without entering cells. Some fungi are **biotrophic**, living within the plant and obtaining nourishment without causing cell death. Some are **necrotrophic**, killing cells and absorbing nutrients from the dead tissues. Others have both a biotrophic and later a necrotrophic phase. Many fungi are important plant pathogens causing severe crop losses whereas others, such as mycorrhizal fungi (page 394), are mutualistic and can have an important role in plant nutrition. The way in which fungi infect plants, and the subsequent biotrophic or necrotrophic relationship, will now be considered.

The Infection Process – How Fungi Penetrate Plant Tissues

Plant morphology and fungal invasion

Plants have two entirely different kinds of surface, each in contact with a very different environment. All the parts above ground – stems, leaves, flowers and fruits – are exposed to the air, and water in the form of dew and rain. A fungus arriving at this surface must do so as a spore, and in this way the process of infection of aerial surfaces is different from root infection, where it is possible for fungal growth to be sustained in the surrounding soil, and for the growing hyphae to attack the root. Thus the success of fungi which live as airborne symbionts of aerial plant surfaces depends on the effectiveness of their mechanisms (Chapter 4) for spore production, liberation, dispersal in air and reinfection. Mechanisms for infecting the new host plant are specially important where the fungus is an obligate symbiont of a narrow range of plant species and thus highly host-specific. The subterranean part of the plant, on the other hand, is in continual contact with the microbes and animals living and actively growing in the soil. Nutrients and water are both normally present near plant roots, and the soil environment is also favourable to microbial life because conditions fluctuate less than above ground, and potentially damaging sunlight is excluded. Fungal hyphae can therefore approach a root by growing through the soil. The fungus does not necessarily have to initiate growth from spores in a nutrient-poor, exposed environment, as airborne pathogens do. The root is continually surrounded by a microbial community, as described in Chapter 6. Most members of this are saprotrophs, but others live symbiotically as mutualists or pathogens. Most plant roots have mutualistic symbioses with fungi, forming the associations known as mycorrhizas. There are also pathogenic fungi which arrive at the root surface by means of motile spores (page 382) which swim through water held in the interstices within the soil.

In view of the different problems encountered by fungi invading aerial and subterranean parts of plants, it is not surprising that most are specialized to enter either one or the other (Table 7.1). Thus, some plant pathogenic fungi are airborne, like rusts (Uredinales), and others are soil-borne like the vascular wilt fungi of the genera *Fusarium* and *Verticillium*. The rest of the present section describes the morphological and physiological features of plants which are likely to dictate the behaviour of a successful invading pathogen.

Table 7.1 Some common airborne and soil-borne pathogens of plants

Pathogenic fungus	Type of disease
Airborne	
Basidiomycetes:	
Uredinales (rusts)	Biotrophic
Ustilaginales (smuts)	Biotrophic
Ascomycetes and mitosporic fungi:	
Erysiphales (powdery mildews)	Biotrophic
Cladosporium	Necrotrophic – leaf spots
Colletotrichum	Fruit rots and leaf spots
Drechslera (*Helminthosporium*)	Leaf spots, wilts
Magnaporthe (*Pyricularia*) *grisea*	Necrotrophic: a destructive pathogen of rice (*Oryza sativa*)
Sclerotinia fructigena	Rots of apples
Sclerotinia fructicola	Rots of plums
Zygomycetes:	
Rhizopus stolonifer	Rots of fruit
Oomycetes:	
Peronosporaceae (downy mildews)	Biotrophic
Soil-borne	
Basidiomycetes:	
Armillaria mellea	Kills trees, rots wood
Heterobasidion annosum	Rots wood inside trunk of tree
Ascomycetes and mitosporic fungi:	
Fusarium spp.	Root rots; vascular wilts
Gaeumannomyces graminis	Take-all of wheat
Sclerotinia sclerotiorum	Root rots
Verticillium spp.	Vascular wilts
Oomycetes:	
Pythium spp.	Damping off of seedlings
Chytridiomycetes:	
Synchytrium endobioticum	Scab of potatoes
Plasmodiophoromycetes:	
Plasmodiophora	Hypertrophy of roots of *Brassica*
Airborne and soil-borne	
Phytophthora spp.	Necrotrophic; many species, most with a limited host range and highly destructive

Aerial surfaces Most fungal pathogens that attack the aerial surfaces of plants do so at places such as stems, leaves (Fig. 7.1A), flowers or fruits where there is an outer layer of living cells, the **epidermis**. Woody surfaces such as the bark of trees are practically impregnable to fungi unless damaged. The living epidermis of land plants is a layer of cells forming a flat sheet. Its surface is covered by a fatty layer, the **cuticle**, which is normally about 1 μm thick, and more in some plants, particularly those that grow in dry places. The cuticle is waterproof and also water repellent. These properties are due partly to its chemical composition, which includes long-chain aliphatic hydrocarbons, wax esters, primary alcohols,

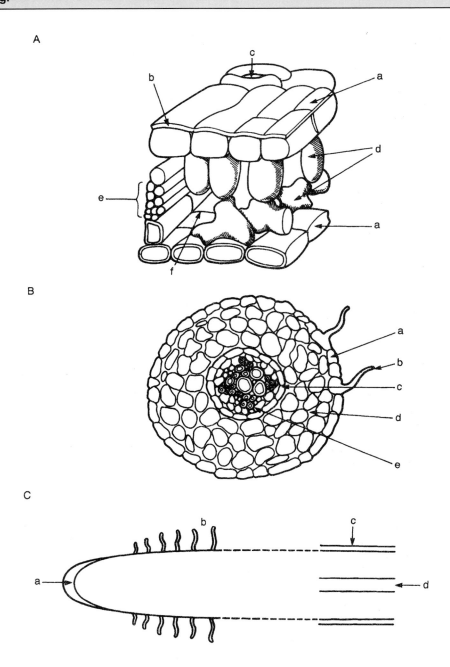

Figure 7.1 The structure of plant tissues. A. Section across a leaf: a, epidermis; b, cuticle; c, stoma; d, mesophyll; e, vascular strand; f, airspace. B. Section across a root: a, epidermis; b, root hair cell; c, endodermis; d, cortex with water-filled intercellular spaces; e, central vascular cylinder contained within the endodermis, consisting of the conducting tissues xylem and phloem, and surrounding cells. C. Longitudinal section of a root: a, root cap, from which cells are shed into the surrounding soil; b, root hair region; c, older region in which the surface is covered in impermeable bark; d, vascular cylinder.

fatty acids and **cutin**, a polymer of C_{16} and C_{18} hydroxy and epoxy fatty acids. The water-repellent nature of cuticle is also partly due to the shapes in which the waxy materials are deposited on the surface. These are characteristic of particular plant species and form patterns of rolls or scales which lie on the surface. A leaf surface may consequently be so hydrophobic that water drops falling on it remain almost spherical. This will have the effect of reducing the area of leaf available for invasion by fungal spores that need liquid water in which to germinate and grow. The epidermis is interrupted by **stomata** (sing. **stoma**), which are holes of variable aperture leading into gas-filled spaces in the leaf. These spaces are also lined with a thin cuticle. A fungal hypha can enter the internal tissues of a leaf either by penetrating the cuticle and cell wall of an epidermal cell (Fig. 7.2B), or by growing through a stoma into the cavity beneath and then penetrating the cuticle and cell wall of an internal cell (Fig. 7.2A). Conditions in the substomatal cavity are probably more favourable to fungal growth than on the open leaf surface; the relative humidity approaches 100% and some of the light is filtered out by overlying leaf cells. Some species of pathogenic fungi regularly infect leaves through stomata, others (e.g. most *Botrytis* species) by direct invasion of epidermal cells. The mode of entry is usually characteristic of the species (Figs. 7.2, 7.3). For those that enter via stomata, there are two problems: to locate a stoma, and to do so when it is open. The stomata of most plants are open during the hours of daylight – when photosynthesis is occurring – provided that the water available to the plant is adequate. When the stomata are open, oxygen produced by photosynthesis diffuses out through them. The tropic response of the germ-tubes of some plant pathogens may be elicited by the gradients of oxygen tension so generated. There is evidence that the point of emergence of the germ-tube from spores of the saprotroph *Geotrichum candidum* is affected by oxygen gradients so that it starts to grow on the side of the spore closer to the oxygen source. A similar effect may operate in spores of pathogens germinating on the leaf surface near stomata which would provide a source of oxygen while the tissues of a leaf were photosynthesizing and the stomata open. Negative phototropism – growing away from the light – is also seen in germ-tubes of the weakly parasitic *Botrytis cinerea* and a number of rust fungi and may help to direct their growth into the leaf.

Root surfaces Plant roots (Fig. 7.1B,C) have been described in Chapter 6, page ; here, features of special concern with respect to infection by fungi will be considered. Like stems, roots in their early stages of development are covered by a living epidermis. Root epidermis has none of the waterproof, hydrophobic properties of leaf epidermis. It is covered in mucilage, and in the root hair region, usually a few millimetres behind the tip, hair-like extensions of the epidermal cells grow into the soil, rather like fungal hyphae themselves although wider. The root tip is covered by the root cap, which consists of cells which die and are sloughed off or abraded away as the root grows past soil particles. These cells, together with mucilage and exudates (page 344), make the root a rich source of nutrients for soil microbes, supporting a rhizosphere microflora which includes saprotrophic and parasitic fungi as well as mutualistic symbionts (pages 394). Exudates may also attract the swimming spores of parasitic fungi such as *Phytophthora* (page 382) and stimulate the production of infection structures

A

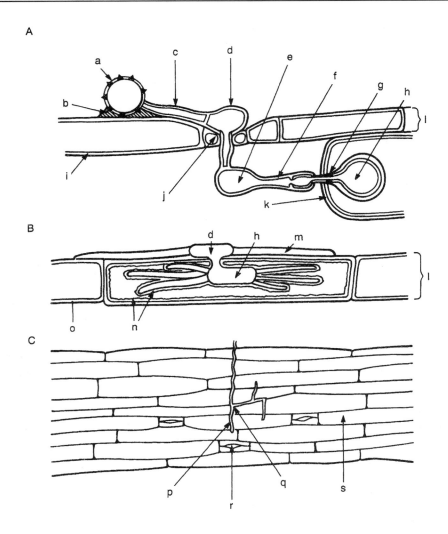

B

C

such as rhizomorphs of *Armillaria mellea*. One exudate component that has both of these effects is ethanol, produced by anaerobic metabolism when waterlogging of soil results in a poor oxygen supply to roots. Although the release of nutrients from roots makes fungal growth in their vicinity possible, the saprotrophic rhizosphere population thus supported may make it harder for parasitic fungi to colonize the root. This is because of competition for nutrients and perhaps because of antibiotic production by some microorganisms. Soil-borne root-infecting fungi are often inferior to saprotrophic soil fungi in their ability to compete for resources, and less able to produce and tolerate antibiotics. Thus the favourable conditions for fungal growth in the neighbourhood of roots are not necessarily advantageous for those which need to gain entry to the root itself.

As roots elongate the older parts, as with shoots (Fig. 7.4), become covered in bark. This is made of waterproof layers of dead cells which develop from a layer of actively-dividing cells (periderm) just beneath the root surface. Each cell acquires a wall thickening of lignin and the waterproof phenolic substance suberin. The successive layers of these cells die and constitute the bark. Single fungal hyphae cannot penetrate this, although a few species of soil-borne tree pathogens, such as *Armillaria mellea*, are able to overcome the bark barrier.

Wounds The term 'wound' is used to describe a breakage in a plant's surface layers which exposes the tissues beneath. The word suggests, to us animals, an injurious and infrequent happening. Wounds are, however, a normal part of plant development. Plants grow continuously, repeatedly producing the same structures throughout their lives. Some of these – fruit, leaves, branches – are shed, leaving wounds or scars. Plants have evolved side-by-side with grazing animals and damage by grazing is a normal part of their biology. Roots are eaten by soil animals, abraded by mineral particles in soil, and the surfaces of older roots are disrupted by the emergence of lateral branches from the internal tissues where they are initiated. Some fungi regularly gain entry to plants via wounds; for example, the brown rot of fruit caused by *Sclerotinia fructigena* occurs as a result of wasps or birds breaking the epidermis of the fruit and introducing the conidia of the fungus. The wounds that occur as part of development – emergence of lateral roots, leaf fall – do not act as infection sites for fungi as do injuries caused by external agents. Probably plants have evolved ways of preventing microbial invasion at these potentially vulnerable points.

Internal tissues Within the plant, different tissues provide different environments for fungi. Most specialized pathogens are confined to one kind of plant tissue. Inside the leaf, and in the outer parts of green stems, cells have spaces between

Figure 7.2 Examples of infection structures produced by fungi on leaf surfaces. A, Diagram of stages in infection of a leaf of bean (*Vicia faba*) by a spore of rust fungus (*Uromyces viciae-fabae*). This fungus characteristically enters via a stoma into the air-filled space beneath, then invades a cell of the mesophyll. a, Uredospore, stuck to the leaf epidermis by adhesive material; b, the site of cutinase and esterase production; c, germ-tube; d, appressorium, site of cellulase production; e, substomatal vesicle, producing pectin esterase; f, infection hypha producing cellulase; g, neck ring (page 393); h, haustorium; i, wall of epidermal cell; j, guard cell at the edge of a stoma through which the hypha is entering the leaf. (After Mendgen & Deising (1993). Infection structures of fungal plant pathogens – a cytological and physiological evaluation. *New Phytologist* **124**, 193–213.) B, The lobed haustorium of *Erysiphe graminis* (a powdery mildew) in a host cell. This fungus penetrates directly into one of the epidermal cells. d, Appressorium; h, haustorium; m, hypha on the leaf surface; n, the haustorial sac, cytoplasm continuous with parietal cytoplasm of the leaf epidermal cell; o, wall of epidermal cell. (After Bushnell, W. R. & Gay, J. (1978). Accumulation of solutes in relation to the structure and function of haustoria in powdery mildews, pp. 183–235. In D. M. Spencer, ed., *The Powdery Mildews*, Academic Press, London.) C, The development of germ-tubes of *Puccinia hordei* on the surface of a barley leaf. p, Primary germ-tubes growing at right angles to the longitudinal axes of the epidermal cells (thigmotropism). q, initiation of short branches at junctions between epidermal cells. Similar patterns of growth are seen on inert replicas of the leaf surface. Monocotyledonous plants, such as barley and other cereals, have this pattern of long lines of epidermal cells with linear arrays of stomata at intervals. The ability of the germ-tubes to orient their growth in the way shown increases their chances of meeting a stoma through which the fungus can enter as in A above. (After Read, N. J., Kellock, L. J., Knight, H. & Trewavas, A. J. (1992). Contact sensing during infection by fungal pathogens. In J. A. Callow & J. R. Green, eds., *Perspectives in Plant Cell Recognition*, pp. 137–172. Cambridge University Press, Cambridge.)

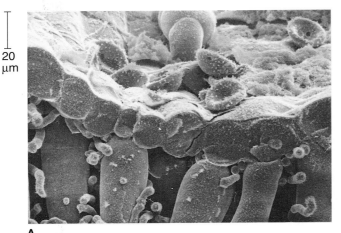

A

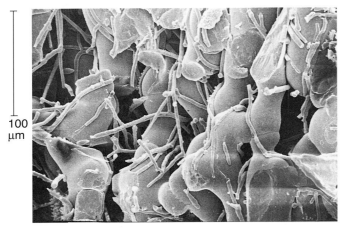

B

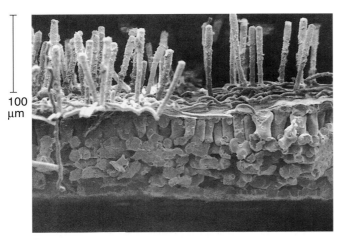

C

Figure 7.3 Biotrophic pathogens of leaves. A, Infection of broad bean (*Vicia faba*) by uredospores of a rust fungus, *Uromyces viciae-fabae*. Spores are germinating on the surface of a leaf. Hyphae are entering the intercellular spaces of the leaf mesophyll via stomata, leaving empty, collapsed spore cases on the surface of the epidermis. Compare with the diagram in Fig. 7.2A showing the stages in development of infection of the leaf by *Uromyces viciae-fabae*. B, The internal mesophyll of the leaf (of ragwort, *Senecio jacobaea*) is shown, with hypha of another rust (species unknown) growing over and among the living cells. C, Infection by a powdery mildew fungus *Erysiphe pisi* on pea (*Pisum sativum*). Fungal growth is confined to the epidermis and there are no hyphae within the mesophyll. There is abundant development of conidia on the surface of the leaf, which are easily brushed off, giving a powdery appearance to infected leaves. (A–C; B. E. Juniper.)

20 µm

100 µm

100 µm

them which are normally filled with gases – water vapour, and air with amounts of carbon dioxide and oxygen that vary due to photosynthesis and tissue respiration. Similar intercellular spaces in the outer layers of roots, the **cortex** (Fig. 7.1B) are filled with gas or with water. These spaces intercommunicate with the external environment. However, the central region (**stele**) containing the conducting tissues which carry water from the roots to leaves and sugar from photosynthetic to non-photosynthetic parts, is surrounded by a continuous layer, the **endodermis**, that does not allow free diffusion of water solutes. The cells of the stele are closely packed and do not have spaces between them (Fig. 7.1B). Thus these tissues are likely to be saturated with liquid water and gaseous exchange within them will be slow. The only fungal pathogens which infect tissues in the stele of young plants are the vascular wilt fungi. Other pathogens grow between the cells of outer, cortical tissues, or may be confined to the epidermis, like powdery mildews.

When a woody plant develops, the stem thickens because new woody conducting tissue, **xylem**, is added each year to the central core (Fig. 7.4). At the same time the sugar-conducting tissue, **phloem**, is replaced by new cells and bark develops to cover the outside of the stem. Thus the central trunk of a mature tree has a central core of dead wood cells, which as they age cease to be active in water conduction and become impregnated with tannins and phenolic substances. Outside this core, called **heartwood**, is a layer of wood cells (the **sapwood**) which are active in conducting water. These are bounded by the **vascular cambium**, the cells of which continue to divide throughout the life of the plant, and to produce derivatives which differentiate into woody xylem cells on the inside and sugar-conducting phloem cells on the outside. Some also generate cells which remain as living, less highly specialized cells within the xylem and phloem. These form the **medullary rays** in wood. Outside the cambium is the living phloem. This remains as a thin layer because the older cells are squashed as the new ones develop. Phloem is the slimy layer found when bark is peeled off a twig. The outermost layer of all is the bark which is continuously generated by a second cambium, the **cork cambium**. Bark is impermeable to water and gases, although in twigs it is perforated at intervals by pores, **lenticels**. Although the cork cambium develops, and begins to produce bark in the normal course of events as a green twig becomes woody, new cork cambium may also appear in a tree invaded by the mycelium of a parasitic fungus, or in response to purely physical damage such as pruning a branch. The result is a localized corky periderm layer sealing off the damaged part from the rest of the tree.

A tree provides a new range of habitats and substrates for fungi, compared with a non-woody, herbaceous plant. Tree pathogens such as *Armillaria mellea* and *Heterobasidion annosum* are able to enter and grow in living trees. Other fungi are saprotrophs on wood as described in Chapter 6, or harmless endophytes in the xylem (page 405). Many tree pathogens can continue to live saprotrophically on the wood after killing the tree, as *A. mellea* does. In this case, the point at which necrotrophy (pages 385) becomes saprotrophy is hard to identify. The longevity of trees enables a symbiotic fungus to maintain a vegetative colony for long periods, and tree pathogens do not have to reach a new host frequently in order to survive, as do pathogens of herbaceous and in particular annual plants.

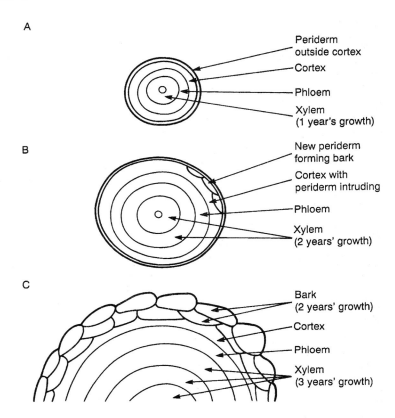

A

Periderm
outside cortex

Cortex

Phloem

Xylem
(1 year's growth)

B

New periderm
forming bark

Cortex with
periderm intruding

Phloem

Xylem
(2 years' growth)

C

Bark
(2 years' growth)

Cortex

Phloem

Xylem
(3 years' growth)

Figure 7.4   The structure and composition of wood. A, B and C are cross-sections of 1, 2 and 3 year-old twigs. The twig thickens mainly because of an increase in the amount of xylem, to which a new layer is added each year. The increase in thickness is accompanied by the extension and renewal of the surface layers by the periderm, bark-producing layers which arise just below the surface and generate the impermeable cells of bark. In a mature woody trunk, the bulk consists of xylem, with phloem occupying a cylindrical space outside the xylem and under the bark. The cortex disappears as periderms arise within it.

Features and activities of fungi facilitating plant infection
Fungi form a variety of structures that enable them to enter plant tissues and achieve intimate contact with cells. The processes and structures involved are different in aerial and subterranean parts of the plant, and moreover each parasitic fungus species invades the plant in a characteristic way.

Establishing by DNA technology fungal attributes crucial for infection Most fungi are unable to invade other organisms, being exclusively saprotrophic. The ability to invade and establish a parasitic or mutualistic relationship is relatively rare, although it occurs in every fungal phylum. Many symbiotic fungal species, however, appear similar in form and life cycle to their saprotrophic relatives, and are closely related to them. For example, *Cladosporium fulvum* is a pathogen of tomato plants whereas its relative *C. herbarum* is exclusively saprotrophic. The question therefore arises as to what characteristics enable some fungi to invade a living host organism. Until the advent of **molecular biology** it was not possible to determine with certainty which of the many cellular processes apparently playing a part in invasiveness were essential. For example, cell wall degradation and toxin production might be observed to accompany invasion of a host, but there was no way of testing each separately, and finding out whether the fungus could invade without it. However, gene cloning and sequencing techniques now make it possible to identify genes critical for pathogenicity, and in many cases to identify the molecules encoded and their cellular functions (Table 7.2).

Two approaches which have been very successful in identifying the cellular processes that are required for establishing plant infections are targeted deletion of particular genes, and the identification of genes expressed preferentially during the infection process.

Two basic strategies are used to test a gene for a critical role in pathogenicity. In the first, a collection of mutant strains of the pathogen is made and tested systematically for their ability to cause disease. Random mutagenesis in fungi can be brought about by treatment with ultraviolet light or chemical mutagens, although the use of insertional mutagenesis procedures, such as restriction enzyme mediated insertion of DNA, is now more widespread. In this process, the fungus is transformed with a linearized plasmid vector and a corresponding endonuclease. The endonuclease causes nicks to be made in chromosomal DNA and allows insertion of the DNA to occur randomly throughout the genome. The inserting plasmid creates mutations by disrupting genes and also acts as a tag to facilitate gene cloning. Large banks of mutants have been made in fungal

Table 7.2   Some genes shown to be required for pathogenicity of fungi to plants. Targeted deletion of single genes (page 380, Fig. 7.7) has shown that some molecules, but not others, are essential for pathogenicity of some fungi. This method can only be applied to pathogens which are haploid and transformable. (See R. Oliver & A. Osbourn (1995), Molecular dissection of fungal phytopathogenicity. *Microbiology* **141**, 1–9.)

Nature of process required in invasion of host	Fungus	Enzyme or other molecule thought to be necessary to carry out process	Does disrupting the gene encoding this molecule affect pathogenicity?
Overcoming physical barriers	*Magnaporthe grisea*	Cutinase	No
	Cochliobolus carbonarum	Endoxylanase	No
Detoxification of preformed inhibitors	*Gaeumannomyces graminis*	Avenacinase	Yes
Toxin production	*Cochliobolus carbonarum*	HC-toxin synthetase 1	Yes
Recognizing and responding to signals from host	*Magnaporthe grisea*	MAP kinase (component of intracellular signal transduction pathway)	Yes
Development of infection structure	*Magnaporthe grisea*	Synthetase in melanin synthesis pathway	Yes
	Ustilago maydis	Chitin synthase	No

pathogens using insertional mutagenesis and a number of genes required for successful plant infection have been isolated in this way.

The second strategy involves generating a targeted gene deletion at a particular locus. This procedure is used to assay the role of a known gene – normally defined by its expression profile, or its homology to other genes – in the infection process. Many genes of fungal pathogens have now been assayed using the one-step gene replacement procedure first used to determine gene function in *Saccharomyces cerevisiae*. In outline, the process begins when a gene is identified and a large genomic clone is isolated, corresponding to the whole gene locus. An area spanning the protein-coding portion of the gene is removed from this clone using restriction endonucleases and an antibiotic resistance marker gene is inserted in its place. The linear gene construct is then transformed into the fungus and a double cross-over event occurs in the flanking regions, replacing the chromosomal copy of the gene with the antibiotic resistance gene marker. The rate of homologous recombination varies among fungal species but is often of the order of 10–30%, allowing gene replacement mutants to be readily generated.

Once generated, gene replacement mutants can be tested for their ability to cause disease and this has proved informative in identifying a number of conserved signalling pathways required for pathogenicity in fungi. In the corn smut fungus for example, the mating type genes have been characterized in detail in this manner and elements of a G-protein coupled signal transduction cascade have been shown to operate in the fungus during the establishment of dikaryotic growth which is necessary for infection of the host. In the rice blast fungus a cyclic AMP response pathway and a mitogen-activated protein kinase pathway have both been shown to be required for the formation and function of infection structures.

The application of genomic approaches to the investigation of plant disease is currently a very active area of research. New techniques such as rapid automated DNA sequencing and genome-wide expression profiling are transforming the study of fungal cell biology. It is now possible, for example, to examine the changes in expression pattern of all the genes of a fungus in response to a specific stimulus. Carrying out such experiments on pathogenic fungi is still some way off, but the prospect of studying the expression patterns of all pathogen genes during the infection process is an exciting one. The generation of a clear picture of the integrated action of whole sets of pathogen genes during infection is now a realistic goal.

Airborne pathogenic fungi

Airborne fungi arrive on the plant surface as spores. The process of spore deposition is discussed in Chapter 4, where mention is made of the high proportion, among plant pathogens, of large-spored fungi, and the possible importance of size and weight of spores in increasing rates of impaction against objects such as stems and leaves.

Efficient spore production, liberation and dispersal are obviously essential for an airborne plant pathogen. A fungus which is dependent on a particular plant host must be able to infect new host plants; the more specific the pathogen and the less available the host, the more efficient does the cycle of sporulation and infection need to be. Infection of the host requires structural and physiological adaptation by the parasite. Structures that enable a spore to infect a leaf cell include **germ-tubes, appressoria** (sing. **appressorium**), **infection pegs** and **haustoria** (sing. **haustorium**). Physiological adaptations enable the development of such structures to be started at appropriate times and places. Developmental responses can be due to environmental stimuli, for example the production of swimming spores by *Phytophthora* when it is raining, or they can be elicited by some feature specific to the host.

Spore germination begins the infection process. The spore first swells and then a germ-tube emerges. Even at this early stage the host plant may exert a physiological effect, although the structures developing are no different from those of saprotrophic fungi. Spores of the highly host-specific pathogen *Erysiphe graminis*, powdery mildew of cereals, show a higher rate of germination on the leaves of the host than on other leaves. There are two ways in which such specificity can be mediated. One is by a 'contact stimulus' where contact with a surface of appropriate shape triggers development of a spore. Thus the swimming zoospores of the downy mildew fungus *Pseudoperonospora humuli* always settle, encyst and germinate at the open stomata of the host leaves. When a film of zoospore suspension was spread over a plastic replica of the leaf surface, this preference for development over 'stomata' was still shown, so that the possibility of a chemical stimulus was excluded. Elsewhere, chemical effects have been demonstrated. For example, spores of the downy mildew of lettuce, *Bremia lactucae*, are stimulated to germinate by the presence of germinating lettuce seeds, and uredospores of the safflower rust *Puccinia carthamni* by several volatile polyacetylenes isolated from the host (Fig. 4.18A). These induce germination at concentrations as low as 0.15 ng ml^{-1}.

There is a clear advantage for a pathogen in having a mechanism by which host-specific substances and no others trigger the development of infection structures since spores that start to germinate in the absence of a host will die and be wasted.

Environmental factors also affect spore germination and hence the behaviour of a pathogenic fungus in nature. The water relations of spores vary. Powdery mildews (Erysiphales) are more often seen in dry weather since their asexual spores can germinate and infect leaves without liquid water; indeed wetting reduces the viability of their conidia. On the other hand, sporangia and conidia of downy mildews (Peronosporales) soon decline in viability unless the relative humidity is high. Infection of hops by the hop downy mildew, *Pseudoperonospora humuli*, is greatly increased if the crop has been soaked by rain. Water in the form of dewdrops on the leaf is essential for leaf penetration by spores of the rice blast fungus *Magnaporthe grisea*, to allow the appressoria to generate hydrostatic pressure osmotically and force an infection peg through the plant epidermal cell wall.

Water has a special role in the life cycle of the downy mildew fungi because many of these Oomycete fungi, like their saprotrophic relatives, have swimming zoospores. Many parasitic Oomycetes attacking land plants show what appears to be adaptation of a zoosporangium to their life as terrestrial parasites. Instead of a submerged zoosporangium being formed and releasing zoospores under water, a sporangium is formed on the end of a hypha which grows up into the air from the stoma of an infected plant. Sporangia break away and are dispersed whole, like large conidia. They may germinate directly, producing germ-tubes, or else release zoospores which swim for a period before developing a cell wall and germinating by means of a germ-tube. Germ-tubes are usually produced directly from sporangia in the genera *Bremia* and *Peronospora*; *Plasmopara* normally has a swimming, zoosporic stage.

A nutrient supply from outside the spore is needed for the germination of spores of some species of fungi but not others. There is evidence that host-specific parasitic fungi are less likely to need exogenous nutrients for germination and infection than saprotrophs. The genus *Botrytis* includes both host specific plant parasites such as *B. fabae* – causing chocolate spot disease of field beans – and the much less specific *B. cinerea*, the cause of grey mould on senescent or frost-damaged plant parts in autumn. Spores of *B. fabae*, *B. tulipae* and *B. squamosa*, all host specific, do not require nutrients to germinate, whereas isolates of *B. cinerea* vary in their need for exogenous nutrients. Nutrients in the spore's environment can, in saprotrophic species, act as a useful signal to trigger germination. Since the leaves of healthy plants do not release nutrients in large quantities as senescent ones do, nutrients do not necessarily signal the presence of a suitable substrate to spores of host-specific pathogens. Instead, attack on a healthy plant may need substantial nutrient reserves, a possible reason for *B. fabae* having much bigger spores than *B. cinerea*. Spores of pathogens, on arrival at the leaf surface, may secrete a sticky material which holds them in place on the leaf. Scanning electron micrographs of spores of the rice blast fungus *Magnaporthe grisea* show exuded material attaching the ungerminated spore to the epidermis.

The emergence of a germ-tube from a spore on a host leaf is only the beginning of the infection process, which to succeed must bring about intimate and continuing contact between fungus and plant. The way in which the hypha then develops varies with the nature of the plant tissue it is colonizing (Fig. 7.2). The structures involved in entering leaves are different from those of fungi which invade wounds. A leaf- or stem-infecting fungus must breach the cuticle and cell wall of the host, and this usually involves the production of first a germ-tube and then an **appressorium**, a structure specialized for firm attachment which is formed at the tip of the hypha, often in a particular position in relation to the cells of the leaf. For example, appressoria of *Uromyces appendiculatus* (bean rust), developing from uredospores on French bean leaves, normally form after the tube makes contact with the lip of the stomatal guard cell (one of the two cells that surround the stomatal opening) and are thus positioned over stomata. In *Botrytis cinerea* hyphae growing on a leaf surface form multicellular infection cushions from which large numbers of appressoria develop. A hydrophobin protein (page 102) has been found by gene disruption to be important for the development of *M. grisea* spores on leaves, and hydrophobins are thought to be involved in leaf invasion by many fungi. *M. grisea* develops appressoria from the tips of germ tubes on any hydrophobic surface, including artificial membranes as well as leaf cuticles. A hydrophobin protein appears to be essential for the recognition of a suitable surface. The protein may act by forming an amphipathic layer with its hydrophobic side against the hydrophobic leaf cuticle and the hydrophilic side of the layer presented to the developing fungus.

Sections through appressoria of species which penetrate directly through the

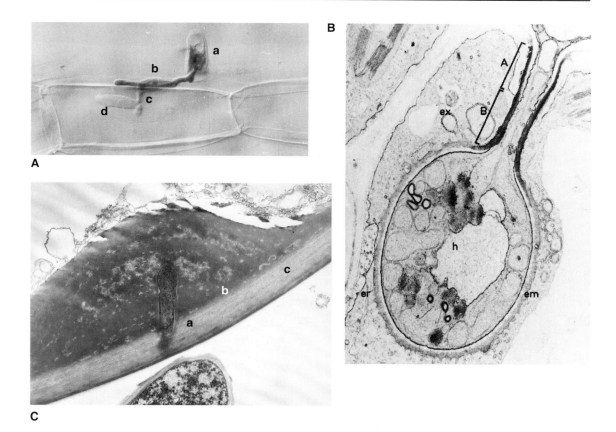

Figure 7.5 The development and structure of haustoria. A, Infection of an epidermal cell of a susceptible strain of barley (*Hordeum vulgare*) by barley powdery mildew (*Erysiphe graminis*) viewed by interference microscopy. a, a spore on the leaf surface; b, the germ-tube; c, the infection peg; d, a lobed haustorium characteristic of *Erysiphe*, developing within an epidermal cell (S. A. Archer). B, An haustorium of the Oomycete biotrophic pathogen *Albugo candida* in a host cell. The cell membranes of both fungus and host remain distinct, except (see text) where they are fused in a neck ring (A–B). (Woods, A. M. & Gay, J. L. (1983). Evidence for a neckband delimiting structural and physiological regions of the host plasma membrane associated with haustoria of *Albugo candida*. *Physiological Plant Pathology* 23, 73–88). C, A penetration peg (a) of *E. graminis* enclosed by a massive papilla (b) of wall material, halting penetration by the fungus. The reaction is an example of a response by a resistant strain of the host (barley) to a non-virulent race of the pathogen. (c) is the normal cellulose wall of the cell. (J. L. Gay).

cuticle and cell wall into the epidermis of leaves (Figs. 7.2, 7.5) usually show a narrow hyphal branch, the infection peg, developing from the region of the appressorium attached to the leaf. Thus the appressorium can provide a firm attachment that enables the infection peg to exert mechanical pressure against the cell wall. For this to be effective a high hydrostatic pressure must be generated in the appressorium, which will need a tough cell wall to withstand the pressure. In *M. grisea* the wall is melanized and inextensible, and genes encoding the enzymes of the melanin biosynthetic pathway have been shown to be needed

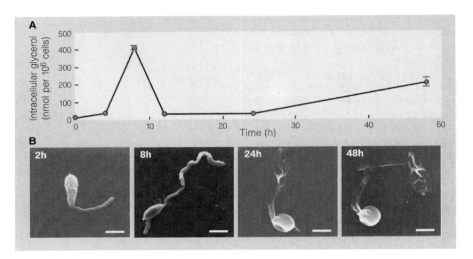

Figure 7.6 The formation of appressoria in *Magnaporthe grisea* accompanied by an increase in intracellular pressure. The internal pressure that drives infection hyphae of the rice blast fungus *M. grisea* through plant cell cuticles results from high osmotic pressure, generated metabolically in appressoria with strong walls. In this experiment conidia were allowed to germinate on a hydrophobic surface in water drops. During the first 8–10 h following germination glycerol was synthesized rapidly. From 10–20 h the glycerol was metabolized, decreasing the internal osmotic potential, and the conidium and germ-tubes collapsed. From 25 h appressoria began to form, separated from the hyphae by septa. Their walls became melanized and impermeable to glycerol, and glycerol level again rose. At 48 h the intracellular glycerol concentration, and the resulting pressure, were at their highest, because the glycerol was contained in a smaller volume within the appressoria, the rest of the hyphae having collapsed. Scale bars, 20 μm. (Reproduced with permission from de Jong, J. C., McCormack, B. J., Smirnoff, N. & Talbot, N. J. (1997) *Nature* **389**, 244–245, MacMillan Magazines Ltd.)

for pathogenicity. Lipid and carbohydrate reserves in the spore are metabolized to produce 3.2 M glycerol in the appressorium. The resulting high osmotic potential results in water uptake from dew to generate a hydrostatic pressure of 8 MPa, the highest so far found in a eukaryotic cell, and forty times the pressure of a car tyre (Fig. 7.6). *M. grisea* is able to drive its infection peg through a variety of artificial materials including polystyrene. In several different leaf surface pathogens, including *M. grisea*, it has been found that both germ-tubes and infection pegs contain more of the cytoskeletal protein actin than do other hyphal apices, suggesting adaptation of the infection structures to their mechanical role. In some fungi – but not *M. grisea* – there is evidence that the breach in the cell wall and cuticle may also be facilitated by enzymic digestion. Electron micrographs show a decrease in the electron density of the plant cell wall in the vicinity of the infection peg in many host–pathogen interactions. Many plant pathogenic fungi secrete a cutinase enzyme in the presence of cutin, the major structural component of cuticle. The presence of the enzyme was demonstrated immunologically around infection pegs of the pea pathogen, *Fusarium oxysporum* f. sp. *pisi*, which were in the process of invading a cell. With infection of papaya fruit by *Colletotrichum gloeosporoides*, specific inhibitors of cutinase prevented infection, but only while the cuticle was intact, demonstrating the role of cutinase in breaching the cuticle.

Beneath the cuticle lies the cell wall, consisting (Chapter 6) of pectic polysaccharides, hemicelluloses, cellulose, protein, and, in specialized cells, lignin. Many fungi can degrade some or all of these polymers.

Pathogenic fungi which are not obligate symbionts can be grown in culture, and enzymes that degrade plant cell walls are then usually detectable in the culture filtrate. Some necrotrophic pathogens such as *Sclerotinia fructigena* release, as described below, large amounts of enzymes which break down pectic substances and cause cells to separate. The role of the cell wall-degrading enzymes is harder to establish in diseases where the fungus is an obligate symbiont which cannot be cultured separately, and where tissue degradation is limited to the point of entry of the infection peg. Here the existence of enzyme activity has been inferred from ultrastructural observations. In transmission electron micrographs loss of microfibrillar structure probably indicates dissolution of cellulose by cellulases, and loss of the middle lamella between cells shows pectic enzyme action. These changes are usually very localized, suggesting that the enzymes may be bound to the hyphal wall, or quickly inactivated. Cutinase and cellulase have been shown by expressed sequence tagging to be synthesized by powdery mildew (*Erysiphe graminis*) when infecting its barley host, and the resulting cutin monomers act as signals to induce the development of infection structures. Cutinases can enhance infectivity, as was shown by transforming the weakly parasitic wound-infecting fungus *Mycosphaerella* with a cutinase gene from *Fusarium*. The transformant acquired the ability to invade a healthy host plant with an intact epidermis.

Whatever their role, appressoria develop as part of the infection process and their development often appears to be initiated by physical contact with particular leaf surfaces or by specific chemical stimulants. 'Contact stimuli' which are common in all fungi, not only pathogenic ones, are not well understood. They are regarded as occurring when development is elicited by physical contact in the absence of a chemical stimulus; the settling of zoospores on replicas of stomata, mentioned above, is an example of the effect of a contact stimulus. In *Uromyces appendiculatus*, the bean rust, contact of the germ-tube tip with a 0.5 µm stomatal guard cell lip is enough to trigger development of appressorium development. The fact that this is a response to a physical stimulus has been shown by mimicking the response on plastic replicas of silicon wafers possessing ridges of a similar height. Specific substances – mostly short-chain unsaturated aliphatic aldehydes and ketones – have also been shown to induce germ-tubes of *Puccinia graminis* to develop appressorium-like structures *in vitro*. Volatile fractions from leaves induce the formation of infection structures, as do several chemical compounds similar to those produced by leaves. However, in some fungi, including *P. graminis*, contact stimuli alone have been found to be sufficient to induce all the necessary infection structures. An example of the orientation of germ-tubes in response to contact stimuli is shown in Fig. 7.2C.

The ability of fungi to recognize the host plant is essential for invasion. Genes encoding molecules which are components of a signal transduction pathway, for example, receptor proteins and protein kinases, are expressed by fungal pathogens specifically during the infection process. When such genes are inactivated by targeted gene disruption (Fig. 7.7) the mutant strain is avirulent,

1. Chromosomal locus of interest

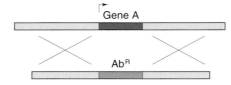

Gene A

2. Introduction of Gene Replacement Vector

3. Resulting Null Allele

Ab^R

Figure 7.7   One step gene replacement in fungi. In this process a genomic clone spanning a gene of interest is first identified from a genomic library. A restriction fragment containing regions of at least one kilobase on either side of the gene of interest is cloned, and an antibiotic resistance gene cassette (Ab^R) is inserted into the space normally occupied by the protein-encoding part of the gene of interest. The antibiotic resistance gene cassette will contain a promoter and terminator sequence to allow high level expression in the fungus. The gene replacement vector is then introduced into the fungus by transformation. A double cross-over event can occur, allowing the selectable marker gene to replace the chromosomal copy of the gene. The resulting locus is known as the **null allele**, as it expresses none of the biological activity associated with the native gene. Targeted gene replacements provide a direct test of the role of specific genes in fungal pathogenicity. Figure and legend by N. J. Talbot.

but virulence can be restored by transformation to reintroduce an undamaged copy of the gene. Fungi in which specific components of cyclic AMP and MAP kinase signalling pathways have been identified and shown to be involved in the infection of a host are *Magnaporthe grisea, Ustilago maydis*, and the insect pathogen *Metarhizium anisopliae* (page 429).

When the fungus has penetrated the cell wall the outcome of the close contact between the two organisms depends on the reaction of the invaded cell. It may die; if it does, the continuing growth of the fungus depends on its ability to utilize the dead cell as a substrate. In most specialized symbioses the invaded cell does not die immediately and an interface develops between the living plant cell and the fungal hypha. Such interfaces are the characteristic feature of biotrophic associations (pages 391–408). Some fungi such as Plasmodiophomycetes (page 552) and pathogenic Chytridiomycetes become entirely intracellular. Most fungal pathogens, however, invade plant cells by means of specialized hyphae while developing most of their mycelium outside the cells of the host, either within the tissues or on the plant surface. A hypha which has penetrated the cell wall then swells at its tip and ceases apical growth, forming a haustorium. Normally the plasmalemma of the host cell is not broken, but forms an invagination so that the hypha is still topologically on the outside of it.

The structure of haustoria of fungi is characteristic of the major taxonomic divisions to which pathogens belong. Erysiphales (Figs. 7.2, 7.5) have lobed or digitate haustoria with filamentous processes in which the surface area/volume ratio is high. They are separated from the mycelium by a septum which has a central pore as is typical of all Ascomycetes. There is one nucleus per haustorium.

Haustoria of downy mildews are without septa and continuous with the intercellular mycelium. Like the hyphae of other Oomycetes they are coenocytic and contain variable numbers of nuclei. Haustoria of rust fungi are also simple in shape if they are developed from mycelium originating from uredospores or teliospores, but basidiospore infections give rise to filamentous haustoria. Although there is much variability in the form of haustoria between different taxonomic groups, there are also some striking similarities between them. Throughout both rusts and powdery mildews, the host plasmalemma, where invaginated round the haustorium, differs from the rest of the host plasmalemma. The intramembrane particles seen in freeze-fracture preparations of normal plasmalemma are absent, and polysaccharide material is abundant on the outer surface. Because of these distinctive features this modified plasmalemma has been termed 'the extrahaustorial membrane'. The other striking similarity between haustoria of rusts and powdery mildews, not seen in downy mildews, is an electron-dense band around the neck of the haustorium which extends to include both plant cell and hyphal plasmalemma. A functional advantage of this may be that when the plant protoplast shrinks due to plasmolysis under water stress, the membrane contact between plant and fungal cell is not disrupted. Another possible function might be to isolate the membrane of the haustorial complex from the non-haustorial membranes of the plant and fungal cells and thereby to maintain differences in membrane composition.

The haustorium is believed to have an important role in the transfer of substances between host and fungus. Nutrients are transferred otherwise the mycelium would not grow. Other substances are also transferred between the partners and evoke the morphological and physiological changes characteristic of the disease. Intercellular hyphae may also act as sites for chemical exchange, and the relative importance of haustoria and intercellular mycelium has been the subject of considerable research.

Some fungi are dependent on nutrient uptake via haustoria. For most powdery mildews, haustoria constitute the only possible site of exchange because most of these parasites develop no intercellular mycelium. Spores of these fungi germinating on the leaf surface establish haustoria in the epidermal cells only, and then the mycelium develops from the appressorium over the surface of the leaf. The mycelium only develops after haustoria have formed; radioactive labelling of sulphur and phosphorus compounds in the host were not transferred to mycelium until haustoria had formed. However, in other symbioses the intercellular hyphae are either significantly or exclusively involved. The genus *Claviceps* which causes ergot in grasses does not form haustoria, and the ergot (sclerotium) which develops in place of a seed is supplied with nutrients from the host via intercellular mycelium which replaces the host cells at the base of the ovary. The mitosporic fungus *Cladosporium fulvum* is a pathogen of tomato with a specific gene-for-gene relationship with its host (page 417) but grows exclusively in intercellular spaces without invading cells.

Most rust fungi have both haustoria, in leaf mesophyll cells, and an extensive intercellular mycelium. The hyphae can be seen closely adpressed to host cells. Experiments with rusts and downy mildews showed uptake of some substances such as sugars by intercellular hyphae whereas others, including amino acids, were only taken up when haustoria had been formed.

Root-infecting fungi

The soil is a reservoir of fungal propagules – spores, sclerotia, mycelium on living plants and dead plant residue – from which pathogenic fungi can grow and invade plant roots. Two processes are involved; the pathogen must first reach, then invade, the host root.

Reaching a host root   For soil-borne pathogens the problems of reaching a host differ from those of airborne fungi. Spores cannot travel so fast or so far in soil as they can in air, so dispersal is more limited. Any movement of spores in soil must be caused by the flow of soil water, by the swimming of motile spores or by the activities of soil animals. Where the disease spreads largely by hyphae growing from root to root through the soil, spread is slow.

Soil is a good medium for fungal survival, and for growth also, when nutrients are available. Pathogenic fungi may thus remain in it, viable but dormant, for long periods, in some cases 10 years or more. Finding a host root then becomes a matter of starting to grow at the right time, when a host root approaches. Soil fungistasis prevents fungal spores germinating unnecessarily. Nutrient substances exuded from roots overcome this inhibitory effect for some spores. However, more specific ways are known in which host roots break fungal dormancy. The white rot disease of onion plants is caused by a fungus, *Sclerotium cepivorum*, which survives the absence of its host in the form of sclerotia. These germinate only in the presence of exudates from roots of plants of the onion (*Allium*) genus. The active substances are unsaturated short-chain thiols (Fig. 4.18D), as has been proved by inducing sclerotia to germinate by their application. Such specificity, combined with the ability of the sclerotia to remain dormant for long periods, is an effective and economical way for a pathogenic fungus to maintain itself at the expense of a stable plant community in which the host can be depended on to recur at intervals.

If growth starts only in the vicinity of a root it is likely that some hyphae will grow on to the root surface. It is known that some roots attract the hyphae of some fungi so that they alter their direction of growth by tropism. The pathogen of cereals *Gaeumannomyces graminis* (Fig. 7.8) attacks seedling roots, growing through soil from crop residues on which it survives, towards living roots. Germ-tubes of the Oomycete genera *Pythium* and *Phytophthora* show tropism towards root exudates. Tropism is effective over distances of a few millimetres and probably enables hyphae which are already growing close to a root to 'home in' on it. This tropism is preceded by the taxis of zoospores which make use of their ability to sense, and swim in response to, a variety of stimuli. Zoospores of pathogens are released into water and can travel passively in water which drains down through the spaces between soil particles. However, the direction of swimming is affected by gravity and their negative geotaxis causes them to swim upwards, which may help them to regain the upper layers of soil where roots are plentiful. They also swim against the direction of flow, which could have the same effect. Zoospores gather on root surfaces, most abundantly in the elongation zone just behind the root tip, and have been shown to be attracted by many of the substances exuded by roots – amino acids, sugars, ethanol and other alcohols, fatty acids and aldehydes. There is little evidence for host specificity in this attraction but it could make a spore more likely to come to rest on a root than on

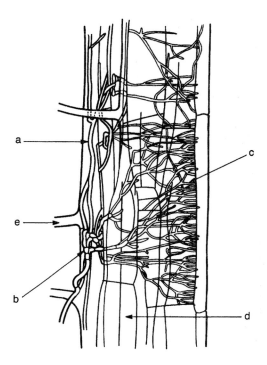

Figure 7.8 Infection of a root of a seedling of wheat (*Triticum aestivum*) by the take-all fungus (*Gaeumannomyces graminis*). Dark-pigmented runner hyphae (a) which may aggregate into strands (b) consisting of several hyphae, growing over the root surface. Colourless hyphae (c) arising from these grow into the root, through the cortex (d) and towards the vascular cylinder, which they eventually invade. Root hair (e). (After Skou, J. P. (1981). See references, page 453.)

an inert particle. It has been found that the electrical field that exists around plant roots can play a part in attracting swimming zoospores (page 229). By the use of very sensitive microelectrodes it has been possible to measure the charge around root tips of pea plants. When suspensions of zoospores of the Oomycete root pathogens *Pythium aphanidermatum* and *Phytophthora palmivora* were placed in electrical fields of comparable magnitude to those measured round roots, *Pythium* spores moved towards the cathode and *Phytophthora* spores towards the anode. If a root is wounded, the site of the wound becomes relatively negatively charged, and it was found experimentally that the wound site attracted zoospores of *Pythium* (Fig. 7.9A). When these were washed off and the root placed in a suspension of *Phytophthora* zoospores these aggregated on the undamaged, positively charged part of the root surface (Fig. 7.9B).

Once on the root surface, swimming spores encyst and this is stimulated by the presence of nutrients, which also promote subsequent germ-tube development in preference to production of more zoospores. *Phytophthora* zoospores have been found to become briefly sticky just after encystment, another possible adaptation

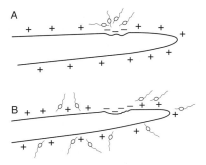

Figure 7.9 Electrotaxis of Oomycete zoospores to plant roots. The direction of swimming of zoospores placed in an electrical field was found to vary between species, with *Pythium aphanidermatum* zoospores moving towards the cathode and *Phytophthora palmivora* towards the anode. Plant roots bear a small electrical charge on their surface, normally positive, but changing to negative in parts of the root that are damaged. In an experiment to test for a role of electrotaxis in the attraction of zoospores to plant roots, damaged roots were placed in suspensions of zoospores of either *Pythium* or *Phytophthora*. *Pythium* spores (A) were differentially attracted to the negatively charged damaged area whereas *Phytophthora* zoospores (B) aggregated all over the positively charged intact parts of the root. (Redrawn from Morris, B. M. & Gow, N. A. R. (1993). *Phytopathology* **83**, 877–882.)

to infection which would prevent the spore from being washed off the root surface before invading it (page 229).

Root penetration Most soil fungi that invade plant roots are only able to do so via younger parts of the root or through wounds. Only a few fungi are able to invade roots protected by bark, but many can enter the younger regions where the outer layers consist of living cells. The root hairs, epidermal cells which grow extensively into the soil, are particularly vulnerable.

The environment generated in the soil by the presence of a plant root – the rhizosphere – was described in Chapter 6. Potential pathogens differ from the saprotrophic fungi of the rhizosphere by being able to invade the root. As in the case of airborne pathogens, special structures are formed to infect and colonize. Invasion is often preceded by a phase of ectotrophic growth with hyphae spreading extensively on the root surface. They may form various kinds of aggregation such as rhizomorphs, or infection cushions, depending on the fungal species. This local concentration of growth appears to allow them to accumulate the resources necessary to overcome the host's resistance at the point of invasion. *Gaeumannomyces graminis*, the cereal disease known as 'take-all', in which roots and stem bases are attacked, at first grows ectotrophically in the form of 'runner hyphae' on the root surface (Fig. 7.8). These become wide and darken, and may anastomose and aggregate, sometimes even encasing the root in a sock-like mycelial cylinder. Narrow, colourless infection hyphae then develop and penetrate the root epidermal cells, sometimes many hyphae attacking each cell. Host cells normally resist invasion by forming lignified caps ('lignitubers') of cell wall material which enclose and wall off tips of penetrating hyphae. An equilibrium phase may occur in which lignitubers are broken through by the hyphae as fast as the cell can produce them. Finally the hyphae overcome resistance, invade the cell and adjacent ones, and spread rapidly in the root cortex. Although invasion is slowed by the endodermis, the fungus can cross it and enter the stele, destroying the phloem tissue and invading the xylem water-conducting cells, causing the plant to die. Active resistance by the plant is evident in the production of lignitubers, the slowing of invasion at the endodermis, and in the observation and that the infection takes considerable time to develop and so is never seen at the root apex.

The root cortex is not the preferred habitat of all root pathogens. The soil-borne fungi known as vascular pathogens have the ability to enter living roots across the cortex and endodermis and grow mainly in the xylem vessels. They are believed to enter via natural or accidental wounds in roots, then grow across the cortex and into the xylem vessels in the central stele, where most of their mycelium is produced. Some, for example *Verticillium albo-atrum* in hops and lucerne and *Fusarium oxysporum* f. sp. *cubense* in bananas, cause important diseases. The symptoms are wilting and dying of aerial parts of the plant. Some apparently similar fungi, however, appear to inhabit the xylem without adverse effects on the plant. These are known as **endophytes** (page 405).

The Plasmodiophorales, for example *Plasmodiophora brassicae*, which causes club-root of cabbages, have a unique mode of invasion of host root hairs. The infecting zoospore encysts on the cell surface and then produces an intracellular, bullet-like structure which appears to be shot quickly and violently into the host cell, to be followed by the contents of the zoospore.

Necrotrophic Attack

Necrotrophic pathogens first kill, then feed upon, the host's cells. Necrotrophic disease is recognizable by patches of dead, often blackened, tissue. These range in size from leaf spots a few millimetres in diameter to extensive dead areas, as occurs in potato blight or Dutch elm disease when the whole plant is killed. Necrotrophs produce toxins which kill plant cells, and enzymes which degrade plant constituents so that the fungus can use them for food. Necrotrophy may occur from the beginning of an infection, as in rots of fruit caused by fungi of the genus *Sclerotinia*. However, some infections begin with a biotrophic phase, with haustoria in living cells, and become necrotrophic later on. This happens with leaf infections by spores of *Phytophthora infestans*.

Many necrotrophs show low specificity to host plants. *Botrytis cinerea*, for instance, is an opportunistic pathogen of a wide variety of weakened, juvenile or senescent plants, causing grey mould in cold, damp conditions (Fig. 7.10). Some necrotrophs have much greater host specificity. Several species of *Phytophthora* (e.g. *P. infestans*, *P. megasperma*) are specific for a few host species, and furthermore, any strain ('race') of the fungus can attack only a limited number of strains ('cultivars') of the host plants.

Unlike most biotrophic pathogens, necrotrophs are usually easy to culture in the absence of their hosts. For this reason necrotrophs like *Phytophthora*

Figure 7.10 Grapes attacked by the necrotrophic pathogen *Botrytis cinerea*. The host tissues are discoloured and broken down, and there are clusters of conidia on the grape surface. Such attack spoils grapes intended for the table, but in some vineyards grapes are deliberately left on the vine until, as tissues age and resistance declines, extensive *B. cinerea* infection occurs. Damage to the grape cuticle then causes water loss and hence increased sugar concentration, permitting the production of fine sweet wines such as Sauternes (Chapter 8) (F. M. Dewey).

megasperma, in which virulence may be determined by single genes, have been much used to investigate molecular interactions between host and parasite, since it is possible to culture mutant strains of the fungus with altered pathogenicity to investigate the disease process, and also to obtain substances of exclusively fungal origin and test their effect on host tissue.

Necrotrophic parasites survive the absence of their hosts in the form of resting structures such as spores or sclerotia, or in dead host material. Some necrotrophs can survive away from the host as saprotrophs. They have a free-living phase during which they have to compete with other fungi which are exclusively saprotrophic. Thus their ability to reinfect a host depends on their competitive saprotrophic ability.

Toxins

Toxins of pathogenic fungi are secondary metabolites, like the antibiotics and mycotoxins discussed in Chapter 6. Most have low molecular weights. They are synthesized from precursors derived from the different pathways of primary metabolism and so are chemically diverse (Fig. 7.11). The chemical type of toxin is usually related to the taxonomic group of fungus, and any species produces one or a few chemically related toxins. For example, the **fusicoccins** formed by

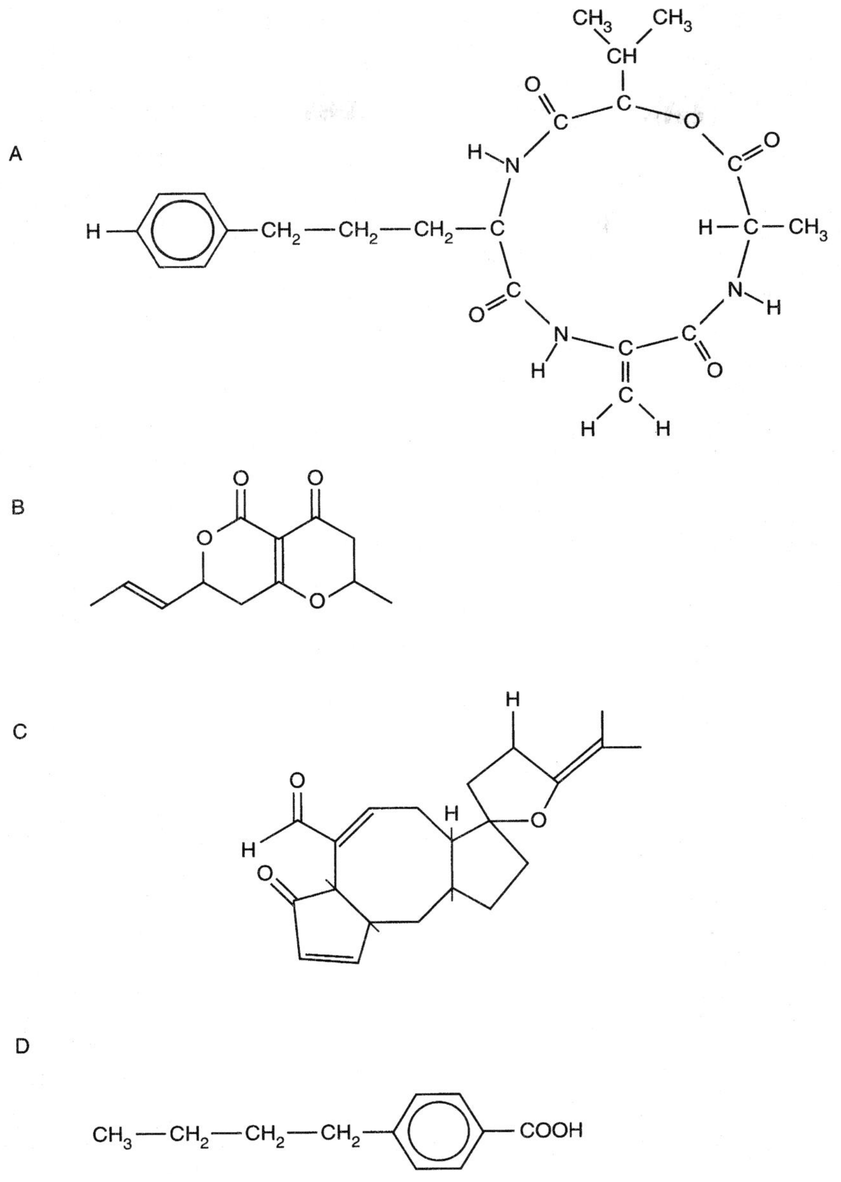

Figure 7.11 Examples of toxins produced by plant pathogenic fungi. They are secondary metabolites (page 513). A, A cyclic depsipeptide, from a pathogenic race of *Alternaria alternata* which infects apples. It is derived from amino acids, and causes the death of leaf cells in susceptible apple cultivars, but not in resistant ones. B, Deoxyradicin, from *Alternaria helianthi*, derived from ketones. The fungus is a pathogen of sunflower (*Helianthus annuus*) but the toxin affects plants of other species as well. C, A sesquiterpene, derived from acetate, produced by *Cochliobolus heterostrophus* (anamorph *Drechslera maydis*). Susceptibility of maize (*Zea mays*) to this toxin, is conferred by a single gene. D, Fusaric acid, produced by several species of *Fusarium*, which binds metal ions, and also increases the permeability of cell membranes.

Fusarium amygdali, the cause of peach and almond canker, are diterpenes with varying substituent groups.

Like production of other secondary metabolites, toxin production is affected by environmental variables, especially nutrient supply. Toxins may be produced only at certain stages in the host life cycle or fungal disease cycle. Determining their role in pathogenesis is difficult. Although many can be purified from fungal cultures, chemically characterized, and applied to plant tissues, it is hard to establish whether their action under these conditions is identical to processes occurring during disease. However, in some cases the symptoms of the disease can be reproduced by application of the toxin. For example, fusicoccin given to uninfected plants causes the symptoms of peach and almond canker, which indicates that the toxin does determine pathogenicity. Moreover it is found in diseased tissue at the concentration required to cause symptoms. Conclusive proof for the essential role of a toxin in pathogenicity has been obtained by targeted gene disruption in the maize pathogen *Cochliobolus* (formerly *Helminthosporium*) *carbonarum*. Mutants were produced which lacked the gene for an enzyme required for the synthesis of its host-specific peptide toxin, HC-toxin. These mutants were non-pathogenic, but could be restored to pathogenicity by the re-insertion of this single gene. Apart from killing cells, toxins affect plants in a variety of ways. Some may cause yellowing (chlorosis) of green parts, whereas others interfere with processes vital to development, resulting in abnormal growth, or with the transpiration stream, causing wilting and death, sometimes of the whole plant as in Dutch elm disease.

The specificity of toxins for plants varies, probably because they act on different plant components, some of which are common to all plants, for example cell membranes, whereas others are unique to a species or strain. Fusicoccin has a toxic effect on all plant species tested. It activates an ion transport system which pumps potassium ions into the cell. This interferes with the normal mechanism for regulating opening of the stomata, mediated through changes in potassium transport, which close stomata in response to environmental changes. Fusicoccin keeps the stomata open and the plant wilts.

A few cases are known where a fungal toxin affects only a few species or cultivars of plants. Susceptibility to a toxin may result from a single gene, as in the case of some strains of cereals which are susceptible to toxins from *Cochliobolus (Helminthosporium)* species. A biologically interesting but economically unfortunate coincidence is that single genes conferring advantages to a grower (resistance to rust fungi in oats, male sterility in maize) also confer susceptibility to particular *Cochliobolus* toxins. Crops into which such resistance genes have been bred have been prone to epidemics of disease caused by *Cochliobolus*. It appears that the gene product which is the receptor for the rust fungus elicitor (pages 417) is the target for the *Cochliobolus* toxin.

The barley pathogen *Rhynchosporium secalis* owes its pathogenicity, at least in part, to its production of several small toxic necrosis-inducing peptides ('NIPs') which cause necrosis in barley. The action of these toxins is not highly specific, and they act equally against different barley cultivars. As well as playing a role in attack on the host, they are also one of the features of the pathogen that are recognized by the host, causing it to respond by expressing resistance. In this capacity, as elicitors of host resistance, they are very highly host-specific, as barley cultivars vary in their ability to recognize different NIP toxins.

Enzymes

Enzymes produced by pathogens enable them to counteract host resistance and degrade host tissues. The tissue-degrading action of enzymes is conspicuous in soft rots, necrotrophic diseases in which breakdown of pectic substances causes maceration of the tissues which then die, with softening and browning in advance of the hyphae of the pathogens. *Pythium* attacks the stems of seedlings and *Sclerotinia* the fruits of apple and plum in this way. The cell wall-degrading enzymes of such fungi are probably the best investigated extracellular enzymes of pathogens, and their mode of attack on substrates, and regulation, is well understood. Other enzymes may determine pathogenicity or affect virulence by breaking down antifungal substances such as phytoalexins or other phenolics (page 410) in the host plant.

Most of our understanding of the significance of cell wall-degrading enzymes in plant disease is based on the specificities and regulation characteristics of enzymes produced by pure cultures of pathogens. Inferences can be drawn from these studies about the probable role of enzymes in disease. There is evidence that in some diseases pathogenicity depends on the ability to produce wall-degrading enzyme. One example is the attack by *Fusarium oxysporum* on pea leaves where pathogenicity is lost in mutants lacking the gene for a particular cutinase enzyme, one of two cutinases that the fungus produces.

Cultures of plant pathogenic fungi produce a variety of pectinases and cellulases, as well as enzymes which degrade hemicellulose and protein. Pectinases are produced by many fungal and bacterial plant pathogens. They break down proteins in the middle lamellae, having both endo- and exopolygalacturonase activity. They also demethoxylate galacturonic acid residues. The effect is to separate the plant cells from each other and soften their walls, a process called maceration. This action alone is sufficient to cause the death of plant cells, perhaps because their osmotic balance is upset when the cell wall is softened, also possibly as a result of electrolyte leakage. It is thus possible that the ability to produce enzymes could confer pathogenicity or enhance virulence. Cellulase, although produced by necrotrophic pathogens, does not appear to kill plant cells.

The amounts of many wall-degrading enzymes are increased by the presence of their specific substrates. They are produced constitutively at low levels, but enzyme induction increases their amounts. Polysaccharidases of the wilt pathogens *Verticillium albo-atrum* and *Fusarium oxysporum* are specifically induced by very low levels of monomers of their polysaccharide substrates. Galactose induces pectinase activity, xylose induces xylanase, and several other enzymes are similarly subject to specific induction. Their activities are all repressed in a less specific way by easily metabolized carbon compounds. Even the sugars which at low concentrations act as inducers become repressors at slightly higher concentrations.

This regulation of the activities of different wall-degrading enzymes probably occurs in the plant. An invading hypha meets different chemical components of the tissue in sequence, because of their spatial arrangement, so that a sequence of different enzyme activities is likely to be displayed during the course of fungal invasion. There is evidence that this occurs. Washed, insoluble cell walls induce cultures of *Verticillium albo-atrum* and other pathogens to produce in sequence first pectic enzymes, then hemicellulases and finally cellulase. The same sequence has been found in plant tissues of several plant species as fungal disease progresses, with pectic enzymes active in the early stages and cellulase only when tissues are moribund. It has been suggested that potentially necrotrophic pathogens could be restricted to biotrophy by repression of wall-degrading enzymes in the presence of sugars produced by the host plant.

Although wall-degrading enzymes are a conspicuous feature of necrotrophic pathogens, they are not the only enzymes associated with pathogenicity. Enzymes which counteract host resistance can also determine pathogenicity. Phytoalexins, antifungal substances produced by plants in response to infection (pages 408–411), are broken down by some fungi. An example where this ability promotes pathogenicity is in the attack on broad bean, *Vicia faba*, by *Botrytis* species, discussed earlier in this chapter (page 376). *Botrytis fabae* is a parasite regularly producing chocolate spot disease, while the related *B. cinerea* is a less virulent, opportunistic invader. Applying a drop of *B. cinerea* spore suspension to a leaf induces rapid accumulation in the leaf of the broad bean phytoalexin, wyerone, to a concentration inhibitory to the fungus. However, when *B. fabae* invades the leaf, wyerone first accumulates and then decreases in concentration as the tissue becomes necrotic and the lesion spreads. The ability to break down wyerone may therefore contribute to the virulence of *B. fabae*. A similar phenomenon has been shown for *Nectria haematococca* in which virulence to peas is correlated with ability to detoxify pisatin, a phytoalexin.

The possession of enzymes for detoxification of phytoalexins affects the host range of strains of the fungus *Gaeumannomyces graminis*, the cause of the cereal disease take-all. Some strains of this fungus cause disease in both oats and wheat, whereas others can only infect wheat. The ability to cause disease in oats depends on the fungus possessing a gene for the enzyme avenacinase, which detoxifies the antifungal substance avenacin which is only present in oat roots. Targeted disruption of the avenacinase gene renders the oat-infecting strain harmless to oats, but does not impair its pathogenicity to wheat which has no avenacin in its roots.

Blockage of vessels – the vascular pathogens

The direct actions of toxins and enzymes on the cells of the host plant have been described above. Some fungi, the vascular pathogens (vascular wilt fungi) can also cause cell death indirectly, by obstructing the flow of liquid in the xylem vessels and tracheids. This is a consequence of mycelial growth and the secretion of viscous substances, or of enzymes that partly degrade host walls, releasing viscous material. Some can also kill their host cells with toxins (page 386). They are a specialized group with a distinctive ecology and appear to be adapted to inhabit the water-transporting xylem tissue of plants. As xylem inhabitants they have a similar habitat to the harmless or sometimes even beneficial fungi which live as endophytes in the tissues of plants without causing any symptoms of disease (page 405). Only a few species of fungi cause vascular wilts. In *Fusarium oxysporum* about 70 'form species' (abbreviated to f. sp.) have been designated which are more or less host-specific for different plants such as peas, beans, tomatoes and bananas. Much less host-specific, but with the same kind of pathogenesis, is *Verticillium albo-atrum*, which affects many plants. Both are mitosporic fungi which occur in the soil, although they do not grow very much there but survive as thick-walled mycelium, sclerotia or chlamydospores. These are kept quiescent by soil mycostasis (page 337) until exudates from passing roots reverse the inhibitory effect of the soil, as occurs, for example, when chlamydospores of *F. oxysporum* f. sp. *cubense* germinate and attack the host, banana. Dutch elm disease is a vascular wilt in which the pathogen, *Ophiostoma ulmi* or *O. novo-ulmi* (page 53), does not live in the soil but is introduced into the xylem by the flying beetles which feed on the tree and carry the spores. When the hyphae of vascular wilt fungi infect a root they grow radially across the cortex towards the xylem, and enter via pits in the vessel or tracheid walls. The fluid in the xylem contains low levels of nutrients, but sufficient for growth of vascular wilt fungi. Growing in the xylem, the fungi produce spores by budding from the mycelium. Such spores are termed **blastospores**. Thousands of blastospores per millilitre of xylem sap have been recorded. These spores are swept along in the xylem stream until arrested at one of the perforation plates, sieve-like structures which occur at intervals in the tubes. Here they return to hyphal growth and grow through the perforation plate, sporulating again on the other side. The xylem sap becomes polluted by toxins and enzymes of the fungus. Its flow is physically obstructed by the mycelium and blastospores, or by the gels and tyloses which are induced to form in the xylem by the fungus. These have the effect of inhibiting the spread of spores, but as a resistance mechanism have the disadvantage of disrupting the host's water supply. The gels are of hemicellulose and pectin and are produced from the plant cell wall. The growth substances ethylene and indoleacetic acid can induce gelation in the absence of fungi. Since fungi also produce these substances they may provide the mechanism of induction in disease as well. A **tylosis** is a barrier to flow developing in a vessel, when the cell wall of an adjacent living parenchyma cell of the medullary ray balloons through a pit into the vessel. It appears to be an active process, and to involve induction of plasticity in the parenchyma wall. Resistance in plants appears to be correlated with an ability to produce abundant tyloses. Tyloses also occur in plants as a result of non-fungal damage.

Biotrophic Associations

Necrotrophic attack, described in the foregoing section, invariably harms the host plant, killing its tissues and using its accumulated food supply. However, in biotrophic associations, where the tissues are invaded but remain alive, there is the possibility of a benign or mutually advantageous relationship. All biotrophic associations are symbiotic in the sense that two organisms live together. The difference between parasitic and mutualistic biotrophy is in the sources of nutrients for each of the partners. Parasites get all their food from their plant hosts, while in mutualism there is a two-way flow of nutrients. Other differences between parasites and mutualistic symbionts, for example differences in host specificity and in compatibility between partners, result from this difference in nutrition, as is discussed below.

Fungal symbionts of aerial parts of plants (Fig. 7.12A) must be parasites since they are not connected to any other food source. Symbionts of roots (Fig. 7.12B) which have mycelial extensions into the soil have the possibility of introducing nutrients into the root as well as removing them out of it. In fact almost all biotrophic fungi in roots appear to be mutualistic and such a fungus–root association, of which there are many forms, is termed a **mycorrhiza**. Mycorrhizal associations are described in a separate section below. Probably the only

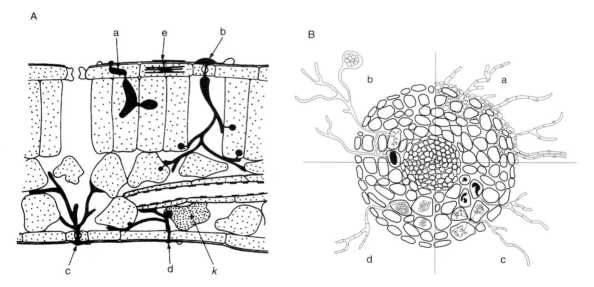

Figure 7.12   A, Types of interaction which fungal pathogens may have with susceptible leaf tissue. a, An intracellular biotrophic fungus; b, a stoma-penetrating, mainly intercellular biotrophic fungus that forms morphologically differentiated, intracellular haustoria; c, an intercellular biotrophic fungus without haustoria; d, a fungus which penetrates between epidermal cell walls and has a short biotrophic phase in each successively invaded cell, later killing at (k); e, a haustorium-producing biotroph with extracellular mycelium developing on the leaf surface. (After Heath, M. C. (1986). Fundamental questions related to plant–fungal interactions: can recombinant DNA technology provide the answers? In J. Bailey, ed., *Biology and Molecular Biology of Plant–Pathogen Interactions*, pp. 15–27. NATO, ASI Series, Vol. H.1.) B, Types of interaction between plant roots and different kinds of mycorrhizal fungi. a, Ectomycorrhiza with fungal sheath outside the root epidermis connected to mycelium in the soil, and to intercellular fungal tissue (the Hartig net) in the root cortex; b, vesicular–arbuscular mycorrhiza, infecting the root from a large spore in the soil, forming intracellular vesicles and branched arbuscules in cells of the inner root cortex; c, orchid mycorrhiza, forming intracellular hyphae which live briefly in each invaded cortical cell before being digested; d, ericaceous mycorrhiza forming bundles of hyphae in cells of the inner cortex, which are connected to mycelium in the soil. After Selosse, M. A. & Le Tacon, F.; see Further Reading and Reference.

biotrophic root fungi which are parasitic are those which are wholly intracellular such as Plasmodiophorales and endobiotic Chytridiales.

In all biotrophic relationships between heterotrophic fungi and autotrophic plants, the nutrients which pass to the fungus are partly or mainly products of current photosynthesis. The fungus is subject to selective pressure which favours the ability to redirect assimilates away from the host and towards the fungus. Also, because the host remains alive the association persists for long periods. During this life together, complex interactions affect physiology and development. In mutualistic associations, where the partners are fitter (i.e. survive better) together than apart, selection has probably favoured cooperative physiology and development. The relationship between parasite and host, on the other hand, is more likely to have resulted from a sequence of evolving resistance by the host which in turn selects for the parasite that can overcome the resistance.

Thus the ecology and development of parasitic and mutualistic biotrophs differ and will be described in separate sections. However, at the cellular level the kinds of interface formed are in many respects similar and are now discussed.

Interfaces between plant and fungus in biotrophic associations

The interface for nutrient transfer is an area where host and fungal cell lie alongside each other. Transfer of nutrients occurs through both plasmalemmas. Specialized structures where the walls of both partners are absent – haustoria of parasites, and arbuscules and vesicles of endotrophic mycorrhiza – occur in many, but not all, biotrophic associations. Where these structures are present they do not penetrate the host plasmalemma but invaginate it. They thus remain functionally outside the cell in the intercellular space and nutrients and water must still enter the fungus across two plasmalemmas. The plant tissue outside the cytoplasmic part of the cells is the apoplast, and water movement and diffusion of small molecules may occur passively in it. The haustorium is functionally located in the plant's apoplast, since it is outside the plant cell membrane.

The presence of an infecting fungus can change the chemistry of the cell walls and permeability of the cell membrane. In haustorial regions the part of the plant cell plasmalemma pushed inwards by the fungus is prevented from forming a cell wall as it does in the rest of the cell. Several substances have been shown by immunocytochemistry to accumulate in the space between the host cell membrane and the fungal cell wall. Among these are hydroxyproline-rich glycoproteins, which are also abundant in the walls of actively dividing tissue at the tips of plant roots. In some arbuscular mycorrhizas wall material is still synthesized around the fungal arbuscules but remains as fibrils and is not organized into a normal wall. However, as the fungal structures age plant cell wall synthesis may be reactivated to seal them off from the rest of the root. There is evidence that, as well as locally inhibiting wall formation, biotrophic fungi can carry out a limited lysis of the pectic components of existing cell walls.

A wall structure which must be of functional significance, since it occurs in the majority of biotrophic parasites but not mutualists, is the 'neck ring' at the base of the haustorium (Fig. 7.5B). It was mentioned in connection with the infection process (page 381). It is a suberized ring, presumably waterproof, formed in the plant cell wall surrounding the point of entry of the haustorium. It appears to join the host and fungus plasmalemmas together and also to seal them to the plant cell wall. Thus the haustorium appears to be isolated in its own apoplastic compartment both from the cell it is in, and from the cell walls and intercellular spaces (apoplast) of the surrounding tissue. Arbuscules of mycorrhiza do not have neck rings and are thus not isolated in this way from the cortical apoplast, the part of the root in which there is free diffusion of water and dissolved substances from the soil.

The structure of host cell membranes, as well as walls, can be altered by a biotrophic partner. However, there are differences in the way parasites and mutualists affect the host's membranes. Although mycorrhizal arbuscules and fungal haustoria look similar in electron micrographs, with a double membrane structure in both cases, the composition of the plant membrane component is different. In parasitic associations, for example with species of *Erysiphe*, *Uromyces*, *Puccinia*, *Albugo*, *Bremia* and *Plasmodiophora*, one class of membrane proteins, H^+ ATP-ases associated with active membrane transport, are absent from the invaginated part of the host membrane, though not from other parts. These proteins are present in normal amounts in host cell membranes associated with fungi in other types of mutualism, for example ectotrophic and orchid mycorrhizas. The implication is that in parasitic associations the fungus alters normal plant cell membrane transport in the region of the haustoria, and that this is not so in mutualistic symbioses.

Mycorrhizal symbiosis and endophytes

The establishment and growth of most plants requires, or is enhanced by, the presence of specialized fungi in the soil which form close associations with their roots. Such associations (Figs. 7.12B–7.14) are known as **mycorrhizas** (Greek *mykos*, fungus; *rhizon*, root). Experiment has shown that mycorrhizas, as compared with uninfected roots, can increase the uptake of plant nutrients such as phosphorus and nitrogen, particularly when these are at low concentrations or in insoluble form in the soil. Fungi assist nutrient uptake by plants from dilute solutions by two mechanisms. Hyphae extending into the soil from the root may play a scavenging role, since the active transport systems in the cell membranes of fungi are capable of concentrating solutes against steep concentration gradients (page 149). The plant nutrients thus concentrated are then released by the hyphae and taken up by the associated root cells. Also, nutrients concentrated by the fungus and held in the vicinity of the root may serve as reserves for future utilization. Such reservoir effects are probably of greatest importance to the plant where the concentration of soil nutrients shows seasonal fluctuations. The mechanisms which cause the fungus to release its stores of nutrients are not understood. Uptake by the plant is assumed to be by normal uptake systems, as is suggested by the continuing presence at the host–fungus interface of H^+ ATP-ase, mentioned above.

Nutrient flow is a reciprocal process in most mycorrhizal associations. The plant supplies the fungus with carbon compounds, mainly hexose sugars. The fungus takes these up and converts them into sugar alcohols such as mannitol, arabitol and erythritol. Thus the hypha becomes a sink for hexose sugars, maintaining a hexose gradient that permits continued passive diffusion of hexose

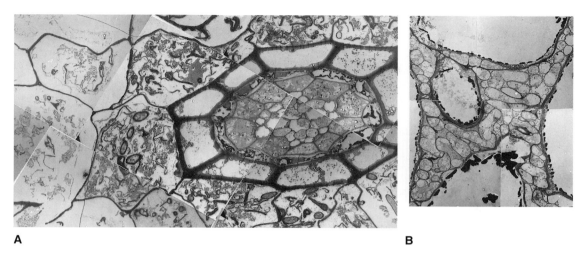

A   B

Figure 7.13   The relationship between fungal hyphae and plant root cells in endotrophic and ectotrophic mycorrhiza. A, Endomycorrhiza of the hair roots of *Pteridium aquilinum* (bracken), showing the intracellular location of the fungus. The electron micrograph of a cross-section of the root shows, in successive cell layers, the uninfected outer layers of the root cortex; the innermost layer of the cortex with cells filled with fungal hyphae; the endodermis, and within it the vascular cylinder, with no fungal tissues. (Turnau, K., Kottke, I. & Oberwinkler, F. (1993). Element localization in mycorrhizal roots of *Pteridium aquilinum* (L.) collected from experimental plots treated with cadmium dust. *New Phytologist* **123**, 313–324.) B, Ectomycorrhiza formed by the fungus *Rhizopogon roseolus* with larch (*Larix*). The electron micrograph shows relatively large and empty cortical cells of the host with intercellular spaces packed with fungal tissue, forming the Hartig net (K. Turnau).

towards the fungus. Plant cells do not normally leak sugars, but the fungus appears able to induce this selective increase in plant cell membrane permeability. Similar processes occur in the lichen symbiosis, where the unicellular photosynthetic symbiont releases its assimilates to the fungal host only when associated with it and not when free-living.

Mycorrhizal fungi become endemic in soils where their hosts grow, and form extensive networks which can connect the roots of plants of different species. This could lead to carbon being transferred between plants. Although it has been demonstrated in pot plants it is not yet known how significant such carbon transfer is in nature. The question is important for plant ecology and productivity, as it may influence competition for resources between plants in a community of different species. The low degree of host specificity of some mycorrhizal fungi is illustrated by the fact that the type of mycorrhiza found in a plant community can be determined by the dominant plant species. For example, ferns normally have VAMmycorrhizas (page 397) but when growing in beech or pine woods they acquire the ectomycorrhizas of the dominant trees.

The potential agricultural importance of mycorrhizas is obvious. It was first revealed when exotic crops could not easily be established in new situations. For example, efforts to establish pine trees in countries where they were not native were unsuccessful until mycorrhizal fungi were introduced into the

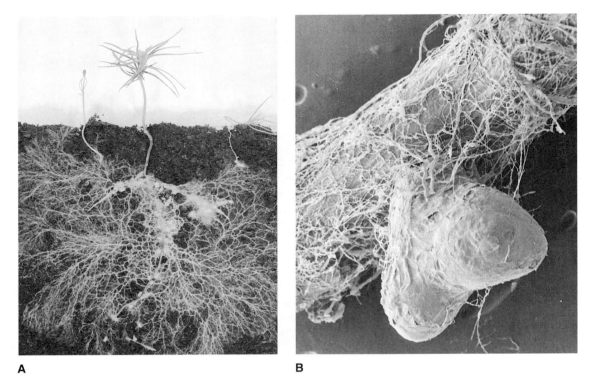

A   **B**

Figure 7.14   The *Suillus bovinus* ectomycorrhiza of *Pinus*. A, The association between mycelium and seedlings revealed by artificial inoculation of seedlings in transparent chambers. Strands of aggregated mycelium extend from the fungal sheath which covers the root. The fan-shaped advancing fronts are composed of separate, diverging hyphae. The mycelium forms cords connecting the three seedlings. When fresh nutrient resources are presented to the mycelium in the form of small lumps of leaf litter, these are colonized, and the use of radioactive phosphate added to the leaf litter resource has shown that phosphorus is transported from the litter into the seedling via the fungal mycelium. (D. Read, (1990). B, Scanning electron micrograph of the ectomycorrhiza. The hyphae form aggregated strands over the root surface and grow through the mucilagenous sheath surrounding the apex of the emerging lateral root. (J. Duddridge.)

nursery soil. Mycorrhizas can be essential, but can also improve crops without being indispensable. Because they can make nutrient uptake from poor soil more efficient, inoculation of crops with appropriate fungi could increase yield and decrease dependence on fertilizers. Most crop plants are naturally mycorrhizal. Strains of fungi differ in the effectiveness of the mycorrhiza they produce and there is interest in developing ways of inoculating crops with the best fungus. Clearly this is easiest if young seedlings in sterile soil can be inoculated. With established plants in field soil there is the problem of pre-existing mycorrhizal infection with sub-optimal strains, as well as competition with the soil microflora. However, the potential gains are promising enough for considerable research on agricultural applications of mycorrhizal infection to be in progress.

There are several different kinds of mycorrhiza. All except those of mycoheterotrophic plants, described below, have the reciprocal nutrient flow described above, but differ in anatomy and physiology.

Vesicular–arbuscular mycorrhizas (VAM) or arbuscular mycorrhizal fungi (AMF)
These mycorrhizas (Fig. 7.12), the commonest, are so-called because their haustorial structures within root cells are branched **arbuscules** (tree-like structures, from Latin, arbor, a tree), as well as vesicles in most species, and it is by these that the presence of the mycorrhiza is recognized. The majority, about 90%, of all vascular plant species as well as many non-vascular lower plants have such fungal symbionts, to the extent that mycorrhizas have been described as the main absorbing organs of plants.

The fungal mycelium of VAM is confined largely to the internal tissues of the root, with no organized fungal tissue outside it, although individual hyphae and spores do develop in the soil. Within the plant the mycelium develops only in the primary cortex and epithelium of the root and does not penetrate the endodermis into the conducting tissues within the stele, nor does it grow into the parts of the plant above the ground.

The group of fungi forming VAMs are placed in a single order, Glomales, which is placed in the Zygomycetes on account of its aseptate mycelium, and the formation by some species of zygospore-like structures preceded by fusion of two hyphal branches. Their association with plants is ancient, perfect recognizable arbuscules having been found in the root cells of fossils of the earliest land plants. Within the Glomales, six genera are distinguished by spore type and mode of formation: *Acaulospora, Entrophospora, Gigaspora, Glomus, Sclerocystis* and *Scutellospora*. The spores are a means of survival and possible spread in soil. All are produced outside the root, in the soil. Spores are up to 800 μm in diameter, large enough to be separated from soil by sieving. The fungus can also survive as fragments of mycelium in plant material. Inoculum for agricultural purposes consists of spores, pieces of infected root, or soil.

The genera named above are responsible for mycorrhizal associations with thousands of different plant species, and single isolates can infect widely different plants. This low host specificity contrasts with that of parasitic biotrophs. A possible explanation for the low specificity is that mutualism, being advantageous to both partners, results in selection for the ability to form associations and against the development of easily triggered resistance responses as occurs with parasitic biotrophs. In spite of generally low host specificity, there are strain differences within species of Glomales which result in differences in morphology and physiology of mycorrhizal associations.

The effects of the Glomales on plant growth in ecosystems have been investigated, for example see Fig. 7.15. Seed mixtures were sown into soil trays inoculated with one to fourteen fungal species or strains per tray. A positive correlation was found between the diversity of the VAM inoculum, and the diversity and productivity of the plants growing from the seed mixture. In addition, there were more hyphae in the soils with more diverse inoculum, and the soils were more effectively depleted of phosphate than the soils with the same amount of inoculum but lower fungal diversity. Comparison of the diversity of

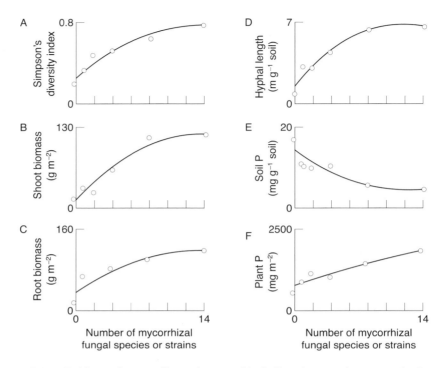

Figure. 7.15 Evidence for an effect of mycorrhizal diversity on plant growth. Sets of replicate trays of sterilized soil placed out of doors in a field site were inoculated with soil containing 1, 2, 4, 8 or 14 different strains or species of mycorrhizal fungi. The inoculated trays were then sown with 100 seeds from a mixture of 15 plant species. After one growing season, the plants growing in each tray were assessed for diversity of species (A), shoot and root biomass (B and C, respectively) and the total length of hyphae of mycorrhizal fungi in the soil (D). The levels of phosphate were also measured both in soil (E) and plants (F). Plants grew better, and the plant community was more diverse, when 8 or 14 different mycorrhizal fungi were present. There were accompanying increases in hyphal growth (shown by measurements of the lengths of hyphae recovered from the soil), and phosphate uptake from soil to plants. Reproduced by permission from Van der Heijden *et al.* (1998), *Nature* **396**, 69–72.

Glomalean fungi of agricultural and semi-natural woodland soils showed a much greater diversity of types in the latter.

Orchidaceous mycorrhizas Some plants form associations with fungi which supply them with carbon compounds. Such plants have been termed 'mycoheterotrophic', because their fungal association enables them to escape the need to photosynthesize and live autotrophically. Among these are orchids. Most orchidaceous mycorrhizal fungi (Fig. 7.12B, c) are confined to plants of the orchid family, which contains tens of thousands of species, mostly tropical. Their physiological relationship with the host plant differs from that in all other types of mycorrhiza in that the fungus supplies carbon nutrients to the plant, obtaining them either by breakdown of insoluble materials in the soil, or from other plants that it has infected. The fungus is most necessary to the plant at the seedling stage because orchid seeds are very small with insufficient food reserves to fuel seedling development, and depend on forming a symbiotic fungal association to obtain the necessary materials. Orchid seeds can be germinated in culture if suitable sugars, and in some cases vitamins, amino acids and growth factors are supplied, but these are not normally present in soil and so the orchid is obligately symbiotic with its mycorrhizal fungus under natural conditions.

Infection occurs after the seed coat has been ruptured by the embryo inside taking up water and swelling. It emerges and produces a few root hairs, and the fungal hyphae invade these first. The intracellular hypha invaginates the plasmalemma and forms coils inside the cell. The mycelium spreads within the plant invading the cells as it goes, but each intracellular coil has a short life, lasting only a few days before it degenerates and is digested by the host cell. The young plant thus becomes connected to mycelium in the soil via an extensive intercellular and intracellular system of hyphae. It has been shown that, in the association of *Dactylorchis purpurea* with *Rhizoctonia solani*, fungal cellulase can break down cellulose, and translocate the soluble carbon thus released into the orchid seedling, enabling it to grow.

The fungi of orchid mycorrhizas can be grown axenically and are not obligately mycorrhizal as are VAMs. Some are sterile, although many of these show the Basidiomycete feature of clamp connections. Many orchids are associated with *Rhizoctonia* species, including *R. solani*, which is also a common plant pathogen. Some of the fungi form fruit bodies enabling them to be identified further. They belong, for example, to the genera *Corticium* and *Marasmius*, and include *Armillaria mellea*, better known as a tree pathogen, and *Coriolus versicolor*, a common timber-decaying saprotroph. Isolates of several free-living wood decay fungi have been shown to enhance germination of orchid seeds, showing that symbionts are not necessarily specialized strains adapted for symbiosis, but can be free-living saprotrophs.

Association with fungi not only permits orchid seeds to develop but also allows some mature orchid plants to live without photosynthesis. *Spiranthes spiralis* may live for a year or more underground and then flower. Other non-green orchids survive by parasitizing other plants via mycorrhizal connections. For example, *Gastrodia elata* derives its carbon from living or dead trees in this way. Photosynthetic orchids may also maintain fungal associations when adult, but the extent to which this occurs is not known. It is also not clear whether there is ever a significant flow of nutrients from plant to fungus.

Mycoheterotrophy has evolved separately in many plant groups, being found in over 400 species of vascular plants in 87 genera. Mycoheterotrophic plants are mainly found in moist forests with much surface accumulation of plant litter, and dense shade which presumably favours plants which can limit their dependence on photosynthesis. They are most abundant and diverse in the humid tropics, the Malaysian peninsula having 10 genera of orchids as well as many other mycotrophic species.

Mycoheterotrophic associations are unusual among plant/fungus symbioses in that the plant partner seems able to induce the transfer of carbon compounds out of fungal hyphae and into plant cells. In all other symbioses involving fungi in which nutrient transfer has been studied – other mycorrhizas, plant and animal pathogens, and lichens – carbon flow is always into the fungal partner.

Ericaceous mycorrhizas These (Fig. 7.12B, d), like orchidaceous mycorrhiza, are characterized by association with a specific group of plants, the Ericaceae. This is another large family, with many species and worldwide distribution, which has mycorrhiza of characteristic structure, type of fungus and physiology. Ericaceae form ecologically important plant communities particularly on moors, swamps and on peat, and include plants such as heathers, rhododendrons and azaleas. It is with plants of this sort that the following account is concerned. The different mycorrhizal associations of allied plants such as *Arbutus* and *Monotropa* are referred to later.

The typical habitat of heathers (*Calluna, Erica*) is on nutrient-poor, acid soil at high altitudes and at colder latitudes. It is likely that this tolerance of poor conditions is made possible by the activities of the mycorrhizal fungus. Heathers have root systems in which the distal parts consist of very fine, hair-like roots and the cortex, composed of only a few or even a single layer of cells, is infected by mycorrhizal hyphae. The hyphae fill the cells and if there is only a single cortical layer, up to 80% of the root volume can be composed of fungal tissue. The symbionts from a number of associations have been isolated and found to be Ascomycetes or mitosporic fungi, for example *Hymenoscyphus ericae* and *Oidiodendrum griseum*, and some sterile unidentifiable septate fungi. The identifiable fungi are capable of being cultured apart from their hosts. Hyphae from spores or existing mycorrhiza colonize newly formed roots by making a network over the surface in the root mucigel layer, and invading the space between epidermal and cortical cells. Intracellular hyphae develop as branches from the surface mycelium, penetrate the root cells, and grow and branch inside them to form intracellular coils bounded by host plasmalemma. The host cell remains alive at first, with active organelles, but later phenols accumulate, the plasmalemma degenerates and the cell dies. This contrasts with the orchid–fungus interaction in which the host cells periodically digest the fungal intrusions. The connections between root and soil are very abundant. Thus in a plant of the heather *Calluna vulgaris* about half of the root system was infected, and the infection was very dense with frequent hyphal connections to the soil mycelium, about 2000 entry points for hyphae per centimetre length of host root.

The nutritional basis of the symbiosis appears to be the supply of carbon compounds to the fungus by the plant, in a way similar to most other biotrophic symbioses, and enhancement of nutrition of the plant by activities of the fungus that appear to be characteristic of this type of mycorrhiza. The fungus enables the plant to utilize protein and amino acids from the substratum. Most plants can only use nitrate, or in some cases ammonium ion. Infection with its fungal symbiont enables the American cranberry, *Vaccinium macrocarpum*, to utilize amino acids as sole nitrogen source, which the uninfected plant cannot do. Protein can also be used, and the symbiotic fungus *Hymenoscyphus ericae* produces a proteinase, an acid carboxypeptidase with a low optimum pH corresponding to that of heathland soils, and also a chitinase. This ability to utilize hydrolysable organic nitrogen compounds gives a clear adaptive advantage to plants growing in acid heathland soils, because in this type of soil conversion of organic forms of nitrogen back into nitrate and ammonium ('mineralization') is slow owing to the low level of microbial activity in the soil. It has been estimated that in heath soil up to 20% of all nitrogen present is locked in fungal mycelium, which in a fertile soil would be rapidly decomposed by bacteria and soil animals.

A further adaptive advantage known to be conferred on plants by ericaceous mycorrhizas is tolerance of heavy metals and other contaminants in the soil. Infected plants can tolerate up to 100 ppm of heavy metals and are among the first plant colonizers of soils containing excessive quantities of metals such as cadmium, copper and zinc. The mycorrhizal roots retain the metal and thereby appear to reduce its uptake by the plant. There is evidence that ericaceous plants with mycorrhizas produce siderophores (pages 155–156), proteins that bind iron and remove it from solution.

There are other types of mycorrhiza that infect plants in the Ericaceae. *Arbutus*, the strawberry tree, has mycorrhizas resembling the ectomycorrhizas described below but with intracellular penetration by hyphae. *Monotropa*, a genus of colourless, parasitic plants, has mycorrhizas which serve a similar function to those of orchids, facilitating transfer of carbon compounds from the roots of photosynthetic host plants.

Ectomycorrhizas These mycorrhizas (Figs. 7.12B, a, 7.13B, 7.14, Plate 3) are mainly formed on roots of woody plants. A thick fungal sheath develops around the terminal lateral branches of roots and is connected to an intercellular network of hyphae known as the 'Hartig net', in the root cortex. A wide variety of fungi form ectomycorrhizas but only about 3% of plant species are involved and these are mostly trees or shrubs. This is presumably because the relatively massive fungal structures involved – the fungal sheath, and, in the case of Basidiomycetes, large fruit bodies, can only be sustained by a long-term connection with the autotrophic partner. Ectomycorrhizas are believed to be more common in temperate zones of the world, where there are seasonal climatic changes, than in the tropics. Possible physiological reasons for this are discussed below. The most studied ectomycorrhizas are those of pine (*Pinus*) and beech (*Fagus*), although ectomycorrhizas have been reported from 130 different genera in 43 plant families. Fungi listed as ectomycorrhizal include 45 Basidiomycete genera of which 10 are Gasteromycetes, 18 Ascomycetes, mainly those with underground fruit bodies (truffles), the Zygomycete genus *Endogone*, and fungi which have not been found sporulating and are therefore difficult to identify.

Ectomycorrhizas start to develop when hyphae infect the root in the region behind the root cap, and grow backwards towards the older parts forming a weft which later may become a bulky sheath. This sheath can occupy up to 30% of the root's volume and in some cases forms a distinct tissue with differentiated types of hyphae in it. Hyphae from the sheath grow inwards between the epidermal and cortical cells, fanning out over the cell walls to form a structure termed the Hartig net. This is thought to be the main interface for exchange of substances between the symbionts. Hyphae do not enter the stele. Late in development some intracellular growth may occur. Infection alters the growth pattern of the root. The rate of cell division at the tip decreases, slowing elongation. Cortical cells elongate radially, so the infected root becomes short and thick compared with uninfected ones. A specialized interfacial matrix develops between fungal and plant cells. Adhesion between them is mediated by glycoproteins. At contact sites between the fungal and plant cell walls there is enhanced secretion of extracellular fibrillar polymers, but this only occurs when the partners are compatible.

Gene expression in roots during the formation of ectomycorrhiza is being studied in the interaction of *Eucalyptus* seedling roots with the ectomycorrhizal basidiomycete *Pisolithus tinctorius*. This association is currently being developed for commercial inoculation of tree seedlings for forestry, providing added impetus for discovering how formation of the symbiosis can be optimized. Although no single gene has been found that is critical for mycorrhiza formation, SR (symbiosis-related) genes for several kinds of molecule show large increases or decreases in transcript level. Overall plant protein synthesis decreases, presumably accounted for by the observed suppression of root cap cell division and root hair growth. In the fungal partner, there is a big increase in the rate of synthesis of symbiosis-specific forms of acidic polypeptides in the fungal cell wall, which coincides with initial adhesion to the root and subsequent building of the fungal mantle. Hydrophobins (page 102) are also synthesized by the fungus within hours of inoculating roots. It is thought that they may play a part in causing hyphal aggregation to form the ectomycorrhizal mantle, since hydrophobin production also accompanies the onset of multihyphal development in fruiting body formation. Study of symbiosis-related gene expression has also provided evidence that roots can be stimulated to initiate mycorrhiza development by a diffusible signal from the fungus. Hypaphorine, a molecule related to a plant growth hormone and secreted by *Pisolithus*, suppresses root hair development in eucalyptus roots and was also found to stimulate the expression of a gene, *EgPar*. The sequence of this SR gene had homology with genes encoding a plant growth hormone binding protein, a glutathione *S*-transferase, which it is thought might be involved in hormone metabolism and the alteration of root morphogenesis which accompanies mycorrhizal development.

Throughout the root system and during the life of the plant there is a dynamic equilibrium between colonization and loss of fungus as the roots outgrow the fungal sheath and as secondary thickening begins. Recolonization continually occurs from fungus in the soil and new roots, growing out from their points of origin within mature roots, become colonized as they pass through the fungal sheath. At any time the root system contains both colonized and uncolonized roots. Some ectomycorrhizas have many hyphal outgrowths forming an interconnected network of mycelium and mycelial cords (pages 340–344) through the soil (Fig. 7.14A), but others have smooth surfaces with few hyphal outgrowths. The ecological significance of this is discussed below.

Colonization is of interest to foresters attempting to grow trees exotic to an area. Ectomycorrhizal colonization has been shown to be necessary for the successful establishment of some trees such as *Pinus*. It has been found that roots become colonized not only from established mycorrhiza in the soil, but also by spores of mycorrhiza-forming fungi. For example, when spores from *Suillus bovinus* were allowed to fall on soil containing seedlings of *Pinus sylvestris*, mycorrhizal colonization was increased twentyfold. Mycorrhiza initiation by airborne spores could be important in allowing tree seeds to colonize new areas.

Spores of mycorrhizal fungi have been found to resemble those of biotrophic parasites in requiring a specific stimulus for germination. They do not germinate on nutrient agar media which support mycelial growth, but require a specific stimulus which has been shown to be produced by growing plant roots and also by mycelium of other fungi. The germination of spores in the presence of existing mycelium could, if cell fusion subsequently occurs, convert monokaryons into dikaryons and contribute to the genetic material in an existing mycelial network.

Mycorrhizal colonization enhances mineral nutrient uptake from the soil into the plant, and thereby improves growth when this is limited by scarcity of such nutrients. Evidence for this comes from many experiments comparing colonized with uncolonized seedlings in pot culture. Uptake of phosphate and nitrogen compounds has been found to be most affected, although sulphate and cations including calcium and potassium are also taken up more efficiently by colonized plants. Since colonized roots are entirely enclosed in fungal tissue, their uptake of all substances is likely to be affected to some extent. Phosphate is actively absorbed into the fungal sheath in the form of water-soluble orthophosphate. Thence it may pass, probably passively, to the root, also in the form of orthophosphate, or it may, under conditions of excess phosphate supply, be converted into polyphosphate (Fig. 3.3D). This is present in the vacuoles of root cells and can be seen as characteristic granules in electron micrographs. The amounts of phosphate stored in this way are sufficient for it to be a possible reservoir of phosphate for the tree. It has been calculated that when phosphate levels in the environment are high, enough could be stored in the mycorrhizal roots to supply the tree's requirements for about 10 days. Thus mycorrhizal plants may be buffered against periodic phosphate shortage. Ectomycorrhizas also absorb nitrogen, in the form of ammonia and amino acids. A few fungi, for example *Hebeloma* and *Laccaria*, also take up nitrate. Ammonium uptake is also an energy-requiring process, which is fuelled by the autotrophic partner, and in excised roots requires a supply of metabolizable carbon compounds. The affinity of fungal uptake systems for ammonia and for phosphate is greater than those of the uninfected root. This difference may contribute to their role in scavenging scarce resources from the soil. Mycelia of ericaceous and ectomycorrhiza have been found to produce enzymes such as proteinase and chitinase which enables them to obtain nitrogen and phosphorus from organic material in the leaf litter layer.

Ectomycorrhizas may affect the ecology of plant communities as well as individual plants. Many have low host specificity, as experiments show that isolates from one plant species can colonize another, although the degree of host specificity varies between fungi. Ectomycorrhizas may connect many of the plants in a community, serving as a conduit for nutrients, and allowing seedlings to become established under conditions such as shading, where otherwise they might be outcompeted by mature plants. Translocation of nutrients between plants and through mycelial networks in the soil has been demonstrated in microcosms such as those shown in Fig 7.14A. An experiment under natural conditions in a forest showed that ectomycorrhizal tree seedlings were more able than non-ectomycorrhizal ones to obtain useful amounts of carbon from other trees, including trees of very different species (Table 7.3). Tree seedlings were planted in sets of three, each set containing one each of paper birch and Douglas fir, which

Table 7.3   Evidence for a net transfer of carbon compounds between ectomycorrhizal plants. These results are from an experiment to measure the amount of photosynthate transferred between plants of different species infected by ectomycorrhiza, under field conditions. Groups of three seedlings were grown close together in a forest, each group consisting of one seedling each of the ectomycorrhizal tree species *Pseudotsuga menziesii* (Douglas fir), *Betula papyrifera* (paper birch), and one seedling of western cedar, *Thuja plicata*, which has only VAMs that do not form extensive mycelial networks. The seedlings became infected by mycorrhizal fungi present in the soil. The fir seedlings were then shaded to reduce the photosynthetic rate to varying degrees (columns 1 and 2). Carbon dioxide labelled with different isotopes of carbon, ^{14}C and ^{13}C, was fed to each of the ectomycorrhizal seedlings, so that transport of photosynthate in both directions could be measured and net transfer calculated. The fir seedlings in which the rate of photosynthesis had been significantly reduced by shading received proportionately more photosynthate from the birch than seedlings with normal rates of photosynthesis. The non-ectomycorrhizal seedlings of *Thuja plicata* received only small amounts of labelled carbon. (After Simard *et al.* (1997), see Further Reading and References.)

Degree of shading of fir seedlings	Intensity of light falling on fir seedlings (photosynthetically active radiation, μmol μm^{-2} s^{-1})	Photosynthetic rate of fir seedlings (as μmol s^{-1})	Net carbon transfer to fir seedlings (% of total labelled carbon in birch)
Deep shade	60	0.05	9.5
Partial shade	250	0.18	4.3
Full ambient light	1330	0.14	2.7

are ectomycorrhizal, and one western cedar, which forms VAM but not ectomycorrhiza. After three years, the birch and fir saplings had formed ectomycorrhiza with fungi already present in the forest soil. The birch and fir were then fed with carbon dioxide labelled with isotopes of carbon, and measurements were made of the amount assimilated by photosynthesis and the net proportion transferred to neighbouring saplings. The use of two isotopes, a different one to each of the two different plants, was necessary to ascertain whether or not there was net transfer of C. The fir saplings were shaded to mimic the effect of growth under a dense forest canopy and it was found that shade sufficiently dense to reduce the fir trees' photosynthetic rate to a third of that in full light, also induced a threefold increase in net carbon transfer from birch to fir. A relatively small amount of carbon was transferred to the VAM cedar seedlings, presumably by VAM-mediated uptake of root exudates from the labelled birch. Ectomycorrhizal fungi produce potentially very large clonal networks capable of connecting tree roots within a single clone over a large area. Many ectomycorrhizal fungi produce translocating mycelial cords or rhizomorphs (pages 340–344, Fig. 7.14A). There is, therefore, the possibility of resource-sharing via mycelial channels for nutrient flow, connecting many trees in a forest – the so-called 'wood-wide-web'. This has led to intensive investigation of ectomycorrhizal clones in natural ecosystems, which has been facilitated by new molecular methods for identifying mycelial and mycorrhizal samples (pages 325–328). *Suillus* is ectomycorrhizal on pine, and produces large sporophores of the kind familiar to many as Ceps, mushroom-shaped but with pores instead of gills. Molecular methods were used to map the extent of clones of *Suillus pungens* in a Californian forest, from samples of sporophores and mycorrhizal roots. It was found to form clonal patches several hundred square metres in extent, with large numbers of trees being colonized by one clone. *S. pungens*, although producing most of the sporophores, forms a relatively small proportion of the ectomycorrhiza on roots. Either it is exceptionally efficient in taking carbon from the root, or its mycelium is able to acquire carbon from some other source, such as leaf litter.

Ectomycorrhizas are also believed to be important in the cycling of nutrients, capturing substances such as phosphate, nitrogen compounds and cations as they are released by exudation from the leaf canopy and leaf fall, and returning them to the root system. Plant remains in the soil can also supply nutrients for some mycorrhizal fungi, for example *Suillus bovinus* grows out from infected roots of *Pinus* through soil in microcosms (Fig. 7.14A, Plate 3), colonizing patches of leaf litter as well as other roots. Ectomycorrhizal roots typically grow in litter layers (Fig. 6.15), where conditions vary more widely than in lower soil horizons with events such as leaf fall, precipitation and the activities of animals, and where consequently the availability of nutrients is discontinuous. Here their ability to store reserves of phosphate may be particularly useful to the tree.

Fungal endophytes

Some fungal inhabitants of plants do not cause disease or form any close contact with the plant cells, but inhabit the apoplastic spaces of their plant hosts. They can be specific to particular plant species. Because they have no obvious effect on the plant they are only found when they are searched for, but they are probably widespread. An example is the fungus *Pezicula aurantiaca*, an Ascomycete with a conidial stage which is found in healthy alder trees (*Alnus*). The plant seems simply to provide a niche suitable for fungal growth. Endophytes may play a part in initiating decay after a plant has died. Decay of dead branches of trees within the canopy usually begins from within the wood, probably due to the presence of endophytic fungi such as *Xylariales* which start to decay the wood when the branch dies.

Endophytes which produce secondary metabolites toxic to animals (mycotoxins, page 440) confer a selective advantage on their plant hosts by preventing grazing or insect attack (Fig. 7.16). Members of the Hypocreales related to the ergot fungus are found in some grasses where they produce alkaloids that deter insect attack. The weed known as darnel is the grass *Lolium temulentum* infected by a toxin-producing fungal endophyte which prevents animals feeding on it. Some fungal endophytes have been shown to produce antibiotics against other fungi, and bacteria. Some fungal endophyte–grass associations, such as that between perennial rye grass *Lolium perenne* and the fungus *Epichloe* (anamorph *Acremonium)*, are common and widespread in natural populations. The association has become very close where the endophyte invades the flowers, is incorporated in the seeds, and passed to the next generation of host grass (Fig. 7.17). In some grass endophytes the production of sexual spores by the fungus has been lost, and it hence exists as large clonal populations transmitted exclusively via the seeds of the host. The infection rate is

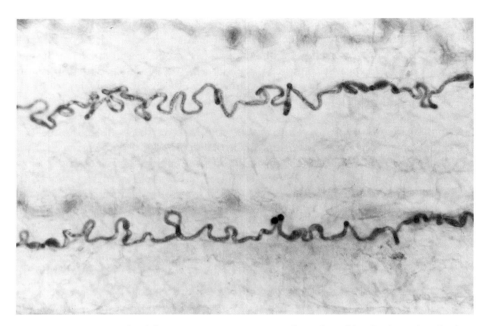

Figure 7.16 Stem pith of the grass *Panicum agrostoides* infected by the fungal endophyte *Balansia henninsiana*. (Leuchtmann, A. & Clay, K. (1988). *Atkinsonella hypoxylon* and *Balansia cyperi*, epiphytic members of the Balansiae. *Mycologia* 80, 192–199. Copyright 1988, The New York Botanical Garden.)

very high in such associations, and the plant carries the endophyte throughout its geographical range. The characteristics conferred by infection, such as increased vigour of vegetative growth, and the presence of secondary metabolites antagonistic to herbivores, have become general features of the grass species. Many species of the common grass family Poaceae are affected in this way by endophytes.

Compatibility and specificity in biotrophic relationships

Symbiotic fungi only successfully infect their host species. Their spores may arrive anywhere but it is only on the right plant or animal that all the further stages of infection can proceed. There are many stages, for example germination, development of special infection structures, and invasion of cells with haustorium formation, at which specificity has been seen in particular fungi. Spores of rust fungi form appressoria on leaves of grasses but not on leaves of other plants. Soil fungi which invade roots and inhabit xylem vessels change the form of their hyphae in response to the different host tissues they grow through on their way from root cortex to vessel interior. Zoospores of *Phytophthora* adhere preferentially to roots of susceptible hosts. Many biotrophic pathogens can avoid triggering plant resistance when they invade cells.

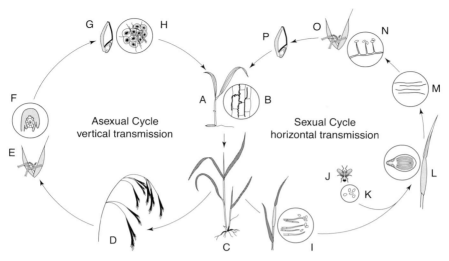

Figure 7.17   The life cycles of endophytic fungi in grasses. **Left:** in some fungus–endophyte associations the fungus is transmitted from parent to progeny by vertical transmission down the generations. The fungus does not sporulate at all on the surface of the plant (D), but its mycelium is passed to the next generation exclusively through infection of the ovule (E, F) and seeds (G, H) by vegetative mycelium (B) present in the tissues of the host grass. This is the sole mode of transmission in some grass–endophyte associations, for example *Neotyphodium lolii* in the perennial rye grass *Lolium perenne*. It results in the inheritance of clones of the endophyte, unchanged by sexual genetic recombination. Molecular phylogenetic studies comparing ribosomal or nuclear gene sequences in grasses and their endophytes show that the fungus and grass species in such associations have often evolved in parallel. **Right:** The fungus is transmitted from one grass plant to another by means of spores produced on the fungal reproductive structure (stroma). A stroma forms (I) on which the fungus's sexual structures, receptive hyphae and spermatia are produced. Insects (J) carry spermatia (K) from one stroma to another, effecting cross-fertilization which is followed by the production of ascospores in perithecia (L) developing in the stroma on the grass. The thread-like ascospores (M) are wind dispersed to grass flowers, where they germinate and produce secondary conidia (N) which may infect ovules (O) and developing seeds (P) of a new host after entering through stigmata. As a result the next generation of plants is infected by new strains of the endophyte. A typical example of such an association is *Epichloë typhina* infecting the cocksfoot grass *Dactylis glomerata*. Several grass endophyte associations, for example *Epichloë festucae* on red fescue, *Festuca rubra*, are capable of both seed and ascospore transmission, the balance depending on the environmental conditions or genotypes involved. (Reproduced by permission from Leuchtmann, A. & Schardl, C. L. (1998). Mating compatibility and phylogenetic relationships among two new species of *Epichloe* and other congeneric European species. *Mycological Research* **102**, 1169–1182.)

The fungal developmental programme responds to physical or chemical conditions which are unique to the host and act as signals to the fungus – the fungus 'recognizes' its host. Following the initial development at the host surface the host may then also recognize the fungus in the sense that the host's developmental programme is uniquely affected by the presence of a particular fungus. It may participate in the development of structures not otherwise formed, such as haustoria, or it may repel the fungus through the activation of the genes controlling the metabolic pathways involved in resistance.

Compatibility between symbiotic fungi and their hosts, being specific, is clearly genetically controlled. In biotrophic plant pathogens such as rusts, and powdery and downy mildews, there is a genetic compatibility between host and fungus which allows the two to function closely together as a physiological and developmental unit. Against the background of this basic compatibility, however, resistant races of the host arise from time to time. In these resistant races, which can differ from susceptible ones by a single gene, some races of the pathogen, which can also differ from successful invaders by only one gene, cannot infect because attempts to do so elicit resistance (page 418). These incompatibility genes and their products are being much studied in the hope of understanding the molecular mechanism of recognition. In mutualistic symbiosis, such as that between mycorrhiza-forming fungi and roots of their host plants, recognition between the potential symbionts is followed by compatible interaction. The genetics of this has received relatively little attention compared with that of pathogens.

The Host's Defences against Fungal Parasites

Plants have features that provide protection against fungal invasion, such as the physical barrier of the cuticle and its water-repellent surface, and the chemical substances found in plant cells. These include glycosides, tannins, phenols, resins and saponins. Many plant proteins are antifungal, including cell wall lytic enzymes, particularly chitinases and β-(1→3)-glucanases, peptides released from germinating seeds, the fungitoxic cell wall proteins of cereals (thionins), chitin-binding lectins, and peptides inhibitory to various enzymes. Among the latter is a polygalacturonase inhibitor that prevents fungal polygalacturonase from completely degrading pectin into its monomers, producing instead oligosaccharides that act as signal molecules to alert plant defences (e.g. Fig 7.18). As well as having preformed defence compounds, plants respond to invasion by induced resistance. New proteins (pathogenesis-related or PR proteins) appear, polymers such as cellulose, callose, lignin and suberin are formed, and phytoalexins accumulate. Cells may die as a result of hypersensitivity. Cellular changes can include an 'oxidative burst' with formation of the powerful oxidizing agents hydrogen peroxide and the superoxide anion, release of hydrolytic enzymes, and toughening of the cell wall by cross-linking of its components. Induced resistance is expressed both locally and systemically, elicited by signals which travel throughout the plant. Longer-term responses include growth responses to seal off and abandon invaded tissues. These responses are common to many different kinds of plant, and normally occur together as a response not only to fungal invasion but also to other stresses imposed on the plant, such as wounding or toxic chemicals. There is therefore a problem concerning the way in which the observed specificity of response to cultivar-specific fungal pathogens is mediated. This is discussed below in relation to separate accounts of the processes involved in resistance.

The hypersensitive response
This is the rapid death of a few cells at the point of invasion, followed by cessation of fungal invasion. As a result small, usually darkened, spots are seen

	Ability to elicit phytoalexin production	Ability to compete with elicitor for membrane binding sites
	Concentration (nM) required for 50% of maximum effect	
(1) hepta–β–glycoside 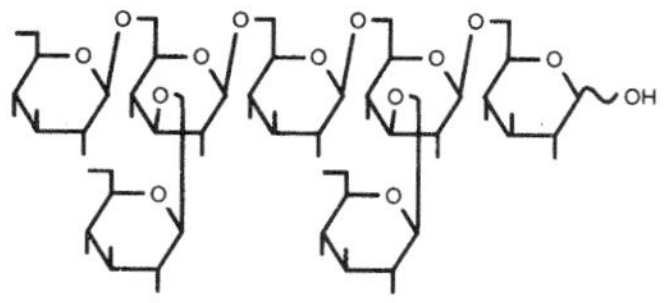	3.9	8.04
(2) α–allyl–hexa–β–glycoside 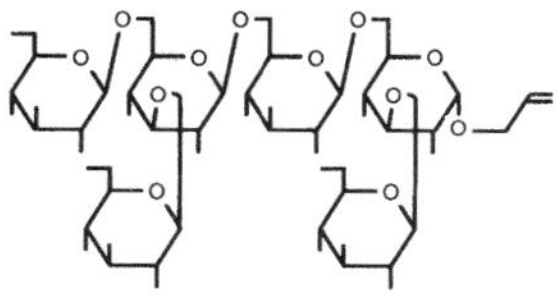	71	21.5
(3) hepta–β–glycoside 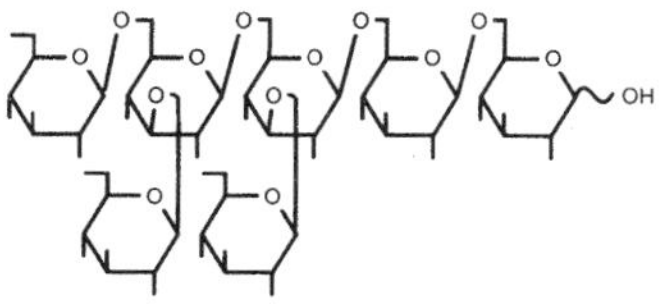	6880	15 700

Figure 7.18 Elicitors of phytoalexin synthesis: Compound (1) is a hepta-β-glycoside, purified from the fungus *Phytophthora megasperma*, and active in inducing phytoalexin accumulation in leaf cells of soya bean (*Glycine max*). Compounds (2) and (3) were synthesized, and compared with 1, for ability both to induce phytoalexin, and to compete for the same binding site on soya bean cell membranes. The analogue with a similar configuration to the natural elicitor was nearly as effective as the elicitor in inducing phytoalexin and in competing for membrane binding sites. The other analogue required far higher concentrations to achieve comparable effects. (After Cheong *et al.* (1991). See references, page 452.) The results suggest that the elicitation might be mediated via stereospecific binding of elicitor to a cell membrane protein.

which are most obvious on leaves or stems. In race-specific resistance, such as is shown by resistant cultivars of barley to avirulent strains of the barley powdery mildew fungus, the cell challenged by a germ-tube approaching from the leaf surface may die before penetration. The cells next to it may also die. The fungus then ceases growth, leaving a small restricted necrotic lesion. This is a common response in race-specific resistance, occurring either before or just after penetration of the cell wall. The faster the response, the smaller the resulting lesions. Biotrophy obviously cannot be established in the face of a hypersensitive response, so hypersensitivity confers resistance to biotrophs. However, plants can also respond hypersensitively to invasion by necrotrophs, and by non-pathogenic fungi. When potato leaves are invaded by either pathogenic or avirulent strains of *Phytophthora infestans* the initial result is the same in both cases, with necrosis of mesophyll cells in contact with hyphae. It is only later that the difference between strains becomes evident in their ability to colonize the leaf and sporulate. So hypersensitivity in itself is not necessarily sufficient to prevent fungal invasion, and cell death must be accompanied by other changes to inhibit fungal growth. Hypersensitivity can vary between tissues and even cells of the same plant. In wheat there is a gene which confers hypersensitivity to the wheat stem rust fungus on mesophyll cells, but not epidermal cells, of the leaf. The fungus can hence infect the epidermis but not the mesophyll, where necrosis of invaded cells stops its growth. Hypersensitivity not only occurs in plants repelling their pathogens, but is also one of their responses to fungi which are non-pathogens ('non-host' resistance). It is accompanied by a variety of chemical changes such as the accumulation of phytoalexins (see below) and other metabolites which may inhibit further fungal growth. The hypersensitive response is widespread in plants and effective in resisting a variety of pathogens.

Phytoalexins

These are low-molecular-weight compounds formed locally in plant cells in response to fungal invasion and also to other stresses, and which inhibit fungal growth. Most phytoalexins are absent from unchallenged, healthy tissue. Several chemical types are well known, for example phenylpropanoids, isoprenoids and acetylenes, each group being derived from different pathways of secondary metabolism. Two of these pathways have been investigated in detail, and their activation in response to fungal elicitors studied. The general pathway of phenylpropanoid metabolism (Fig. 7.19) leads from phenylalanine to precursors of coumarins, flavonoids and stilbenoids, and to synthesis of lignin monomers. Enzymes in the pathway such as phenylalanine ammonia lyase (PAL) are regulated in relation to various stimuli, physical and chemical, as well as those produced by fungi. Another well-studied pathway leads from mevalonate to terpenoids, such as the sesquiterpenes typically produced as phytoalexins by the Solanaceae. As is the case with secondary metabolites, several of the types of phytoalexins produced are characteristic of particular plant families. Most of the family Papillionaceae accumulate isoflavonoids in response to fungal infection. One species of plant can produce several different phytoalexins. Thus tomato (*Lycopersicon esculentum*) has the polyacetylenes falcarinol and falcarindiol, and the sesquiterpene rishitin. Different tissues of the same plant may produce different phytoalexins. The potato has several different ones and the phytoalexin mixture depends on both the organ of the plant and on the inducing stimulus.

Phytoalexins are commonly toxic to fungi, bacteria, and plant and animal cells. Their mode of action is not well understood. Typically concentrations of about 10^{-4} M are needed to inhibit fungi, and this concentration does build up in invaded plant tissues. There is evidence that the ability of a plant to prevent invasion depends on the speed and magnitude of phytoalexin accumulation at the site of invasion. Pathogens may have differential sensitivity and some are able to break down phytoalexins. Tolerance to phytoalexin may also enable a pathogen to overcome this form of host defence. This has been shown with the Oomycete pathogen of peas, *Aphanomyces eutiches*, which develops tolerance to pisatin when cultured in the presence of extracts of pea plants, even though the pisatin is not broken down. Another pea pathogen, *Nectria haematococca*, demethylates pisatin. This presumably facilitates, but is not essential for, plant invasion; a strain in which the gene for pisatin demethylase has been disrupted (pages 372–374) remains virulent.

The production of phytoalexins often accompanies the hypersensitive response and is elicited by a variety of stimuli. These include pathogenic fungi and some substances, known as **elicitors**, derived from them. However, non-specific treatments such as the application of mercury salts can also elicit the response. The molecular processes involved in fungal elicitation have been intensively studied in some fungi which have a host specificity determined by a single gene in the host and a corresponding one in the pathogen. Elicitors from some fungi have been chemically characterized, and include oligosaccharides of defined chain length and conformation (Fig. 7.18). Such elicitors act very quickly, and application of a glucan elicitor from *Colletotrichum lindemuthianum* to cultured cells of its bean host is followed within 5 minutes by the transcription of PAL genes. Not all accumulation of antifungal substances is due to *de novo* synthesis arising from increased gene expression. Thus the concentration of some substances such as phenolic compounds may increase following the hydrolysis of a precursor. This occurs when soya bean is infected by *Phytophthora megasperma* and there is rapid release of glyceollin as glycosylated intermediates are hydrolysed.

Pathogenesis-related proteins

Comparison of infected and uninfected plant cells and tissues shows that resistance is accompanied by the expression of many genes encoding small proteins which accumulate in the intercellular spaces. These pathogenesis-related proteins are most common in hypersensitive responses and appear to contribute to systemic acquired resistance (page 416). They are currently classified into fourteen families, PR-1 to PR-14, members of at least ten of which have direct antifungal activity. Of these, PR-2 represents β-(1→3)-glucanases, PR-3, 4, 8 and 11 represent different types of chitinases and PR-7 represents endoproteinases. These enzymes can be potently antifungal by digesting the walls of potentially pathogenic fungi, especially at hyphal apices. The chitinases and β-(1→3)-glucanases work strongly in synergism (i.e. mixtures of the two activities are much more potent than the sum of either acting alone). In addition, PR-5, 'thaumatin-like' proteins, PR-12, defensins, PR-13, thionins, and PR-14, lipid-transfer proteins, show a range of antifungal activities. There are now many examples where transgenic plants expressing alien pathogenesis-related proteins, particularly chitinases, show increased resistance to fungal pathogens.

Some plant chitinases also have lysozyme activity, so can be active against the peptidoglycan of bacterial cell walls as well as the chitin of fungal cell walls. Others also have chitosanase activity, so are active against cell walls of fungi that have chitosan (deacetylated chitin) as a wall component. The plant defence chitinases may also have activity against invertebrate pests, such as mites and insects, which have chitin as a major component of their exoskeletons and their peritrophic membranes (gut linings). Some of the chitinases and $\beta(1\rightarrow3)$ glucanases are stored in the plant vacuoles, others are secreted to the apoplast. The former probably play the major antifungal role, while the latter (which tend to have weaker antifungal activities) may be involved in releasing oligosaccharide elicitors that activate host defence reactions. Increased production of chitinases and $\beta(1\rightarrow3)$ glucanases is induced by these elicitors and also by ethylene, a plant growth substance that is produced by both plants and fungi during infection. The response of the plant infection may be non-specific, with infection by fungal, bacterial or viral pathogens inducing the same chitinases and $\beta(1\rightarrow3)$ glucanases. In contrast, the legume *Medicago trunculata* showed notably different patterns of induction of its eight known chitinases in response to infection with the arbuscular mycorrhizal fungus *Glomus intraradices*, with the virulent compatible fungal pathogens *Fusarium solani* f. sp. *phaseoli* and *Phytophthora megasperma* f. sp. *medicaginis*, with the avirulent compatible pathogen *Ascochyta pisi*, with the non-infective *Fusarium solani* f. sp. *pisi*, and with the nodulating bacteria *Rhizobium meliloti*.

Changes in cellular and cell wall chemistry
These changes occur in response to challenge by pathogens. The cell walls form the apoplast, the extracellular system of interconnecting walls and spaces which is continuous throughout the plant and over its exterior surfaces (see above, pages 370–371). Plant cell walls are now known to be more than inert inextensible containers for the cells. They interact with the cytoplasm of the cell and their chemical composition and physical structure change in response to cellular changes. In response to fungal invasion, substances may be added to the wall. These include lignin and related phenolic compounds, melanin, carbohydrates including callose and pectins, and several different kinds of protein including hydroxyproline-rich glycoproteins, glycine-rich proteins, and enzymes such as peroxidases which are involved in the synthesis and cross-linking of wall polymers.

The formation of lignin in the walls of cells adjacent to sites of attempted invasion by fungi is thought to be a critical factor for resistance in some plants. For example, synthesis of PAL, and cell wall lignification, as well as hypersensitivity, were elicited in some rust-resistant cultivars of wheat by inoculation with avirulent, but not with virulent, strains of *Puccinia recondita*, a wheat rust fungus. No phenylpropanoid phytoalexin – in the sense of an induced antifungal substance – was produced, so that PAL activity in this instance was

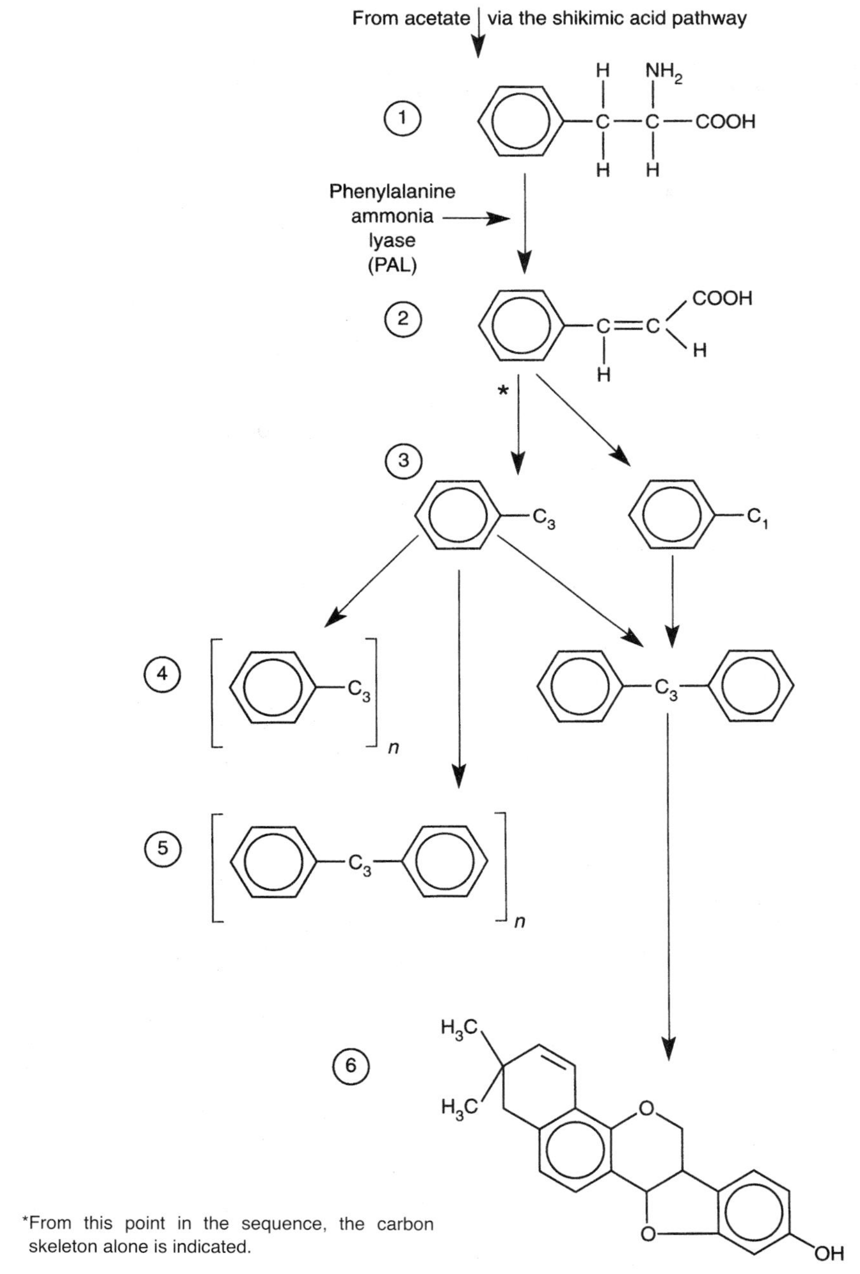

Figure 7.19 The pathway of biosynthesis of plant phenolic compounds. Cinnamic acid, produced from phenylalanine by the action of the enzyme phenylalanine ammonia lyase (PAL), is the precursor of a large number of phenolic compounds such as lignin, tannins, and some phytoalexins, which have roles both in defence and in the development of cell walls. 1, Phenylalanine; 2, cinnamic acid; 3, phenylpropanoid compound; 4, lignin; 5, condensed tannins; 6, glyceollin 1, a flavonoid phytoalexin.

directed to lignification. However, both phytoalexins and lignin precursors are products of the pathway of phenylpropanoid synthesis and may be produced together in response to challenge, as in rust-resistant flax cultivars which produce fungitoxic coniferyl alcohol and coniferyl aldehyde, as well as lignin, when challenged with avirulent cultivars of *Melampsora lini*. Wall lignification can be quickly activated. Cinnamyl alcohol dehydrogenase, which catalyses the reaction producing cinnamyl alcohol, a main precursor of lignin synthesis, increases in bean within 1.5 hours of treatment with fungal elicitor. It can also be prolonged, lignification continuing for several days as secondary walls are formed. When lignin is synthesized in the presence of cellulose, pectins, cutin and cell wall proteins, it forms covalent cross-links with them, strengthening the wall. The pathway involved in lignin biosynthesis and a key enzyme are shown in Fig. 7.19. The activities of these enzymes usually increase preceding the lignin synthesis associated with disease resistance. The synthesis of enzymes is discussed below, in the context of induced protein synthesis.

Tannins, like lignin, are phenol polymers. They occur in various plant tissues, such as bark and the heartwood of trees. Tannins bind to proteins including enzymes, and in so doing can inhibit enzyme activity. This property accounts for the traditional use of bark in tanning leather; animal skin consists of keratin and other proteins, and the tannins of bark prevent enzymic attack by microorganisms on these proteins, thus preserving the leather. Similarly, the tannins in bark and other plant tissues prevent attack by fungal enzymes. Tannin formation is also induced in plant tissues normally lacking them following fungal attack, and accounts for the darkening of plant cell walls that occurs during necrosis.

Saponins, such as avenacin, comprise a family of similar antifungal compounds of plants which are soap-like substances that inhibit the growth of fungi by disrupting their membranes. They exist in a preformed state in the plant cell vacuole, being activated when the precursor form comes into contact with an activating enzyme as a result of damage to the cell by an invading fungus. In the take-all fungus of cereals, the pathogenicity of some strains depends on possession of an enzyme, avenacinase, that destroys the saponin avenacin (page 390), and it is thought that other fungal pathogens also may owe their pathogenicity to an ability to break down saponins.

Callose deposition in cell walls is a common response to wounding, or damage or attempted invasion of a cell by a fungus. Callose is a $\beta(1\rightarrow3)$ glucan, formed by glucose polymerization catalysed by the enzyme $\beta(1\rightarrow3)$ glucan synthetase. Its deposition is highly localized. It is formed on the inner side of the cell wall as fungal hyphae or infection pegs begin to penetrate the cell wall. For example, localized deposits of cell wall material, papillae, form just inside the wall and outside the plasmalemma of barley epidermal cells, when infection pegs of germinating *Erisyphe* spores of an avirulent strain penetrate the wall. Infection may then be arrested and the fungal hyphae contained by the papillae (Fig. 7.5C). The sequence of events indicates recognition by the plant of an invading pathogen. The whole cell wall can also be thickened by callose and cellulose additions, and this can happen in living, uninvaded cells surrounding necrotic lesions. The enzyme $\beta(1\rightarrow3)$ glucan synthetase is localized in the plasmalemma and is calcium dependent. It can be rapidly activated in cell cultures of bean, tobacco and parsley when fungal elicitation of resistance is mimicked by addition of chitosan. The response can be so rapid that it can be detected within 5 minutes of the stimulus, and it is probably mediated in the cell by an increase in membrane permeability, which allows the calcium level to rise in the vicinity of the plasmalemma-bound enzyme. There is some recent evidence that stretching the plasmalemma alone can stimulate local callose synthesis, so perhaps papilla formation is elicited simply by the pressure of an infection peg.

Cell wall proteins synthesized in response to infection include some which appear to change the nature of the wall. **Hydroxyproline-rich glycoproteins** (HRGPs) are found as a normal wall component but also increase on infection. They accumulate locally in the walls of living cells close to sites of infection where fungal growth is inhibited, and in papillae round infection pegs. Fungal elicitors induce HRGP accumulation in many plants that have been investigated including bean, melon and soya bean. Cucumber cultivars resistant to *Cladosporium cucumerinum* increase their cell wall hydroxyproline levels faster on infection than do susceptible cultivars. Thus there is circumstantial evidence that HRGPs confer resistance, although it is not clear how. Possibly they do so simply by strengthening the wall. They are synthesized in the cytoplasm and secreted into the wall where they become cross-linked, probably by means of covalent bonding between tyrosine residues catalysed by peroxidase. Increased synthesis is a result of activation of transcription. There are many slightly differing types of HRGP and there is evidence for differential synthesis of these depending on the nature of the eliciting stimulus. Other cell wall proteins synthesized in response to fungal infection include glycine-rich proteins, whose role is unclear, and enzymes with a role in cell wall synthesis and strengthening. The latter include peroxidase, which catalyses the production of phenoxy radicals from cinnamic acid, which then polymerize to form lignin. Peroxidase probably has a role in other polymerizing and cross-linking reactions which take place in the cell wall.

Development of periderms

Development of secondary tissues is a common response to fungal growth, although it has been less investigated than the immediate responses that characterize resistance to infection in herbaceous plants. However, in fungal infection of the woody parts of trees – bark, sapwood and heartwood – it is the relatively gradual responses such as periderm formation (page 371) that are best described. These tissues are mainly composed of the thickened walls of dead cells and microbial infection becomes evident as decay. The living tissues within bark and wood are activated by wounding or fungal invasion as is shown by the appearance of zones of renewed cell division. In bark of the conifer *Picea sitchensis* (sitka spruce), a new layer of cells appears in the course of several weeks near the site of damage, and becomes impregnated with suberin. In sapwood, the functional, water-conducting part of the xylem containing living parenchyma (page 371), wounding can induce the development of up to fifty new cell layers of parenchyma, the cell walls of which become impregnated with suberin. Gums and tyloses (page 391) may be laid down and also suberized and other, uncharacterized, phenolic compounds are added to cell walls.

Phytoalexins are also produced in sapwood. In gymnosperms the substances best characterized as phytoalexins are stilbenes. Many substances such as resins are produced in wood following infection, but their roles are poorly understood. The resistance to decay characteristic of sapwood of living trees appears to be due to a combination of factors. These are the induced structural and chemical responses described above, as well as constitutive features which include not only structure and chemistry but also the unsuitability of the environment provided by living wood for fungal growth due, for example, to high water content. Once a fungal colony is established within the woody parts of a living tree it is surrounded by a zone of altered structure and chemistry which defines the margin of fungal growth, but it is not clear whether this zone is in fact a barrier to fungal growth. Heartwood (page 371), being entirely dead, does not react to fungal invasion, but becomes decayed by some fungi making the tree hollow. However, timber decay fungi that are saprotrophs usually decay the sapwood of timber much more readily than heartwood because the heartwood is impregnated with antifungal compounds similar in type to those produced in response to damage in living woody tissue. The fungi that are able to decay the resistant heartwood are enabled to do so by their ability to degrade these substances.

Systemic resistance

Some resistance responses occur not only in the immediate vicinity of eliciting stimuli, but also in more remote parts of the plant. A peptide, **systemin**, consisting of an 18-amino acid chain, is mobile within the plant and active in inducing cellular responses including the oxidative burst. It is thought that systemin is formed from its 200 amino acid peptide precursor in response to cell membrane binding of defence elicitors such as oligosaccharides (Fig 7.18). The systemin is transported throughout the plant and binds to receptors on the plasma membrane, activating an intracellular signal transduction pathway that switches on genes encoding defence proteins. Some fungal infections, for example that of the blue mould fungus *Peronospora tabacina* on tobacco plants, result in lasting immunity to a range of pathogens, through **systemic acquired resistance (SAR)**. The initial fungal attack induces localized necrosis, and also activates responses in more distant tissues inducing cell wall changes and synthesis of PR proteins. Subsequent challenge in any part of the plant induces a more rapid activation of defence responses than in plants that have not been subject to an earlier attack. Chemicals have been developed to elicit SAR artificially as a way of controlling plant diseases by increasing plants' immunity to attack by fungi and other pathogens.

How the host perceives the fungus

The recognition of a pathogenic fungus by the host When a fungus makes contact with a plant there are four possible outcomes: disease, mutualistic symbiosis, resistance or no reaction. Which of these occurs in any one plant depends on the kind of fungus. Thus plants distinguish between fungus and non-fungus, as is seen with activation of different PR genes by fungal elicitors and by wounding (page 411). Plants also distinguish between pathogen and non-pathogen, since some species of fungus cause disease and others do not, and between potential symbionts and non-symbionts in the soil. Furthermore, they distinguish between races of pathogens which carry virulence and avirulence genes as described below, since the latter elicit resistance responses which block the normal disease processes. When a plant recognizes a fungus many genes are activated, and this aspect of the plant's response is being intensively studied, particularly in the resistance response, revealing a more complex set of responses than was formerly suspected. What signals to the plant the presence of a fungus, and whether it is a virulent or avirulent pathogen, symbiont, or non-pathogen? DNA technology (page 372) is leading to an understanding of the signal transduction that follows a challenge to the plant and results in the activation of genes involved in resistance. Oligosaccharides from fungal cell walls and from plant cell walls trigger resistance responses in many plants, and appear to signal to the plant the presence of a fungus. The hepta-β-glycoside (Fig. 7.18) was found to elicit phytoalexin production in soya bean. The specificity of this molecule is such that it is the only one with elicitor activity out of about three hundred different heptaglycosides released by partial hydrolysis of cell walls of the pathogen *Phytophthora megasperma*. The structural requirements for elicitor activity were extremely precise. Nanomolar quantities were found to elicit transcription of resistance genes. However, the elicitor was not found to be race-specific, and so does not provide an explanation of the difference between responses of soya bean to virulent and avirulent strains of *P. megasperma*. Another class of oligosaccharide elicitors are the oligogalacturonides produced by the action of pectic enzymes on plant cell walls. Fungi produce a variety of exo- and endohydrolases which depolymerize pectin. For example, an endo $\beta(1{\rightarrow}4)$ polygalacturonase from *Rhizopus stolonifer*, a necrotrophic pathogen, elicits phytoalexin production in the castor oil plant (*Ricinus communis*) by releasing an oligogalacturonide elicitor from the plant cell wall. Since hydrolases are among the proteins produced in the disease resistance response it seems likely that the plant's response to any challenge, by fungi, insects or nematodes, or to abiotic damage, is likely to be amplified by the action of wall-derived elicitors. There is some evidence of the ability of a plant to respond differently to different oligosaccharide elicitors. For example, tomato plants respond to damage by transcribing two genes, among others, one for a proteinase inhibitor protein, the other for an enzyme of ethylene synthesis. The first is activated by oligo-galacturonide of 10–20 residues length, the second by smaller molecules, down to trisaccharides. The effectiveness, stereospecificity and speed of action of oligosaccharide elicitors strongly suggest that the signal transduction pathway which leads to regulated gene activation must be short. Perhaps the elicitor binds to a protein receptor which directly affects gene expression.

Race-specific recognition of pathogens The existence of race-specific resistance genes of host plants and matching avirulence genes of fungal parasites has been known for many years (page 267). However, it is only recently that some of the proteins that they encode have been characterized. Detailed studies of the recognition process required for resistance to fungal disease have been largely carried out in the tomato leaf mould pathogen *Cladosporium fulvum*. This fungus grows exclusively in the intercellular spaces in leaves and is a relatively benign biotrophic fungus. Proteins secreted by *C. fulvum* in the apoplast, such as ECP1 and ECP2, can act as pathogenicity factors, or can be recognized by plants carrying particular resistance genes. In the latter case they are known as avirulence gene products. The first avirulence gene characterized from *C. fulvum* was the *AVR9* gene. *AVR9* encodes a small cysteine-rich protein that is processed within the apoplast by a combination of fungal and plant proteases, and is then recognized by the product of the tomato *Cf-9* gene (Fig. 7.20). Targeted disruption (page 380) of *AVR9* leads to virulence of tomato *Cf-9* cultivar and the converse experiment – where *AVR9* is introduced into an *AVR9* race of the fungus – results in avirulent transformants. This shows that *AVR9* is necessary and sufficient for pathogen recognition by *Cf-9* tomato cultivars. Recently it has been found that even the virulence factors ECP1 and ECP2 can be recognized as avirulence gene products in certain tomato strain backgrounds. Avirulence genes have now also been isolated from a number of other fungi, including the rice blast fungus *Magnaporthe grisea* and interestingly, no homology has yet been detected among them. Taken together these observations highlight the capacity of plants to respond to and to recognize secreted fungal proteins during infection, and thus induce a successful defence reaction.

The recognition of potential mutualistic symbionts This has been much less studied than that of potential pathogens. The fungi which form vesicular–arbuscular mycorrhizas (VAM) (pages 379–398) infect the majority of plants, showing little host specificity. Ability to form VAM is thought to be determined by specific plant genes. In peas two sets of mutants have been

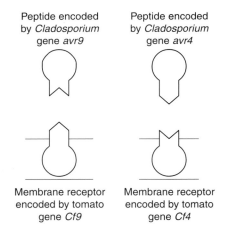

Figure 7.20   Race-specific resistance to a fungal pathogen by a host plant. Each avirulence gene (avr) of the tomato pathogen *Cladosporium fulvum* encodes a peptide which elicits resistance in the tomato host by binding to a complementary plant cell membrane receptor encoded by a *Cf* gene. There is a series of *AVR* genes in *Cladosporium fulvum*, and a corresponding series of *Cf* genes in the tomato. The interactions between the products of genes *AVR9*, *AVR4*, *Cf-9* and *Cf-4* have been investigated in detail. Fungal strains with *AVR9* elicit resistance from cultivars with *Cf-9* but not *Cf-4*, and vice versa. (See DeWit (1997), Further Reading and Reference.)

obtained which determine the ability of the root to form VAM. One set, *'early myc'*, causes fungal appressoria to be poorly developed, and the other, *'late myc'*, limits the branching of fungal arbuscules within cells. These genes are linked with others that determine the ability of pea plants to form symbiotic nitrogen fixing nodules with the soil bacterium *Rhizobium*. This suggests the existence of a region of the plant genome concerned with the formation of symbiotic associations.

Host root exudates such as some flavonoids and volatile substances including carbon dioxide have been found to stimulate germination of VAM spores and growth of germ-tubes, providing a non-specific signal from root to fungus. Fungal invasion of the root follows, apparently without eliciting the defence responses that usually occur when a fungus invades a plant cell. There is some chitinase and peroxidase synthesis which soon falls below the levels of those in uninfected roots. Growth of the fungus within the root must be controlled or the cost to the plant would be too high for mutualism to be sustained. Some root cells may limit fungal development, for example by producing papillae of cell wall material, and suberized cells such as the endodermis (page 366) probably prevent the passage of hyphae. Other root cells react to invasion so as to promote fungal growth. Successful arbuscule development within cortical cells is accompanied by an increase in size of the nucleus and in the amount of cytoplasm and organelles in the invaded cell, and there is a decrease in both when the arbuscule dies and is degraded. Infection can stimulate root branching so as to increase the tissue available for colonization. Thus there is evidence of signalling between host root and mycorrhizal fungus, and of developmental events brought about in both partners by approach and contact, which act to promote the association. So far there is little understanding of the molecular events underlying these responses, partly because, as VAM fungi cannot be grown apart from their hosts, it is difficult to analyse the interaction. The study of plants which are non-mycorrhizal, either naturally or because of mutation or transformation, may help to isolate cellular features determining compatibility.

Plant Pathology

Economic consequences of fungal attack on plants
The most important effect of fungi on human activities is the destruction of crops. A fungal epidemic which destroys the staple crop of a society depending on subsistence farming is a human catastrophe with far-reaching social and economic consequences. The rapid spread which can occur with fungal pathogens is due to their abundant production of both dispersal spores which are carried on the wind and survival spores (page 187) and other structures which lie in the soil. The destructiveness of fungal pathogens results from their ability to break down plant tissues (page 385), to alter physiology so as to reduce yield, and to produce toxins (page 440) poisonous both to the plant and to animals eating it.

The economic consequences of plant disease due to fungi are most likely to be serious when a society depends on a single crop species, or even variety, and a pathogen not normally present is introduced, as in the Irish potato famine of 1845. Potatoes were the sole source of food for subsistence farmers, and a million people died of starvation when the late blight fungus, *Phytophthora infestans*, arrived. Potatoes, originating in South America, had been introduced into Ireland, and grew so well there that a family could live solely on the potato crop from a small piece of land, becoming entirely dependent on the crop. In the uplands of the Andes there are many kinds of potatoes which are in equilibrium with their parasite, *P. infestans*. When first introduced into Ireland, the potato, as a non-native plant, was relatively free of pathogens until the arrival of *P. infestans*, not previously present in Ireland. The fungus was able to spread rapidly in 1845 because the wet weather in that year increased the rate of fungal sporulation and infection, and potatoes were grown throughout the country so that there was rapid secondary infection of one host plant by another. Epidemic spread was also likely since there was little resistance to the fungus in the potato strains that were being grown, and little genetic variation between individual plants as these were all derived by vegetative propagation from a few varieties. *P. infestans* attacks both leaves and tubers necrotrophically, so the crop and the seed potatoes for the next year were all destroyed, and with them the peasant society that depended on them. The great Bengal rice famine of 1942, when two million people died of starvation, was another instance of the destruction of a staple crop by a fungus, *Helminthosporium oryzae*, which causes a disease known as brown spot of rice.

Coffee rust disease due to *Hemileia vastatrix* is economically serious because coffee is an important cash crop for many countries. Like potatoes, coffee is grown in many places where it is non-native, and where *H. vastatrix* does not occur. When introduced the pathogen spreads rapidly and destructively. For example, in the last century it effectively ended coffee growing in Sri Lanka, where coffee is not native. The pathogen probably originated in Ethiopia, the home of coffee, *Coffea arabica*, and its range is still expanding in South America in spite of strenuous efforts by governments to control it, since it is spread by airborne uredospores.

Survival structures of (page 171) pathogenic fungi in the soil, such as spores, sclerotia and hyphae associated with killed host plant remains, may make it difficult or impossible to grow susceptible crops in some areas. Tea plantations in Africa in places where the native forest had been felled, suffered attack by *Armillaria mellea* rhizomorphs growing through the soil from remaining stumps.

Plant pathogens not only affect the growing crop but also plant produce, such as roots, fruits and grain, that have been harvested, before they are consumed. These **post-harvest** pathogens are mainly bacteria and fungi. The fungi commonly responsible are necrotrophs (page 385), including *Sclerotinia* species, *Botrytis cinerea*, *Penicillium* species and the Zygomycete *Rhizopus stolonifer*, which can all invade tissues that have been damaged, and once established, break down cells enzymically. Biotrophic organisms do not appear to attack fruits and storage organs, presumably because there is no advantage for them in establishing a biotrophic relationship with tissues in which there is no continuing photosynthesis. Post-harvest losses are also caused by fungi which produce mycotoxins (page 440) in stored products. The economic costs of post-harvest fungi include not only the losses due directly to infection, but also the cost of maintaining temperature and humidity-controlled storage depots. Loss of commercial timber by fungal attack and spoilage (page 352) is also a form of post-harvest loss.

Controlling fungal attack on plants
Fungal attack occurs when a virulent fungus meets a susceptible host, so quarantine arrangements and cultivation methods may be enough to prevent some fungal diseases. However, problems of control arise when a crop has to be grown in the presence of its pathogens, and the main methods available include the use of fungicidal chemicals, biological control with other organisms which attack the fungus, breeding for crop plants resistant to attack, or **integrated** control with combinations of more than one of these methods.

Fungicides These are usually sprayed on to the aerial surfaces of plants or used to coat seeds, less often added to soil where there are many fungi besides the target pathogen. They vary in their mode of action (pages 175–180). Some, like the copper sulphate used in the traditional Bordeaux mixture, are toxic to a wide variety of different fungi, whereas others affect only certain genera, for example the aminopyrimidines, to which Erysiphales are more susceptible than other fungi. Those fungicides with a wide spectrum of action tend to remain effective over many years, whereas those which act by interfering with a single metabolic step, like the ergosterol biosynthesis inhibitors, are likely to become less effective as the pathogen more easily develops resistant mutants. An example of this tendency of fungicides to select among fungus populations for fungicide-tolerant strains was found in French vineyards with varying frequencies of fungicide spraying against *Botrytis cinerea*, a mitosporic fungus which rots grapes on the vine. The proportion of resistant *B. cinerea* isolates from the field increased progressively from the south of France where the crop is sprayed once a year, to Burgundy with two applications per year, to a maximum in Champagne, where, as the crop is very valuable, four applications are normally made per year. Although surface spraying can be effective, biotrophs such as rusts grow mainly within the leaf tissues, inaccessible to surface sprays. Some fungicides are taken up through the leaves of the plant and into its cells, acting as systemic fungicides. Benomyl, a specific antagonist of fungal tubulin synthesis, is used systemically. It has the disadvantage for food crops that it cannot be removed by washing, as well as having a highly specific site of action so that resistant strains of the pathogen develop. In 1997, the estimated global fungicide sales were in excess of 5.5 billion US dollars. Of these, about 70% were of triazoles (page 178). The sales of strobilurins, however, are rapidly increasing, particularly for cereals (page 524). The major fungicide markets are Western Europe and Japan, accounting for 70% of global sales, largely as a result of prevailing climatic conditions and the highly intensive agricultural practices of these regions. For the different types of crop, nearly half the sales are for fruit and vegetables, followed by about 25% for cereals and 14% for rice.

Fungicides for the control of fungal decay of timber do not have to be non-toxic to animals, so substances like creosote, and mixtures of copper, chromium and arsenic salts are used as timber preservatives. However, concern about the possible environmental effects of these substances, and of the organic solvents in which they are applied, is leading to investigation of other ways of preventing the growth of wood decay fungi.

Biological control Organisms which inhibit the growth of others can be used to control pests and pathogens, a method termed biological control (Table 7.4). Fungal growth can be controlled by means of organisms which attack or compete with fungi, such as nematodes, other fungi and bacteria. Biological control (Plate 9) can reduce the incidence of some plant diseases caused by fungi. Mycoparasites (page 448) which live in soil can destroy resting structures of pathogens. For example, oospores of Oomycetes such as *Phytophthora* are attacked by Chytridiomycetes, and sclerotia are attacked both by facultative parasites such as *Trichoderma harzianum*, and by mycoparasites found only on or near sclerotia, such as *Coniothyrium minitans* and *Sporidesmium sclerotivorum*.

Mycoparasites also attack the sporulating mycelium of pathogenic fungi on aerial plant surfaces, and some, for example *Sphaerellopsis filum* (tel.

Table 7.4   Some examples of the many parasitic fungi actually or potentially used as biological control agents

Fungal genus	Fungal group	Host attacked
Parasites of insects		
Coelomomyces	Chytridiomycetes	Mosquitoes
Conidiobolus *Entomophaga* *Entomophthora*	Zygomycetes	Aphids, cockroaches, flies
Beauveria *Metarhizium*	Mitosporic fungi	Many, including termites, mosquitoes, tsetse flies and locusts
Parasites of nematodes		
Dactylaria *Arthrobotrys*	Mitosporic fungi with trap-forming mycelium	
Harposporium	Mitosporic endoparasite	
Hirsutella	Mitosporic obligate parasite invading adults	Many nematode species
Paecilomyces	Mitosporic fungus with mycelium that invades eggs and egg masses	
Parasites of weeds		
Puccinia chondrillina	Basidiomycetes: Uredinales	*Chondrilla juncea* (skeleton weed)
Phytophthora palmivora	Oomycetes	*Morrenia odorata* (milkweed vine)
Parasites on fungal parasites (hyperparasites)		
Coniothyrium minitans	Mitosporic fungus	Sclerotia of fungi in soil
Eudarluca caricis	Ascomycetes	Rust fungi on leaves

Eudarluca caricis), found on spores of many species of rust fungi, appear to be obligate parasites. Although many examples of mycoparasites have been observed in the laboratory, it has proved difficult to use them to control fungal pathogens in the field. The problem lies in establishing a new microorganism, or increasing the amount of an existing one, in an ecosystem which is already occupied by large numbers of competing species in equilibrium. The best results so far have been obtained by increasing the numbers of a mycoparasite which is already present. *Trichoderma harzianum* spores, applied to seeds in an adhesive dressing before sowing, control seed rots and damping off of seedlings caused by the Oomycete *Pythium*. For control of soil-borne fungal pathogens of mature plants it is necessary to establish the introduced mycoparasite not just in the neighbourhood of seedlings but throughout the soil. To overcome competition from the soil microbial population a very large inoculum must be used, for example the biomass from a fermenter, and it must be added to the soil together with a suitable energy source such as straw. *Trichoderma harzianum* has been used successfully to control *Sclerotinia minor* on lettuce, when added to soil in the presence of a fungicide to which *T. harzianum* is resistant, but *S. minor*, and probably many of the competitors of *T. harzianum*, sensitive. This combined use of more than one method is termed **integrated control**.

Nematodes can cause large crop losses. Fungi attack nematodes by trapping and digesting their bodies, by growing as endoparasites within their bodies, and by invading and parasitizing their egg masses and the cysts which are the form adopted by the females of some nematode species (see Fig. 7. 23). Initial attempts at control used mainly trap-forming fungi, but have not been very successful in the field. Possible reasons may be the difficulty of establishing introduced fungi in the existing soil population, maximum trap formation failing to coincide with peaks of nematode numbers, and the inaccessibility of some nematode life cycle phases such as cysts to the attacking fungi. The mitosporic fungus *Paecilomyces lilacinus*, which invades and destroys the egg masses of *Meloidogyne arenaria*, has been found to reduce the numbers of these nematodes in field soil, and to increase the yield of the oat plants which the nematode attacks. Some of the fungi which parasitize insects (page 426) are being investigated or are in use for the control of insect pests of crops and vectors of human disease (Table 7.4).

Host-specific plant pathogenic fungi can be used to control weeds in crops. The pathogen can be newly introduced into an area and become established and breed there. *Puccinia chondrillina*, a rust fungus, was released into Australia in 1971 to control its host, skeleton weed, *Chondrilla juncea*. A strain from Italy which is specific to the weed was used, and has become established, increasing its range in successive seasons. Fungal pathogens can also be used to spray on to local outbreaks of weeds, without becoming permanently established. Total control of milkweed vine, *Morrenia odorata*, can be achieved with a single application of spores of its pathogen, *Phytophthora palmivora*. Fungi used in this way have been termed **mycoherbicides**. Plant pathogenic fungi with possible uses as mycoherbicides include species of the genera *Alternaria*, *Ascochyta*, *Bipolaris*, *Cercospora*, *Colletotrichum*, *Drechslera*, *Fusarium*, *Microsphaeropsis*, *Phoma*, *Phomopsis*, *Phytophthora* and *Sclerotinia*. Breeding for increased pathogenicity could increase the effectiveness of fungi used as mycoherbicides, and recombinant DNA techniques could be used to produce more virulent transformants. The obvious risk is the accidental creation of new crop pathogens.

Biological control has advantages over chemical control, because it is less likely to damage non-target organisms, and because resistance is slower to evolve. Pests and pathogens can evolve resistance to many specific chemicals with only a small genetic change. Natural predators and parasites, on the other hand, have a long evolutionary history of overcoming the resistance of their fungal hosts. This evolution has resulted in ability not only to overcome resistance, but also to locate their hosts, so they can target their attack. Biological control methods are also preferred over chemical methods because they do not leave any residue of toxic substances, and are therefore preferred to the use of chemicals for sustainable agriculture and forestry. Consequently there is much interest in developing biological control methods. However, methods which work well under controlled conditions in the laboratory are usually difficult to scale up to use in the field, because of environmental variables which may not favour attack and spread of the control organism.

Breeding for disease resistance Disease can be controlled in crop plants by breeding for the characteristics that confer resistance. Traditionally this is done by selecting and propagating varieties of plant which appear to suffer least from disease, as when mildew-resistant varieties of rose are grown and bred from, in preference to susceptible ones. Such resistance is termed **non-race-specific** or **horizontal**. In plants which have a gene-for-gene relationship with a specific pathogen (page 267), specific resistance genes can, by selective breeding, be incorporated into new cultivars of plant. The presence of a specific resistance gene in a cultivar is established by testing its susceptibility to known races of the pathogen. This resistance to disease due to only a few genes, termed **race-specific** or **vertical resistance**, is liable to be overcome by genetic change in the pathogen which enables it to infect the new cultivar. When a crop consists entirely of plants with one specific resistance gene, evolution of resistance by natural selection of mutations in the pathogen population is rapid. However, barley powdery mildew is successfully controlled by maintaining four cultivars of barley each with a different set of resistance genes, and each year sowing seed consisting of a different combination of three of them. A breeding programme of this type takes large resources and is only feasible for very widely grown and profitable crops such as cereals.

The few genes responsible for race-specific resistance are believed to arise later than the probably larger number of genes which make a plant compatible with its pathogen and so allow it to be host to a particular fungal species. Compatibility between a host and its pathogen result from a great many interactions, both physiological and developmental (pages 406, Fig. 7.21). In future it may be possible, using recombinant DNA technology, to identify and prevent the expression of genes responsible for this basic compatibility between plant and host. Plant compatibility genes could be inactivated to produce cultivars to which the pathogen is less likely to be able to adapt, and it might even be possible to infiltrate pathogen populations with individuals carrying altered, non-functional compatibility genes. However, this can only be achieved with greater

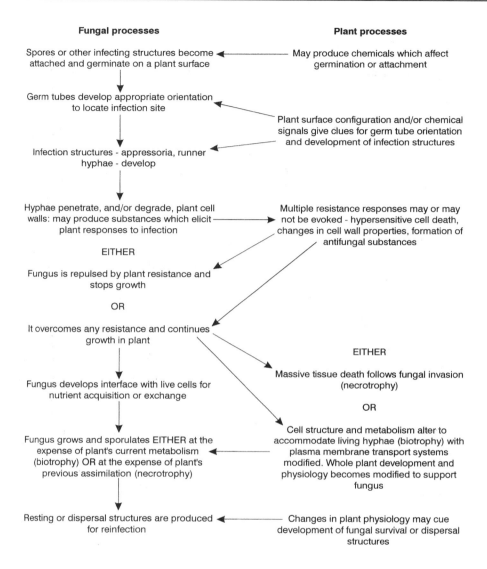

Fungal processes

Spores or other infecting structures become attached and germinate on a plant surface

Germ tubes develop appropriate orientation to locate infection site

Infection structures - appressoria, runner hyphae - develop

Hyphae penetrate, and/or degrade, plant cell walls: may produce substances which elicit plant responses to infection

EITHER

Fungus is repulsed by plant resistance and stops growth

OR

It overcomes any resistance and continues growth in plant

Fungus develops interface with live cells for nutrient acquisition or exchange

Fungus grows and sporulates EITHER at the expense of plant's current metabolism (biotrophy) OR at the expense of plant's previous assimilation (necrotrophy)

Resting or dispersal structures are produced for reinfection

Plant processes

May produce chemicals which affect germination or attachment

Plant surface configuration and/or chemical signals give clues for germ tube orientation and development of infection structures

Multiple resistance responses may or may not be evoked - hypersensitive cell death, changes in cell wall properties, formation of antifungal substances

EITHER

Massive tissue death follows fungal invasion (necrotrophy)

OR

Cell structure and metabolism alter to accommodate living hyphae (biotrophy) with plasma membrane transport systems modified. Whole plant development and physiology becomes modified to support fungus

Changes in plant physiology may cue development of fungal survival or dispersal structures

Figure 7.21 Stages in the interaction between a fungus and a host plant.

understanding of the mode of operation of genes which control non-race-specific resistance. The interaction between plants and their microbial pathogens is one of the most active areas in current mycological research.

Integrated control Fungi cause plant disease through a series of processes – infection, colonization, sporulation, dissemination – each of which requires appropriate conditions (Fig. 7.21). Plant disease can be prevented or its consequences minimized by interfering with any of these stages. Current procedures for controlling coffee rust (*Hemileia vastatrix* parasitizing *Coffea arabica*) in Brazil illustrate how a combination of methods, termed **integrated control**, can be most effective and economical. A study of how each disease process is affected by climatic conditions makes it possible to forecast disease intensity and apply fungicides at the time when spores are developing on the leaves. The parasite evolves fungicide resistance quickly, so copper compounds are used as a general fungicide, as well as occasional systemic fungicides. A coffee breeding programme generates and maintains cultivars with both race-specific and non-race-specific resistance. In the future it may be possible to make use of fungal hyperparasites (fungi parasitic on fungal parasites) which have been found to reduce the numbers of spores produced on leaves. Use of the mathematical techniques of systems analysis may help to show which steps in the disease process it would be most effective to inhibit, to limit the incidence of disease.

Fungi and Animals

The relationships between fungi and animals are based on the fact that both are heterotrophs, with fungi feeding on animals and animals on fungi. Some fungi are saprotrophs on or in animals or animal remains. Others are parasites or mutualistic symbionts which are able to feed on or in living animals. Like fungi that infect plants these vary in the degree of damage and disease they cause, ranging from rapidly lethal to mutualistic relationships. Most species of animals for which fungi are an important source of food are small, such as soil- or wood-inhabiting arthropods and nematodes. Interactions between animals and fungi are important in ecosystems. For example, the rate of plant litter decomposition is affected by nitrogen cycling between plants, fungi and arthropods (page 334; Fig 6.15). Tripartite interactions involving animals and fungi jointly feeding on plants also occur, for example scale insects parasitic on plants are themselves parasitized by specific fungi.

Fungi that Attack Animals

Most healthy animals are resistant to attack by fungi and normally each species is only invaded by a few specialized pathogens. Only commercially or medically important fungal pathogens have been thoroughly investigated, although fungi have been found growing in, and sometimes isolated from, diseased populations of many different animal phyla and presumed to be the causal agents of disease. Protozoans, crustaceans and molluscs have highly specialized fungal parasites, and probably every species of animal has specific fungal parasites.

Fungi parasitic on insects
Most insects die from disease, and their diseases are mainly caused by fungi. The ability to attack and invade insects (Fig. 7.22) has apparently evolved repeatedly in fungi, with some insect parasites known from most taxonomic groups. Most

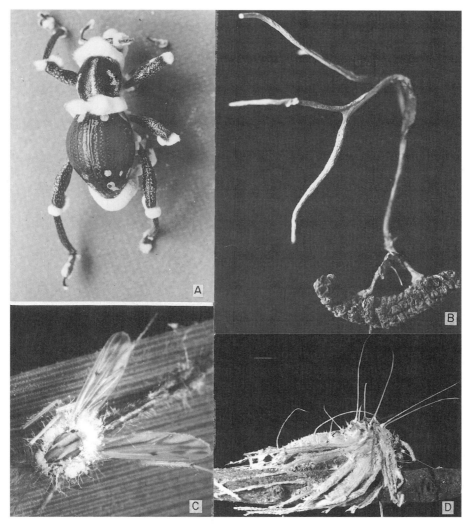

Figure 7.22 Examples of insects attacked by parasitic fungi. A, Cocoa weevil, *Pantorhytes plutus*, infected by *Beauveria bassiana*. B, Coleopteran larva, originally buried in a log, with *Cordyceps* stroma emerging from the insect tunnel and forming brightly coloured fertile heads with immersed perithecia. C, Dipteran fly infected by *Erynia* (Entomophthorales), clinging to a rice leaf and attached to it by fungus. D, Moth, fastened to a tree branch by fungal mycelium, with *Cordyceps* perithecia forming on the dorsal surface, and cylindrical synnemata of *Akanthomyces*. (Figure and legend from Evans, H. C. (1988). In Pirozynski, K. A. & Hawksworth, D. L. eds. *Coevolution of Fungi with Plants and Animals*, Academic Press, London.)

insect pathogens, however, are either from the Entomophthorales, a Zygomycete order, or from the Ascomycetes and related mitosporic fungi. Insects can be attacked during larval or pupal stages which often live in soil or within substrates, or in the adult, often aerial form. Most species are specialized to attack one stage. *Coelomomyces* is a Chytridiomycete genus, members of which live in the coelomic cavity of mosquito larvae, some species being specific to a single host species. They form thick-walled resting spores which germinate to produce zoospores. The fungus does not have a true cell wall but produces 'hyphal bodies' in the host's blood which exploit the adipose tissue and kill the larva.

Entomophthorales occur mainly in temperate regions, unlike the Ascomycete insect pathogens. They invade mature insects by hyphal growth through joints between plates of the exoskeleton. Different species of this group each attack one or more insect species. *Entomophaga grylli* is a parasite of many grasshoppers and locusts, whereas *Entomophthora muscae* is widespread and common on many insects. A number of species live on aphids, and one genus, *Massospora*, is confined to cicadas. Most types of Entomophthorales are highly invasive and destroy the host's tissues within a matter of days. Invasion and destruction, rather than toxins, cause death. A common feature is the alteration in behaviour produced by infection, so that insects crawl upwards just before death, and become anchored by the drying of sticky exudates or fungal mycelium to exposed places, such as the tops of grass stems. The conidiospores are then forcibly discharged from the surface of the mummified dead body (Plate 10). The modification of the insect's behaviour may be of advantage to the fungus, facilitating the dispersal of spores. Some fungal species induce even more remarkable effects. For example *E. grylli* causes synchronized death of locust populations, with most individuals dying in the afternoon so that spores are released at the most humid time of day. A few Entomophthorales do not cause rapid death but colonize only a limited part of the insect's body, discharging spores through a small hole in the exoskeleton over a longer period. In this way the spores are dispersed by the insect itself acting as vector, with social interaction between insects aiding spread of the disease. Fungi of the Entomophthorales survive outside host tissues as resting sporangia, and in the mummified bodies of dead insects.

Ascomycete fungi, and some fungi so far found only as conidial anamorphs but which are probably also Ascomycetes, include hundreds of obligate insect parasites. Many of these are tropical and therefore little investigated. For example, the genus *Hypocrella* consists of over forty species, all very effective parasites of scale insects and whitefly on leaves in humid tropical forests. The best known Ascomycete genus of insect parasites is *Cordyceps*, with several hundred species. It commonly infects larvae or pupae, producing yeast-like cells in the haemocoele, and usually killing its host some weeks after initial infection. Continued hyphal growth within the body produces a kind of sclerotium which acts as a survival stage for the fungus. Finally, when conditions are favourable, a phototropic aerial fruit body emerges from the sclerotium, and bears perithecia and conidia. *Nectria* species attack scale insects, producing coloured perithecia on the host and an anamorphic conidial stage, placed in the genus *Fusarium*.

The mitosporic fungi include a number of genera with the ability to colonize living insects. *Beauveria* and *Metarhizium* are the most studied of fungal insect parasites, and occur mainly in soil, but can be grown in culture. Some species of *Aspergillus*, *Paecilomyces* and *Hirsutella* are facultative insect pathogens. Insects become infected by conidia, which germinate on the host's surface to give germ-tubes and appressoria, from which hyphae penetrate into the tissues mechanically and by means of chitinases, lipases and proteinases. The fungus is dimorphic, and inside the host's blood it grows as single yeast-like cells. Later, as the host's tissues are killed, the fungus reverts to the hyphal form and grows throughout the body. Hyphae grow through the layers of the exoskeleton and anchor them together, preventing moulting which otherwise might enable the insect to slough off the infected layer. Toxins (page 440) have been found to be produced during the yeast phase of *Metarhizium anisopliae*. These **include** cyclic depsipeptides (e.g. Fig. 7.11A) known as destruxins, which induce paralysis when injected experimentally into healthy insects. Insects can be killed within a few days of infection.

Some common fungal pathogens of insects that are known to cause epidemic disease in natural populations, such as species of *Metarhizium* and *Beauveria*, have been investigated as possible agents of biological control of insect pests. Unlike other biological control agents, *Metarhizium* and *Beauveria* infect by directly penetrating the cuticle and do not have to be ingested. Therefore they provide the only means of biological control of insects that feed by sucking plant or animal juices. These fungi can be cultured easily and large volumes of spores can be produced with which to infect the pests. This allows a formulation containing fungal spores to be used as a 'mycoinsecticide', applied in a similar way to a chemical insecticide to control an outbreak of a pest. The molecular genetics of host recognition and infection is being investigated in *M. anisopliae* to discover genes conferring virulence and rapidity of infection. These could then be used to engineer more virulent and host specific transgenic insect pathogens. So far, forty gene sequences have been found that are specifically expressed during fungal growth on insect hosts. Some encode molecules involved in host recognition and initial growth, for example protein kinases and other signal transduction pathway components. Others are required for penetration of the insect cuticle which is composed of protein and chitin, and these include proteinases and chitinases. Yet others encode insect toxins. No single gene conferring virulence has been identified, and the effectiveness of the fungus as an insect pathogen appears to result from the activity of products of many genes.

Although fungi are potentially effective as mycoinsecticides for short-term control, viral and bacterial diseases of insects appear likely to be more effective means of **long-term** control than fungi. One reason for this is that fungal infection is not passed down from parents to offspring, and each individual insect can only be infected by spores from its environment, so that the spread of infection depends on variables such as temperature and humidity that affect sporulation.

Nematode-trapping fungi

Parasitism is a relationship in which one species infects another and obtains nutrients from it, and which may continue for a long time. Parasites are nearly always much smaller than their hosts – a host could not readily provide sustained support for an organism as large as itself. Many animals do, however, obtain nourishment from organisms that are smaller than, or about the same size as themselves, by killing and eating them, an activity termed **predation**. There are also predacious fungi which feed on soil animals such as amoebae, rotifers, nematodes (eelworms) and mites. Nematode-trapping fungi (Fig. 7.23) belong mainly to the mitosporic fungi, but some to the Zoopagales, a Zygomycete order

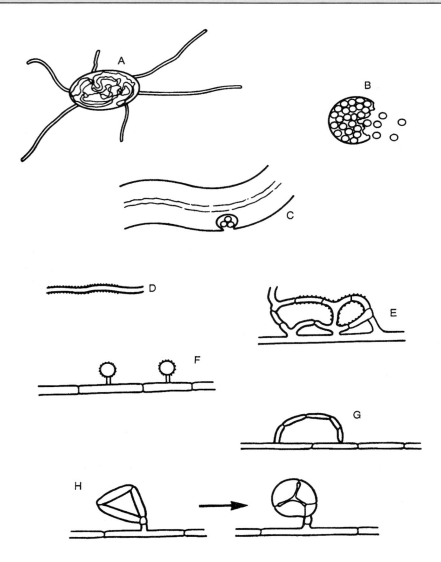

Figure 7.23 Ways in which fungi attack nematodes. Endoparasites: A, *Verticillium chlamydosporium*, a conidial fungus which invades nematode eggs, showing mycelium within the egg and extending from it into the soil. B, *Nematophthora gynophila*, a zoosporic fungus producing spores inside the body of a cyst-forming nematode, in which the female takes the form of a cyst. C, *Catenaria anguillulae* developing sporangia in the tissues of the adult. Trap-formers: (dots represent adhesive material). D, Adhesive hyphae. E, *Arthrobotrys oligospora* with networks of adhesive hyphae. F, Adhesive knobs, found on mitosporic and Basidiomycete mycelium, e.g. *Pleurotus*. G, Non-constricting rings. H, constricting rings found on *Arthrobotrys dactyloides* open (left) and closed (right). (After Stirling (1991). See references, page 458.)

of obligate parasites, and some to Basidiomycetes. Most nematode traps involve adhesive hyphae to immobilize the nematode, but some produce loops and snares of non-adhesive hyphae. In the Zoopagales, the touch of a nematode against the mycelium induces exudation of a sticky droplet at the point of contact, to which the nematode adheres. Within a few hours hyphae invade its body, and assimilate the contents. In the presence of nematodes some predacious Hyphomycetes produce adhesive lateral hyphae or hyphae that form either a constricting or non-constricting ring according to species. A nematode crawling through a non-constricting ring becomes jammed and is invaded by hyphae. With constricting rings, the ring inflates within one-tenth of a second after the nematode touches it, trapping the animal, which is later invaded and digested. The mechanism of this very rapid inflation is not understood. Perhaps the most remarkable example of an infection structure yet described in a fungal parasite of animals is the 'gun cell' in the biflagellate zoosporic genus Haptoglossa, which parasitizes nematodes and rotifers. The zoospore encysts, and the cyst germinates to form an infection cell, which matures to form an intracellular harpoon-shaped 'missile', within a looped, invaginated 'injection tube'. When a nematode brushes past the triggered infection cell, this tube everts within a fraction of a second, penetrating the host's cuticle and injecting the parasite's cytoplasm.

In addition to the soil fungi that behave as predators with respect to small animals, there are others showing more conventional parasitic activity. These are endoparasites, in which a spore ingested by a nematode, amoeba or other small animal germinates to bring about an infection. Other fungi which live in soil invade nematode cysts and egg masses, and one such fungus, *Paecilomyces lilacinus*, has been used successfully in biological control of plant disease caused by nematodes (page 423). It is likely that the Zoopagales are obligate parasites or predators. However, the nematode-trapping mitosporic fungi, which are readily grown in pure culture, may be saprotrophs obtaining supplementary nourishment by predation.

Fungi parasitic on amphibians and fish
Over the last decade there have been unprecedented declines in populations of frogs and toads, resulting in extinctions of some species, especially in Central America and Australia. A major international effort resulted in the unexpected identification of a fungus of the order Chytridiales as the causative agent of this fatal disease. It has been named *Batrachochytrium dendrobatides*. The fungus is strictly keratinophilic, growing on the keratinized skin of adults and killing them. The cause of death may be the disruption of the animal's osmoregulation, as the fungus especially colonizes the skin of the pelvic patch ('drink patch') of the adults. Infected tadpoles are not killed, as the fungus is confined to their mouth parts, their only keratinized body region, but there is a high mortality rate after metamorphosis. This is the first report of the parasitism of a vertebrate by a member of the Chytridiomycota. The sudden and catastrophic appearance of this cutaneous chytridiomycosis raises a number of questions about how this came about. A worrying suggestion is that the infection has been unwittingly spread by biologists studying amphibians to populations that had hitherto not encountered the pathogen.

The majority of fungal species reported to infect fish are facultative pathogens, such as *Saprolegnia* (page 26) which are widespread cutaneous pathogens, for example of salmon when they enter rivers. An exception, however, is *Icthyophonus hoferi,* an obligate fish pathogen which has been implicated in major losses (sometimes more than 50%) of plaice, herring and haddock, especially in Scottish waters and the Norwegian and Baltic Seas. *Icthyophonus* is a genus until recently considered to be fungal, although of uncertain relationship. It is now considered to be a member of a group of organisms formed just after the true fungi had split off, so it is a 'cousin' to the fungi rather than an ancestor. Infection of a fish is by ingestion of spores, from previously infected fish tissue. The spores germinate in the gastrointestinal tract to give amoeba-like cells which invade the intestinal wall and enter the bloodstream to invade other body tissues, where they eventually encyst. Death is commonly by predation of a weakened individual.

Medical Mycology

Fewer than two hundred species of fungi are capable of causing disease in otherwise healthy human beings. The diseases that they cause are called **mycoses**. A wider range of fungi cause us problems by producing **mycotoxins**, giving rise to **mycotoxicoses**, or by being **allergenic**.

Mycoses

Fungal diseases can be classified into three groups, of increasing severity: superficial, subcutaneous and systemic infections. Another classification is whether the fungus is a true pathogen, able to invade and develop in tissues of an otherwise healthy host with no recognizable predisposing factor, such as *Coccidioides immitis* (page 435), or an opportunist pathogen, such as *Aspergillus fumigatus*, only able to invade the tissues of a debilitated or immunosuppressed host. Medical practices developed over the last half century, especially the widespread use of antibacterial antibiotics, steroid drugs, cytotoxic drugs for cancer chemotherapy, and immunosuppressive drugs for transplantation, have greatly increased the incidence of disease from opportunist fungal pathogens. A further factor is increasing international travel, leading to worldwide dissemination of what had been local infective agents. A very important factor over the last twenty years has been the emergence and worldwide spread of AIDS (acquired immunodeficiency syndrome) as a consequence of infection by HIV (human immunodeficiency virus). The occurrence of fungal infections among AIDS patients is very high (see below) and is a major cause of mortality. Worldwide, a WHO estimate is that there were 2.3 million deaths from AIDS in 1999, so fungi currently are taking a very heavy toll on human life. Unfortunately the development of antifungal treatments has not kept pace with the enormously increased prevalence of fungal infections resulting from a combination of all of these causes, and much effort needs to be put into this area of medical research.

The major fungal pathogens and the diseases that they cause will now be described. There has been much interest in elucidating virulence factors, i.e. features that allow attachment, entry and dissemination though the body. The rationale is that understanding these could suggest novel ways for treatment. Obvious features of importance are the ability to grow at physiological temperatures and pH values. Other features are: production of specific adhesins for attachment to host tissues, production of aggressive enzymes, mechanisms of evading host defence mechanisms, and, for the dimorphic fungi, the transition from non-pathogenic saprobic form to pathogenic form (or from less-invasive to more-invasive form).

As with plant pathogens, molecular biological techniques are now being widely used to pinpoint key attributes of the fungus in its pathogenesis, and the relative pathogenesis of deletion mutants can be studied. This course is not to be undertaken lightly, in view of current attitudes to animal experimentation, and the expense of maintaining laboratory animals.

Many fungal pathogens have a relatively limited geographical distribution and are only rarely encountered elsewhere. Not unexpectedly, a greater variety is found in the tropics, where high temperatures and high humidities favour fungal growth and survival.

Candida albicans is the most ubiquitous opportunistic fungal pathogen. There is evidence for a 'general purpose' genotype which is highly prevalent throughout the world, in all types of patients and types of infections. *Candida* is a large genus, with about 165 species recognized, of yeasts lacking sexual cycles (Fig. 2.27F). A few other species apart from *C. albicans* can also occasionally cause human disease, but only *C. albicans* is dealt with here. It is a commensal organism in the mouth and gastrointestinal tract of around 30–60% of the healthy population, but can cause a wide series of types of disease, candidosis, of a great range of severity. Most commonly, it can invade mucosal surfaces of the mouth and vagina, causing 'thrush'. Oral candidosis tends to occur in infants and also in old age, for example associated with the wearing of false teeth. Up to 75% of women will suffer from at least one episode of troublesome vaginal thrush during their lives. There is a very long list of 'pre-disposing factors', favouring the change from benign commensal to pathogen for this 'Dr Jekyll and Mr Hyde fungus', notably treatment with antibacterial antibiotics, or with steroids, and conditions such as diabetes and HIV infection (Fig. 7.24A, and see below). More disseminated life-threatening infections, with invasion of deep organs, occur predominantly in severely immunocompromised patients, especially as a result of organ transplants, but also among other debilitated surgical and intensive care patients, including premature babies. A further complication is that intravenous catheters can become infected, resulting on continuous inoculation of the bloodstream. In its most commonly encountered form, *C. albicans* grows as a budding yeast (Fig. 8.7B), microscopically almost indistinguishable from *S. cerevisiae*. In the laboratory, in a rich broth, it has a very fast growth rate, doubling in less than an hour. It is diploid and has no known sexual cycle (but intriguingly has genes very similar to mating type genes of *S. cerevisiae*). It has a non-standard codon usage whereby CUG encodes a serine residue instead of a leucine. These features both contribute to difficulties in investigating it with molecular biological techniques (for example, both copies of a particular gene have to be deleted to assess its role). It can undergo the still incompletely understood phenomenon of phenotypic switching. This is the high frequency (about 10^{-4}) reversible switching of colony types on agar, giving a range of appearances differing from the basic smooth

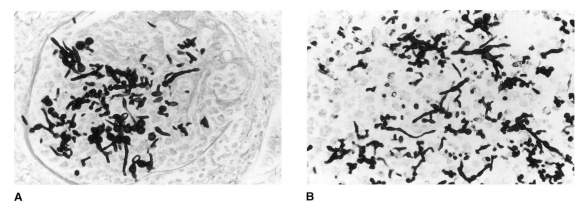

A   **B**

Figure 7.24   Fungal infections of man. The tissue sections have been stained with light green to show human cells and hexamine-silver to render the fungus black. A, *Candida albicans* in glomerulus of kidney of patient with the human immunodeficiency virus (HIV) which subsequently resulted in AIDS. The fungus shows mould–yeast dimorphism. Some yeast cells are budding but others are giving rise to hyphae. B, *Aspergillus fumigatus* in post-mortem lung tissue of an AIDS patient. Fungal spores are present, and some have germinated to give hyphae. A process resembling microcycle sporulation (page 193) appears to be occurring, with some hyphae, after very little growth, bearing phialides from which spores are being budded. (Both preparations by Jacqueline Wong.)

form, with names such as opaque, fuzzy, star, stipple and ring. The frequency of switching can be increased about 30-fold by treatment with low doses of u.v. light. The different colony phenotypes are accompanied by differences in properties such as adhesion, enzyme secretion, drug susceptibility and surface antigenicity, all features associated with virulence, suggesting that phenotypic switching itself can be of importance in pathogenicity. Another feature of importance is the 'dimorphic switch'. This is characteristic of *C. albicans*, and is used in diagnosis. In some culture conditions, for example in mammalian serum at 37°C, yeast cells, instead of budding, form 'germ-tubes' which elongate as true hyphae, and branch to give a mycelium, which can penetrate tissue in ways that yeast cells cannot. The hyphal cells differ from yeast cells in ways other than their apical polarity; they have an altered cell wall composition, with a threefold increase in chitin content. The germ-tubes and hyphal tips show the phenomenon of **thigmotropism** (page 239), i.e. directional growth in response to surface irregularities, also shown by some plant pathogens (Fig. 7.2C). If grown on a membrane with pores, they will grow into them, a response which should aid their penetration of tissue. Initial attachment is required before penetration, and *C. albicans* has the ability to adhere to a variety of host surfaces including epithelial and endothelial cells, fibrin-platelet matrices and neutrophils. The most important adhesins are a variety of surface mannoproteins. Some are specific to yeast or hyphal cells, or to particular switching phenotypes, and this variation may be involved in evading host immune defences. As with plant pathogens, secretion of specific enzymes is of importance in pathogenesis. In *C. albicans*, most interest has centred on a family of aspartyl proteases encoded by at least nine genes (*SAP1–9*). These different enzymes, together with lipases and phospholipases have distinct roles during colonization and invasion of the host, involving modification of cell surfaces and promoting attachment and penetration. This has been shown by assessing the properties of a range of deletion mutants; for example deletion of *SAP2* gave a strain that had lost its virulence.

There are four systemic fungal pathogens, described below, for which a dimorphic switch is of key importance, with notable differences to the yeast–mycelium transition of *C. albicans*. In each case the yeast or spherule is the pathogenic state, and the mycelium is the saprobic form; the dimorphism is regulated by temperature, with the growth as yeast in tissue or enriched media at 37°C, and as mycelium in nature or laboratory at 25–30°C; more chitin is in the cell walls of the yeast than hyphae; the route of infection is via the lungs, but then can become widely disseminated through the body; and the source of infection is exogenous, from propagules in the environment, rather than commensal cells.

Coccidioidomycosis is caused by *Coccidioides immitis*, and is endemic in parts of southwestern USA, Central and South America. In the soil it grows as septate hyphae, alternate cells of which form hardy arthroconidia, which are released to the air when the soil is disturbed. These can germinate to form more hyphae, or if inhaled, enlarge in the lung tissues to give large multinucleate spherules (80 µm), which form many small (2–5 µm) uninucleate endospores, which can spread the infection in the surrounding tissue. In most individuals the infection is relatively mild and transient.

Blastomycosis, caused by *Blastomyces dermatitidis*, and histoplasmosis, caused by *Histoplasma capsulatum*, are similar diseases. Both are endemic chiefly in parts of America and Africa, primarily causing pulmonary infections. *B. dermatidis* has rarely been isolated from soil, but *H. capsulatum* can readily be isolated from soil enriched with bird or bat droppings, and numerous epidemic outbreaks have been reported in the USA associated with building construction or demolition in contaminated sites. The two species have been shown to be anamorphs of the same teleomorph genus, *Ajellomyces*, and isolates of opposite mating type of each species produce ascospores in cleistothecia when grown in pairs in the laboratory.

Paracoccidioidomycosis is caused by *Paracoccidioides brasiliensis*. It is very largely confined to South and Central America. At 28°C it grows as mycelium, at 37°C as yeast cells. It has scarcely ever been reported from soil samples, so the source of infection by inhaled spores is unclear, despite the fact that skin testing in endemic regions shows that a high proportion of the population of both sexes have come into contact with it by the age of 20. There is a rare acute juvenile form of the disease of both sexes, which may lead to death from fungal growth in liver, spleen and bone marrow, but most cases are of pulmonary infections of adults, nearly all males between 30 and 50. This remarkable susceptibility (varying in different regions, but often >50:1, men:women) results from a very specific inhibition of the mycelium/conidium to yeast transition by physiological concentrations of the female sex hormone oestradiol. It is only this morphogenetic transition which is inhibited; growth of the yeast cells is unaffected. The cytosol of yeast and hyphal cells contains a high-affinity binding protein with similar properties to mammalian receptor proteins responding to steroid hormones, and so there may be a signal transduction pathway operating in the fungus similar to those in mammals. There are marked cell wall differences between the two growth forms: in hyphae, β-(1→3)-glucan is the major component, being replaced by α-(1→3)-glucan in yeast. The α-(1→3)-glucan may play a part in evasion of host immune defences.

A further systemic yeast infection acquired by inhalation of airborne cells is cryptococcosis, caused by *Cryptococcus neoformans*, with two varieties. Yeast cells of *C. neoformans* var. *neoformans* have been repeatedly isolated throughout the world from pigeon droppings, but the birds are not naturally infected, and those of *C. neoformans* var. *gatti* have only been found in association with the red gum tree in the tropics. The sexual phase is the basidomycete *Filobasidiella neoformans*. The yeast cells are notable for producing a thick acidic mucopolysaccharide capsule. Another virulence factor is the production of polyphenoloxidase, which converts phenolic substrates into melanin. Both capsule and melanin inhibit phagocytosis. Pulmonary cryptococcosis can develop rapidly in immunocompromised patients, but meningitis is the most common consequence, following dissemination from the lungs, particularly in AIDS patients (see below).

Pneumocystis carinii has emerged from obscurity to be a major fungal pathogen of AIDS patients (see below). This enigmatic organism was described in 1909 as the rare cyst stage of a protozoan parasite, and until recently was the sole province of parasitologists and is treated by antiparasitic drugs. In the last decade, it has become clear that it is a true fungus, albeit a very unusual one. It has yet to be grown *in vitro*, which considerably hampers the study of its natural history. It appears to be passed from person to person by breathing in spores, which germinate and invade the alveoli, resulting in extensive damage to the alveolar epithelium. Different features of its molecular phylogeny, using information from different proteins and RNA species, had pointed in different directions, variously towards affinity with Zygomycota, Ascomycota or basidiomycetous yeasts, but its 18S rRNA gene sequence and other more recent evidence suggest its affinity with ascomycetous yeasts. In this light, its cyst should be referred to as an ascus. Its fungal features include the presence (initially elusive) of the fungal-specific elongation factor 3 for protein synthesis, and chitin, β-(1→3)-glucan and mannoproteins in its cell wall. The presence of β-(1→3)-glucan is in accord with successful treatment of animal infections with experimental drugs having glucan synthesis as target – echinocandins and papulocandins. A feature unlike fungi is the presence of cholesterol and not ergosterol in its cell membranes, which is in accord with the lack of activity against it of amphotericin B and azole antifungal drugs. Strains from humans, rats, mice, ferrets, rabbits and other animals are genetically distinct, and show a high host species specificity, as shown by cross-infection experiments.

Aspergillosis results from opportunistic infection by *Aspergillus* species (Figs. 2.19D, 7.24B), most importantly *A. fumigatus*, but also other species, such as *A. flavus*. In its most serious form there is widespread growth of the fungus mycelium in the lung and dissemination to other organs often follows. Infections follow inhalation of spores into the lungs and paranasal sinuses, and a wide range of different conditions can develop, for example the formation in the lung of a 'fungus ball' called an aspergilloma. Mucormycosis (zygomycosis) is a rare but potentially very serious opportunistic infection by fungi of the order Mucorales, such as *Rhizopus oryzae*. There can be a range of clinical syndromes following infection by inhalation of spores, the most common of which is rhinocerebral mucormycosis, i.e. growth from the paranasal sinuses to involve the palate, face, behind the eyes and brain.

Most people suffer at some time in their lives from superficial or cutaneous fungal infections, i.e. of their skin, nails or hair. A disease of the skin is pityriasis versicolor, caused by overgrowth of *Malassezia furfur,* a lipophilic yeast found as a benign commensal on oily regions of the skin. The 'versicolor' refers to discoloured patches of skin, unsightly but otherwise harmless. The yeast is also implicated in causing dandruff. The major causes of skin disease, however, are the large group of dermatophytes, more than 40 species of closely related fungi of the genera *Microsporum, Trichophyton* and *Epidermophyton.* These are classified into three groups, based on their primary source: anthropophilic, human; zoophilic, animal (see section on veterinary mycology); and geophilic, soil. The tinea (ringworm) diseases that they cause are classified by the site on the body. Thus tinea pedis is ringworm of the feet, most commonly caused by *T. rubrum*, *T. mentagrophytes* or *E. floccosum.* What these fungi share is the ability to utilize keratin, the characteristic protein of skin, nails and hair. They do not invade the underlying tissue, but may penetrate between cells where there is keratin. They can be identified by the characteristics of their macroconidia or microconidia, and morphological and nutritional characteristics. *Microsporum* and *Trichophyton* are anamorphs of the Ascomycete genus *Arthroderma*. Some species are common throughout the world, others have a restricted distribution. A further cutaneous infection, which may however spread to give subcutaneous and lymphatic infections, is sporotrichosis, caused by the dimorphic fungus *Sporothrix schenckii*. It is widely found throughout the world as a mycelium in soil and plant materials. Infection usually follows abrasion of the skin by infected material, for example a thorn. People at risk include farmers, gardeners, florists and miners. An outbreak of 3000 cases occurred in a South African gold mine in the 1940s, contracted from infected pit props.

As mentioned earlier, the occurrence of fungal infections among AIDS patients is very high. In the developing world, this has led to a dramatic increase in the number of deaths from systemic fungal infections. Infection by *Pneumocystis carinii* played a key role in the elucidation of AIDS as a new 'plague of our time'. In early 1981, five patients in Los Angeles, two of whom had died, were diagnosed as having *Pneumocystis* pneumonia. This was so unusual that it was immediately investigated by the US Center for Disease Control, and, together with consideration of the concurrent occurrence of other rare diseases, this led to the first description of AIDS. Another previously exceptionally rare human pathogen, *Penicillium marneffei,* has emerged in the last 10 years as one of the most frequent opportunist infections encountered in AIDS patients in South East Asia. Up to 25% of AIDS patients in the north of Thailand are infected with it. This species was first described in 1959, isolated from bamboo rats by a mycologist who inadvertently infected himself with a contaminated needle, and the first natural human infection was recorded in USA in 1973. However, the most important life-threatening fungal infection of AIDS patients today is cryptococcosis. The incidence varies from one continent to another, from 5–10% in Europe and North America to 15–30% in Africa. More than 80% of these patients have meningitis (infection of the brain and meninges by the fungus). Patients with AIDS with cryptococcosis can seldom be cured, and mortality is 30–60% within 12 months. Oral candidosis often spreads to the oesophagus in AIDS patients, and is used as an 'AIDS-defining illness', i.e. to aid diagnosis of AIDS in an HIV-positive patient. More deep-seated infections of these patients by *C. albicans* is uncommon. Lung infections of *Coccidioides immitis* have become important in areas where this fungus is endemic, and many of these patients die, despite antifungal treatment. Histoplasmosis occurs in 2–5% of AIDS patients living in the endemic regions of USA, mostly as disseminated infection, either newly acquired or reactivation of a quiescent infection, and again the death rate is high.

There has been a steady rise in the incidence of nosocomial (hospital-acquired) fungal infections since the development of immunosuppressive therapies. Ironically, these therapies are based on the discovery of the fungal metabolite, cyclosporin (page 520). The most frequently acquired systemic infections are candidiasis and aspergillosis, due mainly to *A. fumigatus* and to a lesser extent *A. flavus*, followed by mucormycosis. As discussed earlier, *Aspergillus* species are ubiquitous in nature (pages 55–56) and their small spores (Fig. 4.13j) are readily transmitted in air, and inhaled to reach the pulmonary alveolar spaces, where they germinate. These conidia will be present in the non-filtered, non-ventilated air in hospital wards, but ventilation systems themselves may be sources of infection, by providing spaces where *Aspergillus* can grow, and then disseminating the resultant spores. Large quantities of spores are likely to be released during construction work on hospital sites, particularly if this involves demolition of older buildings. Prevention of infection is based on elimination of *Aspergillus* conidia from the environment of high-risk patients, but this is expensive. In contrast to *Aspergillus* species and the Mucoraceae, *C. albicans* is part of the normal gastrointestinal tract and cutaneous flora of most humans. During hospitalization, patients exposed to broad-spectrum antibiotics, which suppress gastrointestinal bacterial flora, may become extensively colonized by *C. albicans*. It may be invasive, and spread from the gastrointestinal tract in surgical patients and patients receiving cytotoxic drugs which often disrupt the normal integrity of the gut mucosal epithelia. Skin colonization by *Candida* may serve as a source of infection through intravenous catheters. If infection becomes established in the blood (candidaemia) the mortality rate is very high.

Clearly the growing problem of nosocomial fungal infections requires multidisciplinary approaches to its solution, with inputs across the board from architects, builders, clinical and nursing staff, as well as research efforts by medical mycologists to improve methods of diagnosis and treatment.

Treatment of mycoses
Before treatment can be started, an accurate diagnosis has to be made. This can be surprisingly difficult with the wide range of clinical manifestations that some fungi may cause. Microscopic examination of samples from scrapings or swabs, or of tissues themselves, can be informative in skilled hands, but often cultures need to be grown, for identification and sometimes to determine drug sensitivity. Radiography can be useful in the diagnosis and assessment of outcome of

treatment of some fungal infections, for example pulmonary aspergillosis. Diagnosis may be aided by serological detection of specific fungal antibodies or antigens, using commercial kits.

Once the infection has been diagnosed, there are very few antifungal drugs available to the clinician, particularly in comparison to the large number of effective antibacterial drugs. Those in current use and others that have promise for future use are among the antifungal agents described in Chapter 3 (pages 177–180). For all medically important fungi (apart from *Pneumocystis carinii*, see above) the 'gold standard' drug for treatment of systemic mycoses remains amphotericin B, which has been used for about 40 years. Acquired resistance is not a problem, and it is very potently fungicidal *in vivo*. Unfortunately, it is toxic at blood levels very little higher than for its therapeutic use, so it needs careful administration by slow intravenous infusion and constant monitoring of the patient to ensure that toxic levels are not reached. To limit its adverse effects, in particular kidney damage, lipid formulations of this very hydrophobic drug have been designed to control availability of the drug in the body. The major one is a liposomal preparation, where amphotericin B is incorporated into tiny (<100 nm) microspheres of pure phospholipid, cholesterol and drug in the ratio 2.8:1.0:0.4. These liposomes are an efficient way to deliver the drug to the fungus, but unfortunately this treatment is very expensive. Topical amphotericin B preparations can be used to treat mucosal and cutaneous forms of candidosis, as the drug is not absorbed. Another polyene antifungal antibiotic in regular use is nystatin, discovered in 1949. It is used for treatment of oral, cutaneous and mucosal candidosis.

The triazole drugs, fluconazole, itraconazole, voriconazole and others, have a broad spectrum of antifungal activity. Fluconazole and itraconazole have been used for many years for the treatment of a wide range of fungal infections, chiefly orally, with few side effects. Repeated courses of low-dose fluconazole, particularly against oral and oesophageal candidosis in AIDS patients, have led to the emergence of strains of *C. albicans* with various forms of acquired resistance to this drug. Upon examination, these strains have several different mechanisms of acquired resistance. Some have altered 'multidrug resistance' efflux pumps in their cell membranes, some have mutations in the target enzyme, 14α-demethylase, and others have an overproduction of this enzyme by gene amplification. Many of these fluconazole-resistant strains, however, appear to retain their susceptibility to other azole drugs, for example voriconazole. Ketoconazole, miconazole, clotrimazole and other imidazole drugs all have their uses, chiefly against a range of superficial infections. Despite the fact that all the imidazole and triazole drugs have the same mode of action, there are distinct differences in their spectra of activities, and in their pharmacokinetics (details of their fate in the body), giving the clinician a choice for the particular situation. The allylamine, terbinafine, is used topically against dermatophytes and cutaneous *Candida* infections, and may also be used orally. Griseofulvin, the antifungal antibiotic from *Penicillium griseofulvum* (see page 177) has been used for many years for the treatment of dermatophytes. Its antifungal activity is almost restricted to this group of fungi. It is given orally, and has the remarkable pharmacokinetic property of concentrating just where it is needed, in the keratin of skin, hair and nails, i.e. just where the dermatophytes are growing. A further drug, which also has been in use for many years is 5-fluorocytosine. It has a limited spectrum of activity, including *Candida* species and *Cryptococcus neoformans*. Resistance is rapidly acquired, so it is usually used in combination with amphotericin B or an azole drug, for example in treatment of deep-seated *Candida* infections or cryptococcal meningitis in AIDS patients.

Mycotoxicoses

As described in detail in Chapter 8, fungi produce an enormously wide range of biologically active secondary metabolites. It is not surprising that many of these chemicals are toxic to humans. They impinge on our lives in two ways; by our unwittingly eating or drinking food contaminated with a wide range of mycotoxins, or by our mistakenly eating poisonous mushrooms (Table 7.5).

Mycotoxins in food can come directly, when the food has been contaminated by toxigenic fungi while growing or after harvest, or indirectly through the food chain, for example in milk from cows fed with contaminated food. Mycotoxins contaminate up to 25% of the world's food supply. Over 300 are known, of which about 20 are serious contaminants of crops used in human foods and animal feeds. The most notorious mycotoxins are the aflatoxins, from *Aspergillus flavus* (hence 'A-fla-toxin'), *A. parasiticus* and *A. nomius*. These were not discovered until 1960, when there were mass deaths from liver disease of turkeys in Norfolk, UK, followed by deaths of other farm animals. After eliminating other causes, the common feature in the different outbreaks proved to be feedstuff containing one consignment of groundnut meal, which was contaminated with *A. flavus*. Aflatoxins are a family of secondary metabolites, of which the most important is aflatoxin B_1. They are acutely toxic to a wide range of vertebrates, but more insidiously, they are very potent carcinogens. The high incidence of liver

Table 7.5   Some mycotoxins and mushroom toxins

Toxin	Typical producing species	Pathological effects
Mycotoxins		
Aflatoxins	*Aspergillus flavus*	Liver damage, liver cancer
Ochratoxins	*Aspergillus ochraceus*	Kidney damage
Patulin	*Penicillium expansum*	Kidney damage
Trichothecenes: T-2	*Fusarium sporotrichioides*	Alimentary toxic aleukia
Vomitoxin	*Fusarium graminearum*	Vomiting, antifeedant
Zearalenone	*F. graminearum*	Gynaecological disturbances
Fumonisins	*Fusarium moniliforme*	Oesophageal cancer
Ergot alkaloids	*Claviceps purpurea*	Vasoconstriction, gangrene
Mushroom toxins		
Amanitin	*Amanita phalloides*	Liver damage
Phalloidin	*A. phalloides*	Liver damage
Muscarine	*Amanita muscaria*	Sweating, vomiting
Gyromitrin	*Gyromitra esculenta*	Liver and kidney damage
Orellanine	*Cortinarius speciosissimus*	Kidney damage
Coprine	*Coprinus atramentarius*	Alcohol poisoning
Psilocybin	*Psilocybe cubensis*	Psychotropic effects
Ibotenic acid	*A. muscaria*	Psychotropic effects

cancer in some parts of the tropics has circumstantially been attributed to continual ingestion of low levels of aflatoxins, in particular synergistically with simultaneous infection by hepatitis B virus. The carcinogenicity and toxicity of aflatoxin B_1 are attributed to two metabolites produced in liver cells, the epoxide and then the dihydroxy derivative, respectively. Ochratoxins are also produced by *Aspergillus* species, notably *A. ochraceus* in the tropics growing on coffee and cocoa, and *Penicillium verrucosum* in temperate regions growing on cereals such as barley. Ochratoxins have been implicated in high local incidences of kidney disease in the Balkans. Patulin is produced by species of *Penicillium*, *Aspergillus* and *Byssochlamys*. A notable example is its production by *P. expansum*, a blue mould causing a soft rot of apples. This has resulted in its being found in samples of commercial apple juice, which has caused concern as it is regarded as being mutagenic. Another important genus for the production of mycotoxins is *Fusarium*. Thus *Fusarium sporotrichioides* and *F. poae* producing the trichothecene T-2 while growing under the snow on millet left in the fields over the winter in the Soviet Union in the 1940s led to terrible epidemics of alimentary toxic aleukia (a disease involving degeneration of the bone marrow, and leading to haemorrhaging, necrosis of the gastrointestinal tract and a progression of blood abnormalities). Another trichothecene, deoxynivalenol, produced by *F. graminearum* and *F. culmorum*, induces vomiting, hence its alternative name, vomitoxin. Zearalenone is an oestrogenic (i.e. oestrogen-mimicking) mycotoxin produced by *Fusarium graminearum*, *F. culmorum* and related species. Intriguingly, this metabolite may play a role in regulating sexual reproduction in the fungi producing it. The most recently characterized mycotoxins of major significance to human health are the fumonisins, produced by *Fusarium* species such as *F. moniliformae*, for example when growing on maize. These are implicated in local high incidences of oesophageal cancer in China and southern Africa. A further group of mycotoxins, the ergot alkaloids, are discussed in detail in Chapter 8 (page 521), in terms of their value as therapeutic agents. It must be emphasized that, as befits secondary metabolites (see page 513), production of these mycotoxins is strain-dependent. Thus the strain of *F. graminearum* used for mycoprotein (page 497), does not produce the mycotoxins mentioned above for other strains of this species.

In many countries there is legislation governing the permitted levels of mycotoxins in foods. For example, current UK limits include a guideline level of 4 µg kg^{-1} for aflatoxin B_1 in human foods used for direct consumption, and 50 µg kg^{-1} for patulin in apple juice. The implementation of such limits, however, requires a complex system of sampling, chemical analysis and reporting. This is impracticable in the many parts of the world where large populations are supported by subsistence farming and where there is not the infrastructure to support such legislation. The problem is exacerbated in such countries which also do not have the expensive facilities for harvesting crops with minimum damage, or for good storage of these crops, and where ambient temperature and humidity encourage fungal growth. For example, a pile of ground nuts, many with damaged husks, stored under a tarpaulin at 35°C and 95% humidity, forms an ideal breeding ground for *Aspergillus flavus*. Thus much of the world's population will be continually at risk of mycotoxicosis for the foreseeable future.

Luckily for people who gather wild mushrooms, there are only a few toxic ones (Table 7.5). The most toxic, the 'death cap' *Amanita phalloides*, produces two closely related families of bicyclic peptide toxins, the most abundant of which are α-amanitin and phalloidin, with completely different actions. α-Amanitin specifically inhibits RNA polymerase II and so prevents synthesis of mRNA, and phalloidin irreversibly binds to filamentous actin and so disrupts cell structure. Both toxins cause major liver damage. *Amanita muscaria*, the 'fly agaric', and especially species of *Inocybe* and *Clitocybe*, produce muscarine. This is a small molecule that is an analogue of acetylcholine, so the toxicity of muscarine results from its binding to acetylcholine receptors on synapses of the nerve endings of smooth muscles and endocrine glands, causing continuous stimulation. Despite its specific name meaning 'edible', *Gyromitra esculenta*, the 'false morel', is deadly if eaten raw due to the presence of gyromitrin, a compound that gives rise to toxic hydrazines, which may also be carcinogens. In parts of Eastern Europe it is eaten after boiling, discarding the water and re-cooking, but even then may produce cumulative toxic effects. Some species of *Cortinarius* produce the bipyridyl compound orellanine, a metabolite of which causes severe kidney damage. An unusual toxicity is caused by coprine, a metabolite of one of the ink caps, *Coprinus atramentarius*. By itself, this mushroom is not toxic, but a metabolite of the coprine inhibits the liver enzyme aldehyde dehydrogenase. If alcohol is consumed at the same time or up to 48 hours later, this results in accumulation of acetaldehyde, leading to gastrointestinal and cardiovascular disturbances. This action of coprine is the same as that of Antabuse® (disulfiram), a drug used to treat alcoholism by aversion therapy. Finally, and notoriously, some mushrooms produce hallucinogenic psychotropic drugs. The best known are *Psilocybe* species, producing psilocybin, which were used in rituals by Mexican Indians. Symptoms from ingesting this drug, which start within 30 minutes and may last for several hours, vary between individuals and may include visual distortions, with feelings of relaxation, nausea or anxiety, or panic attacks. As well as muscarine, *Amanita muscaria* also produces an amino acid, ibotenic acid, which with its decarboxylated derivative, muscimol, may induce hallucinations, dizziness and erratic behaviour. There has been speculation that *A. muscaria* may represent 'Soma', the divine mushroom of immortality of the ancient Eurasian Vedic civilization, and that it was consumed by Siberian shamans to get into an ecstatic stupor.

Allergies

The human body is well equipped with a wide range of defence mechanisms, most notably the immune system, a complex series of cellular and humoral elements. The immune system has evolved to recognize molecular structures, antigens, and to distinguish between them. Thus for example potential pathogens are recognized and destroyed. Occasionally, however, there is an overactive immune response, that may cause more damage than the potential pathogen. This is known as hypersensitivity. There are different types of hypersensitivity. Type I, immediate hypersensitivity, is mediated by immunoglobulin E (IgE) recognizing the fungal antigens, which are then termed allergens. The binding of the allergen to the IgE triggers allergic responses which can include asthma, eczema, hay fever, rhinitis (inflammation of nasal mucous membranes) and urticaria ('nettle-rash').

Susceptible individuals can become sensitized by continual low-dose exposure to environmental allergens, usually via mucosal surfaces. We are constantly exposed to fungal antigens, in particular from aerially dispersed spores (see page 221), and so these are prime candidates, along with plant pollen, to become allergens. The genera most frequently involved are *Alternaria*, *Cladosporium*, *Helminthosporium*, *Penicillium*, *Didymella* and *Aspergillus*. There is a distinct seasonal cycle, with high counts of aerial spores in the autumn and a lesser peak in spring. Another type of hypersensitivity, Type III, can result from persistent inhalation of particular fungal antigens. This can lead to the chronic complaint of farmer's lung. This involves the inhalation deep into the lung of fungal and actinomycete spores, for example from mouldy hay. There their antigens interact with antibodies, primarily immunoglobulin G (IgG), to form immune complexes which are deposited in the alveoli, leading to inflammation. This may result in allergic alveolitis, involving destruction of the alveoli and irreversible fibrosis, and may even eventually lead to death. Similar symptoms have been reported from people deliberately inhaling puffball spores as snuff or as a folk remedy for persistent nose bleeds.

Veterinary Mycology

Domestic, farm and zoo animals suffer from a range of fungal infections, many involving the same or similar species to those causing human infections. Ringworm diseases are widespread among animals and birds, chiefly involving the zoophilic dermatophytes, but usually cause little problem to the animal. Hedgehogs in particular tend to be infested with a range of species. Superficial and mucosal candidosis outbreaks, chiefly involving *C. albicans*, have been reported among poultry and other animals, but these are not common. *Malassezia pachydermatis* is associated with ear infections of dogs, growing in the waxy secretions. Of more deep-seated infections, one that deserves mention is avian aspergillosis, caused by spores of *A. fumigatus* from feed and bedding being inhaled by poultry, caged birds and, notably, penguins in zoos. Under these conditions, death rates can be high. *A. fumigatus*, together with several Zygomycetes, especially *Mortierella wolfii*, is the major cause of mycotic abortion, a phenomenon found in cattle, and to a lesser extent horses, sheep and pigs, in which the fetus and placenta become infected, leading to abortion. It is commonest where cattle are housed in winter, but the disease strikes sporadically.

Luckily there are few fungal zoonoses (i.e. human infections acquired directly from animals). One group that is common, however, are ringworm infections acquired from pets and farm animals, in particular *Microsporum canis* from cats and dogs, *Trichophyton equinum* from horses and *T. verrucosum* from cattle.

Mycotoxins (Table 7.5) are well known to veterinarians, as they cause many problems to animals, both when feeding on infected feedstuffs (e.g. the groundnut meal with aflatoxins), and when grazing in fields where the grass may be infected with toxic pathogens or toxic endophytes. There have been many historical accounts of ergotism, due to *Claviceps purpurea* (pages 54, 522), in horses and other farm animals, for example causing lameness and gangrene of the feet. Some animals may be particularly sensitive to particular toxins; for example zearalenone readily causes vulvovaginitis and abortion in sows. An unusual example is the indirect cause of facial eczema in sheep, especially in New Zealand, because of intensive sheep management practices and climatic conditions there. This occurs after sheep have fed on dead grass that has become infected with the saprotroph *Pithomyces chartarum*, which produces the toxin sporidesmin. This is a hepatotoxin to which sheep are especially sensitive, and its effect is that their blood accumulates incomplete breakdown products from chlorophyll that are photodynamic, hence their facial eczema. It will be the liver damage that will be the cause of death.

Animals can also suffer hypersensitive allergic effects, a troublesome example being chronic obstructive pulmonary disease (COPD) of horses, a loss of lung capacity resulting from long-term inhalation of fungal spores in mouldy hay and bedding in stables.

Fungi and Animals that Live as Mutualistic Symbionts

Some species of fungi and insects form close and lasting associations which enable them to live on plant materials such as wood. Fungi have biosynthetic and degradative abilities that are lacking in most insects. They synthesize vitamins including those of the B group, and sterols, and secrete lytic enzymes which break down cellulose and lignin, so that wood colonized by fungal mycelium becomes a suitable food for insects. Fungal enzymes such as cellulase may also continue to be active in the gut of an insect which has eaten the fungus. Insects promote fungal growth by importing nitrogen compounds to wood and other nitrogen-poor plant remains. Insects have ways of carrying their symbiotic fungi with them when they colonize new sites, so aiding fungal dispersal. Some associations between fungi and insects or other animals which have been investigated in detail are described below.

Bark beetles and ambrosia fungi

These are members of a subfamily of *Coleoptera*, the *Scolytidae*, also known as ambrosia beetles, presumably because they eat a special food (ambrosia was the food of the gods in Greek mythology). This consists of fungal mycelium which develops in the tunnels bored in the sapwood, immediately under the bark, by the female beetle before egg laying. The ambrosia fungi include Basidiomycetes, Ascomycetes, yeasts and mitosporic fungi. An assemblage of different species is usually found in one tunnel, growing and sporulating. The female beetle carries the ambrosia fungi in **mycetangia**, glandular invaginations of the body surface, and inoculates walls of new tunnels with the spores of yeast-like fungal cells which exude from the mycetangia. The larvae and adult beetles feed exclusively on the fungus. Ambrosia beetles normally attack weakened or dead trees. Related beetles without fungi bore into phloem tissue, where they feed on sap. Ambrosia beetles are thought to have evolved from these, being able to utilize dead wood lacking sap, as a result of the association with fungi.

Fungi are also inoculated into wood by wood wasps (*Siricidae*). The female has mycetangia near the ovipositor, and some fungal cells are exuded from this, and attached to each egg as it is laid in the wood. The larva also has mycetangia, and young adult females acquire the fungus extruded from the larval mycetangia, by taking up fungal material from them into the pouches at the base of the ovipositor. The fungus on each egg invades the wood, but is not abundant enough to form the sole food for the wasp larva.

Leaf-cutting ants, termites and Basidiomycetes
These ants live in colonies which forage over wide areas to obtain vegetation and accumulate it in underground chambers. There they inoculate it with a fungus which they maintain in pure culture. Each ant species maintains a different species of fungus. When a queen leaves the nest to start a new colony she takes the fungus culture with her as mycelium, carried in a special pouch. The symbiosis thus depends on highly developed social behaviour. The fungi are Basidiomycetes. They do not form fruit bodies in an active ants' nest, but may do so in an abandoned nest. In an active nest the fungi develop nutrient-rich swellings of the hyphae called bromatia, which the ants eat. The fungus is not known in the wild apart from its association with ant nests, so it presumably benefits from being maintained by the ants. Not only do they disperse it, but also they remove contaminant fungi to maintain the pure culture. In addition, they secrete a proteinase which they exude on to the leaves they have gathered, but the reason for this is not clear – a Basidiomycete fungus would normally have extracellular proteinases of its own.

The leaf-cutting ants are confined to the New World, where their large, vegetation-filled excavations improve soil fertility in forests. They are, however, serious pests in cropped regions owing to the scale of their activities – tens of tonnes of soil can be excavated to make one nest into which several tonnes of surrounding plant material are removed.

Leaf-cutting ants are comparable to termites in their social organization and ability to destroy plant material rapidly. The termite subfamily *Macrotermitinae* have fungal symbionts in their nests, with each termite species having its own fungal species. The fungi are Basidiomycetes of the genus *Termitomyces*, which is found nowhere else. The termites cultivate the mycelium in a fungus garden, called a comb, consisting of a pile of decaying plant material with fungal mycelium growing on the top. Enzymes, especially cellulases, from the fungus digest the plant material, and the comb is eaten by the termites, which reinoculate fresh plant material with fungal conidia produced by the mycelium. The fungus fruits when the rains come, producing distinctive large mushroom-like fruit bodies. *Termitomyces titanicus* produces the largest known fungus fruit bodies, weighing up to 2.5 kg and with caps over 60 cm in diameter. The termites forage widely and the symbiosis with the fungus enables them to use a variety of different plant material as food. Termites and their fungi are found only in tropical climates, in forest or savannah ecosystems.

Animal gut and anaerobic Chytridiomycetes
The gut of an animal normally contains a large number of microbes of many different kinds. This microflora is, for a given animal species, fairly constant in composition. The microbes are provided with a sheltered environment and an

abundant supply of food, usually broken into fragments with a large surface area as a result of chewing. The host may also benefit, there being good evidence that the normal microflora often prevents pathogenic microbes from being established. The gut microflora makes a more direct contribution to the host's welfare with herbivores, ranging from insects to mammals. A high proportion of plant biomass is cellulose, but very few animals produce cellulase, molluscs such as slugs and snails being among the few exceptions. Herbivores which do not themselves produce cellulase are dependent on cellulose-decomposing components of their microflora for the utilization of cellulose. The herbivorous mammals that utilize cellulose most efficiently are the ruminants, animals having a rumen, a region of the gut specialized for cellulose digestion. Because of the importance of ruminants – cattle, sheep and goats – to humans, the microbiology of the rumen has been studied intensively. Conditions in the rumen are anaerobic, with very dense populations of anaerobic bacteria and protozoa. Some of the bacteria are cellulose decomposers, and in pure culture will metabolize cellulose to a range of organic acids, some of which (e.g. acetic, propionic and butyric acids) are assimilated by the herbivore, providing a major nutrient source. Other acids, such as formic and succinic acids, are further metabolized by other bacteria. Carbon dioxide and hydrogen are also produced, but the hydrogen and part of the carbon dioxide are utilized by methanogenic bacteria, yielding methane. The herbivore eliminates the methane and residual carbon dioxide by belching.

In the last two decades it has become increasingly clear that obligately anaerobic rumen Chytridiomycetes (page 37) are an important component of the rumen microflora. Their rhizoids can penetrate plant tissue fragments, including cell walls, and their extracellular enzymes can degrade wall components such as cellulose and xylans (xylose polymers), as well as storage materials including starch. Studies with pure cultures show that sugars are metabolized to formate, acetate, lactate, ethanol, carbon dioxide and hydrogen. The evolution of the latter gas being a way of eliminating reducing equivalents, a metabolic necessity in an environment in which materials are highly reduced – the redox potential of the rumen is very low, about −300 mV. Hydrogen evolution results from the activity of hydrogenosomes, cell organelles that also occur in anaerobic protozoa. The subsequent removal of the hydrogen by methanogens is beneficial for the fungi and their cellulolytic activity. For example, in pure culture *Neocallimastix frontalis* achieved a 16% solubilization of crystalline cotton cellulose, a highly refractory material, in 72 hours. In co-culture with *Methanobrevibacter smithii*, however, 98% of the cotton cellulose was solubilized in the same time. This is extremely efficient cellulolysis, better even than that achieved by *Trichoderma reesii*, used in the commercial production of cellulase.

The attraction of zoospores to plant exudates (page 229) explains how rumen Chytridiomycetes can persist in the rumen by establishing themselves on newly arrived plant material. How they reach and infect young animals is less clear, since the known phases of the life cycle – thallus and zoospores – soon die on exposure to air. Anaerobic Chytridiomycetes indistinguishable from rumen species can, however, be cultured from dried faeces many months old. It seems probable that rumen Chytridiomycetes have survival spores that have not so far been observed or produced in culture.

Anaerobic Chytridiomycetes of the same species as those found in the rumen occur in the gut of many other herbivorous mammals. Fungi also occur in the gut of other animals, yeasts being widespread in the gut of insects, but in comparison with the rumen Chytridiomycetes, relatively little is known of their activities.

Fungi as Food for Animals

Mycelium

Fungal mycelium growing on plant remains is eaten by a variety of plant-litter-inhabiting animals. Animals grazing on mycelium probably have much more important effects on ecosystems than those eating mushrooms. Although the large fruiting bodies are more conspicuous than mycelium, they are relatively short-lived. It is the ability of the mycelium to colonize resources in the plant litter which determines the success or otherwise of a fungal species in an ecosystem. Some fungi appear to have more palatable mycelium than others, as soil insects show consistent preferences. Mycelium of *Trichoderma* and *Aspergillus* species was generally not eaten in experiments where insects were offered a choice between various fungi in culture. Evidence that such preferences may affect the roles of fungal species in ecosystems comes from experiments on fungal growth in spruce litter colonized by the two Basidiomycete fungi, *Mycena galopus* and *Marasmius androsaceus*. The relative success of the two competing species was decided by the presence or absence of the collembolan *Onychiurus*, which preferred to graze on the *Marasmius*; without *Onychiurus*, the *Marasmius* was the predominant species, but adding the collembola resulted in predominance of the *Mycena*. Grazing on fungi which are decomposing plant remains has the advantage for animals of providing soluble carbon sources, protein and vitamins. Symbioses in which animals and fungi are nutritionally interdependent have been described earlier (page 382).

Mycelium is also eaten by man, as part of fermented foods such as tempeh (page 504), and, more recently, in the form of fungal biomass produced in fermenters, dried and compacted to make a meat-like protein-rich food (page 497) marketed as 'Quorn®'.

Fruit bodies

When mycelium is aggregated into a fruit body, containing an accumulation of reserve substances of the mycelium such as fat and protein, it provides a food source which is accessible to a variety of animals. These include the larvae of beetles and flies which use the fruit body as a habitat as well as a food supply. The mushrooms of Agaricales last for only a few days, compared with those of bracket-forming polypores, limiting their insect inhabitants to those with a short larval stage which is completed within the life of the fruit body. The genus *Drosophila* includes some species, the larvae of each of which are found in many different agaric fruit bodies, and other species which are only found in hard, long-lived fruit bodies of polypores. Some associations between beetles and fungus fruit bodies are species-specific, larvae of a beetle species being confined to a single fungal species. Insect specialization to fungivory can include morphological adaptations, particularly of mouthparts, and physiological features. Fungus-eating *Drosophila* species are resistant to the toxin amanitin found in some poisonous toadstools including the death cap *Amanita phalloides*, whereas other *Drosophila* species are susceptible to it. The cultivation of fruit bodies (page 492) such as those of the edible mushroom, *Agaricus bisporus*, and shiitake, *Lentinus edodes*, for human consumption is an economically important method of food production (Chapter 8). It is still the only economically profitable technology for the conversion of lignocellulosic wastes.

Mycoparasitism

There are many species of fungi which are parasites of other fungi (Fig. 7.25, Plate 5). When the host is itself a parasite (page 363), such fungi are termed **hyperparasites**. Some mycoparasites are biotrophs, growing on living mycelium of other fungi. Most biotrophic mycoparasites are members of the Zygomycetes, for example the genera *Dispira*, *Dimargaris*, *Piptocephalis* and *Tieghemomyces*. They invade the mycelium of their host and have haustoria or fine hyphae growing within the hyphae of the host. They may be grown as dual cultures with their hosts, but in the absence of a host fungus they grow poorly. Other mycoparasites are necrotrophic, invading and killing their hosts. Necrotrophic mycoparasites include members of the mitosporic genera *Gliocladium* and *Trichoderma*, and the Oomycete *Pythium*, as well as the Zygomycete genera *Dicranophora*, *Spinellus*

Figure 7.25 Mycoparasitism. One of the fruit bodies of the agaric *Mycena galopus* is being attacked by *Mucor* sp. The slender sporangiophores are visible against the dark background, and the black sporangia where it is bright. (John and Irene Palmer.)

and *Syzygites*. Several genera common in soil can invade and destroy hyphae and sclerotia of their hosts, and these have attracted most interest as possible agents of biological control (page 422) of soil-borne plant disease which could act specifically to remove the sclerotia of a pathogen from soil. Grown in paired culture, mycoparasites have been shown to locate and grow towards hyphae of their hosts, responding to the presence of host hyphae with chemotropic growth and increased hyphal branching. Some Zygomycete mycoparasites have been found to show very specific attraction to exudates, probably proteins, of their host species. On contact with the host hypha, the parasite's hyphae coil round it and may invade it after penetrating the wall. Some necrotrophic mycoparasites kill the host hypha before contact, others after. Host specificity of necrotrophic species varies. Some, such as *Trichoderma* and *Gliocladium* species have a wide host range, whereas *Sporidesmium sclerotivorum* appears to be limited to attacking sclerotia of the genus *Sclerotium*. Exudates from sclerotia stimulate conidia of *S. sclerotivorum* to germinate, and the germ-tube then invades the host sclerotium, causing degradation of its glucan reserves, which are then used by the parasite to supply hyphae which grow out of the sclerotium into the soil to infect other neighbouring sclerotia. Mycoparasites in the soil can invade the large spores of endomycorrhizas (page 397). For example, *Pythium oligandrum* has been found in spores of *Glomus macrocarpum*. Although mycoparasites found in soil have been most investigated, mycoparasitism also occurs between the fungal inhabitants of leaf surfaces. For example, the yeast *Tilletiopsis minor* is a hyperparasite of powdery mildews.

Fungi and Other Microorganisms

Fungi as Parasites and Mutualistic Symbionts of Other Microorganisms

In aquatic environments fungi are important parasites of algae. Among such fungi are members of the Chytridiales (page 348) that infect diatoms, for example *Rhizophidium* which infects *Asterionella*, and can have a major effect on diatom population densities in lakes. In soils, fungi are important parasites of protozoa. Some predacious fungi can trap protozoa such as amoebae, and endoparasitic fungi develop inside protozoa after their spores have been ingested. One type of mutualistic relationship that involves fungi has great ecological significance, i.e. the relationship between fungi and photosynthetic microorganisms that has evolved in many Ascomycete orders to give the lichens.

Lichens
Lichens are of interest both with respect to fungal diversity (page 76), and as a symbiotic relationship. Like fungi forming mycorrhiza and endophytes, lichen fungi depend for their carbon and energy supply on photosynthetic organisms. In the majority (85%) of lichens these are unicellular or filamentous green algae, and in the remainder they are cyanobacteria or a combination of these and green algae. The fungus forms a tissue within which the photosynthesizing photobiont cells occupy a relatively small volume in particular positions within the thallus.

Lichens can live on exposed surfaces such as rock wherever there is light and water. Being resistant to water stress and in many species to freezing, they are the dominant form of vegetation together with mosses at latitudes and altitudes which are too cold and dry for higher plants – about 8% of terrestrial ecosystems. Many species are particularly tolerant of desiccation and can grow, though slowly, under conditions of intermittent water supply. This is possible because of their unique cellular structure and physiology.

In the large structurally complex lichen, *Sticta sylvatica* the hyphae of the outer top and bottom layers (upper and lower **cortex**) are more or less isodiametric, embedded and stuck together in a hydrophilic matrix. Within this, the central part (**medulla**) provides a gas-filled zone of filamentous hyphae covered in a hydrophobic protein with a rodlet structure reminiscent of that seen in hydrophobin layers of other fungi (page 102). The photobiont cells form a layer at the lower side of the upper cortex. The interface for nutrient exchange is unlike that in many biotrophic associations, in that no intracellular haustorial structures are formed. Instead, the medullary hyphae grow into the gelatinous sheath that surrounds the photobiont cells, and the associated photobiont and hyphal cells become sealed together within the hydrophobic coating material. When the lichen is moistened by rain or humid air, photosynthate is released from the photobiont and taken up by the closely associated hyphae. These are able to transport nutrients within the thallus to growing regions. It is not known how the fungus induces the photobiont to leak its photosynthate. The carbon compounds transferred from the photobiont are polyols from green algae, and glucose from cyanobacteria. The latter also contribute nitrogen compounds to the symbiosis by virtue of their nitrogen fixation capability. In dry weather, the cortical layers lose water, shrink and become brittle, although cell damage is minimal even though air spaces can occur within the fungal cells.

The lichenized way of life is very common. In Ascomycetes 46% of all known species occur as lichens, and such fungi comprise 98% of all lichens, although there are also a few lichens formed by Basidiomycetes and mitosporic fungi. Few photobionts have been assigned to species, but it appears that their diversity is much less than that of lichen fungi, with the majority belonging to the green algal genus *Trebouxia*. Gene sequence data confirm that lichen symbiosis has evolved in several separate classes of fungi, and several groups contain both lichenized and saprotrophic species, as well as species with other nutritional strategies such as plant parasitism and mycorrhizal symbiosis. Recently, in a comparison of gene sequences of some lichenized fungi with those of closely related but non-lichenized sister groups, it was found that the rate of DNA evolution appeared to have increased as a result of adopting the lichen habit. Although it is possible in many cases to culture the mycobiont and photobiont separately, the lichen association does not re-form when they are brought together *in vitro*. The fungal partner alone does not develop into the characteristic lichen form unless the photobiont is present. Appropriate differentiation clearly requires both partners, and there are several examples of lichens where one fungus can form a lichen distinctly shaped or pigmented depending on which of two different photobionts it associates with (Plate 1). Some lichens with both cyanobacterial and green algal photobionts keep them in different regions of the thallus which show different morphology, the cyanobacteria being limited to surface nodules called cephalodia.

Reproduction and dispersal is by ascospores (which must locate the photobiont and re-form the lichen wherever they land and germinate) or by fragments of the thallus, or special dispersed propagules called **soredia** and **isidia** that are composed of tissue from both partners. Fungal colonies arising from spores probably acquire their photobionts from degenerating soredia. Lichens vary in their preferred ecological niche, some being aggressive competitors and pathogens on other lichens. They are sensitive to air pollution, especially sulphur dioxide, but to varying degrees, so that lichen diversity can be used to monitor air pollution. Lichens bind metal cations and accumulate metal-rich particles within the thallus. *Cladonia*, subgenus *Cladina* (Fig. 2.28C) species, mainly *C. rangiferina* and *C. stellaris*, that consitute 'Reindeer Moss' or more accurately 'Reindeer Lichen', a major component of the arctic vegetation, accumulates radionuclides from fall-out from H-bomb testing and accidents such as that at Chernobyl. This constitutes a health hazard for indigenous people who depend on meat of the caribou and reindeer that graze lichen (page 157).

The close interaction of the partners in the lichen symbiosis is expressed not only in their collaborative morphology but also in their secondary metabolites, the 'lichen acids', phenolic compounds which crystallize on the surfaces of the medullary hyphae and are only formed by the partners in association. They are used traditionally to produce a range of pleasing muted colours when used to dye cloth, and can also be used in the identification of species. Some of them have antibiotic activity, and one of these, usnic acid, has been incorporated into antiseptic shampoos. Others are mycotoxins, and one of these, vulpinic acid, is produced in *Letharia vulpina* which has been used traditionally as a poison for wolves and foxes.

Attack on Fungi by Other Microorganisms

Fungi are liable to attack by other microorganisms as well as mycoparasitic fungi. Bacteria and Protozoa are known to attack fungal hyphae and have been investigated for use as agents of biological control of fungal pathogens. The best studied examples are interactions that occur in soil or on leaf surfaces, and these are discussed in relation to biological control (pages 422).

Acknowledgements

We are grateful to Professor T. D. Bruns and Professor M. Seaward for advice on text relating to mycorrhiza and lichens respectively. Figure 7.17, showing the life cycles of endophytic fungi in grasses, and the accompanying legend, were kindly supplied by Prof. A. Leuchtmann. The sections headed *Identifying by DNA technology attributes crucial for infection* and *Race-specific recognition of pathogens* were contributed by Prof. N. J. Talbot, who also supplied Fig. 7.7 and the accompanying legend illustrating the experimental procedure of targeted gene replacement, and read the whole chapter.

Further Reading and Reference

General

Ayres, P. G., ed. (1992). *Pests and Pathogens: Plant Responses to Foliar Attacks*. BIOS Scientific Publications, Oxford.

Campbell, C. L., Peterson, P. D. & Griffith, C. S. (1999). *The Formative Years of Plant Pathology in the United States*. American Phytopathological Society Press, St Paul, Minnesota.

Fitter, A. H. & Stribley, D. P. eds. (1996). *Plant–Microbe Symbiosis: Molecular Approaches*. Cambridge University Press, Cambridge.

Isaac, S. (1992). *Fungal–Plant Relationships*. Chapman & Hall, London.

Lockwood, J. L. (1987). Evolution of concepts associated with soil-borne plant pathogens. *Annual Review of Plant Pathology* 26, 93–121.

Pirozynski, K. A. & Hawksworth, D. L., eds. (1988). *Coevolution of Fungi with Plants and Animals*. Academic Press, London.

Selosse, M. A. & LeTacon, F. (1998). The land flora: a phototroph–fungus partnership? *Trends in Ecology and Evolution* 13, 15–20.

Smith, D. C. & Douglas, A. E. (1987). *The Biology of Symbiosis*. Arnold, London.

Plant and Fungal Structures and Substances Involved in Interactions

Agarwal, A. A., Tuzun, Z. & Bant, E., eds. (1999). *Induced Plant Defences against Pathogens and Herbivores: Biochemistry, Ecology and Agriculture*. The American Phytopathological Society, St Paul, Minnesota.

Andrews, J. H. & Hirano, S. S., eds. (1990). *Microbial Ecology of Leaves*. Springer-Verlag, New York.

Boller, T. & Meins, F., eds. (1992). *Genes Involved in Plant Defence*. Springer Verlag, Berlin.

Bolwell, G. P. & Wojtaszek, P. (1997). Mechanisms for the generation of reactive oxygen species in plant defence – a broad perspective. *Physiological and Molecular Plant Pathology* 51, 347–366.

Bonfante-Fasolo, B. & Scannerini, S. (1993). The cellular basis of plant–fungus interchanges in mycorrhizal associations. In M. F. Allen, ed., *Mycorrhizal Functioning: an Integrative Plant–Fungal Process*, pp. 65–101. Chapman & Hall, New York.

Borges-Walmsley, I. M. & Walmesley, A. R. (2000). CAMP{ signalling in pathogenic fungi: control of dimorphic switching and pathogenicity. *Trends in Microbiology* 8, 133–141.

Brown, J. S. & Holden, D. W. (1998). Insertional mutagenesis of pathogenic fungi. *Current Opinion in Microbiology* 1, 390–394.

Buczacki, S. & Harris, K. (1998). *Pests, Diseases and Disorders of Garden Plants*, 2nd edn. Harper Collins, London.

Callow, J. A. & Green, J. R., eds. (1992). *Perspectives in Plant Cell Recognition. Society for Experimental Biology Seminar Series*, vol. 48. Cambridge University Press, Cambridge.

Cheong, J. -J., Birberg, W., Fugedi, P., Pilotti, A., Garegg, P. S., Hong, N., Ogawa, T. & Hann, M. G. (1991). Structure–activity relationship of oligo-ß-glucoside elicitors of phytoalexin accumulation in soybean. *Plant Cell* 3, 127–136.

DeWit, P. G. M. (1997). Pathogen avirulence and plant resistance: a key role for recognition. *Trends in Plant Science* 2, 452–458.

Ebbole, D. J. (1997). Hydrophobins and fungal infection of plants and animals. *Trends in Microbiology* 5, 405–408.

Epstein, L. & Nicholson, R. L. (1997). Adhesion of spores and hyphae to plant surfaces. In G. C. Carroll & P. Tudzynski, eds., *Plant Relationships*, pp. 11–25. Vol. VA *The Mycota*, eds. K. Esser & P. A. Lemke. Springer-Verlag, Berlin.

French, R. C. (1992). Volatile chemical germination stimulators of rust and other fungal spores. *Mycologia* 84, 277–288.

Godiard, L., Grant, M. R., Dietrich, R. A., Kiedrowski, S. & Dangl, J. L. (1994). Perception and response in plant disease. *Current Opinion in Genetics and Development* 4, 662–671.

Henson, J. M., Butler, M. J. & Day, A. W. (1999). The dark side of the mycelium: melanins of phytopathogenic fungi. *Annual Review of Phytopathology* 37, 447–471.

Howard, R. J. (1997). Breaching the outer barriers – cuticle and cell wall penetration. In G. C. Carroll & P. Tudzynski, eds., *Plant Relationships*, pp. 43–60. Vol. VA *The Mycota*, eds. K. Esser & P. A. Lemke. Springer-Verlag, Berlin.

Howard, R. J., Ferrari, M. A., Roach, D. H. & Money, N. P. (1991). Penetration of hard substrates by a fungus employing enormous turgor pressures. *Proceedings of the National Academy of Science, USA* 88, 11281–11284.

Kronstadt, J. W. (1997). Virulence and cyclic AMP in smuts, blasts and blights. *Trends in Plant Science* 2, 193–199.

Lamb, C. (1996). A ligand–receptor mechanism in plant-pathogen recognition. *Science* 274, 2038–2039.

Martin, F., Laurent, P., de Carvalho, D., Voiblet, C., Balestrini, R., Bonfantes, P. & Tagu, D. (1999). Cell wall proteins in the ectomycorrhizal basidiomycete *Pisolithus tinctorius*: identification, function and expression in symbiosis. *Fungal Genetics and Biology* 27, 161–174.

McCrae, M. A., Saunders, J. R., Smyth, C. J. & Stow, N. D., eds. (1997). *Molecular Aspects of Host–Pathogen Interactions. Symposium of the Society for General Microbiology*, Vol. 55. Cambridge University Press, Cambridge.

McGee, P. A., Smith, S. E. & Smith, F. A., eds. (1989). *Plant–Microbe Interface: Structure and Function*. CSIRO, Melbourne, Australia.

Mendgen, K. & Deising, H. (1993). Infection structures of fungal plant pathogens – a cytological and physiological evaluation. *New Phytologist* 124, 193–213.

Nicholson, R. L. & Hammerschmidt, R. (1992). Phenolic compounds and their role in disease resistance. *Annual Review of Phytopathology* 30, 369–389.

Osbourn, A. E. (1996). Saponins and plant defence: a soap story. *Trends in Plant Science* 1, 4–9.

Pearce, R. B. (1987). Antimicrobial defences in secondary tissues of woody plants. In G. F. Pegg & P. G. Ayres, eds., *Fungal Infection of Plants. Symposium of the British Mycological Society* 13, 219–238. Cambridge University Press, Cambridge.

Read, N. D., Kellock, L. J., Knight, H. & Trewavas, A. J. (1992). Contact sensing during infection by fungal pathogens. In J. A. Callow & J. R. Green, eds., *Perspectives in Plant Cell Recognition*, pp. 137–172. Cambridge University Press, Cambridge.

Skou, J. P. (1981). Morphology and cytology of the infection process. In M. J. C. Asher & P. J. Shipton, eds. *The Biology and Control of Take-all*, pp. 175–197. Academic Press, London.

Staples, R. C. & Hoch, H. C. (1997). Physical and chemical cues for spore germination and appressorium formation by fungal pathogens. In G. C. Carroll & P. Tudzynski, eds., *Plant Relationships*, pp. 27–40. Vol. VA *The Mycota*, K. Esser & P. A. Lemke, eds. Springer-Verlag, Berlin.

Taiz, L. & Zieger, E. (1998). *Plant Physiology*, 2nd edn. Addison-Wesley, Redwood City.

Talbot, N. J., Kershaw, M. J., Wakley, G. E., de Vries, O. M. H., Wessels, J. G. H. & Hamer, J. E. (1996). *MPG1* encodes a fungal hydrophobin involved in surface interactions during infection-related development of *Magnaporthe grisea*. *Plant Cell* 8, 985–999.

Van Etten, D. H., Soby, S., Wasmann, C. & McCluskey, K. (1994). Pathogenicity genes in fungi. In M. J. Daniels, J. A. Downie & A. E. Osbourn, eds., *Advances in Molecular Genetics of Plant–Microbe Interactions*, pp. 163–170. Kluwer Academic, Dordrecht.

Walton, J. D. (1996). Host selective toxins: agents of compatibility. *Plant Cell* 8, 1732–1733.

Zhu, Q., Droge-Laser, W., Dixon, R. A. & Lamb, C. (1996). Transcriptional activation of plant defence genes. *Current Opinion in Genetics and Development* 6, 624–630.

Necrotrophs and Necrotrophic Attack

Collmer, A. & Keen, N. T. (1986). The role of pectic enzymes in plant pathogenesis. *Annual Reviews of Phytopathology* 24, 383–409.

Hardham, A. R. (1992). Cell biology of pathogenesis. *Annual Review of Plant Physiology and Plant Molecular Biology* 43, 491–526.

Hohn, T. H. (1997). Fungal phytotoxins: biosynthesis and activity. In G. C. Carroll & P. Tudzynski, eds., *Plant Relationships*, pp. 129–144. Vol. VA *The Mycota*, K. Esser & P. A. Lemke, eds. Springer-Verlag, Berlin.

Pegg, G. F. (1985). Life in a black hole – the microenvironment of the vascular pathogen. *Transactions of the British Mycological Society* 85, 1–20.

Van Alfen, N. K. (1989). Reassessment of plant wilt toxins. *Annual Reviews of Phytopathology* 27, 533–550.

Yoder, O. C., Macko, V., Wolpert, T. & Turgeon, B. G. (1997). *Cochliobolus* spp. and their host-specific toxins. In G. C. Carroll & P. Tudzynski eds, *Plant Relationships*, pp. 145–166. Vol. VA *The Mycota*, K. Esser & P. A. Lemke, eds. Springer-Verlag, Berlin.

Biotrophic Relationships and Mutualistic Interactions

Agerer, R. (1987–95). *Colour Atlas of Ectomycorrhizae*. Einhorn Verlag, Schwabisch Gmund.

Allen, M. J., ed. (1992). *Mycorrhizal Functioning: an Integrative Plant–Fungal Process*. Chapman & Hall, New York.

Ashford, A. E., Allaway, W. G., Peterson, C. A. & Cairney, J. W. G. (1989). Nutrient transfer and the fungus–root interface. *Australian Journal of Plant Physiology* 16, 85–97.

Bacon, C. W. & White, J. F. (2000). *Microbial Endophytes*. Marcel Dekker, New York.

Barbosa, P., Krischik, V. A. & Jones, C. G., eds. (1991). *Microbial Mediation of Plant–Herbivore Interactions*. Wiley, New York.

Barea, J. M. (1991). Vesicular–arbuscular mycorrhiza as modifiers of soil fertility. *Advances in Soil Science* 15, 2–31.

Brundrett, M. (1991). Mycorrhizas in natural ecosystems. *Advances in Ecological Research* 21, 171–313.

Brundrett, M., Bougher, N., Dell, B., Grove, T. & Malajczuck, N. (1996). *Working with Mycorrhizas in Forestry and Agriculture*. Australian Centre for International Agricultural Research, Canberra.

Bruns, T. D. & Gardes, M. (1993). Molecular tools for the identification of ectomycorrhizal fungi: taxon-specific oligonucleotide probes for suilloid fungi. *Molecular Ecology* 2, 233–242.

Cairney, J. W. G. & Chambers, S. M. (1999). *Ectomycorrhizal Fungi: Key Genera in Profile*. Springer-Verlag, Heidelberg.

Carroll, G. C. (1992). Fungal mutualism. In Carroll, G. C. & Wicklow, D. T., eds., *The Fungal Community, its Organisation and Role in the Ecosystem*, 2nd edn, pp. 327–354. Marcel Dekker, New York.

Clay, K. (1991). Parasitic castration of plants by fungi. *Trends in Ecology and Evolution* 6, 162–166.

Debaud, J. -C., Marmeisse, R. & Gay, G. (1997). Genetics and molecular biology of the fungal partner in the ectomycorrhizal symbiosis *Hebeloma cylindrosporum* × *Pinus pinaster*. In G. C. Carroll & P. Tudzynski, eds., *Plant Relationships*, pp. 95–115. Vol. VB *The Mycota*, K. Esser & P. A. Lemke, eds. Springer-Verlag, Berlin.

Gianinazzi-Pearson, V. (1996). Plant cell responses to arbuscular mycorrhizal fungi: getting to the roots of the symbiosis. *Plant Cell* 8, 1871–1883.

Gloer, J. B. (1994). Applications of fungal ecology in the search for new bioactive natural products. In D. T. Wicklow & B. E. Soderstrom, eds., *Environmental and Microbial Relationships*, pp. 249–268. Vol. IV *The Mycota*, K. Esser & P. A. Lemke, eds. Springer-Verlag, Berlin.

Harley, J. H. & Harley, E. L. (1987). A check list of mycorrhiza in the British Flora. *Supplement to New Phytologist* 105, 1–102; 1990, addenda and errata, *Supplement to New Phytologist* 115, 699–711.

Harrison, M. J. (1998). Development of the arbuscular mycorrhizal symbiosis. *Current Opinion in Plant Biology* 1, 360–365.

Helgason, T., Daniell, T. J., Husband, R., Fitter, A. H. & Young, J. P. W. (1998). Ploughing up the wood-wide web? *Nature* 394, 431.

Hetherington, A. M. (1998). Mycorrhizas – structure and function. *New Phytologist: The Tansley Review Collections 2*. Cambridge University Press, Cambridge.

Horton, T. R. & Bruns, T. D. (1998). Multiple host fungi are the most frequent and abundant ectomycorrhizal types in a mixed stand of Douglas fir (*Pseudotsuga menziesii* D. Don) and bishop pine (*Pinus muricata* D. Don). *New Phytologist* **139**, 331–339.

Ingleby, K., Mason, P. A., Last, F. T. & Fleming, L. V. (1990). *Identification of Ectomycorrhizas*. HMSO, London.

Korde, R. T. & Schreiner, R. P. (1992). Regulation of the vesicular–arbuscular mycorrhizal symbiosis. *Annual Review of Phytopathology* **43**, 557–581.

Leake, J. R. (1994). The biology of mycoheterotrophic ('saprophytic') plants. *New Phytologist* **127**, 171–216.

Leake, J. R. & Read, D. J. (1994). Mycorrhizal fungi in terrestrial habitats. In D. T. Wicklow & B. E. Soderstrom, eds., *Environmental and Microbial Relationships*, pp. 281–301. Vol. IV *The Mycota*, K. Esser & P. A. Lemke, eds. Springer-Verlag, Berlin.

Leuchtmann, A. & Clay, K. (1997). The population biology of grass endophytes. In G. C. Carroll & P. Tudzynski, eds., *Plant Relationships*, pp. 185–202. Vol. VB *The Mycota*, K. Esser & P. A. Lemke, eds. Springer-Verlag, Berlin.

Martin, F., Laurent, P., de Carvalho, D., Voiblet, C., Balestrini, R., Bonfante, P. & Tagu, D. (1999). Cell wall proteins of the ectomycorrhizal basidiomycete *Pisolithus tinctorius*: Identification, function, and expression in symbiosis. *Fungal Genetics and Biology* **27**, 161–174.

Marx, D. H., Maul, S. B. & Cordell, C. E. (1992). Application of specific mycorrhizal fungi in world forestry. In G. S. Leatham, ed., *Frontiers in Industrial Mycology*, pp. 78–98. Chapman & Hall, London.

Newman, E. I. (1988). Mycorrhizal links between plants; their functioning and ecological significance. *Advances in Ecological Research* **18**, 243–266.

Newsham, K. K., Fitter, A. H. & Watkinson, A. R. (1995). Multifunctionality and biodiversity in arbuscular mycorrhizas. *Trends in Ecology and Evolution* **10**, 407–412.

Norris, J. R., Read, D. J. & Varma, A. K., eds. (1992). *Techniques for the Study of Mycorrhiza. Methods in Microbiology*, vols 23 and 24. Academic Press, London.

Podila, G. P. & Douds, D. D., eds. (2000). *Current Advances in Mycorrhizae Research*. American Phytopathological Society Press, St Paul, Minnesota.

Read, D. (1990). Mycorrhizas in ecosystems. In D. L. Hawksworth, ed., *Frontiers in Mycology*, pp. 101–130. CAB International, Wallingford.

Read, D. J., Lewis, D. H., Fitter, A. H. & Alexander, I. J., eds. (1992). *Mycorrhizas in Ecosystems*. CAB International, Wallingford.

Redlin, S. C. & Carris, L. M., eds. (1996). *Endophytic Fungi in Grasses and Woody Plants: Systematics, Ecology and Evolution*. APS Press, American Phytopathological Society, Minnesota.

Saikkonen, K., Faeth, S. H., Helander, M. & Sullivan, T. J. (1998). Fungal endophytes: a continuum of interactions with host plants. *Annual Review of Ecology and Systematics* **29**, 319–343.

Simard, S. W., Perry, D. A., Jones, M. D., Myrold, D. D., Durall, D. M. & Molina, R. (1997). Net transfer of carbon between ectomycorrhizal tree species in the field. *Nature* **388**, 579–582.

Simon, L. (1993). Origin and diversification of endomycorrhizal fungi and coincidence with vascular land plants. *Nature* **363**, 67–69.

Smith, F. A. & Smith, S. E. (1997). Structural diversity in (vesicular)–arbuscular mycorrhizal symbioses. *New Phytologist* **137**, 373–388.

Smith, S. E. & Read, D. J. (1997). *Mycorrhizal Symbiosis*, 2nd edn. Academic Press, London.

Smith, S. E. & Smith, F. A. (1990). Structure and function of the interfaces in biotrophic symbioses as they relate to nutrient transport. *New Phytologist* **114**, 1–38.

Stierle, A., Strobel, G. & Stierle, D. (1993). Taxol and taxane produced by *Taxomyces andreanae*, an endophytic fungus of Pacific Yew. *Science* **260**, 214–216.

Van der Heijden, M. G. A., Klironomos, J. N., Ursic, M., Moutoglis, P., Streitwolf-Engel, R., Boller, T., Wiemken, A. & Sanders, I. R. (1998). Mycorrhizal fungal diversity determines plant diversity, ecosystem variability and productivity. *Nature* **396**, 69–72.

Wright, D. P., Scholes, J. D., Read, D. J. & Rolfe, S. (2000). Changes in carbon allocation and expression of carbon transporter genes in *Betula pendula* Roth. colonised by the ectomycorrhizal fungus *Paxillus involutus*. *Plant Cell and Environment* **23**, 39–49.

Plant Disease Caused by Fungi

Agrios, G. N. (1996). *Plant Pathology*, 4th edn. Academic Press, San Diego.

Barley, J. A. & Jeger, M. J., eds. (1992). *Colletotrichum: Biology, Pathology and Control*. CAB International, Wallingford.

Bishop, C. (1993). *Plant Resistance to Microbial Pathogens*. Chapman & Hall, London.

Blakeman, J. P. & Williamson, B., eds. (1994). *Ecology of Plant Pathogens*. CAB International, Wallingford.

Bowyer, P. (1999). Plant disease caused by fungi: pathogenicity. In R. P. Oliver & M. Schweizer, eds., *Molecular Fungal Biology*, pp. 294–321. Cambridge University Press, Cambridge.

Bruehel, G. W. (1987). *Soilborne Plant Pathogens*. Macmillan, New York.

Butin, H. (1995). *Tree Diseases and Disorders*. Oxford University Press, Oxford.

Campbell, C. L. & Madden, L. V. (1990). *Introduction to Plant Disease Epidemiology*. Wiley, New York.

Coley-Smith, J. R., Jarvis, W. R. & Verhoeff, K., eds. (1980) *The Biology of Botrytis*. Academic Press, London.

Crute, I. R., Holub, E. B. & Burdon, J. J., eds. (1998). *The Gene-for-Gene Relationship and Plant–Parasite Interactions*. CAB International, Wallingford.

Duncan, J. (1999). *Phytophthora* – an abiding threat to our crops. *Microbiology Today* **26**, 114–116.

Fox, R. T. V. (1994). *Principles of Diagnostic Techniques in Plant Pathology*. CAB International, Wallingford.

Giese, H., Hippe-Sanwald, S., Somerville, S & Waller, J. (1997). *Erysiphe graminis*. In G. C. Carroll & P. Tudzynski, eds., *Plant Relationships*, pp. 55–78. Vol. VB *The Mycota*, K. Esser & P. A. Lemke, eds. Springer-Verlag, Berlin.

Govers, F., Drenth, A. & Pieterse, C. M. J. (1997). The potato late blight pathogen and other pathogenic Oomycota. In G. C. Carroll & P. Tudzynski, eds., *Plant Relationships*, pp. 17–36. Vol. VB *The Mycota*, K. Esser & P. A. Lemke, eds. Springer-Verlag, Berlin.

Gurr, S. J., McPherson, M. J. & Bowles, D. J. (1992). *Molecular Plant Pathology*, Vol. 2. Oxford University Press, Oxford.

Holliday, P. (1998). *A Dictionary of Plant Pathology*, 2nd edn. Cambridge University Press, Cambridge.

Honee, G. (1999). Engineered resistance against fungal pathogens. *European Journal of Plant Pathology* **105**, 319–326.

Ingram, D. S. & Robertson, N. F. (1999). *Plant Disease: a Natural History*. Harper Collins, London.

Joosten, M. H. A. J., Honee, G., Van Kan, J. A. L. & De Wit, P. J. G. M. (1997). The gene-for-gene concept in plant-pathogen interactions: Tomato – *Cladosporium fulvum*. In G. C. Carroll & P. Tudzynski, eds., *Plant Relationships*, pp. 3–16. Vol. VB, *The Mycota*, K. Esser & P. A. Lemke, eds. Springer-Verlag, Berlin.

Judelson, H. S. (1996). Genetics of Oomycete plant pathogens. *Molecular Plant–Microbe Interactions* **9**, 443–449.

Knogge, W. (1996). Fungal infections in plants. *Plant Cell* **8**, 1711–1722.

Kranz, J. (1990). Fungal diseases in multispecies plant communities. *New Phytologist* **116**, 383–405.

Lonsdale, D. (1999). *Principles of Tree Hazard Assessment and Management*. H.M.S.O., London.

Lucas, J. A. (1998). *Plant Pathology and Plant Pathogens*, 3rd edn. Blackwell, Oxford.

Lucas, J. A., Shattock, R. C., Shaw, D. S. & Cooke, L. R., eds. (1991). *Phytophthora. Symposium of the British Mycological Society*. Vol. 17. Cambridge University Press, Cambridge.

Mace, M. E., Bell, A. A. & Beckman, C. H., eds. (1981). *Fungal Wilt Diseases of Plants*. Academic Press, New York.

Manners, J. G. (1993). *Principles of Plant Pathology*, 2nd edn. Cambridge University Press, Cambridge.

Melchers, L. S. & Struiver, M. H. (2000). Novel genes for disease-resistance breeding. *Opinion in Plant Biology* **3**, 147–152.

Mendgen, K. (1997). The Uredinales. In G. C. Carroll & P. Tudzynski, eds., *Plant Relationships*, pp. 79–94. Vol. VB, *The Mycota*, K. Esser & P. A. Lemke, eds. Springer-Verlag, Berlin.

Oliver, R. & Osbourn, A. E. (1995). Molecular dissection of fungal pathogenicity. *Microbiology* **141**, 1–9.

Parry, D. W. (1990). *Plant Pathology in Agriculture*. Cambridge University Press, Cambridge.

Pipe, N. D., Buck, K. W. & Brasier, C. M. (1995). Molecular relationships between *Ophiostoma ulmi* and the NAN and EAN races of *O. novo-ulmi* determined by RAPD markers. *Mycological Research* **99**, 653–658.

Prell, H. H. & Day, P. (2001). *Phytopathogenic Fungi: Interactions with Plants*. Springer-Verlag, Berlin.

Punja, Z. K. (1985). The biology, ecology and control of *Sclerotium rolfsii*. *Annual Reviews of Phytopathology* **23**, 97–127.

Salaman, R. (1985). *The History and Social Influence of the Potato*. Reprint of 1949 edn. Cambridge University Press, Cambridge.

Salzer, P. *et al.* (2000). Differential expression of eight chitinase genes in *Medicago truncatula* roots during mycorrhiza formation, nodulation, and pathogen infection. *Molecular Plant-Microbe Interactions* **13**, 763–777.

Scheffer, R. P. (1997). *The Nature of Disease in Plants*. Cambridge University Press, Cambridge.

Schots, A., Dewey, F. M. & Oliver, R., eds. (1994). *Modern Assays for Plant Pathogenic Fungi: Identification, Detection and Quantification*. CAB International, Wallingford.

Schwarze, F. W. M. R., Engels, J. & Mattheck, C. (2000). *Fungal Strategies of Wood Decay in Trees*. Springer, Heidelberg.

Shah, D. (1997). Genetic engineering for fungal and bacterial diseases. *Current Opinion in Biotechnology* **8**, 208–214.

Smith, I. M., Dunez, J., Lelliott, R. A., Phillips, D. H. & Archer, S. A. (1988). *European Handbook of Plant Diseases*. Blackwell, Oxford.

Spencer, D. M., ed. (1978). *The Powdery Mildews*. Academic Press, London.

Spencer, D. M., ed. (1981). *The Downy Mildews*. Academic Press, London.

Staskawicz, B. J., Ausubel, F. M., Baker, B. J., Ellis, J. G. & Jones, J. D. G. (1995). Molecular genetics of plant disease resistance. *Advances in Plant Pathology* **24**, 90–167.

Strange, R. N. (1993). *Plant Disease Control: Towards Environmentally Acceptable Methods*. Chapman & Hall, London.

Strouts, R. G. & Winter, T. G. (1994). *Diagnosis of Ill-health in Trees*. H.M.S.O., London.

Valent, B. (1997). The rice blast fungus. In G. C. Carroll & P. Tudzynski, eds, *Plant Relationships*, pp. 37–54. Vol. VB *The Mycota*, K. Esser & P. A. Lemke, eds. Springer-Verlag, Berlin.

Wolfe, M. S. & Caten, C. E., eds. (1987). *Populations of Plant Pathogens: their Dynamics and Genetics*. Blackwell, Oxford.

Woodward, S., Stenlid, J., Karjalainen, R. & Hütlermann, A. (1998). *Heterobasidion annosum: Biology, Ecology, Impact and Control*. CAB International, Wallingford.

http://www.scisoc.org/ The American Phytopathological Society

http://www.cabi.org Commonwealth Agricultural Bureau International

http://www.bspp.org.uk/mppo1 British Society of Plant Pathologists; molecular plant pathology online

Biological Control

Adams, P. (1990). The potential of mycoparasites for the biological control of plant diseases. *Annual Review of Phytopathology* **28**, 59–72.

Andrews, J. H. (1992). Biological control in the phyllosphere. *Annual Review of Phytopathology* **30**, 603–635.

Bills, S., Podila, G. K. & Hiremath, S. (1999). Genetic engineering of a mycorrhizal fungus *Laccaria bicolor* for use as a biological control agent. *Mycologia* **91**, 237–242.

Boland, G. J. & Kuykendall, L. D. (1998). *Plant–Microbe Interactions and Biological Control*. Marcel Dekker Inc., New York.

Charnley, A. K. (1994). Entomopathogenic fungi and their role in pest control. In D. T. Wicklow & B. E. Soderstrom, eds., *Environmental and Microbial Relationships*, pp. 185–201. Vol. IV *The Mycota*, K. Esser & P. A. Lemke, eds. Springer-Verlag, Berlin.

Chet, I., Inbar, J. & Hadar, Y. (1994). Fungal antagonists and mycoparasites. In D. T. Wicklow & B. E. Soderstrom, eds., *Environmental and Microbial Relationships*, pp. 165–184. Vol. IV *The Mycota*, K. Esser & P. A. Lemke, eds. Springer-Verlag, Berlin.

Feng, M. G., Poprawski, T. J. & Khachatourian, G. G. (1994). Production, formulation and application of *Beauveria bassiana* for insect control: current status. *Biocontrol Science and Technology* **4**, 3–34.

Hawkins, B. A. & Cornall, H. V. eds. (1999). *Theoretical Approaches to Biological Control*. Cambridge University Press, Cambridge.

Kerry, B. R. & Jaffee, B. A. (1994). Fungi as biological control agents for plant parasitic nematodes. In D. T. Wicklow & B. E. Soderstrom, eds., *Environmental and Microbial Relationships*, pp. 203–218. Vol. IV *The Mycota*, K. Esser & P. A. Lemke, eds. Springer-Verlag, Berlin.

Kooyman, C., Bateman, R. P., Langewald, J., Lomer, C. J., Ouambama, Z. and Thomas, M.B. (1997). Operational-scale application of entomopathogenic fungi for control of sahelian grasshoppers. *Proceedings of the Royal Society B* **264**, 541–546.

Lansen, M. (2000). Prospects for controlling animal parasitic nematodes by predaceous microfungi. *Parasitology* **120**, S121–S131.

Maniania, N. K. (1998). A device for infecting adult tsetse flies, *Glossinia* spp., with an entomopathogenic fungus in the field. *Biological Control* **11**, 248–254.

McCoy, C. W., Samson, R. A. & Boucias, D. G. (1988). Entomogenous fungi. In C. M. Ignoffo & N. B. Mandava, eds, *Handbook of Natural Pesticides*. Vol 5: *Microbial Insecticides*, Part A, *Entomogenous Protozoa and Fungi*, pp. 151–226. CRC Press, Boca Raton, Florida.

Naseby, D. C., Pascual, J. A. & Lynch, J. M. (2000). Effect of biocontrol strains of *Trichoderma* on plant growth, *Pythium ultimum* populations, soil microbial communities and soil enzyme activities. *Journal of Applied Microbiology* **88**, 161–169.

Paulitz, T. C. (2000). Population dynamics of biocontrol agents and pathogens in soils and rhizospheres. *European Journal of Plant Pathology* **106**, 401–403.

Stirling, G. R. (1991). *Biological Control of Plant Parasitic Nematodes: Progress, Problems and Prospects*. CAB International, Wallingford.

TeBeest, D. O., ed. (1991). *Microbial Control of Weeds*. Chapman & Hall, New York.

TeBeest, D. O., Xang, X. B. & Cisar, C. R. (1992). The status of the biological control of weeds with fungal pathogens. *Annual Review of Phytopathology* **30**, 637–657.

Van Driesche, R., Kenmore, P. E., Waage, J. K. & Bellows, T. S. (1993). *Biological Control: a Guide to its Applications*. Chapman & Hall, London.

Fungi and Animals other than Man

Anderson, J. M., Rayner, A. D. M. & Walton, D. W. H., eds. (1984). *Invertebrate–Microbial Interactions. Symposium of the British Mycological Society*. Vol. 6. Cambridge University Press, Cambridge.

Arora, D. K., Ajello, L. & Mukerji, K. G. (1991). *Humans, Animals and Insects. Handbook of Applied Mycology*, Vol 2. Marcel Dekker, New York.

Barron, G. L. (1977). *The Nematode-destroying Fungi*. Canadian Biological Publications Ltd, Guelph, Ontario.

Barron, G. L. (1982). Nematode-destroying fungi. In R. G. Burns & J. H. Slater, eds., *Experimental Microbial Ecology*, pp. 533–552. Blackwell, London.

Batra, L. R. (1979). *Insect–Fungus Symbiosis*. Wiley, Chichester.

Courtney, S. P., Kibota, T. T. & Singleton, T. A. (1990). Ecology of the mushroom-feeding Drosophilidae. *Advances in Ecological Research* **20**, 225–274.

Hinkle, G., Wetterer, J. K., Schultz, E. R. & Sogin, M. L. (1994). Phylogeny of the attine ant fungi based on analysis of small subunit ribosomal RNA gene sequences. *Science* **299**, 1695–1697.

Hungerford, L. L., Campbell, C. L. & Smith, A. R. (1998). *Veterinary Mycology Manual*. Iowa State University Press, Iowa.

Khachatourian, G. G. (1996). Biochemistry and molecular biology of entomopathogenic fungi. In D. H. Howard & J. D. Miller, eds, *Human and Animal Relationships*, pp. 331–363. Vol. VI *The Mycota*, K. Esser & P. A. Lemke, eds. Springer-Verlag, Berlin.

Mueller, U. G., Rehner, S. A. & Schultz, T. R. (1998). The evolution of agriculture in ants. *Science* **281**, 2034–2037.

Murrin, F. (1996). Fungi and insects. In D. H. Howard & J. D. Miller, eds, *Human and Animal Relationships*, pp. 365–388. Vol. VI *The Mycota*, K. Esser & P. A. Lemke, eds. Springer-Verlag, Berlin.

Pickering, A. D. & Willoughby, L. G. (1982). *Saprolegnia* infections of salmonid fish. In Roberts, R. J., ed., *Microbial Diseases of Fish. Special Publications of the Society for General Microbiology*, vol 9, pp. 271–297.

Samson, R. A., Evans, H. C. & Latgé, J. P. (1988). *Atlas of Entomogenous Fungi*. Springer Verlag, Berlin.

Theodorou, M. K., Lowe, S. E. & Trinci, A. P. J. (1992). Anaerobic fungi and the rumen ecosystem. In G. C. Carroll & D. T. Wicklow, eds., *The Fungal Community: its Organisation and Role in the Ecosystem*, 2nd edn, pp. 43–72. Marcel Dekker, New York.

Tunlid, A., Jansson, H. -B., & Nordbring-Herz, B. (1992). Fungal attachment to nematodes. *Mycological Research* **96**, 401–412.

Van Cutsem, J. & Rochette, F. (1991). *Mycoses in Domestic Animals*. Janssen Research Foundation, Belgium.

Wheeler, Q. & Blackwell, M., eds. (1984). *Fungus–insect Relationships: Perspectives in Ecology and Evolution*. Columbia University Press, New York.

Wilding, N., Collins, M., Hammond, P. M. & Webber J. F., eds. (1989). *Insect–fungus Interactions. Symposium of the Royal Entomological Society of London in collaboration with the British Mycological Society*, vol 14. Academic Press, London.

Medical Mycology

Arora, D. K., Ajello, L. & Mukerji, K. G. (1991). *Humans, Animals and Insects. Handbook of Applied Mycology*, Vol 2. Marcel Dekker, New York.

Cutler, J. E. & Han, Y. (1996). Fungal factors implicated in pathogenesis. In D. H. Howard & J. D. Miller, eds., *Human and Animal Relationships*, pp. 3–29. Vol. VI *The Mycota*, K. Esser & P. A. Lemke, eds. Springer-Verlag, Berlin.

Day, J. H. (1996). Allergic respiratory responses to fungi. In D. H. Howard & J. D. Miller, eds., *Human and Animal Relationships*, pp. 173–191. Vol. VI *The Mycota*, K. Esser & P. A. Lemke, eds. Springer-Verlag, Berlin.

Evans, E. G. V. & Richardson, M. D., eds. (1989). *Medical Mycology: a Practical Approach*. IRL Press at Oxford University Press, Oxford.

Gale, C. A., Bendel, C. M., McClellan, M., Hauser, M., Becker, J. M., Berman, J. & Hostetter, M. K. (1998). Linkage of adhesion, filamentous growth, and virulence in *Candida albicans* to a single gene, *INT 1*. *Science* **279**, 1355–1358.

Kwon-Chung, K. J. & Bennett, J. E. (1992). *Medical Mycology*. Lea & Febiger, Philadelphia.

Lacey, J. (1990). Aerobiology and health: the role of airborne fungal spores in respiratory disease. In D. L. Hawksworth ed., *Frontiers in Mycology*, pp. 157–185. CAB International, Wallingford.

Schuh, J. C. L., Harrington, K. & Fanslow, W. C. (1997). *Pneumocystis carinii* Delanoe (1912) has been placed in the Archiascomycetales, a class of the Ascomycota. *Infection and Immunity* **65**, 4365–4366.

Warnock, D. W. & Campbell, C. K. (1996). Medical Mycology. *Mycological Research* **100**, 1153–1162.

Fungi and Other Microorganisms

Ahmadjian, V. (1993). *The Lichen Symbiosis.* Wiley, Chichester.

Gargas, A., DePriest, P. T., Grube, M. & Tehler, A. (1995). Multiple origins of the lichen symbiosis in fungi suggested by SSU of rDNA phylogeny. *Science* **268**, 1492–1495.

Honegger, R. (1997). Metabolic interactions at the mycobiont–photobiont interface in lichens. In G. C. Carroll & P. Tudzynski, eds., *Plant Relationships*, pp. 209–221. Vol. VA *The Mycota*, K. Esser & P. A. Lemke, eds. Springer-Verlag, Berlin.

Jeffries, P. (1994). Mycoparasitism. In D. T. Wicklow & B. E. Soderstrom, eds., *Environmental and Microbial Relationships*, pp. 149–164. Vol. IV *The Mycota*, K. Esser & P. A. Lemke, eds. Springer-Verlag, Berlin.

Jeffries, P. & Young, T. W. K. (1994). *Interfungal Parasitic Relationships.* CAB International, Wallingford.

Lumsden, R. D. (1992). Mycoparasitism of soilborne plant pathogens. In G. C. Carroll & D. T. Wicklow, eds., *The Fungal Community: its Organisation and Role in the Ecosystem*, 2nd edn, pp. 275–293. Marcel Dekker, New York.

Lutzoni, F. & Pagel, M. (1997). Accelerated evolution as a consequence of transitions to mutualism. *Proceedings of the National Academy of Sciences USA* **94**, 11422–11427.

Nash, T. (1996). *Lichen Biology.* Cambridge University Press, Cambridge.

Richardson, D. H. S. (1999). War in the world of lichens: parasitism and symbiosis as exemplified by lichens and lichenicolous fungi. *Mycological Research* **103**, 641–650.

http://lichen.com The North American Lichen Project.

Questions

With each question one statement is correct. Answers on page 561.

7.1 Carbon compounds are transferred mainly from fungi to plants in which of these types of mycorrhizal symbiosis?
a) orchid mycorrhiza
b) ectomycorrhiza
c) ericaceous mycorrhiza
d) VAM (AMF)

7.2 A technique by which it is possible to identify the genes involved in pathogenesis by fungi is:
a) enzyme-linked immunosorbent assay (ELISA)
b) dilution plating
c) ITS-RFLP matching
d) targeted gene disruption

7.3 Structures formed by symbiotic fungi inside plant cells are:
a) appressoria
b) haustoria
c) Hartig nets
d) rhizomorphs

7.4 A plant parasite which infects its host by means of rhizomorphs is:
a) *Magnaporthe grisea*
b) *Erysiphe graminis*
c) *Armillaria mellea*
d) *Phytophthora infestans*

7.5 Which of the following is not a ringworm fungus?
a) *Epidermophyton*
b) *Malassezia*
c) *Microsporum*
d) *Trichophyton*

Fungi and Biotechnology

8

The Economic Significance of Fungi

Fungi cause immense economic losses. Their harmful activities as saprotrophs include damage to timber, fuel, food and manufactured goods (Chapter 6). As parasites they cause heavy crop losses and diseases of humans and domestic animals (Chapter 7). The beneficial activities of yeasts and other fungi, however, are also of great economic significance (Table 8.1). They have long been exploited as food, in processing food, and in brewing. In the present century, as the fermentation industry has developed, they have yielded an increasing range of valuable products, including antibiotics and other drugs of great pharmaceutical value, agricultural fungicides and plant growth regulators, vitamins and enzymes. Recently, with the advent of genetic manipulation, fungi are being used to produce hormones and proteins hitherto available only from mammals. The utilization of fungi to benefit humans can be regarded as a part of biotechnology, and is the subject of the present chapter.

'Biotechnology' is a term that became popular at about the same time that genetic manipulation first received wide publicity. Biotechnology was then seen as applied genetic engineering and molecular biology. Subsequently it was recognized that the fermentation industry had already achieved considerable technical sophistication, and was entitled to be regarded as a branch of biotechnology. This led to various broader definitions of biotechnology, for example 'the application of scientific and engineering principles to the processing of materials by biological agents to provide goods and services'. Among the 'biological agents' of outstanding importance in biotechnology are the fungi. In some instances the utilization of fungi in providing 'goods' (e.g. alcoholic beverages) is very ancient, in others (e.g. antibiotics) recent. Some processes, such as antibiotic production, have benefited from 'the application of scientific and engineering principles' throughout their short history, whereas others, such as brewing, are ancient crafts to which such principles are now being increasingly applied. In the present chapter some of the basic methods of biotechnology will be discussed, and then the processes involving fungi to which they have been or are likely to be applied.

Table 8.1 Products of fungal biotechnology

Application	Notes
Enzymes	Mainly for food processing, e.g. producing high fructose syrup
	Also industrial uses, e.g. cellulases
Alcoholic beverages	Mainly by *Saccharomyces cerevisiae*
Mushrooms	*Agaricus bisporus* and several other species
Single-cell protein	Mycoprotein from *Fusarium graminearum*
Yeast biomass	*Saccharomyces cerevisiae*, for brewing and baking
Fermented foods	From soya beans, cereals, meat
Cheeses	Mainly blue-vein and some soft cheeses
Industrial alcohol	Mainly by *Saccharomyces cerevisiae*
Citric acid	Mainly by *Aspergillus niger*
Vitamins	E.g. riboflavin
Antibiotics	Mainly penicillins and cephalosporins
Pharmaceutical drugs	Secondary metabolites, e.g. for controlling cholesterol synthesis, immunological responses, blood pressure
	Via enzymic bioconversions, especially of steroids
Agrochemicals	Fungicides, plant growth regulators
Heterologous gene expression	Mainly using *Saccharomyces cerevisiae*, also with *Aspergillus* species
Biocontrol agents	Versus weeds, invertebrate pests, plant pathogenic fungi

Fermentation Technology

Fermentation technology is concerned with the large-scale culture of microorganisms in fermenters, and the recovery of useful products contained within the microbial cells or released into the surrounding medium. It should be noted that the use of the terms fermentation and fermenter does not necessarily imply anaerobic metabolism, or fermentation in the physiological sense. Indeed, most of the organisms now utilized by the fermentation industry are aerobes, and the adequate aeration of fermenters is a very important part of fermentation technology. Fermentation was, however, a crucial part of some of the earliest processes involving the large-scale culture of microorganisms, such as the production of alcohol by yeast and acetone and butanol by bacteria, leading to an anomalous use of the term fermentation for subsequent aerobic processes.

The successful operation of a fermentation requires laboratory studies on the organism used, suitable materials (feedstocks) on which to grow the organism, the optimum operation of fermenters of appropriate design, and finally, 'downstream processing' to recover the desired product efficiently. These aspects of fermentation technology will now be discussed.

Research and Laboratory Activities Associated with Fermentation Processes

Preliminary investigations

The starting point for a fermentation process is the discovery that a microorganism makes a useful product. Such a discovery does not by itself constitute a sufficient basis for a successful industrial fermentation – the organism may grow poorly, the medium used may be expensive, and yields of the desired product may be very low. These problems are overcome by improving the performance of the microorganism genetically and by providing it with the optimum environment for growth and product formation. The genetic improvement of fermentation microorganisms is dealt with later (page 526). The provision of the optimum environment requires physiological studies to determine the physical and nutritional conditions needed for growth (Chapter 3) and high yields of the desired product. A satisfactory medium or 'broth' should give high growth rates, since slow-growing cultures will occupy expensive fermenters for too long. It should also give good final yields of the organism and of the desired product per unit of biomass. Finally the medium must be cheap and utilize readily available raw materials (page 465).

Scale up

The commercial production of fungal metabolites commonly involves fermenters that contain tens or even hundreds of thousands of litres of medium that is stirred with impellers and sparged with air. Laboratory research on fungal metabolism is usually conducted with flasks that contain about 100 ml of medium and may be static or shaken. If the flasks are static most fungi grow as a mat on the surface, whereas with shaken liquid culture they occur as dispersed hyphae or pellets submerged in the medium (Chapter 3). There are important metabolic differences between surface and submerged cultures (page 137), and achieving good yields in submerged culture of a metabolite discovered in surface culture is an important step towards efficient commercial production. Penicillin, for example, was first produced at great expense and inadequate amounts in surface culture, and a crucial step in obtaining abundant and cheap penicillin was the discovery of a strain that produced the antibiotic in submerged culture. Having obtained a good performance in submerged culture, the first step in scale up is growth in laboratory fermenters which contain only a few litres of medium but which are stirred and sparged. In this they resemble production fermenters, and hence are suitable for studying some of the problems of large-scale culture. Next, studies are conducted at a pilot plant scale, with fermenters ranging in volume from about fifty litres to a few thousand litres. When efficient growth, product formation and product recovery have been obtained on the pilot plant scale, and the economics of the process assessed and found to be satisfactory, commercial production becomes feasible.

Many of the difficulties that arise during scale up result, directly or indirectly, from the problem of aeration. As volumes increase, adequate aeration necessitates more vigorous stirring and sparging. This can cause the medium to foam, reducing oxygen transfer, so antifoam agents that do not harm the process have to be added. The shear generated by the impellers may fragment the fungus, so studies aimed at retaining the required growth form may be needed. Other problems are associated with the decreasing surface to volume ratio as larger vessels are used – for example, as cultures get larger more metabolic heat is generated and more cooling is needed.

Culture preservation

A fungal strain that gives high yields of a valuable product is a precious asset. Such a strain devotes a substantial proportion of its carbon source to the formation of a metabolite in quantities which are unnecessary for survival. Repeated subculture is hence likely to result in the replacement of the high-yielding strain with variants that produce less of the desired metabolite and devote more of their resources to achieving higher growth rates. Excessive subculture must therefore be avoided, and the material needed for initiating fermentations kept in a non-growing state. The ideal starting point for preparing cultures for preservation will be a master culture that has originated from a single uninucleate haploid cell and is hence genetically uniform, and which has been shown, by testing subcultures, to have the desired properties. From such a master culture a set of subcultures is prepared and from them material for preservation.

Fungal cultures can be stored after they have grown on agar media in screw-capped vials. Such cultures will survive for a few weeks or months, spores remaining viable for longer than vegetative cells. Longer survival, usually for over a year, can be obtained by covering the cultures with sterile mineral oil (medicinal paraffin) which probably acts by further reducing water loss, or by storing small plugs of culture plus agar under water in small sealed vials. There are other more sophisticated methods of preserving fungal cultures. Probably the best is liquid nitrogen refrigeration. A suspension of cells, such as asexual spores, is made in a liquid, such as dimethyl sulphoxide (DMSO), which penetrates the cells and prevents ice crystal formation on cooling. The suspension is then placed in sterile ampoules which are sealed, cooled and stored in or over liquid nitrogen in specially designed containers. Fungi preserved in this way remain viable for many years, with no attention except for maintaining the level of liquid nitrogen in the container.

Inoculum preparation

The final inoculum for a production fermentation must be of adequate size, which usually means having a biomass of about 10% of that to be grown in the production fermenter. This requirement for a very large inoculum is so that the largest fermenter, which is an expensive facility and costly to operate, is occupied by a single fermentation for the shortest possible time. Material from the storage ampoule is usually first transferred to an agar slope, and the resulting culture used to inoculate a flask of liquid medium. The contents of such a flask, towards the end of the growth phase, are used to inoculate a laboratory-scale fermenter. By means of a series of successively larger fermenters, sufficient material is obtained to inoculate the production fermenter. Early in the process it is necessary to determine that cultures are behaving correctly with respect to growth and metabolite formation and, throughout inoculum production, checks to ensure that contamination has not occurred are essential.

Feedstocks for Fermentation Processes

For a successful fermentation, the fungus must behave in a consistent, predictable manner, giving similar growth rates and metabolite yields in successive fermentations. A crucial factor in achieving this is to provide a nutrient medium that contains utilizable sources of carbon, nitrogen and other essential elements in the appropriate amounts and ratios. This is easily done in the laboratory by making use of pure chemicals, but on an industrial scale this would be far too expensive. The selection of the right raw materials (feedstocks) for medium preparation is a key factor in the success of an industrial fermentation. Some of the factors in feedstock selection will now be considered.

The nutrient required by a fungus in the greatest amount is the carbon source. The main feedstock must hence contain a utilizable carbon source and, because of the quantities needed, must be cheap. Most fungi are able to utilize glucose or sucrose, so good feedstocks include sugar-cane juice, unrefined sugar, molasses (waste liquor from sugar refineries, which contains uncrystallizable sugar) and hydrolysed starch. Whey, the waste product produced from milk during cheese manufacture, contains lactose, but only a few fungi can use this sugar. Starch itself, and groundnut and maize meals, which contain starch, are good feedstocks for the many fungi that can use this carbohydrate. Cellulose, especially if impregnated with lignin, is attacked by relatively few fungi, and only slowly, so is unsuitable as a feedstock. In view, however, of the abundance and cheapness of lignocellulose, a great deal of research is devoted to the problem of converting this material into a utilizable form. Carbon sources other than carbohydrates include vegetable oils, such as soya bean, palm or maize oil. In addition to a carbon source a fungus must be provided with a nitrogen source. Some feedstocks, such as groundnut meal, beet molasses and whey, are nitrogen sources as well as carbon sources. When, however, the major feedstock contains little nitrogen, a nitrogen-containing feedstock is also needed, for example cornsteep liquor, a by-product of the corn wet milling industry. Other nutrients that may be needed in medium preparation are minerals, trace elements and vitamins, although when complex materials are used as feedstocks these are often present in adequate amounts. If vitamins are needed, they are often added as yeast extract.

When small fermenters are used, the cost of raw materials is only a small part, perhaps 10%, of the cost of operating the process. When large fermenters are needed, the cost of raw materials will then be a large part of the operating cost, for example 50–60% for the production of industrial ethanol. Under these circumstances the cost of raw materials is crucial in determining whether it is economic to operate the process, and the cheapest materials must be sought. With massive amounts of raw materials needed, **transport costs** are important, and it may be necessary to site the fermentation plant near the source of the materials. With any fermentation, raw material costs will be a factor in determining profitability, but where they are a smaller proportion of total costs, features other than cheapness become important in choosing feedstocks. Highly concentrated feedstocks, and those available throughout the year, will have lower **storage costs** than dilute feedstocks available only seasonally. **Ease of handling** is also important. **Constancy of composition** helps to give consistent results in successive fermentations. Some raw materials contain substances that persist through the fermentation and are difficult to separate from the product, increasing **downstream processing costs**. Some may contain materials that are not utilized in the fermentation and remain in the spent medium, causing problems of **waste disposal**.

The selection of feedstocks involves balancing considerations of convenience against cost. Wastes from an industry, such as whey, may be very cheap but have disadvantages, although some, such as cornsteep liquor used in penicillin fermentations (page 516), have proved valuable. Raw materials with other uses, such as unrefined sugar and vegetable oils, may be convenient but more expensive. Such commodities may also vary greatly in cost as a result of the international situation, harvest failures or gluts, and government subsidies or taxes. It is hence useful if a fermentation can be operated with a number of alternative feedstocks to take advantage of fluctuations in commodity prices.

Fermenter Design and Operation

Most industrial fermentations are carried out in **stirred tank fermenters**, and unless otherwise indicated, discussion in this section refers to such fermenters. A stirred tank fermenter (Fig. 8.1) is a cylindrical vessel designed to contain a volume of sterile medium that can be inoculated, aerated, stirred, monitored by sensing devices, heated or cooled, and sampled or fed with additional materials, all without bringing about contamination. It is constructed of high-grade stainless steel, to avoid corrosion by media or leaching of toxic metals into the medium. The basic design of the stirred tank fermenter has changed little since its introduction for penicillin manufacture in the 1940s, although there have been many improvements in detail, and variations that depend on the size of the fermenter.

Fungi can be grown in the laboratory either by **batch culture** or **continuous culture** (Chapter 3), and this is true also on an industrial scale. Continuous culture has some features which, when compared with those of batch culture, seem advantageous. Whereas with batch culture conditions are continually changing, as cell density increases, nutrients are utilized and products formed, with continuous culture optimal conditions for growth can be maintained throughout. Furthermore, with continuous culture a fermenter can be operated for many weeks, whereas with batch culture the fermenter is out of action at frequent intervals for cleaning, sterilizing and reinoculation. Hence with continuous culture it should be possible to obtain the required output of a product with a smaller plant. Continuous culture has, however, disadvantages that have led to batch culture predominating in the fermentation industry. It is a potent method for selecting for variants such as strains with higher growth rates, which devote less of their resources to the synthesis of the desired product. Many of the products sought in fermentations are secondary metabolites, formed only when nutrient levels are low. Here a continuous process would require the organism to be subjected successively to conditions suitable for growth and then for product formation, necessitating a sophisticated and potentially expensive

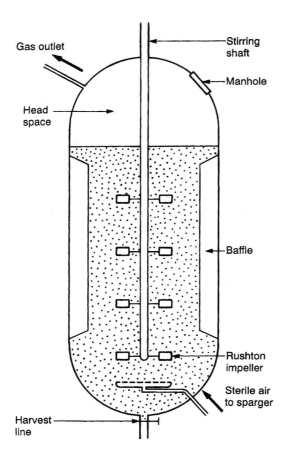

Gas outlet

Stirring shaft

Manhole

Head space

Baffle

Rushton impeller

Sterile air to sparger

Harvest line

Figure 8.1 Diagram of a stirred tank fermenter. The height of the cylindrical fermenter is one to four times its diameter. The nutrient medium with suspended microorganisms (stippled) occupies most of the fermenter, but above the medium is the head space, containing air which has passed through the medium and is hence depleted in oxygen. Sterile air enters the fermenter through a pipe that terminates in a single aperture or a sparger with many small apertures. It leaves the fermenter head space via a gas outlet. A stirring shaft carries several Rushton impellers, horizontal discs carrying vertical plates. Turbulence is increased by baffles. A manhole provides access for cleaning and has a glass window to allow viewing during a fermentation. The culture is harvested via a harvest line at the base. The temperature jacket, cooling coils, steam lines, inlets for inoculation, sampling and additions during the fermentation, and probes for monitoring are not shown.

system. With batch culture a stationary phase suitable for secondary metabolite formation inevitably follows the growth phase. Furthermore, many products are needed in relatively small amounts, so that a single batch fermentation can satisfy demand for a period, and the fermenter used to make other products. Continuous culture is most useful when a product is needed in large amounts, and is a metabolite associated with growth, for example industrial ethanol from the anaerobic growth of yeast, or is the direct result of growth, for example single cell protein. Thus mycoprotein is produced by continuous culture, but the fermentations have to be prematurely terminated as a result of selection of strains with more highly branched mycelium (page 498). An important variant of batch culture is **fed batch culture**, in which a nutrient is continuously or intermittently added to a fermenter during the course of a fermentation. Fed batch culture enables **catabolite repression**, in which high glucose levels inhibit secondary metabolism, to be avoided; glucose is supplied as needed and inhibitory concentrations do not occur. Fed batch culture is extensively used for the production of penicillin and other antibiotics.

Fermenter and medium sterilization
Before a fermentation can begin the fermenter and the medium must be sterilized. This is done by heating. Bacterial spores, the most resistant of contaminants, survive 100°C, so a period at a higher temperature (120–150°C) is needed. Heating sufficient to lower the probability of contaminant survival to one cell per thousand batch fermentations is generally regarded as acceptable.

When a facility has only a few fermenters, medium sterilization is usually carried out within the fermenters. Steam is passed through the temperature control jacket that surrounds a fermenter. It may also be injected into the fermenter. When the temperature reaches 100°C, the gas exit is closed so that a positive pressure develops and higher temperatures are reached. Following sterilization the fermenter is cooled by passing water through the jacket. Vacuum formation is avoided by admitting sterile air to the head space above the medium. The medium is stirred throughout to facilitate heat transfer.

Where there are many fermenters, the medium is often sterilized separately and then transferred along sterile pipework to already sterilized fermenters. This shortens the period during which expensive fermenters are idle, since an empty fermenter takes a much shorter time to sterilize than one filled with medium. The medium may be sterilized in a pressure vessel which is simpler in design and hence cheaper than a fermenter. Alternatively, medium sterilization can be a continuous process, with the medium being pumped through a heat exchanger in which rapid sterilization occurs. Even when the medium is sterilized in a fermenter, it may be desirable to sterilize some component separately and to add it aseptically to the fermenter. This may, for example, be done with sugar, since sugars and amino acids can interact when heated to form compounds that may inhibit growth.

Aeration and agitation
Most industrial fermentations are aerobic, and hence must be supplied with air. The air is pumped through a filter to eliminate contaminants and into the fermenter, commonly at a rate of about half a volume of air per volume of medium per minute. The air may enter the fermenter through a sparger with many small apertures, so as to yield many small air bubbles. This facilitates oxygen diffusion into the medium, since small bubbles have a higher surface to volume ratio than large bubbles. Spargers with small apertures are satisfactory with unicellular organisms such as yeasts, but can be blocked by the mycelium of moulds. With these it may be necessary to introduce air by means of a pipe of

large aperture that is not so readily blocked. The rate of oxygen transfer between bubble and medium depends not only on bubble size but on the residence time of the bubble in the medium. Agitation by means of a stirrer results in bubbles following a tortuous path from pipe or sparger to medium surface, and hence in more effective oxygen transfer. It also breaks large bubbles into smaller ones and hinders small bubbles from coalescing into larger bubbles. The residence time of bubbles will be greatest when the distance from the point of air entry to the medium surface is greatest, so fermenters are usually taller than they are wide, with their height up to four times their diameter. The escape of exhaust gases from the fermenter is restricted in order to obtain a pressure greater than atmospheric in the space above the medium. This positive pressure means that if leakage accidentally occurs it is outwards, so that contamination does not occur. A high positive pressure also increases oxygen solubility and hence favourably influences its transfer from bubbles to medium. Pressure increases with medium depth, so the rate at which oxygen dissolves is greatest with tall fermenters. Forced aeration is not only important in supplying oxygen to the medium but also in removing gaseous or volatile products of metabolism that may be inhibitory to growth.

Mechanical agitation (stirring) helps to distribute dissolved air uniformly throughout the fermenter. It also reduces the width of the static layers of liquid that surround cells and through which oxygen must diffuse to reach the cell. This reduction in width also facilitates the passage of metabolic products from the cells into the medium. These may include soluble products sought in the fermentation, and will include wastes such as carbon dioxide which must be eliminated to obtain optimal growth. Agitation will also hinder the formation of cell clumps and mycelial aggregates, which grow less well than free cells and individual hyphal filaments. Finally it will facilitate the transfer of excess metabolic heat to the cooling surfaces of the fermenter (see next section).

Agitation in fermenters is commonly by means of three or four Rushton impellers mounted on a vertical shaft. A Rushton impeller consists of flat vertical plates, usually six on a pilot scale fermenter but up to twelve on a large production vessel, equally spaced around a horizontal disc. The shaft enters the fermenter through a steam-sterilized double mechanical seal, with bearings lubricated by condensed steam. Several vertical baffles (commonly four) mounted on the fermenter wall increase turbulence and mixing. An uninoculated medium, a yeast suspension or one of mould pellets will normally be of low viscosity and act as a Newtonian fluid, in which viscosity is unaffected by shear. If, however, a microorganism produces a large amount of extracellular polysaccharide or consists of individual hyphal filaments the culture may be very viscous. It will also behave as a non-Newtonian fluid, in which apparent viscosity decreases with shear rate. Close to the impeller shear rate will be high, apparent viscosity low and mixing good. Further away shear rate will be less and mixing poor. With non-Newtonian fluids the relationship between the power supplied and the shear generated will be difficult to calculate. The amount of power needed to stir large fermenters, especially with viscous cultures, is considerable. Hence whereas with laboratory fermenters shaft rotation speeds of 1000 r.p.m. are feasible, with a filamentous organism in a large fermenter power economy may dictate shaft speeds of 50 r.p.m. or less.

Temperature control

Metabolism generates heat, which in laboratory-scale cultures is readily dissipated. The volume of a culture, however, increases as the cube of the vessel's linear dimensions, but the surface through which heat is lost only as the square. Hence the larger the fermenter, the more readily overheating occurs. Small fermenters have a temperature control jacket through which cooling water flows. With larger fermenters the cooling surface needs to be greater, so in addition there are cooling coils within the fermenter. If the temperature within the vessel rises above or falls below the optimum for the fermentation, this is detected by a sensor which controls a valve adjusting the flow rate of the cooling water. Availability of cooling water is a criterion in the selection of sites for large fermenters. Where feasible the use of thermophilic organisms for large-scale fermentations can reduce cooling costs.

Medium monitoring and adjustment

During the course of a fermentation conditions change. Cell density and hence oxygen demand increases, nutrients are utilized, and products are released into the medium. With fungi these usually include acids, lowering the pH of the medium. For a successful fermentation, departures from optimal conditions must be detected by appropriate sensors, and are automatically corrected. The parameters that are monitored commonly include growth, nutrient levels, dissolved oxygen, oxygen consumption (measured by analysis of inlet and outlet gas streams), gas pressure and foaming of the medium which is controlled by the addition of an antifoam agent.

Alternative fermenter designs

An upper limit to the size of stirred tank fermenters is imposed by the cost of the power needed for agitation. This has led to the development of a range of tower fermenters in which aeration is carried out in such a way as to provide adequate agitation (Fig. 8.2A). Tower fermenters are used for mycoprotein production. Fermenters without mechanical agitation are, however, used in brewing. These fermenters ('cylindroconical fermenters') are cylindrical with a tapering base (Fig. 8.2B). At the centre of the fermenter yeast cells with associated carbon dioxide bubbles rise to the surface. At the circumference liquid cools and descends through increased specific gravity. A circulation of cells and nutrients is hence maintained.

Downstream Processing

When a fermentation is complete it will contain cells (biomass) in a large volume of spent medium. In brewing the spent medium is itself the desired product, and in single cell protein production the biomass. Usually, however, the desired product is a minor component of the cells or broth. A litre of broth with biomass may, for example, contain only 1 g of a required enzyme or 10 g of an antibiotic. A small amount of product hence has to be separated from a large volume of

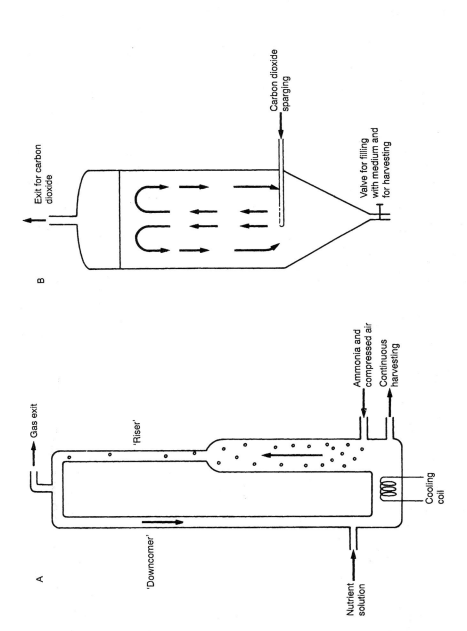

Figure 8.2 Alternative fermenter designs. A, The 30 metre tall airlift fermenter used for mycoprotein production by the continuous culture of *Fusarium graminearum* (page 497). Compressed air and ammonia are injected into the system, the former providing oxygen and bringing about circulation via 'air-lift', and the latter acting as the nitrogen source. Spent gas, containing respiratory carbon dioxide, leaves the fermenter at the top. A nutrient solution (glucose, biotin, mineral salts) is pumped into the fermenter at a constant rate near the bottom of the 'downcomer'. A heat exchanger in the form of a coil maintains a constant temperature of 30°C, and the culture is continuously harvested at the same rate as that at which the nutrient solution is supplied. B, A cylindroconical fermenter as used in brewing. Circulation occurs either as a result of carbon dioxide production by the yeast or, as shown here, through sparging by carbon dioxide. (A, After Trinci (1992); see References, page 541.)

waste material. A fermentation will be economic only if the downstream processing that accomplishes this is efficient.

A series of steps (unit operations) will be required in the concentration and purification of a product. Each unit operation involves expense in terms of equipment, manpower and energy or chemicals, and at each step there will be some loss of product. The number of steps has hence to be kept to a minimum. The operations involved in downstream processing will be considered below, and in later sections in relation to specific products.

Initial broth handling – separation of solid and liquid phases

When a fermentation ends, aeration ceases. If the organism is an obligate aerobe, cells will die and autolytic enzymes may destroy the required products. If the organism is a facultative anaerobe, a switch to anaerobic metabolism may result in the required product being utilized. The broth is also likely to contain residual nutrients, which could support the growth of contaminants. Undesirable changes can be delayed by chilling the broth, but this is too expensive except for small-volume fermentations. It is hence essential that steps that prevent product deterioration are undertaken promptly.

In a few instances it is possible to separate a product directly from untreated broth. Industrial alcohol, for example, is recovered from the fermentation broth by vacuum distillation. Usually, however, a series of preliminary steps, including

Figure 8.3   Downstream processing equipment. A, Rotary vacuum filter. Culture medium and suspended organisms are fed into a trough (t). The aqueous phase passes through a filter cloth on the surface (d) of a slowly rotating drum (0.1–2 rpm) and through numerous hollow spokes – only four shown – to the hollow axle (a) of the drum, through which suction is applied. The solid phase (p), including filter aid if used (see below), is held on the outside of the drum by suction, and is removed by a knife scraper (k). Washing (w) by a spray ensures that products in solution are removed from the solids and pass into the drum. If it is the aqueous phase that is of interest, a filter aid is usually added to the medium being fed into the trough. This can be diatomite (the silica skeletons of microscopic algae, diatoms, mined from extensive deposits) or, in the food industry, relatively expensive food-grade starch. If it is the solid phase that is needed, then filter aid is not used, because of the difficulty of separating it from the product. Rotary filters are relatively cheap to install and operate, and can cope with large volumes of liquid. If, however, a high degree of microbiological containment is needed, either for operator or product, then other types of filter, or centrifugation with containment, must be used. Although currently widely employed, rotary filters may be less used in future, because of concern over possible respiratory hazards associated with diatomite disposal. B, Continuous flow centrifuge. There are a range of types, but a widely used form has, as an essential feature, stacked conical discs (d) (more numerous and closer together than shown). Liquid with suspended particles is pumped into the rotating centrifuge and takes a path ($\rightarrow$) that carries it to the outer wall and then inwards between the discs. Since the discs are close together and steeply sloped, particles moving outward by centrifugal force have only a short distance to travel to reach the under surface of a disc. The particles (shown as stippling) then slide along the discs to reach the wall of the centrifuge. Here, depending on centrifuge design, solids may accumulate or be discharged intermittently, or continuously through a fine nozzle (n).

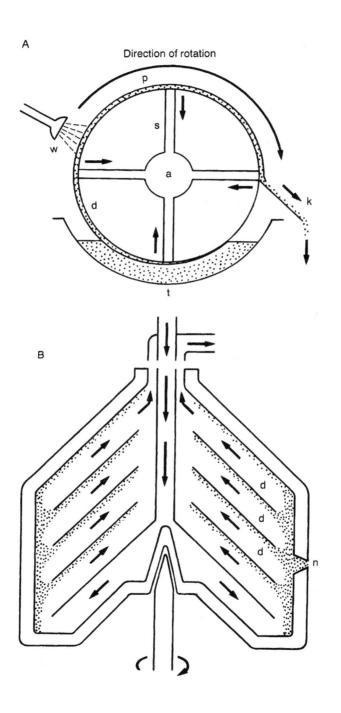

the separation of biomass and broth, is required. Fungal mycelium is usually separated from broth by means of a **rotary vacuum filter** (Fig. 8.3A), in which the filter forms the surface of a rotating drum. Mycelium and broth are fed onto this surface, the broth passes through the filter into the interior of the evacuated drum, and the mycelium is removed from the surface as the drum rotates. Yeast cells, depending on strain, may tend to rise to the top of a fermentation or to sediment. The former occurs in the production of traditional beers (page 488), enabling the yeast to be skimmed from the surface of the vat – cell removal by **flotation**. Where **sedimentation** occurs, settling out can be accelerated by agents that neutralize the charge at the cell surface and permit cells to flocculate, i.e. aggregate to form clumps that sediment readily. An example of a flocculating agent is the protein isinglass, used in the clarification of traditional lager, cider and wine (page 488). **Centrifugation**, using continuous flow centrifuges (Fig. 8.3B), is a more rapid method of cell separation than sedimentation.

Handling of biomass
Biomass may be required either as intact cells, or as a source of intracellular enzymes. When intact cells are required, removal of water, **dewatering**, is needed to obtain material sufficiently dry to avoid deterioration. The obvious method of removing water, drying by various forms of controlled heating, is expensive. Water content is hence first lowered as far as possible by mechanical forms of dewatering, such as rotary filtration for mycelium and centrifugation for yeast cells. Recovery of intracellular enzymes requires **cell breakage** to release enzymes into the aqueous phase. A variety of methods of cell breakage are employed, such as liquid shear (use of a press), solid shear (grinding and milling), wall lysis (for example, with alkali) and osmotic shock.

Handling of the aqueous phase
Various methods are used for the concentration and purification of products from the aqueous phase of fermentations. **Precipitation** followed by filtration is widely employed; citric acid, for example, is precipitated as the calcium salt, and enzymes by the addition of ammonium sulphate. Penicillin is recovered by **solvent extraction** into a phase immiscible with water, and a variety of forms of **chromatography** are used to separate and concentrate other products. The final phase of product recovery is usually drying. **Spray drying** in which a liquid concentrate passes through an atomizer into a stream of hot air is often used. **Freeze drying**, although expensive, is employed with heat-labile products of high value.

Disposal of waste
A fermentation produces a large volume of waste. Commonly about half of the carbon in a nutrient medium is converted into carbon dioxide, which it may be economic to recover from exhaust gases. Solid waste is usually incinerated or buried. Liquid waste, such as spent broth with residual nutrients, presents a major disposal problem. A medium size antibiotic plant, for example, is estimated to have an output of effluent with a potential for polluting waterways equivalent to

that of sewage from a town of 30 000 inhabitants. A plant must hence either pay the municipal authority for effluent treatment or have its own facility. There is hence an interest in recovering useful products, such as methane, fertilizer or animal feed, from fermentation wastes.

Solid Substrate Fermentations

The traditional Asian food fermentations (page 506) are solid substrate fermentations. Moulds are used to attack indigestible plant materials, such as soya beans or wheat grains. Fungal extracellular enzymes break down refractory polymers, especially cellulose, and further microbial processing yields nutritious and palatable products such as soy sauce, *miso* and *tempeh*. The raw materials usually receive pretreatment to facilitate fungal attack. Examples are grinding or milling to increase surface area, soaking to soften, and steaming to kill and eliminate the resistance associated with living seeds. The moulds used are aerobes, and depend for oxygen on air-filled spaces between substrate particles. The water content of the substrate has hence to be kept low so that the air-filled spaces persist. The resulting low moisture levels mean that the moulds have to tolerate low water potentials. Hence *Aspergillus*, a xerotolerant genus (page 55) and a potent producer of extracellular enzymes, is prominent in solid substrate fermentations.

Traditionally the Asian food fermentations were carried out in vertically stacked wooden trays. This gives high enzyme activities and a final product of uniform quality, but requires the provision of large surface areas for relatively small volumes of substrate. With large volumes of substrate, problems arise in distributing oxygen throughout the fermentation and in dissipating the carbon dioxide and heat generated by fungal metabolism. A variety of bioreactors for handling large volumes of substrate with adequate aeration and agitation have been devised. These include reactors in which humid air is forced through the fermenting mass, and rotating drum and fluidized bed reactors. Some of the systems can be operated continuously as well as on a batch basis.

Although traditionally associated with Asian fermented foods, solid substrate fermentation can be used to give a range of other products. Mushroom growing (page 493), the production of the composts used in mushroom growing and the composting of organic wastes are all essentially solid substrate fermentations. The theory and technology of solid substrate fermentation is most highly developed in Japan, and has there been applied to the production of citric and other organic acids and industrial enzyme production. Solid substrate fermentations have many merits. The raw materials are commonly cheap, sterilization is not usually needed, energy requirements are low and the wastes needing disposal are compact and semi-solid. On the other hand, the fermentation process is often slow and the range of organisms able to tolerate the low water potentials involved is limited.

Enzyme Technology

Fungi produce some enzymes of great commercial importance. The nature, production and utilization of such enzymes will now be considered, as will the exploitation of the enzymic activities of immobilized fungal cells.

Fungal Enzymes of Commercial Importance

Many fungi produce extracellular enzymes that enable them to break down polysaccharides and proteins into sugars and amino acids that can be assimilated easily. These enzymes are of considerable industrial importance. They have, for example, largely replaced acid hydrolysis as a method for converting starches into sugars; enzymes, unlike acid, do not cause undesirable side reactions or corrode reactor vessels. Several thermostable enzymes, that can be added to hot mashes, are used in the enzymic hydrolysis of starch (Fig. 8.4). A bacterial amylase attacks $\alpha(1\rightarrow4)$ linkages within the starch molecule, breaking it into oligosaccharides, small soluble polymers. The resulting 'liquefied starch' can then be converted into

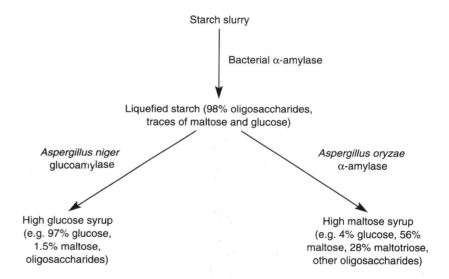

Figure 8.4   The enzymic hydrolysis of starch. A starch slurry is treated with a bacterial α-amylase at a high temperature for a short period (e.g. 95°C, 2 hours) to give 'liquefied starch' , a soluble mixture of oligosaccharides, mainly with three to seven glucose units. The liquefied starch can then be treated with a variety of fungal or bacterial enzymes at suitable temperatures and periods to give syrups containing the required mixture of sugars. Here the production of a high glucose and a high maltose syrup is indicated. A high glucose syrup can be converted into a high fructose syrup with glucose isomerase. Fructose is sweeter than sucrose, and a smaller amount can be used to impart sweetness to soft drinks and confectionery; sales of high fructose syrup are now large.

a required mixture of sugars by treating it with further enzymes, singly or in combination. A 'high glucose syrup', with up to 97% glucose and a little maltose and higher oligosaccharide, can be obtained with a glucoamylase, also called amyloglycosidase, commonly from *Aspergillus niger*. Glucoamylase is an exoenzyme that, starting from the end of polymer chains, attacks both $\alpha(1\rightarrow4)$ and $\alpha(1\rightarrow6)$ linkages. High glucose syrups are readily utilized by yeast and hence useful in baking and brewing. Alternatively the liquefied starch can be treated with a fungal α-amylase, usually from *Aspergillus oryzae*, to give a 'high maltose syrup', with a little glucose, over 50% maltose and considerable higher oligosaccharides. High maltose syrups are less hygroscopic and less prone to crystallize than high glucose syrups, and are suitable for jam and confectionery. Although starch-degrading enzymes are of outstanding importance commercially, a number of other fungal enzymes are used in the food and drink industry. A $\beta(1\rightarrow3)$-glucanase is used to remove haze from beer by degrading glucans. Pectins from the middle lamella of plant cell walls can cause haze in fruit juices and wines. These can be clarified with a fungal pectinase. Lactase, converting lactose into glucose and galactose, can render whey more easily utilized by microorganisms, and ice cream sweeter. Cellulases have also found some use in food processing, but are now receiving intensive study in connection with the utilization of lignocellulose, the cheapest and most abundant of organic carbon sources, but one of the most intractable. Most of the enzymes mentioned above are from *Aspergillus niger* or *A. oryzae*, although lactase is also obtained commercially from the yeast *Kluyveromyces* and cellulase from *Trichoderma*.

Proteolytic enzymes (proteases) vary with respect to their pH optima. In general alkaline proteases are produced by bacteria and acid proteases by fungi. There is a large demand for proteases as detergent additives; as these have to be effective under alkaline conditions most of those employed are from bacteria. An important acid protease is chymosin, used in cheese manufacture to coagulate the milk proteins. The traditional source of chymosin is rennet from the stomach of slaughtered calves, but acid proteases produced by *Rhizomucor miehi*, *Rhizomucor pusillus* and *Cryphonectria parasitica* have been widely used since the 1950s. Microbial rennets produce palatable cheeses, but with flavours slightly different from those produced with animal rennets, and they can be marketed as vegetarian foods. The gene responsible for calf chymosin was cloned into several microorganisms, for example the yeast *Kluyveromyces lactis*, in the 1980s, and microbially produced chymosin is now widely used commercially. The gene for *R. miehi* proteinase has been cloned and expressed in *Aspergillus oryzae*, to give high yields of over 3 g per litre. The product is commercially available, with better properties than the parent enzyme preparations, due to the absence of contaminating enzymes. Other fungal proteases are used to remove protein hazes from beer, and the wheat protein gluten from flour, leading to a softer dough more suited to biscuit making. Other extracellular fungal enzymes that have been used in the food industry are nucleases and lipases. Nucleotides are largely responsible for meaty flavours, and in appropriate foods can be released from RNA with a nuclease from *Penicillium citrinum*. The lipases liberate fatty acids from fats, influencing cheese flavour. Lipases are also used in detergents, including an enzyme from a *Humicola* species expressed in *Aspergillus oryzae*. An enzyme used as an additive to animal feedstuffs is an *Aspergillus* phytase, which hydrolyses the plant storage molecule phytic acid to give inositol and six phosphate anions. Addition of phytase to feedstuffs of monogastric animals such as pigs, which are unable to utilize phytic acid, increases the nutritional value, eliminates the need for additional phosphate in the diet, and decreases problems of phosphate in the manure. The phytase is produced only in very low yield by the parent fungus, but at high yield by heterologous expression in a recombinant strain of *Aspergillus niger*.

Whereas extracellular enzymes from fungi are used extensively in the food industry there are relatively few intracellular fungal enzymes of commercial importance. Glucose oxidase, from *Aspergillus niger* or *Penicillium chrysogenum*, may be included in bottled fruit juice to eliminate traces of oxygen and thus act as a preservative. It is also a component of kits for the detection and estimation of glucose, especially in blood and urine. Catalase, also from *A. niger*, can be used to remove traces of hydrogen peroxide from milk and other materials. Alcohol dehydrogenase from *Saccharomyces cerevisiae* is used in ethanol determinations. Asparaginase, used in the treatment of leukaemia, is obtained from *Aspergillus* and *Penicillium* as well as from bacteria.

The enzymes that break down polysaccharides and proteins are required by industry on a large scale. They are commonly not needed pure, and may be marketed in preparations in which the enzyme constitutes 5% or less of the solid material. Such preparations are relatively cheap to produce, with a low cost per kilogram of active enzyme. Many tons per year of these high volume, low cost enzymes are produced, accounting for at least 80% of enzyme sales. Enzymes for analytical, diagnostic and clinical use are needed in small amounts but highly purified. Here the annual production of an enzyme may be a kilogram or less, but with a value per gram orders of magnitude above that of an industrial enzyme. An increasing range of such low volume, high cost enzymes is being produced, mostly by heterologous expression in bacteria and fungi, notably *Aspergillus* species.

Although the enzymes that are used by industry on a large scale need not be pure, they need to be harmless. Allergies were formerly caused, in both production workers and customers, by the proteases used in detergents. Such problems can be avoided if the enzyme preparations are at no time in the form of fine powders that can become airborne and be inhaled. Preparations that are used as food additives or in food processing must be free from any materials that are poisonous, such as mycotoxins. New products are subjected to extensive toxicity tests.

Production of Fungal Enzymes

Aspergillus is the most important fungus for commercial enzyme production. The microorganisms used for enzyme production are generally grown in fermenters in batch culture, although some use of solid substrate fermentation is made, especially in Japan. The first step in subsequent processing is the separation of cells from broth. When the enzyme is intracellular, the cells must then be disrupted to release the enzyme. However, it is usually the broth that requires processing, since most commercially important fungal enzymes are extracellular, and heterologously expressed enzymes are usually designed to be excreted. Enzymes for industrial use are often sold in aqueous solution. Then all that is needed is concentration to reduce the volume of liquid to be handled and transported. This can be achieved by the application of hydraulic pressure to force water from the solution through a membrane with pores *ca* 10 nm, a process termed ultrafiltration. Precipitation with agents such as ammonium sulphate ('salting out') followed by filtration can be used when a solid enzyme preparation is needed. Where high purity is essential, an expensive sequence of chromatographic procedures may be necessary.

Immobilized Enzymes and Free and Immobilized Cells

The proteases and starch-degrading enzymes used for large-scale conversions in industry are usually added to the substrate in stirred tank reactors. A batch of enzyme is needed for each batch of substrate processed, but since the enzymes are cheap this does not matter. However, some microbial enzymes used in large-scale processes are expensive, so means were sought for immobilizing the enzymes so that they could be retained and used for a long period. Immobilization can be achieved by attaching the enzyme to carriers which may be inorganic (e.g. ceramics), or natural polymers (e.g. cellulose) or synthetic polymers (e.g. nylon). The attachment can be by ionic bonding due to charge differences between enzyme and carrier, or by covalent bonding achieved with appropriate reagents (Fig. 8.5). Depending on the enzyme, the carrier and the attachment method, there can be a stabilization or a loss of enzyme activity, and a narrower or a wider enzyme specificity. An alternative to attachment is entrapment of the enzyme within a porous matrix of cross-linked polymers, within hollow fibres or within semi-permeable microcapsules. Entrapment allows substrates to reach the active site of the enzyme, but prevents enzyme escape. The choice of attachment or entrapment method will depend on the process envisaged and the enzyme to be employed.

The availability of immobilized enzymes allows a stirred tank reactor to be operated on a continuous flow as well as on a batch basis. It has also resulted in the use of other types of reactor, such as enzyme-packed columns through which the substrate solution flows, as well as fluidized beds. In the latter an upward flow of the substrate solution keeps in suspension the particles to which the enzyme is bound. These processes have so far been utilized mainly with enzymes from bacteria. Immobilized enzyme technology, however, has also permitted the development of such analytical methods as the enzyme electrode. Here an immobilized enzyme interacts with a substrate to cause a change in the concentration of a substance measurable with the electrode. For example, with the 'glucose electrode' immobilized glucose oxidase causes disappearance of oxygen from solution at a rate proportional to the glucose concentration, and the change in oxygen concentration is measured by the electrode.

As indicated above, extracellular enzymes, either free or immobilized, are utilized in some important industrial processes. The same is true for intracellular enzymes that retain high activity when extracted from the cell. There are,

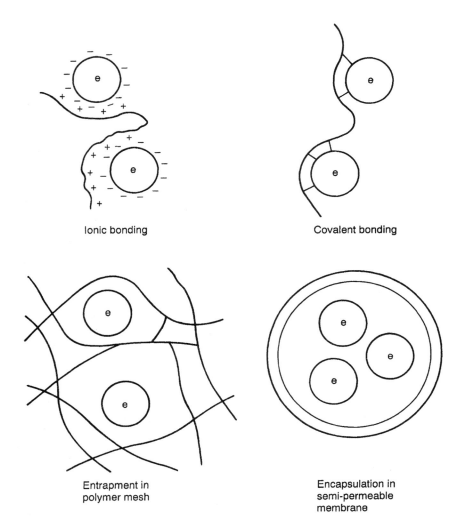

Ionic bonding

Covalent bonding

Entrapment in
polymer mesh

Encapsulation in
semi-permeable
membrane

Figure 8.5   The immobilization of enzyme molecules (e) through attachment to a carrier by ionic or covalent bonding and by entrapment within a polymer mesh or a semi-permeable capsule.

however, reactions of industrial interest that require intracellular enzymes that lose most of their activity when extracted, or need expensive co-factors, or can be accomplished only by a sequence of enzymes. Such reactions can be carried out economically only by intact cells, and microbial cells have been extensively used to carry out transformations that would be difficult to accomplish by purely chemical means. Some fungi are able to bring about precise changes in steroid molecules (Fig. 8.6), and are used to carry out some of the steps involved in the conversion of readily available plant sterols into therapeutically useful compounds. The mycelium of such fungi can be grown in fermenters, washed in a buffer solution and packed into a column through which a solution of the compound to be treated is passed. When unicellular cells such as yeasts or fungal

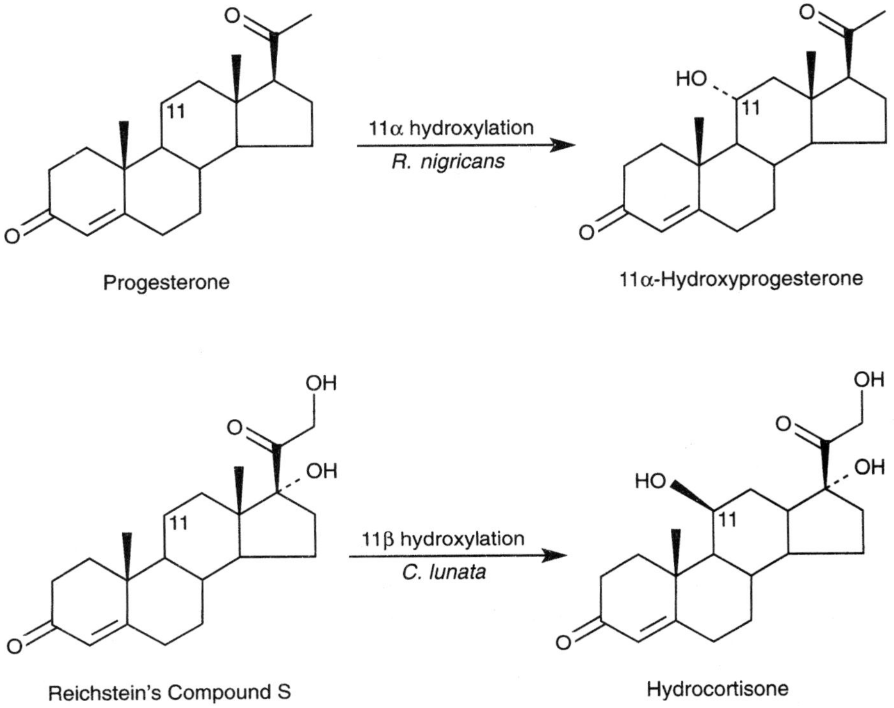

Figure 8.6 Steroid transformations by fungal cells. The Zygomycete *Rhizopus nigricans* carries out an 11α-hydroxylation and the mitosporic fungus *Curvularia lunata* an 11β-hydroxylation on the steroid molecule. The resulting hydroxyl groups differ in their configuration with respect to the planar molecule, – – – representing a bond projecting away and ▲ one towards the observer. Progesterone is obtained by a series of chemical steps from stigmasterol, extracted from soya beans, and Reichstein's Compound S from diosgenin, which occurs in a Mexican yam. Further chemical or microbially catalysed steps convert 11α-hydroxyprogesterone and hydrocortisone to therapeutically useful compounds such as cortisone and prednisolone.

spores are used to effect conversions, immobilization is useful, and this can be accomplished by methods similar to those effective with enzymes, in particular by entrapping the cells in beads of calcium alginate, a seaweed polysaccharide. Alternatively, cells can be cultivated in such a way that they adhere to suitable surfaces or particles. Immobilized cells may be alive, killed by procedures that leave the required enzymic activities unimpaired, or treated to increase cell membrane permeability to the substrate and products.

The Production of Alcoholic Beverages

Under aerobic conditions yeasts metabolize sugar to carbon dioxide and water. If oxygen is scarce or absent, or if the sugar concentration is high (page 72), fermentation occurs with the production of ethanol and carbon dioxide. This alcoholic fermentation is the basis of the production of the great variety and huge quantities of alcoholic beverages consumed by humans. The yeast responsible is, in the vast majority of fermentations, *Saccharomyces cerevisiae* (Fig. 8.7), although important varieties were sometimes given specific names (e.g. *S. carlsbergensis*, *S. ellipsoideus*). Alcoholic beverages can, on the basis of the procedures involved in their production, be grouped into three classes. First, there are those, like **wine** and **cider**, which are made by fermenting plant juices rich in sugars. Secondly there are those, such as **beer**, made from plant materials rich in starch which must be converted into sugars before fermentation can begin. Finally there are **spirits** and **fortified wines**; in the production of these, distillation is used

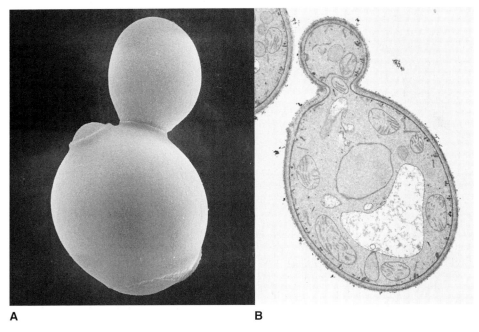

A   **B**

Figure 8.7 Electron micrographs of yeasts. A, Scanning electron micrograph of *Saccharomyces cerevisiae*, the yeast responsible for nearly all alcoholic fermentations, late in the cell cycle. The mother cell, with a prominent bud scar on the upper left hand side and a less prominent birth scar at the bottom, is below, and the bud, soon to be a daughter cell, is above (see page 109). The diameter of the mother cell is *ca* 5 μm. B, Transmission electron micrograph of a thin section of a yeast cell of *Candida albicans,* a benign or pathogenic component of the human body microflora, at a similar stage of the cell cycle. A section through a bud scar is seen at the upper left surface of the mother cell. The cell wall is seen as a multi-layered structure, continuous over the surface of the mother cell and bud, enclosing the dark-stained plasma membrane. Note also the central nucleus surrounded by a double-layered nuclear membrane with a nuclear pore at the upper left, above a large vacuole surrounded by a single membrane, and several mitochondria with cristae (invaginations of the inner mitochondrial membrane). The oval single-membraned structure in the centre of the bud is a microbody. The diameter of the mother cell is *ca* 4 μm. Prepared by chemical fixation; compare with Fig. 4.8, a section of a yeast cell prepared by freeze-substitution. (Reproduced with permission from Osumi, M. (1998). The ultrastructure of yeast: cell wall structure and formation. *Micron* **29**, 207–233.)

to obtain alcohol concentrations higher than can be achieved by fermentation alone.

Beverages made from Sugar-rich Plant Juices

Wine

The major wine-producing countries have laws that define wine as the product of the alcoholic fermentation of the juice of fresh grapes, and strictly limit the materials that may be added during production. A wide variety of black and white grapes are used, and the protection of the grapevines from fungi and other pests is an important branch of plant pathology. There is, however, one plant pathogen, that has an important role in the production of some fine white wines, such as Sauternes. This is the grey mould *Botrytis cinerea*, that the French call *pouriture noble* and the Germans *Edelfäule*, the 'noble rot' (page 385). The hyphae of this mould penetrate the skins of the ripe grapes and cause water loss, resulting in shrivelled grapes with an unusually high concentration of sugar and other constituents. Fermentation of the juice of these grapes results in a sweet wine of high quality.

The differences between wines depend on the variety of the grape, the local soil and climate, the properties of the yeasts and the details of the production method. Most white wines are made from white grapes, whereas to make red wines a substantial proportion of the grapes must be black, since the colour of the wine comes from the pigments in the grape skin. After the grapes are harvested they are crushed to break the skins and release the juice. In making white wines gentle pressing follows to release further juice from the pulp, after which the skins and pips are discarded. In making red wines, fermentation is allowed to proceed in the presence of pips and skins, resulting in the extraction, by alcohol from the fermentation, of pigments from the skins and tannins from the skins and pips. Red wines hence have a higher tannin content that white wines. Rosé wines usually result from a brief exposure of the fermentation to pips and skins, but sometimes from blending of red and white wines.

The fermentation is traditionally carried out in barrels or open vats (Fig. 8.8) but fermenters giving a greater control of conditions are now widely used, especially for white wines. The yeast principally responsible for the alcoholic fermentation was formerly designated *Saccharomyces ellipsoideus*, but is now regarded as a variety of *Saccharomyces cerevisiae*. Inoculation with a specific strain of *S. cerevisiae* is now widely practised, but traditional methods depend on the introduction of a suitable strain with the grapes. Yeast cells are numerous on the surface of grapes, but only a very small proportion of the cells are *S. cerevisiae*. Within a few days from the beginning of the fermentation, however, it becomes the dominant species. Grape juice is rich in sugars and other nutrients, which permit rapid yeast growth which soon results in anaerobic conditions. The juice is also strongly acidic (pH 2.9–3.9) and, especially when red wines are being produced, rich in tannins, discouraging the growth of bacteria and yeasts that might adversely affect the fermentation. Undesirable organisms may be further inhibited by sulphuring. This is the addition of sulphur dioxide, or of potassium or sodium metabisulphite which, by interaction with the grape acids, yields

A   **B**   **C**

Figure 8.8   Wine and beer fermentations. A, Wine fermentation in barrels; sampling. B, Wine fermentation in open vat, with copper vessel for transporting and warming the 'must' – unfermented grape juice. C, Traditional British beer fermentation in an open vat, with a pipe line for filling the vat, and a vacuum line for removing the excess foam produced by a top yeast. A variety of closed vessels are now used for wine and beer fermentations, cylindroconical fermenters (Fig. 8.3B) being widely used for the latter. (A–C, Bernard Lamb.)

sulphur dioxide. The wine yeasts tolerate this treatment. In addition ethanol is produced as fermentation proceeds, and this too is inhibitory to many microorganisms. *S. cerevisiae* tolerates the ethanol, and a high rate of metabolism can continue for several weeks, until an alcohol concentration of 10–12% (v/v) is reached. This results in partial inhibition of the yeast, and cell numbers and metabolic activity fall. Depending on how much sugar remains unutilized, a sweet, medium or dry wine results. Although carbon dioxide and ethanol are the main products of yeast metabolism, hundreds of other compounds are produced in smaller amounts. These, which include higher alcohols, organic acids and ethyl esters, are responsible for the distinctive character of different wines. Glycerol may be present at concentrations as high as 1%, giving 'body' to the wine. The volatile ethyl esters are major components of the aroma of wine. Malic acid and tartaric acid, each of which is present in grape juice at levels of up to 2%, survive the alcoholic fermentation.

When the first phase of the fermentation is over, the wine is transferred to wooden casks or other containers for maturation and storage. Microbiological activity continues at a reduced level during maturation. Lactic acid bacteria that can tolerate high acidity and alcohol concentrations commonly carry out the malolactic fermentation, converting much of the strongly acidic malic acid to the blander lactic acid, and raising the pH. This reaction is desirable where the grape juice is strongly acidic, as it is near the northern limits of vine cultivation. The reaction can be terminated if necessary by the addition of sulphur dioxide.

Microbiological activity that eliminates tartaric acid is less common, and is regarded as undesirable. Excess tartrates are often removed by chilling to −5°C for a few days, to avoid possible precipitation after bottling.

Maturation generally takes about a year, after which white wines generally do not improve. Some red wines and some sweet white wines, however, can improve in storage for many years. Sterile filtration is usual at the time of bottling, to eliminate bacteria that could cause spoilage by, for example, converting the ethanol into acetic acid.

Cider and perry

The production of cider from apples has much in common with the production of wine from grapes. There are many varieties of cider apple, which can be classified into sweet (low in acid and tannin), bittersweet (low in acid, high in tannin), sharp (high acid, low tannin) and bittersharp (high in both acid and tannin). Some use is also made of culinary and dessert apples. The apples are crushed and the pulp pressed to extract the juice. Traditionally the cider fermentation depends on yeasts introduced with the fruit, but the addition of a selected strain of *Saccharomyces* is now widely practised, as is sulphuring to prevent the growth of unwanted yeasts and bacteria. Since apple juice generally has a lower sugar content than grape juice, the alcohol content of cider is usually considerably less than that of wine. Following the alcoholic fermentation, lactic acid bacteria may carry out the malolactic fermentation during storage, converting some of the malic acid into lactic acid. Some varieties of cider apple can by themselves yield a drinkable cider, but many of the more palatable ciders are made by the blending of ciders from, for example, sweet, acid and bitter varieties of fruit. Ciders, like wines, are liable to microbial spoilage, especially by bacteria that convert ethanol into acetic acid. Hence, at the time of bottling, pasteurization or filter sterilization is carried out. Since the apple harvest may yield quantities of apple juice in excess of immediate fermentation capacity, concentrates may be prepared for easy storage, and then diluted as required for fermentation – a practice not permitted in wine making. Perry is prepared from the juice of perry pears or unripe dessert pears. The method of production is similar to that for cider. Ciders and perries are made in a wide range of sweetnesses and amounts of 'sparkle'. Both are bottled, but ciders are also served 'draught' from kegs or barrels.

Other beverages

The use of the term 'wine' without further qualification is restricted to the alcoholic beverage produced by the fermentation of grape juice. A variety of other fruits are used to produce fruit wines, either by amateurs, or commercially. Most of the fruits used have a far lower sugar content than grapes, so sugar has to be added to obtain a satisfactory alcohol content. In other respects the processes involved in producing a fruit wine parallel those involved in making wine from grapes.

In some tropical countries palm wines (Fig. 8.9), made from the sap of a variety of palms, are of considerable nutritional and social importance. The palm tree is tapped near the apex, and the sap, rich in sugars and amino acids, is collected in a container such as a gourd. A rapid alcoholic fermentation is brought about by

A   **B**

Figure 8.9   Production of palm wine from the coconut palm in Sri Lanka. The inflorescence peduncle, the shoot which bears flowers and ultimately coconuts, is tapped and the sap collected in earthenware pots. The sap ferments to yield a palm wine, toddy, which can be distilled to give a spirit, arak. A, Coconut palm. The spherical objects are collecting vessels, not coconuts. B, Collecting vessel with sap. (A,B, Bernard Lamb.)

yeasts originating from the plant surface, the tapping tools, or the container. The wine has to be consumed within a day, or bacterial spoilage renders the wine undrinkable. The Mexican equivalent of palm wine is *pulque*, made from the sap of agave, which like cactus can grow in dry areas and on poor soils.

Honey, derived from the nectar from flowers, is in a sense a plant product, even though the collection and initial processing has been carried out by bees. The sugar concentration and hence the osmotic pressure of honey is so high that until diluted it is immune from microbial attack. On dilution, however, it can be fermented by yeasts to yield mead. The alcohol concentration of mead will depend on the extent of the dilution, but can be as high as 15% (v/v). In the Middle Ages, before the development of sweet grape wines, mead was widely made throughout Europe, but now it is a rarity.

Beverages made from Starchy Materials

Sugar-rich plant juices are available in abundance only in some areas and only for part of the year. Seeds, roots and tubers rich in starch are on the other hand, available in almost any region where plants can grow, and are readily stored for many months. Very few yeasts, however, can utilize starch, so **saccharification**, the conversion of starch into sugar, must be carried out before fermentation can commence. Saccharification can be brought about in several different ways. In the production of beer, the best known beverage based on starchy raw materials, the saccharification process is termed **malting**.

Beer

Beer is basically an alcoholic beverage made from malted barley and flavoured with hops. In Germany, and in some other countries, the materials used in brewing are limited by law to malted barley, water, hops and yeast. Elsewhere the malted barley may be supplemented by other sources of sugar, and other additives may be utilized. Beers are of two main types, **ale** and **lager**. Ale is produced by a top yeast (see below), and is served at room temperature or slightly cooled. It is the typical beer of the British Isles, where it is often referred to merely as 'beer', with the term 'ale' being limited to light-coloured beers or regarded as archaic or genteel. Lager, which is served well chilled, is produced by a bottom yeast (see below) which probably evolved in Bavaria a few centuries ago. Lager is now produced throughout the world, with ale being unusual outside the British Isles.

Malting may be carried out by brewers or by independent maltsters. Barley grains are steeped in water for up to a day, and then allowed to germinate under moist, aerobic conditions. During germination, enzymes are produced which mobilize the nutrient reserves in the seed. Proteinases and amylases begin the breakdown of reserve proteins to amino acids and the abundant starch to sugars. In nature the nutrients would support the early growth of the barley plant but, in malting, germination is terminated after about a week by kilning, drying in a stream of air at 60–79°C. The seeds are killed but the enzymes are not destroyed. The dried grains, known as malt, are packages of enzymes and nutrients that can be stored for months. Such malt is used to make light-coloured beers. Higher temperatures in the kiln give a darker malt with additional flavours but little or no enzyme activity. This can be added to enzymically active malt to give a darker beer.

Malt is milled to give a coarse flour, the grist. This is added to warm water to give a porridge-like mash. Except where German concepts of beer purity prevail, additional starch, in the form of milled maize or rice (corn or rice grits), may be added to the mash. The mash is maintained at about 65°C for a few hours. The barley enzymes continue the breakdown of starch and protein in the malt, as well as attacking the starch in added grits. After mashing, residual solids are separated to leave a liquid, the wort, rich in fermentable sugars and the nutrients required for yeast growth. High glucose syrups (page 476), usually derived from maize starch, may be added at this stage as an alternative to the earlier addition of grits.

The wort is then boiled. This stops enzyme action, sterilizes the wort, and coagulates some of the proteins present. Prior to boiling, dried female flowers of the hop plant, *Humulus lupulus*, are added, or pellets or extracts made from hop flowers. Hop flowers contain cyclic organic compounds termed humulones, which when the wort is boiled isomerize to soluble isohumulones, responsible for the characteristic bitterness of beer. The hop flowers also contain terpene hydrocarbons and other essential oils, important for flavour and aroma. After boiling the wort is cooled, and hop residues and other solid material (such as precipitated protein) are removed. The wort is then aerated, which promotes subsequent yeast growth, probably because biosynthesis of the sterol and unsaturated fatty acid required for plasma membranes (page 147) are sparse in wort, but can be synthesized by yeast if oxygen is present.

Traditionally the fermentation takes place in large rectangular open vats (Fig. 8.8C), but now closed vessels are commonly used, one of the most successful types being the cylindro-conical fermenter (page 470). Closed vessels allow a close control of conditions, and reduce the chance of contamination with inferior yeast strains or of bacteria causing spoilage. Inoculation is usually with a genetically uniform strain of yeast grown in pure culture. The inoculum is grown in a fermenter under aerobic conditions, which allow the carbon source to be converted efficiently into biomass instead of ethanol. The inoculum volume is usually one-tenth of that of the wort to be inoculated, a practice similar to that in fermentation technology (page 464). In ale production a top yeast is used, so called because in the course of a traditonal fermentation the yeast mixed with gas accumulates as a foam at the top of the vat. Lager is produced with a bottom yeast which, probably as a result of more hydrophilic cell walls, tends to sediment instead of rising to the top of the fermentation. In the brewing industry the lager yeast is usually termed *Saccharomyces carlsbergensis* and the ale yeast *Saccharomyces cerevisiae*, although current taxonomic opinion is that both ale and lager yeasts are strains of *S. cerevisiae*. The difference between ale and lager yeasts with respect to flotation and sedimentation disappears in modern fermentation vessels with good circulation of wort. Fermentations producing lager are conducted at 8–15°C and those for ale at 15–25°C. The oxygen initially present in wort is soon exhausted by growth. This, and the high sugar concentrations, result in metabolism being almost wholly fermentative. As a result, sugar is converted into ethanol rather than biomass – with respiration 100 g of sugar will give about 50 g biomass, but with fermentation about 50 g ethanol and only 5 g biomass. As well as producing ethanol, yeast cells produce a wide variety of compounds important for flavour and aroma, especially ethyl esters and organic sulphur compounds, such as dimethyl sulphide. A desirable feature in a brewing yeast is a tendency to flocculate and clump towards the end of a fermentation. This is thought to be due to changes in the cell surface and to the formation of calcium bridges between carboxyl groups on adjacent cells: it facilitates the removal of yeast cells at the end of the fermentation. The harvested yeast cells can be recycled, but after being used several times they are discarded owing to an increasing likelihood of microbial contamination. The discarded yeast may be used as a component of animal feed, or as a source of vitamins. Another major product of fermentation, carbon dioxide, may be collected and compressed for sale. Otherwise it has to be dispersed efficiently, as concentrations of more than a few per cent in air that is breathed are dangerous.

After the primary fermentation is complete, a few weeks' maturation in casks or other vessels follows. During this period a secondary fermentation is brought about through the metabolism of some of the residual sugars by the low concentrations of yeast remaining in the beer. After maturation, **fining** (clarification) of the beer with isinglass or other agents is usual. Isinglass is derived from the buoyancy bladders of fish, and is almost pure collagen. The long positively charged molecules of collagen interact with the negative charges on yeasts and some proteins and the resulting complexes sediment. Auxiliary fining with negatively charged alginate may follow to remove positively charged proteins. Clarification is completed by filtration. Flash pasteurization – a few seconds at 75°C – is then used to reduce the chance of spoilage by lactic acid or acetic acid bacteria, or by bacteria producing unpleasant flavours or odours or causing cloudiness.

Other beverages

Beers of the type described above, although originating in Europe, are now produced on a large scale throughout the world. Other starch-based alcoholic beverages are, however, numerous and still important. They use materials that include cereal grains such as rice, maize and sorghum, and root crops such as cassava. Some are made by peasants in their homes and some in plant rivalling a modern brewery in scale and sophistication.

There are three basic methods of saccharification. It may be brought about, as in European beers, by enzymes formed in germinating grain. The production of Bantu beer in South Africa, for example, uses sorghum malt. Secondly, saccharification may result from enzymes made by moulds originating on the surface of the grain. This is usual in Asian fermentations. In Japan, pure strains of *Aspergillus oryzae* are used for saccharification in the production of saké, a rice wine, whereas in China starch breakdown in rice wine manufacture is usually due to members of the Mucorales, especially *Rhizopus oryzae*. With the traditional beers of tropical Africa, such as pito in Nigeria, enzymes from both the germinating grain and moulds on the grain surface contribute to starch breakdown. Saccharification by moulds is not without hazards, since members of the *Aspergillus flavus-oryzae* series can produce aflatoxin, a carcinogenic mycotoxin. The *A. oryzae* strains used for saké production in Japan, however, do not produce aflatoxin. A third, very primitive, method of saccharification occurs, very rarely now, in parts of South America, in which the starchy raw material is chewed, and amylase in saliva converts starch into sugars, as with the maize-based beverage chica. As with European beers, *Saccharomyces cerevisiae* is the yeast responsible for alcohol formation with these other starch-based beverages. They almost all, however, share a feature in which they differ from European beers. This is a phase of souring, preceding or accompanying alcoholic fermentation, and brought about by lactic acid bacteria, which in brewing European beers are regarded as an unmitigated nuisance. The resulting beverages often contain about 1% lactic acid. Alcohol concentration can vary greatly. With rice wines the yeast strains employed are very alcohol-tolerant, and saké contains 15–16% (v/v) ethanol. The alcohol concentration in most of the other beverages is in the range 1–5%. Rice wines are clear, whereas the other beverages are often cloudy or have a sediment of yeast and unfermented solids. In many areas the low alcohol beverages are of considerable nutritional importance, being a major source of vitamins and amino acids. Their acidity and alcohol content are unfavourable for pathogen survival, resulting in a safe beverage in areas where water supplies are often dangerous.

Beverages made Utilizing Distillation

Beverages containing 10–12% alcohol are readily produced by fermentation. At alcohol concentrations a little higher the metabolism of most strains of yeast is partially or totally inhibited. Hence beverages with very high alcohol concentrations, spirits, can be made only by distilling the product of a fermentation. Although alcohol concentrations a little above 12% can be obtained by fermentation, as with the 15–16% ethanol of saké, even these concentrations are most easily obtained by the addition of distilled alcohol to the products of a fermentation, as with fortified wines, such as port.

Spirits

A wide range of spirits are made from sugary or starchy materials. Among the former are brandy (from grape juice), fruit brandies, such as the plum brandy slivovitz, calvados or applejack (apple juice), rum (juice from the sugar cane), tequila (juice from agave) and a wide range of spirits corresponding to the various palm wines. Among those from starchy materials are whisky from Scotland and (different spelling) whiskey from elsewhere. A variety of cereals are used in their manufacture, including barley, wheat, rye and maize. Gin is also made from rye or maize, and vodka from non-cereal starchy materials, such as potatoes. In Asia there are a range of rice-based spirits.

Where starchy materials are used in spirit manufacture, a saccharification stage is necessary. In the production of Scotch malt whisky, the sole carbon source is barley malt. During kilning this is exposed to peat smoke, important in producing the flavour and aroma of the final product. In making other whisky (which spelling will be used henceforth), barley malt is also used, but with the addition of unmalted grain from barley or other cereals. Where the persistence of flavour from raw materials is unimportant, as with gin and vodka, microbial enzymes instead of malt may be used to convert the starch source into a high glucose syrup.

The fermentations involved in spirit production are varied. That used in making brandy is essentially a wine fermentation. The grapes used, however, yield a wine relatively low in alcohol and high in acid. The yeasts are from the grape surface, and sulphuring, which would adversely affect the distillation apparatus and final product, is not employed. In the production of whiskies, souring of the mash, either with pure strains of lactic acid bacteria or bacteria introduced with the mash, is important. The alcohol production is due to inoculation with pure strains of *S. cerevisiae*. This yeast is also used for the fermentation phase with gin, vodka and the light aroma varieties of rum produced from fresh cane juice. Most rum, however, is of heavy aroma varieties, for which the raw material is molasses, the uncrystallizable but very concentrated mother-liquor left from sugar refining. This is treated to remove unwanted chemicals from the refining process, diluted and fermented with the fission yeast *Schizosaccharomyces*. Another unusual feature of the rum fermentation is that anaerobic bacteria of the genus *Clostridium* have an important role in flavour formation through producing traces of butyric and other fatty acids.

Distillation can be carried out with pot stills (Fig. 8.10). An aqueous solution of alcohol is heated in a pot and the resulting vapour conveyed to a cooling coil for condensation. As alcohol has a lower boiling point than water, the vapour and the condensate will have a higher alcohol concentration than the original solution. Distillation of a 'wine' or 'beer' with an ethanol concentration of 8–10% can give a condensate with an ethanol concentration of 25–30%. Higher ethanol concentrations can only be obtained by distilling the condensate. When this is done, the earliest fractions to condense are discarded, since they are rich in unpleasant or toxic compounds such as methanol. The ethanol concentration in

A **B**

Figure 8.10 Distillation with pot stills. A, Pot still (at right), condenser and collecting vessel for grappa production in Italy. Grappa is made by distilling the product of the fermentation of the residue left after grapes have been pressed for wine production. B, Pot still at a whisky distillery, with a safety valve and access port, also seen in close-up on still in right foreground. (A,B, Bernard Lamb.)

the fractions that are collected is about 70%. Distillation is terminated before fractions containing too much water and excessive amounts of low-volatility higher alcohols are produced. Pot stills are used for making brandy, malt whisky and some other fine whiskies. Pot stills are, however, inefficient in terms of energy consumption and productivity, and continuously operating stills ('patent stills') are widely used in producing other spirits. They are varied in design but similar in concept. A stream of 'beer' descends a column, its descent being slowed by a series of perforated plates. As it descends, it encounters an upwardly flowing current of steam, which strips the beer of its volatile components. The steam and volatiles then pass through condensing equipment, subdivided into successively cooler compartments, in which the various vapour fractions condense. These fractions can be collected separately from the ethanol fraction, which can have a concentration as high as 96%.

Sometimes the aim of distillation is to produce a 'neutral spirit', free from any flavour, aroma or colour originating from the raw materials. Vodka is one such neutral spirit, in which the distillate is purified by treatment with charcoal, and is essentially pure aqueous alcohol. Gin is a neutral spirit that has been flavoured before or during distillation by interaction with juniper berries and various herbs. In contrast to neutral spirits are those in which flavour and aroma originate from materials present before fermentation, but only become satisfactory after the distillate has undergone a long period of maturation. Brandy is matured for many years in wooden casks, with interaction between wood and spirit being responsible for colour and for modifying flavour. Whisky undergoes similar maturation, although here the colour is largely due to the addition of caramel. Apart from 'pure malts', Scotch whisky is the product of skilled blending of malt whisky and grain whisky, the latter produced from a mash containing grain additional to malted barley. Blending, sometimes with neutral spirits, is also practised in the production of other spirits. Marketed spirits have an ethanol concentration of about 40% (v/v), whereas the immediate product of distillation contains 70–96% ethanol. Dilution with distilled, demineralized or soft water is hence necessary. With brandy and some malt whiskies, however, some of the requisite fall in alcohol concentration occurs by evaporation during years of maturation in wooden casks.

Fortified wines

The two best known fortified wines are sherry and port. In the making of sherry, grape juice is fermented and then neutral grape spirit added to bring the ethanol concentration to 15% (v/v), and with port a brandy is used to arrest fermentation, leaving some sugar unfermented, and to adjust ethanol to 17%.

In making Fino and Amontillado sherries, maturation is conducted in casks about 80% full so as to provide contact with a large volume of air. At the air–liquid interface a thick film of yeast, 'flor', develops and has an important role in maturation. One of the changes brought about is the oxidization of a little of the ethanol to acetaldehyde, conferring a distinctive flavour on these sherries. A further adjustment of alcohol concentration follows, to 15% (Fino and Amontillado), 18% (Oloroso sherry) or 20% (port). With all these fortified wines there is then a prolonged period of further maturation, in wooden casks with sherry, and in the bottle with vintage port.

The Cultivation of Fungi for Food

Fungi may be used to modify food to make it more nutritious or palatable (page 500). They may also be used to produce materials of nutritional value such as vitamins, amino acids and lipids (page 510). In addition the macrofungi may be cultivated to yield fruit bodies (mushrooms), and yeast cells and mould mycelium grown in fermenters for food (single cell protein). These direct uses of fungi as food are dealt with in the present section.

Mushrooms and other Macrofungi

The common cultivated mushroom (Agaricus bisporus)

In nature *Agaricus bisporus* occurs on manure heaps, on garden waste and at roadsides. In seventeenth-century France it was found that it would develop on suitably prepared beds of horse manure in gardens. Later it was extensively cultivated in caves and old mine workings, where a constant environment meant that successful cropping was no longer dependent on seasons and weather. Now it is produced in purpose-built mushroom houses, with an annual world output of about two million tonnes a year (Table 8.2).

There are essentially two phases in mushroom production (Fig. 8.11A, B) each of which is a solid substrate fermentation. First there is composting, the preparation of a suitable growth medium, which takes 10–14 days. Then there is the actual growth of *Agaricus* and harvesting of mushrooms, which occupies 9–11 weeks. A mushroom farm, however, is usually organized so that the production of a crop of mushrooms is started every week.

The materials used for preparing a compost vary, but there will usually be a cheap carbon source and a cheap nitrogen source, for example wheat straw and horse manure. Since several tons of material will be processed daily, bulk-handling machinery is needed. The materials are mixed, made into stacks and watered so as to provide damp but not waterlogged conditions. A dense flora of Actinomycetes and other bacteria, up to 10^{10} propagules per gram, and fungi, up to 10^6 propagules per gram, develops. Locally temperatures may reach 70°C, but excessive heating and anaerobic conditions are avoided by frequent turning. The composting process is completed in special chambers at 60°C. This favours thermophilic species, and effects a pasteurization, killing mesophilic organisms that might later infect or compete with *Agaricus*. It also drives off ammonia, toxic for *Agaricus*. Composting results in the disappearance of low-molecular-weight nutrients that can be utilized by many microorganisms. High-molecular-weight materials such as cellulose and lignin, which can be utilized by *Agaricus*, remain, and are indeed supplemented by microbial cell wall material. A medium selective for *Agaricus* is hence produced, which after cooling, can be inoculated. In the UK surplus wheat straw is utilized in mushroom production. The production of macrofungi for food is so far the only really profitable way of utilizing lignocellulose waste.

Table 8.2 Annual production of cultivated macrofungi

Fungus	Class	Quantity (thousands of tonnes)
Agaricus bisporus	Agaricales	1956
Lentinus edodes	Poriales	1564
Pleurotus spp.	Poriales	876
Auricularia spp.	Auriculariales	485
Flammulina velutipes	Agaricales	285
Volvariella volvacea	Agaricales	181
Tremella spp.	Tremellales	130
Hypsizygus marmoreus	Agaricales	74
Pholiota nameko	Agaricales	56
Grifola frondosa	Poriales	33
Others		521
Total		6161

Information provided by Professor S.T. Chang, 1999.

Strains of *Agaricus bisporus* with commercially attractive attributes are maintained by specialist companies. These supply the grower with packages of mushroom spawn, mycelium in pure culture on a cereal grain substrate. The use of pure cultures for inoculation avoids the transmission of pathogens of the fungus, such as the watery stipe virus, bacteria causing fruit body blotch such as *Pseudomonas* species, and fungi that may infect and damage, such as *Verticillium* species. Subsequent handling, however, must still be hygienic, to avoid later infection, or infestation with competitive weed fungi, such as *Chaetomium* and *Trichoderma*. Water supplies are usually chlorinated, and fungicides for which *Agaricus* has a high resistance may have to be used.

Mushroom spawn, at about 0.5% by weight, is thoroughly mixed into the compost, which is then maintained at 25°C, the optimum for mycelial growth. The compost is fully colonized in 10–14 days. It is then put into trays and casing, covering with a 3–5 cm layer of soil or peat and chalk, is carried out. The trays are then placed in a cropping room at 16–18°C, a temperature that allows fruit body initiation and development. The casing is then colonized and the first flush of fruit bodies occurs at about 3 weeks. It is thought that microbial activity in the casing layer removes volatile inhibitors of fruit body initiation, as its effect can be simulated with activated charcoal, which is very effective in adsorbing such compounds. Further flushes of fruit bodies occur at 7–10 day intervals for about a further 5 weeks. The cropping room is then treated with steam at 60°C to kill pests and pathogens, the rooms cleared and cleaned, and the spent compost disposed of, often to horticulturists and gardeners.

Shiitake (Lentinus edodes)

The large-scale commercial growth of shiitake mushroom (pronunciation shii-ta-ke) was pioneered in Japan, and it is now widely grown in very large quantities (Table 8.2). In nature it grows on the dead wood of oaks, chestnuts and related trees. A common way of cultivation in Japan (Fig. 8.11C,D), is in young oak trunks, diameter 5–15 cm, which are felled in late autumn, when the sugar content of the wood is at its highest. The wood is cut into 1 m lengths and stored for 1–2 months. Fifteen to twenty holes are then drilled into the logs to receive an inoculum of *L. edodes* from pure culture. This is in the form either of wood wedges or a sawdust spawn. The holes are then sealed with hot wax to prevent evaporation or infection by competitors. The logs are then laid on racks on a hillside in coniferous woodland, which provides a good environment for the development of *L. edodes* mycelium. Mycelial development in the logs takes one or more years, depending on the wood and the inoculum used. The logs are then transferred to a cool, humid, well-sheltered glade, and placed upright against bamboo fences. This transfer to an environment suitable for fruiting is done in late autumn. A good crop of high quality mushrooms is obtained in the spring, and a poorer one follows in the autumn. The mushrooms may be sold fresh, canned or dried. Shiitake is also produced in greenhouses with a greater environmental control of both growth and fruiting phases. Sawdust and rice bran mixtures are also used for cultivation. This gives faster production but with lower yields and apparently poorer quality fruit bodies. Compared with *Agaricus*, shiitake has a stronger taste and less is needed to flavour a meal.

Figure 8.11 Mushroom cultivation. A, Preparation of compost for growth of *Agaricus bisporus*. B, Young fruit bodies. C, Fruiting of Shiitake, *Lentinus edodes*, on logs. D, The arrangement of the logs in a woodland glade. (A,B, David Wood, reproduced by permission of Horticulture Research International. C,D, Alastair Campbell.)

Other macrofungi in commercial cultivation

Agaricus bisporus and *Lentinus edodes* are not the only Basidiomycetes grown on a large scale (Table 8.2). *Volvaria volvacea* (syn. *Volvariella diplasia*), the padi (rice) straw fungus, has been cultivated for about 2000 years in China and Southeast Asia. The substrate is a compost of rice straw, cotton waste or both, and production procedures range from traditional out-of-door methods to culture in mushroom houses in a controlled environment. *Flammulina velutipes* (syn. *Collybia velutipes*), the winter mushroom, can be grown in plastic bottles with a sterilized sawdust and rice bran substrate. *Pleurotus ostreatus*, the oyster mushroom, is grown mainly in Central Europe on a variety of straw-based substrates. *Pholiota nameko*, the nameko or viscid mushroom, is grown on sterile sawdust and rice bran substrates. *Auricularia* and *Tremella* are also cultivated on a large scale in China and Taiwan. Production methods are essentially similar to those employed for shiitake. Increasingly, other species are also being produced, to cater for a growing worldwide demand, for example for vegetarian dishes with 'exotic mushrooms'.

Prospects for the commercial production of other macrofungi

The fruit bodies of a wide range of macrofungi are collected and eaten. A few species have yielded fruit bodies in cultivation and, as indicated above, only about a dozen have been grown commercially on a large scale. All the latter are saprotrophs (although *Flammulina* can attack living wood) and most are wood-decomposers. It is likely that further research will render profitable the production of various related species, at present grown on a small scale. There are, however, other macrofungi with fruit bodies that are a highly prized and expensive commodity. Among Basidiomycetes there are the chanterelle, *Cantharellus cibarius*, and some species of *Boletus*. There are also a few edible Ascomycetes, especially the highly esteemed morel, *Morchella esculenta*, and truffle, *Tuber melanosporum*. It is likely that such fungi have symbiotic relationships with trees; with *Boletus* a mycorrhizal relationship is well established, and with the truffle there is a close association with the oak. It is possible that current research on the nutrition and fruit body development of such species may lead to cultivation, with or without their associated trees. Meanwhile progress is being made in managing oak plantations in such a way as to enhance truffle production.

Edible Biomass from Yeasts and Moulds – Single Cell Protein

Much of the world's population is poorly nourished, and famines are frequent. The dietary component that is most usually in short supply is protein. Plants have a relatively low protein content, being largely carbohydrate. Plant protein also tends to be low in some essential amino acids, such as methionine and tryptophan. Animals convert plant biomass into high-grade protein, but with poor efficiency, so meat is too expensive for much of the world's population. Finally, both plants and animals alike are affected by climate, and unseasonable weather, floods, droughts and pests can drastically reduce yields.

Many microorganisms are able to use cheap sources of nitrogen, such as ammonium salts and nitrates, and abundant carbon sources such as starch, natural gas and petroleum hydrocarbons. The resulting biomass has a high protein content. Microbial growth rates are high, and a compact fermentation plant can produce as much protein as a large area of agricultural land in any climate. These considerations have led to massive industrial research and development aimed at the production of microbial biomass for human consumption, either directly or indirectly through its use as animal feed. Since the main objective is supplying protein, the biomass is usually termed 'single cell protein' or 'SCP', although other cell components will be present, even after processing. Bacteria, fungi and algae have all been used in SCP production, but here consideration will be restricted to fungi, both yeasts and moulds.

Mycoprotein

In the UK the mitosporic fungus *Fusarium graminearum* is used for SCP manufacture for human consumption. Originally marketed as a 'novel' protein-rich food, it is now sold chiefly as a healthy meat analogue, for example as sausages and burgers, with the trade name Quorn® mycoprotein. The mycelium is processed to simulate the texture of meat, and contributes to the dietary requirement for fibre (Fig. 8.12). Worries in the UK about eating meat, exacerbated by the BSE problem, have led to increased demand. Quorn® has a high protein content with a satisfactory amino acid composition. The lipid content is lower than in meat, and the animal sterol cholesterol is absent. From the original idea, it took 20 years to develop the product, which was first sold to the public in 1985. Retail sales in UK for 1998 were £74 million, with a production of 562 tonnes. It is manufactured by continuous flow fermentation, in air lift fermenters (Fig. 8.2). This system was designed to give a high production

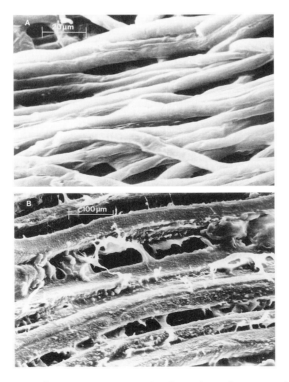

Figure 8.12 Single cell protein (SCP). A, Scanning electron micrograph of the mycoprotein Quorn® prepared from *Fusarium graminearum*. B, A similar image of beefsteak. The fungal hyphae in mycoprotein provide a similar texture to the muscle fibres of meat. (A,B, R. Angold, Lord Rank Research Centre, High Wycombe.)

rate. The 155 m³ fermenters are inoculated with 5 litres of a culture containing about 50 g biomass and it takes about four days before this culture can be switched to continuous mode. The nitrogen source, ammonia, is fed into the fermenter with the sterile compressed air, and the glucose, biotin and mineral salts nutrient medium is fed to the culture to give a dilution rate of about 0.2 h⁻¹, well below the organism's μ_{max} of 0.28 h⁻¹. The culture is maintained at about 30°C by a heat exchanger, and the raw material harvested continuously at the rate of 30 tonnes h⁻¹. An essential part of the processing for human consumption is the lowering of the RNA content, which at about 9% of the *F. graminearum* biomass is far above the WHO recommended limit of 2% (w/w). High levels of nucleic acids in the diet would lead to a build-up of uric acid in the blood, in turn leading to gout and other medical problems. Therefore the mycelium is given a heat shock of 68°C for about 20 minutes, allowing endogenous enzymes to reduce the RNA content to about 1% (w/w) with minimum loss of protein and fibrous structure. The mycelium is harvested by centrifugation to give a product containing about 30% (w/w) total solids. A mechanical process aligns the filaments of the mycoprotein, to give the required 'chewy' fibrous texture, and a coating of egg white is added to stabilize. Flavourings and colourings may be added to give different pre-prepared meals. The mycoprotein is exceptionally adsorptive, and retains flavourings and colourings very well during cooking. Fermentations are terminated 6 weeks or less after the onset of continuous culture. This is due to the appearance of highly branched mutants in the culture, which are not amenable to processing to give the desired texture. The production of Quorn® mycoprotein is a major success story for fungal biotechnology.

Single cell protein from organic wastes
The industries based on agriculture and forestry produce large amounts of organic waste. Many studies have been carried out on the conversion of such waste to SCP, and to a lesser extent pilot plants and actual processes operated. Where the wastes are rich in sucrose or glucose the 'food yeast' *Candida utilis* can be grown. This is done on a large scale with molasses in Cuba and on a small scale with confectionery waste in the UK. Whey (page 504), a waste from cheese manufacture, is used in France and the USA for SCP production by *Kluyveromyces fragilis*, a yeast that can utilize lactose. Wastes rich in starch have been used for SCP manufacture in Sweden with *C. utilis*. The necessary preliminary hydrolysis of starch to sugar is carried out by *Endomycopsis fibuliger*, one of the few yeasts able to do this. There has been limited production of SCP from cellulose wastes with the Ascomycete *Chaetomium*. Cellulose, however, is rather slowly utilized, and most cellulose sources have a substantial lignin content. This is not only very difficult to degrade, but if breakdown does occur, toxic phenolic compounds may be formed. Hence lignocellulose, in spite of its abundance, is a less attractive substrate than sugar or starch for SCP production.

Single cell protein from petroleum hydrocarbons and related substrates
Crude oil contains *n*-paraffins, unbranched hydrocarbon molecules, which can be utilized by some yeasts. From about 1955 to 1970 processes were developed using

these yeasts for the production of SCP with *n*-paraffins as the carbon source. One raw material was gas-oil, a crude petroleum fraction with a boiling temperature 300–380°C and 10–25% *n*-paraffins, mainly in the C_{14}–C_{16} range (i.e. 14–16 carbon atoms per molecule). Fermentations were developed, using *Candida tropicalis*, in which these *n*-paraffins were utilized and the residual gas-oil further processed. Other fermentations were developed using *Saccharomycopsis (Candida) lipolytica* and C_{11}–C_{13} *n*-paraffins separated from crude petroleum. During the early 1970s a major oil company operated a plant capable of producing 16 000 tons of SCP per year from gas-oil and another utilizing purified *n*-paraffins and having a production capacity of 4000 tons of SCP per year.

Natural gas consists mainly of methane which can be utilized by some aerobic microorganisms. Mixtures of methane and air are, however, explosive and dangerous to handle on a large scale. A catalytic process was therefore developed for the conversion of methane into the much safer methanol, which can be utilized by some bacteria and yeasts. A process was developed for the bacterial production of SCP from methanol, and a giant fermenter capable of producing 55 000 tons of SCP per year was operated for a few years. SCP has also been produced from methanol on a pilot plant scale using yeasts in the genus *Pichia*.

Current problems and future prospects for SCP

Attempts to introduce SCP have encountered many problems. Those that are purely technical can be overcome, and SCP produced from a wide variety of raw materials and on a large scale. The other problems are those of safety, acceptability and profitability, and are more intractable.

Crude petroleum contains carcinogenic materials, the possible persistence of which was a cause for concern with the production of SCP from *n*-paraffins and gas-oil. Any microorganism with a high rate of growth and protein synthesis contains abundant ribosomes and hence a high RNA content, so the biomass requires treatment to reduce nucleotide levels. Concern has also been caused by the ability of many moulds, including some species of *Fusarium*, to produce mycotoxins. Hence any new form of SCP intended for human consumption has to undergo extensive – and costly – animal feeding trials and toxicological tests. Requirements for SCP intended for animal feed are less stringent, but even so it must be established that traces of toxic materials do not accumulate and hence occur in meat. The safety issue can, however, be exploited by vested interests to frustrate the introduction of SCP. This occurred in the 1970s in Italy, where a hostile press campaign and political pressures led to the abandonment of SCP production from gas-oil and *n*-paraffins. This was after two companies had constructed plants capable of producing 100 000 tons per year of SCP. Losses to the companies were of the order of £100 000 000.

When a new food is intended for human consumption, it must not only be safe, but people must wish to consume it. Conservatism with respect to food (which must often have a high survival value) is powerful in the Third World, where there has been a reluctance to consume unfamiliar food even in time of famine. In developed countries advertising can succeed in promoting a new food, but the media can be used skilfully by the opponents as well as the protagonists of a novel product. A new food will only succeed if it is clear to people that it is cheaper, or more attractive, or healthier than an existing product.

Finally, a process can only succeed in a market economy if it is profitable. The cost of raw materials, the capital investment involved, running costs and the value of the final product are all of importance in determining profitability. The idea of producing SCP from petroleum or natural gas arose at a time when petroleum prices were low and it is now rarely economically attractive. SCP has, however, been produced from such materials on a very large scale in Eastern Europe, where SCP production was seen as a way of remedying a protein shortage arising from agricultural inefficiency, without using scarce foreign currency. The most readily utilizable agricultural wastes, such as molasses, increasingly have profitable alternative uses, such as feedstocks for the fermentation industry and direct incorporation into animal feed, and even the less attractive wastes can often be composted (e.g. for mushroom production) or used to make biogas. They also tend to be seasonal, and plant must be run continuously to be profitable. SCP intended for animal feed has to compete with soya bean products and fishmeal, which are usually cheaper. Fermentations to produce SCP for animal feed are only likely to be viable if they use a feedstock of high negative value – one for which disposal costs would otherwise be high. SCP for human consumption, such as mycoprotein (page 497) can be sold for high prices, and may prove to be more profitable, even though development costs are high.

Food Processing by Fungi

The previous section dealt with the production of fungal biomass, in the form of mushrooms or single cell protein, for food. Fungi also have a crucial role in the processing of many foods, improving the texture, digestibility, nutritional value, flavour or appearance of the raw materials used. The present section will consider the production of foods in which fungi bring about such improvements, without contributing substantial biomass to the final product.

Bread

The first loaves, as distinct from flat, unleavened bread, were probably made about 6000 years ago. By that time humans were making extensive use of cereal seeds as food. One procedure was to grind the seeds, mix with water to form a dough, and bake on hot ashes or heated stones to give a flat bread. Under such conditions dough could well become contaminated by yeasts, and if left for some time before baking could well 'rise' due to carbon dioxide formation by the yeast. Such dough would give a lighter and more attractive product on baking – a loaf instead of flat unleavened bread. In due course it was probably found that such a result could be obtained more consistently by adding to newly made dough a leaven, a little material from an earlier batch that had risen successfully. When the Israelites left Egypt they had no time to leaven their dough in this way (Exodus

12), so Passover is celebrated with unleavened bread. By Roman times it had been found that dough could be made to rise with material skimmed from the top of a beer fermentation. By the early nineteenth century thousands of tons of brewer's yeast obtained in this way were being used annually in bread making, while the use of a leaven from previous batches of dough still persisted, especially for the production of rye bread. However, during the nineteenth century lager beer became popular over much of Europe, and top yeasts were superseded by bottom yeasts for brewing. These proved less satisfactory for making bread than did top yeast, leading to the production of yeast specifically for baking.

The production of baker's yeast
The yeasts used for bread making are now almost always pure strains of *Saccharomyces cerevisiae*. This yeast can yield about 50 g of cells per 100 g of sugar consumed when respiring, but only about 5 g of cells when fermenting. The production of baker's yeast is hence carried out in well-aerated fermenters. High sugar concentrations bring about fermentation even under aerobic conditions, so fed batch culture is used, with the sugar concentration in the fermenter being retained at about 0.1%. This is most effectively achieved by delivering the carbon supply, usually in the form of molasses, at a rate that increases exponentially to keep pace with growth. At the end of the fermentation the cells are recovered by vacuum filtration, with the amount of dewatering depending on the form in which the yeast is required. Delivery to very large bakeries may be by refrigerated tankers carrying yeast cream, with yeast cells accounting for about 20% by weight. For smaller bakeries the yeast is compressed, extruded as a strip of rectangular cross-section with about 30% solids, and cut into yeast cakes, which can be kept at about 5°C for about a week. Where the need for yeast is occasional, active dry yeast is useful. This is sold as granules with about 95% solids in vacuum or inert gas packs, and can be kept for about a year.

Yeast and bread making
The principal grain used in making bread is wheat. Second in importance is rye, the use of which will be discussed in the next section. Wheat is milled and sieved to remove the coarser fractions, giving wheat flour. The flour is thoroughly mixed with water, yeast and salt to give dough. Other materials may be included such as milk, sugar, fat, eggs, malt, dried fruit and spices to give various types of bread and other baked goods. Wheat flour contains 1–2% sugar, mainly oligosaccharides of glucose, fructose or both. Proteins are also present, gluten, a characteristic wheat protein, being of special importance. In making wheaten bread the amount of yeast included in the dough is usually about 2%, but with other baked goods lower or higher amounts may be used, in the range 1–6%. The salt concentration for bread making is usually 1.5–2%. This concentration partially inhibits some yeast enzymes, and gives an optimal dough for bread. Lower concentrations give a more moist and runny dough suitable for some other baked goods.

In preparing dough, the mixing process (kneading) is prolonged and thorough. This not only ensures uniform distribution of constituents, but increases the elasticity and extensibility of gluten, to give a dough that will rise well. After kneading, the dough is left for a few hours to mature. The yeast ferments much of the sugar present, producing the carbon dioxide which inflates the dough, and ethanol which is driven off during baking. Traces of many other, less volatile, compounds are produced, and these contribute to the flavour and aroma of the bread. Yeast also affects gluten structure, increasing extensibility by oxidizing disulphide bridges between protein chains. Finally baking kills the yeast and makes its own contribution to taste and texture.

Sour doughs and rye bread

Bread can be made from pure rye flour, although most rye bread is produced from a mixture of wheat and rye flours. Rye flour differs considerably in its properties from that of wheat, and a dough with 20% or more of rye flour will not rise satisfactorily unless it is soured. This can be done with edible organic acids, but a more satisfactory dough results from microbiological souring. Such souring may be spontaneous, but is more usually brought about by the addition of material from a previous batch of sour dough or with a commercial sour dough starter culture. Such a culture, or a successful sour dough, contains a wide variety of lactic acid bacteria, which cause souring, and yeasts, which are mainly responsible for leavening. Several yeast species are usually present, including *Saccharomyces exiguus*, *S. cerevisiae*, *Candida krusei* and *Pichia saitoi*, and are of strains more acid and heat tolerant than baker's yeast. The traditional wheaten bread leavens, prior to their replacement by baker's yeast, would have contained lactic acid bacteria, and sour doughs are still used for the production of a few wheaten breads, for example San Francisco Sour Dough French Bread. They remain essential, however, for the production of rye bread. Sour dough bread keeps better than that made with baker's yeast, and pure rye breads are useful in the diet of individuals who cannot tolerate the wheat protein gluten.

Soya Bean Products

The soya bean plant, cultivated in China for thousands of years, and now on a large scale in other countries, has many satisfactory features. The beans have a high protein content and also yield an oil useful for cooking and many other purposes. The beans also have some less attractive features. They contain a variety of unpleasant compounds, some toxic, some antagonizing nutrient utilization, and some causing excessive gas production in the gut. Cooking destroys some of these factors, but the beans do not soften well and remain difficult to digest and generally unappealing. A wide range of fermentations has, however, developed in China and in neighbouring countries yielding harmless, nutritious and palatable products from soya beans. Although these fermentations are ancient, some are now carried out on a large scale in modern plant, especially in Japan. Examples of such fermentations are discussed below.

Soy sauce

Soy sauce is a condiment extensively used in China, Japan and Indonesia to render a monotonous diet more appetizing. There are many varieties of soy sauce, but the procedures used in their production are basically similar. Soya beans are soaked for 16 hours and, after the seed coats have been removed, cooked. An aerobic, solid substrate fermentation follows in which an appropriate strain of *Aspergillus* grows on and within the substrate, the mixture of substrate and fungus being known as *koji*. Inoculation is with spores of either *A. oryzae*, *A. tamarii* or *A. sojae*, all members of the *Aspergillus flavus-oryzae* series, or with *koji* from a previous successful fermentation. The *Aspergillus* produces a wide range of hydrolytic enzymes, breaking down proteins, polysaccharides and other components of the substrate. Traditional *koji* fermentations could continue for as long as 3 months, but with an appropriate controlled environment and properly timed stirring this is reduced to 3 days in modern plant. The *koji* stage is terminated when adequate mycelial growth and enzymic activity has occurred but before sporulation gives mouldy flavours. The material is then mixed with brine having a salt concentration of about 20%, and the *moromi* or salt mash stage follows. This is an anaerobic fermentation in deep unstirred vats, the organisms involved being salt-tolerant yeasts and lactic acid bacteria. Pure cultures of *Zygosaccharomyces rouxii* and the lactic acid bacterium *Pediococcus* can be used to initiate the *moromi* stage. The lactic acid bacteria produce lactic acid, lowering the pH and limiting the range of microorganisms that can grow. The yeasts, which favour acidic conditions as well as tolerating salt, produce up to 2% ethanol and a wide variety of compounds influencing flavour and aroma. The *moromi* stage lasts at least 3 months and in some processes up to 3 years. At the end of the fermentation the liquid is drained from the remaining solids, clarified, pasteurized and bottled.

Miso

Miso is a fermented soya bean paste produced in Japan and, under a variety of names, in other countries in the Far East. It has many culinary uses including the preparation of soups and sauces. Usually the main raw materials are soya beans and rice. The production process has much in common with that for soy sauce, with *koji* and *moromi* stages. The rice is soaked, steamed and inoculated with *Aspergillus oryzae*, and a solid substrate fermentation conducted for two days. It is then mixed with soaked, steamed soya beans, the components of which, like those of the grain, are acted upon by the *koji* enzymes. Salt is then added and the *moromi* stage is initiated either with pure cultures of *Zygosaccharomyces rouxii* or with a starter of fully fermented *miso*. Liquids are drained from the *miso* as fermentation proceeds, and these can be used as a sauce. When a high proportion of rice is used, the *moromi* stage lasts about 2 weeks, and the final product is light coloured, sweet and not very salty. When a high proportion of soya beans is used, the *moromi* state lasts longer, and the product is dark, meat-flavoured and salty. Miso is consumed in sufficient amounts to be nutritionally important. It is a good source of vitamins and proteins, the latter being in a readily assimilated form and high in some amino acids, such as lysine, which are often deficient in predominantly vegetarian diets.

Tempe kedele

Tempe (or *tempeh*) is an Indonesian fermented food that can be cooked in a variety of ways. It is a cake covered in white mould and when prepared from soya beans is known as *tempe kedele*. The initial treatment of soya beans involves soaking, removing the seed coat and cooking. After draining, the material is mixed with *tempe* from an earlier batch or other appropriate inocula, pressed into cakes and incubated for 1 to 2 days. Members of the Mucorales, mainly *Rhizopus*, grow through and on the surface of the cakes. Proteases are produced that render the soya bean protein more palatable, and lipases that release fatty acids from the soya bean oil. On an industrial scale a shorter soaking in 1% lactic acid may replace the bacterial fermentation, and subsequent inoculation is with pure cultures of *Rhizopus oligosporus*.

Sufu

The milking of animals and the consumption of milk are practices that until recently were alien to the Chinese culture area. Instead a vegetable milk is prepared by soaking, grinding and straining soya beans. Heating is carried out either before or after straining. This improves flavour and destroys trypsin inhibitors which could adversely affect some uses of the milk. A soya bean curd, *tofu*, which in China takes the place of cheese, can be prepared from the milk by curdling with calcium or magnesium sulphate. *Tofu*, which is widely used in cooking or eaten flavoured with soya sauce or *miso*, is bland but can by fermentation yield *sufu*, a more highly flavoured food. *Tofu* is cut into cubes, soaked for 1 h in 6% salt and 2.5% citric acid, which prevents subsequent bacterial growth, and pasteurized. The cubes are then surface inoculated and incubated for 2–7 days. The fungi responsible for the fermentation are members of the Mucorales, such as *Mucor racemosus, Rhizopus chinensis* and *Actinomucor elegans*. Proteases from the fungi release peptides and amino acids from the soya bean protein, contributing to flavour. After the fermentation the *sufu* can be matured in a variety of ways, such as ageing in brine and rice wine. It is used as a condiment and in cooking.

Cheese and Fermented Milk

The preparation of cheese and fermented milks such as yoghurt is a fermentation industry second only to brewing in scale and in the value of its products. The main microorganisms involved are lactic acid bacteria which ferment the lactose in milk to lactic acid and produce a range of other metabolites responsible for flavour. Fungi have, however, an important but subsidiary role in the production of some cheeses and fermented milks.

In cheese production, milk is inoculated with starter cultures of lactic acid bacteria. The enzyme rennin is then added to coagulate milk proteins and form a curd. Here there can be indirect fungal involvement, as there is an increasing use of rennin from fungi (page 477) instead of calf rennet. The way the curd is then handled depends on the type of cheese being made, but with all cheeses whey has to be drained from the curd. Various microorganisms in addition to lactic acid bacteria have a role in subsequent cheese ripening, and instances where fungi make a major contribution will be considered below, as will fermented milks in which yeasts as well as lactic acid bacteria are important.

Surface-ripened cheese

The best-known surface-ripened cheeses are Brie and Camembert. The curd is drained and shaped to give the final disc or segmented disc form. The surface is then dry-salted and sprayed with spores of the white *Penicillium* species *P. camemberti*. Salt- and acid-tolerant yeasts and the mitosporic fungus *Geotrichum candidum* develop at the surface although deliberate inoculation is not practised. It is thought that utilization of lactic acid by the yeasts may raise the surface pH and encourage the germination of *Penicillium* spores and subsequent mycelial growth. The white mycelium gives the surface of Camembert and Brie its characteristic appearance but does not penetrate the interior. Proteolytic enzymes from the fungus, however, diffuse into the curd and break down proteins into peptides and amino acids, causing softening and finally runniness. Deamination of amino acids by *Geotrichum* produces traces of ammonia and contributes to flavour. The *Penicillium* makes lipases which release fatty acids, which can be converted into methyl ketones, both of which contribute to flavour, as do a range of other fungal products.

Blue-vein cheese

The best-known blue-vein cheeses are Roquefort, Gorgonzola, Stilton and Danish Blue. The starter bacteria include a heterolactic fermenter which produces carbon dioxide as well as lactic acid. The gas production results in irregular cavities in the cheese. An inoculum of *Penicillium roqueforti* spores is either included in the starter or added to the fresh curd. After the curd has been compressed into a cheese, salt is dusted on to the surface and diffuses into the interior, establishing a concentration gradient. The cheese is then spiked, and the air that enters allows the germination of *Penicillium* spores, and mycelial growth along the perforations and through the cavities in the cheese. Growth is greatest at the intermediate salt concentrations at moderate depth in the cheese, and it is there that the blue coloration resulting from the production of *P. roquefortii* spores is most marked. The fungus produces both proteolytic and lipolytic enzymes, resulting in the release of amino acids and peptides from protein and fatty acids from lipid. The fatty acids to some extent undergo β-oxidation which is a normal part of fatty acid catabolism. This, by removing acetyl units, results in fatty acids of shorter chain length. These fatty acids can be oxidized to the corresponding β-keto acid and decarboxylated to a methyl ketone. Heptan-2-one is the major methyl ketone in blue-vein cheese, in contrast to the C_9 and C_{11} methyl ketones in the surface-ripened cheeses such as Camembert. Fatty acids, methyl ketones and lactones are all important in determining the flavour of blue-vein cheese.

Fermented milks

In the preparation of cheese, curdling is brought about by the addition of rennin as well as the activity of lactic acid bacteria, and whey is drained from a semi-solid curd. The fermented milks, differ in that rennin is not employed, and separation of curd from fluid does not occur. In **yoghurt** lactic acid bacteria alone are responsible for curdling. There are, however, many other fermented milks and in some, moulds and yeasts as well as bacteria have a role. The most esteemed varieties of a Finnish fermented milk called *villi* have a surface growth of *Geotrichum candidum* which lowers acidity by metabolizing some of the lactic acid. Yeasts as well as lactic acid bacteria are involved in the preparation of some other fermented milks. Russian **koumiss** contains 1–2% lactic acid produced by lactic acid bacteria, but lactose-fermenting yeasts produce carbon dioxide, which imparts a slight effervescence, and ethanol to a concentration of 1–3%. **Kefir**, a Russian fermented milk, is produced with *kefir* grains, which vary in length from about 5 to 20 mm. These grains are intimate symbiotic associations of lactic acid bacteria and yeasts, held together by the unique polysaccharide kefiran. This mixed culture yields a product similar to koumiss. Meanwhile the grains grow and divide, and at the end of the fermentation can be separated, washed, dried and stored for up to a year. *Laban* (or *leben*) in the Middle East and *dahi* in India are other fermented milks in which yeasts have a role.

Other Fermented Foods

A very wide range of fermentations have been developed, especially in Asia and Africa, that render rather unappealing foods more nutritious, digestible or attractive. They may involve moulds, yeasts or bacteria, or any combinations of these organisms. A few examples of such fermented foods, illustrating the diversity of the organisms concerned and the purposes accomplished, are now considered.

Ontjom (sometimes spelt *oncom*) is made in Indonesia from the cakes left after groundnuts (peanuts) have been pressed to extract oil. The press cake is soaked, drained, steamed and inoculated with *Neurospora intermedia* or *Rhizopus oligosporus*. The fungi produce hydrolytic enzymes that render the protein more soluble, break down oligosaccharides that can cause flatulence, and release fatty acids from the remaining lipids to produce flavour. Sporulation at the surface of the cakes gives them a pink or grey colour, respectively.

Indonesian *tapé ketan* fermentations use carbohydrate-rich glutinous rice, which is cooked and inoculated with *ragi*, a preparation equivalent to the leaven of the West and the *koji* of Japan. That used for preparing *tapé ketan* contains various fungi, among which *Amylomyces rouxii*, a member of the Mucorales, and *Endomycopsis burtonii*, a yeast able to produce amylases, are important. Two to three days' fermentation converts the rice into a sweet–sour, mildly alcoholic paste with an enhanced protein and vitamin content.

Cassava (manihot, tapioca) is a root crop rich in starch. Many varieties contain cyanogenic glycosides, which probably protect the plants from insect pests, but also render the roots poisonous, unless suitably treated. About 70% of the cassava produced in Nigeria is converted into **gari**. The roots are peeled and grated and packed into jute sacks. Some of the juice is then squeezed out by putting weights on the sacks. The sacks are then left for 3–4 days during which a fermentation occurs, in which *Corynebacterium manihot* and the mitosporic fungus *Geotrichum candidum* appear to be important. The bacterium hydrolyses a little of the starch, producing organic acids and lowering the pH. This causes hydrolysis of the cyanogenic glycosides with the production of hydrogen cyanide which evaporates. The lower pH also encourages the growth of the *Geotrichum*, which produces aldehydes and esters responsible for the flavour of *gari*.

Ang-kak (Chinese red rice) is used to produce red rice wine and to tint a range of foods. Rice is soaked, drained and cooked, and inoculated with material containing the Ascomycete *Monascus purpureus*, which produces about six chemically related red and yellow pigments. A non-glutinous rice variety is used, since it is important that the rice grains do not stick together – the fungus produces pigments best with good aeration. The pigments diffuse into the rice, which at the end of the fermentation is dried and often powdered. As well as being used in food and drink preparation, *ang-kak* features in the traditional Chinese pharmacopoeia.

Fermented meat products are widespread in eastern parts of Europe and the Mediterranean countries, where there are a range of traditional mould-ripened sausages and hams. In these, fungal metabolism plays an important role in imparting flavours and colours, and the surface growth of mould is welcomed. The fungi most commonly encountered are species of *Penicillium* and *Aspergillus*, notably *Penicillium nalgiovense*. Usually these processes involve the development of spontaneous mycofloras. This, however, runs the risk (albeit small) of possible mycotoxin contamination from toxigenic strains of *Penicillium* and *Aspergillus*. To prevent this the use of starter cultures has been advocated, for example non-toxigenic strains of *P. nalgiovense* and the yeast *Debaryomyces hansenii*. The fungi mainly grow on the surface of the product, but have a range of effects throughout the sausage or ham. These include reducing the oxygen concentration, which delays rancidity and improves reddening, and increasing the pH by consuming lactic acid and acetic acid produced by accompanying bacteria, and by proteolysis and deamination to produce ammonia.

The Production by Fungi of Primary Metabolites of Economic Importance

Primary metabolites are those that have to be produced for growth to occur. Nucleic acids, proteins, carbohydrates and lipids have to be synthesized, and their precursors too, unless these can be obtained from the growth medium. The metabolic pathways that produce energy have to be active, and under some circumstances intermediates or end products may accumulate. Vigorous primary metabolism is inevitably associated with the phase of rapid growth (page 127), and in batch culture the maximum accumulation of primary metabolites tends to be towards the end of that phase, since in the stationary phase they may be further metabolized. Continuous culture (page 130), in which the organism is maintained in the phase of exponential growth, is ideal for the production of primary metabolites, but requires greater investment and maintenance than batch culture.

Most of the pathways of primary metabolism are widespread, so an economically interesting primary metabolite may occur in a wide range of microorganisms. Hence, for the production of a specific metabolite, the biotechnologist may be able to select from a wide range of bacteria and fungi. Since many bacteria have far higher growth rates than fungi, it is often from the bacteria that organisms are chosen for the commercial production of primary metabolites. Yeasts and moulds are, however, used for the large-scale production of two such metabolites, ethanol and citric acid, as well as for some others that are needed in smaller amounts.

Industrial Alcohol

Ethanol, as industrial alcohol, is a commercially important chemical. A mixture of 80–90% petroleum and 10–20% ethanol, Gasohol, can be used as a fuel for standard spark-ignition internal combustion engines, and engines can be designed that will run on fuel with a higher ethanol content, even 100%. Ethanol is also a major feedstock in the chemical industry. Until about 1950 most industrial ethanol was produced by fermentation. Ethanol can, however, be made by the hydrogenation of ethylene, cheap and abundant supplies of which became available from petroleum cracking about that time. Most of the world's industrial alcohol is now made in this way, but there are areas in which political or economic considerations have led to the retention and expansion of ethanol production by fermentation. Brazil has a limited output of petroleum and a shortage of foreign exchange for petroleum imports, but the capacity to grow large amounts of fermentable biomass, from which over four million tons of ethanol were produced in 1983. A similar situation prevails in some developing countries, with the fermentation production of ethanol being promoted. The USA, with frequent grain surpluses, has tax legislation promoting fermentation ethanol, and an output second only to Brazil. About 20% of the world output of industrial alcohol is by fermentation, but this is sufficient to render industrial alcohol fourth in value among fermentation products, after alcoholic beverages, cheese and antibiotics.

Fermentations for the production of industrial alcohol are generally carried out with *Saccharomyces cerevisiae*, although some closely related yeasts and also the fission yeast *Schizosaccharomyces pombe* are also used. The substrates are generally not sterilized, hence saving energy. A major substrate is sugar cane juice, with the cane debris being burnt to generate power for operating the fermentation plant. Other substrates are molasses and starchy materials such as surplus grain, potatoes and cassava. The starch has to be converted into sugars before alcoholic fermentation can proceed, and this is done with acid hydrolysis or fungal amylases. One method is to carry out a solid substrate fermentation with the starchy material, using an amylase-producing *Aspergillus*. This initial fermentation can, however, generate a massive inoculum of lactic acid bacteria, which subsequently compete with the yeast and make lactic acid instead of ethanol. Hence an on-site production of *Aspergillus* amylase in sterile fermenters is more usual. There is a limited use of whey as a substrate, using the lactose-fermenting yeast *Kluyveromyces fragilis*. A variety of batch and continuous fermentation systems are used, with the latter being more economic on a large scale. In some systems the yeast is recycled and used to generate more alcohol. This reduces the time and substrate expended for making biomass instead of alcohol. The yeast strains favoured for alcohol production are those with a high ethanol tolerance and an ability to metabolize at, for yeast, relatively high temperatures such as 35°C. A high fermentation temperature reduces cooling requirements and later, when the broth has to be heated to distil off the alcohol, the heating requirement. Distillation yields 95% alcohol. Where anhydrous alcohol is needed, as with Gasohol production, redistillation with continuously recycled benzene follows.

Citric Acid

Citric acid is a component of the tricarboxylic acid (TCA) cycle and hence is produced by most aerobic organisms. Normally, however, it is promptly metabolized via *cis*-aconitic acid to yield other components of the TCA cycle. High concentrations of citric acid occur in citrus fruits, especially lemons, which were the original source of citric acid. Accumulation of citric acid was found about a century ago in *Penicillium* and later in *Aspergillus niger*. Commercial production of citric acid from *A. niger* began in 1923, and soon totally replaced that from lemons. About 300 000 tons a year are now produced by fermentation, mostly for use in food, soft drinks and pharmaceutical preparations, although some is used by industry. *A. niger* remains the main organism used for producing citric acid.

The main carbon source used in citric acid production is molasses from the refining of cane or beet sugar, although in the USA high glucose syrup from grain is also used. The amounts of trace metals in the medium are crucial in determining the success of a fermentation. Some trace metals are essential for growth, but if amounts are excessive or if the balance between nutrients is wrong, then citric acid may not accumulate or may be partially or wholly replaced by other acids, such as the toxic oxalic acid. It is especially important that the level of iron, a co-factor for aconitase, responsible for the further metabolism of citric acid, is kept low. This is usually done by treatment of the medium with ferrocyanide, which also removes manganese, another metal adversely affecting citric acid accumulation. Fermentations were at first carried out by surface culture in shallow trays, and some plants still operate in this way. The medium is initially at pH 5–7, necessary for the germination of the inoculum of *A. niger* spores. Later the medium is acidified to about pH 2, which is optimal for citric acid accumulation and also discourages the growth of contaminants. The fungus grows as a surface mat and citric acid is released into the medium below. Submerged culture is now also used for citric acid production, the fungus being grown in batch culture in stirred and sparged fermenters. The medium is formulated and the propagule number in the inoculum adjusted so as to favour pellet growth (page 135). This reduces energy requirements for stirring and gives good citric acid yields. Recently the production of citric acid in submerged culture has also been carried out using *Candida* spp. Citric acid is recovered from media by adding chalk or lime leading to the precipitation of insoluble calcium citrate. This is washed and citric acid released by treatment with sulphuric acid. Precipitated calcium sulphate is removed and concentration and crystallization of citric acid carried out.

Other Primary Metabolites

Microorganisms are used for the production of primary metabolites of almost every type, although bacteria are often the organisms favoured by the fermentation industry. Representatives of most classes of primary metabolite have, however, been produced commercially by moulds or yeast in the past or at the present time. A brief systematic survey of such metabolites will illustrate the present utilization and future potential of moulds and yeasts for their production. **Nucleotides** and **amino acids** as nutritional supplements and for conferring meaty flavours on bland foods are currently produced by bacteria. **Proteins** with enzymic activity (page 476) and single cell protein (page 496) are produced by moulds as well as bacteria. Several extracellular **polysaccharides** of actual or potential economic significance are made by bacteria and fungi. One of these, pullulan from *Aureobasidium pullulans*, is produced in Japan for making biodegradable plastics and as a food coating which, having low oxygen permeability, minimizes oxidative deterioration of food. **Fats** – glycerol esterified with fatty acids – are accumulated in large amounts by some moulds and yeasts. In Germany *Candida* and *Fusarium* were used to make edible fat during the Second World War, and in the First World War, glycerol was obtained from the anaerobic metabolism of *Saccharomyces cerevisiae*. This was done by including bisulphite in the medium. This combines with acetaldehyde and prevents the formation of ethanol by the step that regenerates NAD^+; by the oxidation of NADH. Under these circumstances the NAD^+ that is required in the path from glucose to pyruvate is regenerated in a pathway by which part of the product of glucose metabolism is diverted to glycerol. Moulds are used for the production of several **organic acids** in addition to citric acid (Fig. 8.13). *Rhizopus* was at one time used to make another TCA cycle intermediate, fumaric acid. Itaconic acid, resulting from the decarboxylation of *cis*-aconitic acid, is produced by *Aspergillus terreus*. Strains of *A. niger* are used to make gluconic acid, resulting from the action of glucose oxidase on glucose. Kojic acid, also derived from glucose, is produced by several *Aspergillus* spp. Yeast extract, from which several vitamins can be purified, is itself widely consumed as a vitamin source, and riboflavin (vitamin B_2) is produced using the Ascomycete *Ashbya gossypii*. The Zygomycete *Blakeslea trispora* and some pink yeasts have been studied as possible sources of carotenoids, which are vitamin A precursors and, consumed by farmed salmon, impart a natural pink coloration to the flesh.

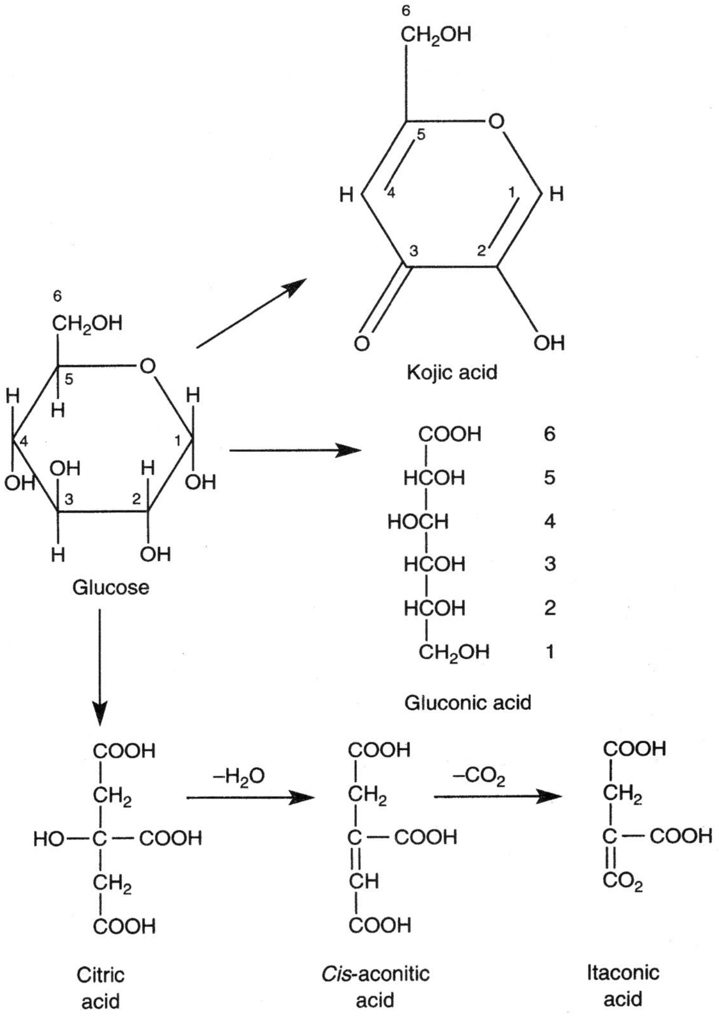

Figure 8.13 Economically important organic acids produced by fungi. The carbon skeleton of glucose, as indicated by the numbering of the carbon atoms, is conserved in the formation of kojic acid and gluconic acid. Gluconic acid, in the form of phosphogluconic acid, is an intermediate in the oxidation of glucose by the pentose phosphate pathway. Citric acid, economically the most important acid produced by fungi, is a key member of the tricarboxylic (TCA) cycle, and accumulates in strains of *Aspergillus niger* with an incomplete TCA cycle. *Cis*-aconitic acid is an enzyme-bound intermediate formed from citric acid during the normal operation of the TCA cycle. In some *Aspergillus* spp. with an incomplete cycle, *cis*-aconitic acid is converted into itaconic acid.

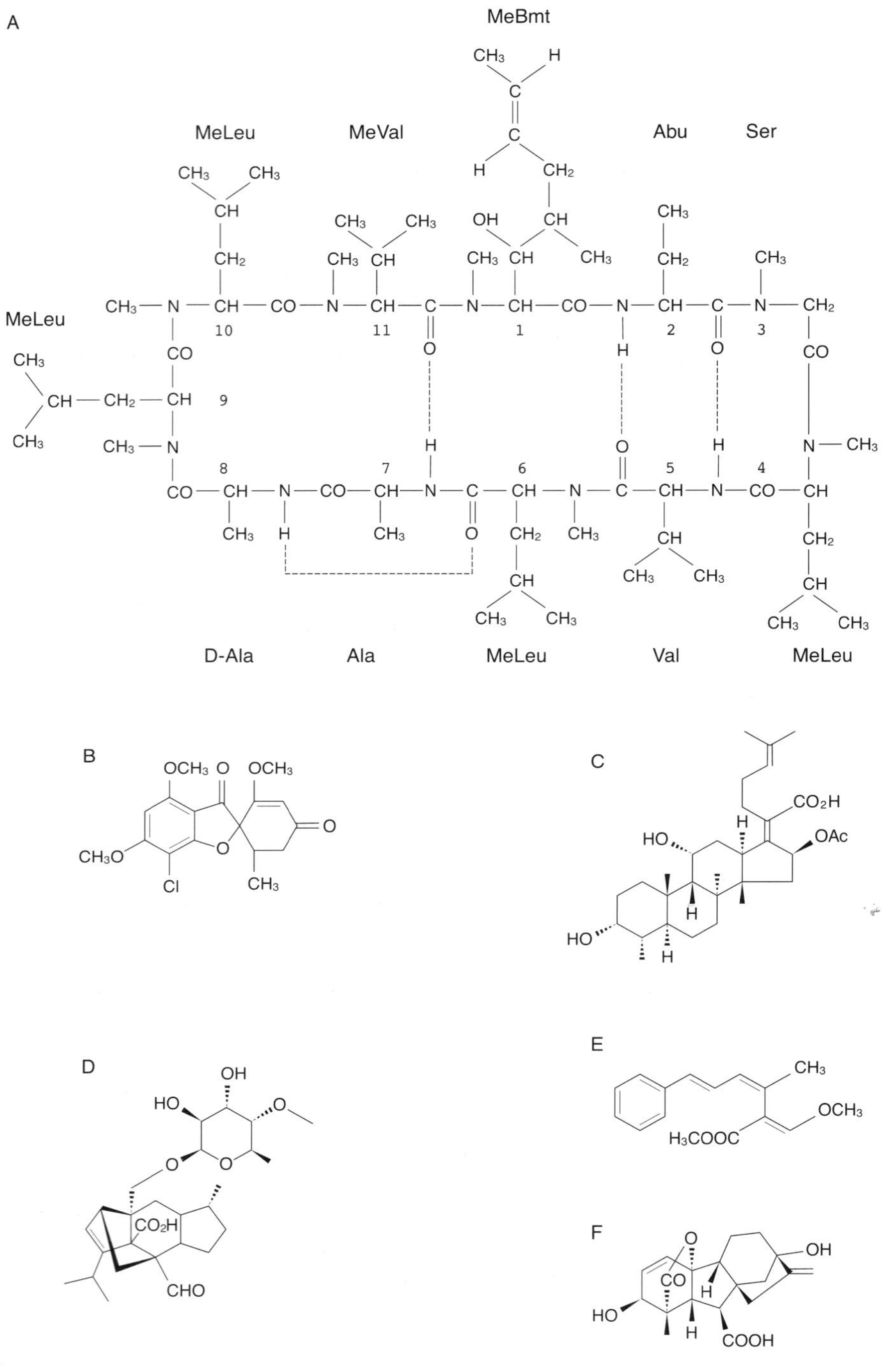

Figure 8.14 Examples of fungal secondary metabolites, illustrating the wide variety of chemical structure. A, cyclosporin A, a cyclic peptide. Of the 11 amino acids, only L-alanine (Ala) and L-valine (Val) are found in proteins. The others are D-alanine (D-Ala), L-methyl-valine (MeVal), L-methyl-leucine (MeLeu), sarcosine (methylglycine) (Sar), α-aminobutyric acid (Abu), and the unusual amino acid butenyl-methyl-L-threonine (MeBmt). The dashed lines signify hydrogen bonds stabilizing the structure. B, Griseofulvin. C, Fusidic acid. D, Sordarin. E, Strobilurin A. F, Gibberellin GA$_3$.

The Production by Fungi of Secondary Metabolites of Economic Importance

The fungi provide us with an enormous variety of strange and wonderful 'secondary metabolites', some of which have profound biological activities that we can exploit. Secondary metabolites are those that are not essential for vegetative growth in pure culture. Secondary metabolism occurs as growth rate declines and during the stationary phase, and often is associated with differentiation and sporulation. This phase of growth has been termed the 'idiophase' by John Bu'Lock (i.e. idiosyncratic phase), in contrast to the 'trophophase' (i.e. feeding phase) characterized by primary metabolism. Thus batch rather than continuous culture is usually favoured for secondary metabolite production. The following generalizations can be made about most fungal secondary metabolites.

First, they show an enormous variety of structures and biosynthetic origins (Figs. 8.14–16, 8.18), and often are very complex with unusual or unique components. For example, of the 11 amino acids in the cyclic peptide cyclosporin (Fig. 8.14A), only two are found in proteins. There have been numerous examples where secondary metabolites have been found to have structural features previously unknown in chemistry.

Secondly, the production of any one of them tends to be restricted to one or a few organisms, which may not be closely related, and it varies considerably between different isolates of the same species. Thus cyclosporin A is produced by several strains of three species of *Tolypocladium*, but also by strains of species of *Beauveria*, *Fusarium*, *Neocosmospora* and *Stachybotrys*.

Thirdly, they are produced in families of closely related molecules. This is a result of their biosynthetic enzymes tending to be less strictly specific in respect to substrate than those of primary metabolism, and tending to be less strictly regulated. Usually members of a family are distinguished by letters of the alphabet, in order of their discovery. Relative proportions of the production of different family members can be manipulated by changing constituents of the medium. Individual members of a family often show very different biological properties. Thus of the first 25 cyclosporins identified as metabolites of *Tolypocladium*, only two have worthwhile immunosuppressive activity, and most others have none, and 17 have antifungal activities.

There is good evidence that genes encoding at least some of the enzymes in many of these complex biosynthetic pathways occur in clusters on a single chromosome. Examples include ergot alkaloids in *Claviceps purpurea*, with 15 enzymic steps from tryptophan to ergotamine; gibberellins in *Giberella fujikuroi*, from farnesyl phosphate; aflatoxins in *Aspergillus parasiticus*, penicillins in *Penicillium chrysogenum* and *Aspergillus nidulans*; and cephalosporins in *Cephalosporium acremonium*. The significance of this is that it may play a role in coordinated regulation of the expression of these genes, via appropriate transcription factors. In contrast, in some cases a single polypeptide gene product may have multiple catalytic sites involved in the biosynthesis of a particular secondary metabolite. The largest enzyme protein known is cyclosporin synthetase, from *Tolypocladium niveum*, which has a molecular mass of

1 689 243 Da, with 11 amino acid-activating domains and 40 catalytic functions involved in the biosynthesis of this cyclic undecapeptide (i.e. composed of 11 amino acids) (Fig. 8.14A). This is encoded by a gene with a giant 45.8 kilobase open reading frame with no introns. The sequence of each of the 11 domains is similar to those of domains of other multifunctional peptide synthetases from bacteria and fungi. Each of these 11 domains has been ascribed to the activation of a particular methylated or unmethylated amino acid, and their order in the protein corresponds to the biosynthetic order of the amino acids.

Why are secondary metabolites produced? In the case of antimicrobial antibiotics such as penicillin, cephalosporin and fusidic acid against bacteria, and griseofulvin, strobilurin and sordarin against fungi, it is generally accepted that they probably play an aggressive role in competition, allowing the producer to invade territory occupied by the victim, and to utilize nutrients released upon its death. Some plant pathogenic and insect pathogenic fungi produce toxic cyclic depsipeptides (peptides also containing ester bonds) that are implicated in their pathogenicity, such as enniatins produced by *Fusarium* species, and beauvericin by *Beauveria bassiana*. But then why are fungi such fruitful sources of potent pharmacologically active molecules, such as the immunosuppressive cyclosporin, the cholesterol-lowering mevinic acids and the insulin mimetic demethyl-asterriquinone? An intriguing possible answer, at least for some of them, lies in the commonality shown by biochemical mechanisms. It has become increasingly clear, for example through results of the yeast and human genome projects, that many fundamental systems in our cells and tissues have homologous systems in fungi. Thus may it be that some of these pharmacologically active molecules are playing roles in the regulation of aspects of fungal metabolism, involving similar mechanisms to those in mammals that have been conserved through evolution, giving rise to the cross-reactivity that we observe? Another puzzle is how do fungi that produce toxic secondary metabolites avoid killing themselves? In some cases the biological target is less sensitive than that of other organisms, for example the RNA polymerase II of *Amanita phalloides* (the death cap mushroom) is a million-fold less susceptible to inhibition by α-amanitin, the major mycotoxin of *A. phalloides*, than is the enzyme of mammals. In contrast, the peptidyl-prolyl isomerase, cyclophilin, of *Tolypocladium inflatum* is as sensitive as that of mammals to inhibition by cyclosporin A, so there must be another explanation for *T. inflatum* being resistant to the antifungal activity of the toxin that it is producing. It may be that there is a physical separation of the cyclosporin and the cyclophilin in the fungal cell.

The majority of useful fungal secondary metabolites have been discovered by routine screening, i.e. the testing of culture broths from a large number of fungal isolates in specific biological or biochemical assays, ranging from the use of whole organisms to the use of single pure enzymes. Extracts that show activity in one screen are then commonly tested with a range of other screens, sometimes with unexpected results. For example, cyclosporin was originally discovered in an antifungal screen, but showed insufficient activity for development as an antifungal drug, but it then showed great potency in a screen for immunosuppressive agents.

As shown in the examples discussed below, screening of fungal isolates for the production of secondary metabolites has been very profitable, and has had (and is having) major effects on the development of medicine and agriculture. It is, however, a costly business. It takes 8 to 15 years to develop a new drug or antibiotic. Only 1 in 15 000–50 000 from the original screen will be developed to the marketplace. It costs more than 50 million dollars to bring a novel compound to market, involving initial research and development, testing for efficacy and safety, and clinical trials. But the rewards can be high. The pharmaceutical antibacterial market is in excess of 15 billion dollars, and the antifungal market close to 5 billion dollars.

Secondary Metabolites in Medicine

Antibiotics

When two different microorganisms are cultured on the surface of an agar medium, one is often seen to affect adversely the growth of the other. Such an antagonism may occur in a variety of ways but often it is the production of secondary metabolites by one organism that interfere with the metabolism of the other organism. These secondary metabolites are known as antibiotics. Use of the term is limited to antimicrobial secondary metabolites. It is not used for simple organic acids, enzymes or toxic proteins responsible for antagonistic effects between species.

As mentioned above, antibiotic production is thought to have a selective value for microorganisms that live in environments where there is intense competition for resources. Antibiotics are produced in particular by some Ascomycetes and related mitosporic fungi that live in the soil and penetrate and utilize dead plant material (page 334). They are also produced by many soil bacteria, particularly Actinomycetes, especially members of the genus *Streptomyces*. These are hyphal bacteria that live in the soil and attack refractory plant material and chitin from fungal walls and insect cuticles. Some other spore-forming soil bacteria, such as members of the genus *Bacillus*, also produce antibiotics. Antagonisms between microorganisms have been known for over a century, and by 1940 some of the factors involved had been studied and in a few instances compounds purified. Then, however, penicillin G, an antibiotic produced by the mould *Penicillium notatum*, was shown to be of enormous clinical value. Administered by injection, it could combat potentially lethal infection by *Staphylococcus* and other bacteria, and rarely caused adverse reactions in humans. The success of penicillin led to the screening of thousands of soil microorganisms for antibiotic production. Several thousand antibiotics have now been isolated, but only about 50 have been used clinically, most being too toxic to humans. Although the majority of antibiotics in current clinical use come from Actinomycetes, the first useful antibiotic, penicillin, and the related cephalosporins, came from fungi, and their derivatives remain the most important, with annual sales in excess of 13 billion dollars (Table 8.3).

Penicillins and cephalosporins The story of the serendipitous discovery of penicillin in 1928 by Alexander Fleming, by his astute observation of a zone of lysed bacteria around a contaminant colony of *Penicillium notatum*, has been

Table 8.3   Annual value (1998) for sales of medicines derived from fungal metabolites

Product	Value (billions (10^9) of dollars)
Penicillins, cephalosporins[a], amoxicillin-clavulanic acid	13.6
Mevinic acids: simvastatin, pravastatin, atorvastatin	7.5
Cyclosporin	1.2

[a] Although originally obtained from fungi, *Streptomyces* is now the major source of cephalosporins.

told many times, and is an excellent example of Pasteur's adage 'fortune favours the prepared mind'. The importance of this discovery was recognized by the award of the 1945 Nobel Prize for Medicine to Fleming, together with Howard Florey and Ernst Chain, who rapidly developed penicillin for clinical use at Oxford, from initiating the research project in 1938, through successful animal experiments in 1940, and first clinical trials in 1941. It was shown to be amazingly effective in treating infections with Gram-positive bacteria, and to have few adverse side effects. Yields were, however, very low, about 2 mg l^{-1}, and consisted of a mixture of penicillins. Production of these first batches of penicillin was by surface culture in any available container, the best being hospital bedpans. There was an urgent need to produce sufficient stocks for intensive clinical trials, but as resources for this major effort were not available in war-time Britain, the culture of *P. notatum* and the accumulated knowledge on penicillin production were taken by Florey to the USA. An immediate aim was to increase yield, and a tenfold increase was obtained by using corn-steep liquor (a by-product of turning corn into starch) as nutrient source, and further increases by developing deep fermentation techniques. In addition a large number of strains of *Penicillium* were tested. A strain of *Penicillium chrysogenum*, isolated from a rotting cantaloupe melon in 1943, gave a yield of 150 mg l^{-1}, and successively improved descendants of this strain (page 528) have been the mainstay of industrial production, with current strains yielding in excess of 30 g l^{-1}. Using corn-steep liquor in the medium, the major product was benzyl penicillin, known as penicillin G, with phenylacetic acid as side chain (Fig. 8.15). It was found that the inclusion of phenylacetic acid in the medium resulted in increased yields of penicillin G, which has remained the most important penicillin produced entirely by biosynthesis. It is, however, unstable in acid solutions. This means that it is usually administered by injection, since if given by mouth, a large proportion of the dose would be destroyed by acids in the stomach. The inclusion of phenoxyacetic acid instead of phenylacetic acid in the medium was found to give penicillin V, phenoxymethyl penicillin (Fig. 8.15). This is not so active as penicillin G, but is more acid-stable and hence can be given by mouth. There are other naturally occurring penicillins, but G and V are the only ones that remain clinically important.

Within a decade of the introduction of penicillin, resistance among bacteria that had at first been susceptible to the antibiotic was becoming a serious problem. This resistance was due to the production by the resistant bacteria of an enzyme rendering the penicillin inactive and hence termed penicillinase. It was established that penicillinase acted by the hydrolytic cleavage of the β-lactam ring and was hence a β-**lactamase**. By 1957 it seemed likely that penicillin would diminish in importance due to the spread of resistance. At this stage, however, it was found that a strain of *P. chrysogenum* produced large amounts of the chemical nucleus of penicillin, 6-amino penicillanic acid (6-APA). This β-lactam lacks antibiotic activity but it was realized that it could be used for the production of a wide range of new 'second generation' penicillins by the chemical addition of a variety of side chains by reaction of 6-APA with the appropriate organic acid. Examples of such **semi-synthetic penicillins** are methicillin and ampicillin (Fig. 8.15). Methicillin is resistant to β-lactamase due to steric hindrance, the bulky side chain preventing access by the enzyme molecule to the β-lactam ring. Ampicillin, unlike the other penicillins mentioned, is a broad-spectrum antibiotic, active against Gram-negative as well as Gram-positive bacteria. Although 6-APA was at first produced by fermentation, it was soon found more convenient to prepare it from penicillin G by removal of the side chain. This is done with a hydrolytic enzyme, penicillin amidase (also termed penicillin acylase), or by chemical cleavage. The development of the semi-synthetic penicillins has resulted

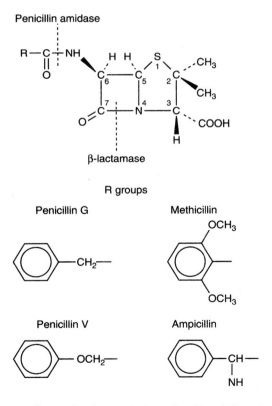

Figure 8.15   The penicillin molecule, consisting of a thiazolidine ring (1–5), a β-lactam ring (4–7) and a side chain. The sites of action of β-lactamase and of penicillin amidase, which removes the side chain to give 6-amino penicillanic acid (6-APA), are indicated. The various penicillins differ in the side chain, as indicated for penicillin G, penicillin V, methicillin and ampicillin.

in the penicillins becoming a versatile group of antibiotics and retaining their importance.

The closely related cephalosporins were discovered quite soon after the development of penicillins for clinical use, and although they took much longer to be developed, in the long run they have proved to be of even more value. In 1945 a strain of the mitosporic fungus *Cephalosporium acremonium*, isolated from a sewage outfall in Sardinia, was shown to inhibit the growth of both Gram-positive and Gram-negative bacteria. After several years' delay, detailed studies were carried out on the nature of the antibacterial agents produced by the fungus. The first antibiotic isolated was named cephalosporin P, because it was only active against Gram-positive bacteria, and was found to be a steroid. A second antibiotic was called cephalosporin N, because it was active against Gram-negative as well as Gram-positive bacteria, and proved to be a new penicillin, so it was re-named penicillin N. A third, cephalosporin C, was one of three impurities (A, B and C) observed when purifying penicillin N by chromatography. A mutant of *C. acremonium* that gave higher yields of cephalosporin C was subsequently obtained, and the structure of the antibiotic elucidated. It proved, like penicillin, to be a β-lactam, but differed in having a six-membered dihydro thiazine ring instead of a five-membered thiazolidine ring (Fig. 8.16), and penicillin N proved to be the substrate for the 'expandase' enzyme that expanded the ring from five- to six-membered. Cephalosporin C proved to be a relatively weak antibiotic, but it had broad antibacterial specificity, and in addition was highly resistant to β-lactamase. These attractive features led to plans for large-scale use, but these were brought to an end in 1960 by the advent of the semi-synthetic penicillins. It was found, however, that cephalosporin C could be cleaved chemically to give 7-amino cephalosporanic acid (7-ACA), after which a very wide range of different side chains could be added. In addition the side chain at the 3-position could be modified, removed or replaced (Fig. 8.16). These developments have made available a wide range of novel second, third and now fourth generation semi-synthetic cephalosporins, some of which, such as the orally administered cephalexin, are very widely used.

Cephalosporin and penicillin biosyntheses share their first two steps. The first is the formation of a tripeptide, L-aminoadipyl-L-cysteinyl-D-valine, catalysed by a multifunctional enzyme. This is cyclized to give isopenicillin N, which is the branch point for the two families of antibiotics. The enzyme involved, isopenicillin N synthetase, shows close similarity in fungi producing the two antibiotics; the enzymes from *P. chrysogenum* and *Aspergillus nidulans*, and *C. acremonium*, producing penicillin and cephalosporin, respectively display about 80% identity. For the biosynthesis of penicillin, the isopenicillin N is hydrolysed to give 6-APA, which is then reacted with a molecule of an organic acid-coenzyme A to give a specific penicillin. For the biosynthesis of cephalosporins, the isopenicillin N is epimerized to give penicillin N, which is then acted on by the expandase, to give deoxycephalosporin C, which is further modified to give cephalosporin C and cephamycin C. The genes for penicillin biosynthesis in *P. chrysogenum*, and most of the genes for cephalosporin biosynthesis in *C. acremonium* are in clusters. Cephalosporins are also biosynthesized by the actinomycete, *Streptomyces clavuligerus*, and bacterial isopenicillin N synthetase has about 60% identity with the fungal enzymes. This has been taken as strong

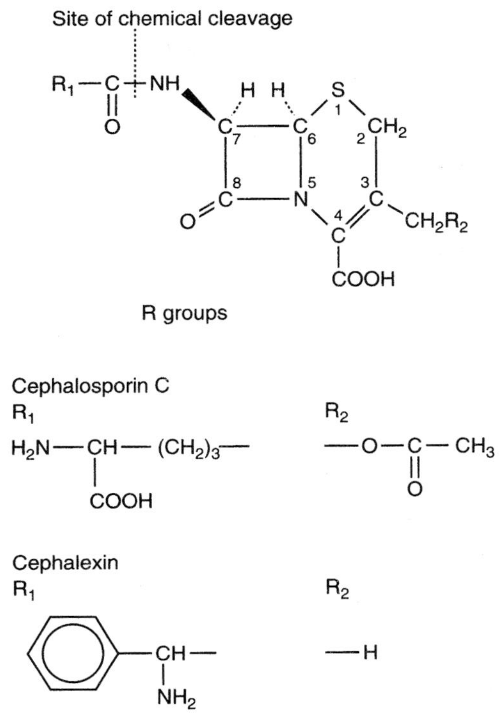

Site of chemical cleavage

R groups

Cephalosporin C

R$_1$

H$_2$N—CH—(CH$_2$)$_3$—
|
COOH

R$_2$

—O—C—CH$_3$
‖
O

Cephalexin

R$_1$

⬡—CH—
|
NH$_2$

R$_2$

—H

Figure 8.16 The cephalosporin molecule, consisting of a dihydrothiazine ring (1–6), a β-lactam ring (5–8) and two side chains (R^1 and R^2). The site at which chemical cleavage acts to yield 7-amino cephalosporic acid (7-ACA) is indicated, as is the nature of the R$_1$ and R$_2$ groups in cephalosporin C and in the semi-synthetic cephalexin.

evidence for horizontal gene transfer of pathway genes from bacterium to fungus, but this appealing analysis has been questioned and it is possible that this observation represents conservation of these genes from a common ancestor.

Other antibiotics Fungi produce a variety of antibiotics in addition to penicillin and cephalosporin. Their diversity will be illustrated by commenting on some that have clinical significance.

Griseofulvin, from *Penicillium griseofulvum*, is one of the relatively few natural organic compounds that contain chlorine (Fig. 8.14B). It was isolated in 1939 as a result of an investigation into the disappearance of chloride ions from the culture medium. A decade later an antibiotic active against plant pathogenic fungi was found to be griseofulvin. It was then decided that employment against plant pathogens would be uneconomic, and work on the topic was abandoned. Then in 1960 it was found that, when administered orally, griseofulvin concentrated in the keratinous layers of the skin. There it inhibits the growth of some troublesome dermatophyte fungi, and is hence clinically useful. It acts by inhibiting assembly of fungal microtubules, and so inhibits mitosis in susceptible species. **Fusidic acid**

(Fig. 8.14C), from the mitosporic fungus *Fusidium coccineum*, is a steroid active against Gram-positive bacteria, and is useful against strains that have acquired a high resistance to β-lactam antibiotics. **Fumagillin**, from *Aspergillus* spp., is active against protozoa. It is used by bee-keepers to combat hive infections by the troublesome *Nosema apis* and in veterinary practice against amoebal infections in dogs. A novel antifungal antibiotic, **sordarin** (Fig. 8.14D) from the Ascomycete *Sordaria araneosa*, is showing promise for development as an antifungal drug. Its potent antifungal activity is by specifically inhibiting protein synthesis.

Pharmacologically active fungal secondary metabolites
Rivalling the β-lactams in value of pharmaceutical sales are the fungal metabolites and their derivatives with very specific pharmacological activities (Table 8.3). For many years the pharmaceutical industry has screened microbial strains for compounds with biological activity, using an ever widening set of bioassays. A series of compounds have been discovered, and many are in widespread use as drugs for a broad set of medical conditions. The most important examples of fungal compounds are now discussed.

Cholesterol-lowering drugs Commercially, the most important group of these fungal compounds are the mevinic acids, which are potent cholesterol-lowering agents. A series of related metabolites have been discovered and continue to be discovered, starting with mevastatin from *Penicillium citrinum*, then lovastatin from *Monascus ruber*. They are small organic acids with an acidic side group which very specifically and potently interacts with the active site of 3-hydroxy-3-methylglutaryl coenzyme A reductase, the key enzyme on the biosynthetic pathway to cholesterol in the liver. They are potent competitive inhibitors of enzyme activity, so that the enzyme has over 10 000-fold affinity for them than for its natural substrate, 3-hydroxy-3-methylglutaryl coenzyme A.

Cyclosporin A One of the great successes of medicine in the recent past has been the ability to transplant organs, notably kidney, heart, lung, liver, pancreas, bone marrow, small bowel and skin. These transplant operations have been made possible by the discovery of the potent immunosuppressive drug, cyclosporin A (also known as cyclosporine A), as a metabolite of an isolate of the mitosporic fungus, *Tolypocladium inflatum*, from a soil sample from the Norwegian mountains. Cyclosporin A acts on T-lymphocytes of our immune system, inhibiting their production of interleukin-2 so that they cannot respond to antigens, thus preventing rejection of the transplanted organ. Interleukin-2 also mediates inflammatory responses in a range of diseases, and cyclosporin is used in the treatment of resistant forms of eczema and psoriasis. Cyclosporin A was first found during a screen for antifungal agents, but it was then chosen for testing in a further 50 bioassays. It was positive in only one test, a newly developed cellular assay designed to detect immunosuppressive drugs, by measuring the haemagglutination count of sera of mice that had been immunized with sheep red blood cells. This positive result in the test showed that the mice had been unable to produce antibodies to the foreign cells. In the early 1970s there was a desperate need for an immunosuppressive drug, as the one year survival rate for kidney grafts was only about 50%. Cyclosporin A, first reported publicly in 1976, filled this need, as seen in its sales in 1998 of about 1.2 billion dollars. It took 12 years of pharmacological, toxicological and medical research to bring cyclosporin to the market. As is common for secondary metabolism, *T. inflatum* produces a family of cyclosporins, from A to Z. Of these only two are strongly immunosuppressive, while 17 have antifungal activity. Several other fungi produce cyclosporin A, but *T. inflatum* is the major producing species. The original isolate produced only a few milligrams of compound per litre of fermentation broth but, with strain improvements by mutation and selection, and optimization of culture conditions, yields are now several grams per litre. Cyclosporin A is a very complex cyclic peptide with many of the constituent amino acids methylated (Fig. 8.14A). Features of its structure and of the remarkable cyclosporin synthetase have been described earlier. It is too complex for ready chemical synthesis, and certainly it would have been impossible to predict that such a structure could have its profound pharmacological effect. More than 1000 analogues of cyclosporin A have been made and tested, but none are as good as the natural compound. Thus its discovery and exploitation, and the subsequent medical advances in transplantation, have relied totally on traditional fungal biotechnology.

Ergot alkaloids
Preparations from ergot, the sclerotia produced by members of the genus *Claviceps* (page 54; Fig. 8.17) have been used in medicine for hundreds of years. The beneficial effects of such preparations, as well as the poisoning of man and animals due to accidental consumption of ergot, are due to alkaloids, nitrogen-containing basic compounds. Detailed studies on the chemistry, mode of action and ways of producing ergot alkaloids began early in the twentieth century and still continue.

The ergoline nucleus (Fig. 8.18A) of ergot alkaloids is derived from tryptophan and mevalonic acid. The most simple ergot alkaloids, the clavine alkaloids, such as elymoclavine (Fig. 8.18B), are a little more complex. They are produced in large amounts by *Claviceps fusiformis*, and although not themselves used in medicine, are precursors for the chemical synthesis of other, useful alkaloids. Lysergic acid (Fig. 8.18C) and its amide are produced by *C. paspali*. Neither is therapeutically useful, but lysergic acid is the starting material for the chemical synthesis of a range of more complex alkaloids. One synthetic derivative, lysergic acid diethylamide (LSD), achieved notoriety as a hallucinogen. The propanolamide of lysergic acid, ergometrine (Fig. 8.18D), is produced by *C. purpurea* and is therapeutically important. *C. purpurea* produces a range of peptide derivatives of lysergic acid, including the medically important ergotamine (Fig. 8.18E). Although ergometrine and ergotamine are the only naturally occurring ergot alkaloids of medical importance, a wide variety of others have been produced by chemical modification of the naturally occurring alkaloids.

At least a dozen ergot alkaloids are used in medicine. Here a few examples of their use and an indication of their mode of action will be given. Ergometrine and its derivative methyl ergometrine act upon the uterus. They are administered in childbirth if contractions of the uterus are weak and irregular and, by stimulating

Figure 8.17 Ergot of rye. The ovary of rye (*Secale cereale*) is infected by *Claviceps purpurea*, and a sclerotium (ergot) instead of a grain develops. Grain contaminated with ergot may be highly toxic, but crops of rye can be deliberately infected with C. *purpurea* to produce ergot as a source of pharmacologically active substances. The fermentation industry, however, is now the main source of ergot alkaloids (Peter Mantle).

uterine smooth muscle, they facilitate delivery. They are also given after childbirth if there is excessive bleeding. This they control by narrowing blood vessel diameter (vasoconstriction), through causing contraction of smooth muscle around the blood vessels. Ergotamine and its derivative dihydroergotamine are used to treat migraine. Here again the alkaloids act through vasoconstriction, counteracting the dilation of veins in the brain that occurs in migraine. Bromocriptine, a derivative of the naturally occurring ergocryptine, is used to stop excessive or untimely milk production. This it does by suppressing the formation by the pituitary gland of prolactin, the hormone that stimulates lactation. The way in which the ergot alkaloids act at the molecular level is ill-understood. There is, however, some structural resemblance between the ergot alkaloids and the neurotransmitters noradrenaline, serotonin and dopamine. These substances, released at nerve endings, stimulate smooth muscle and the glands that produce hormones. It is possible that much of the activity of ergot alkaloids will ultimately be explicable in terms of interaction between the alkaloids and the receptors for neurotransmitter molecules.

The original source of ergot was rye that had been infected naturally with *Claviceps purpurea*. In some areas where rye is grown climatic conditions are

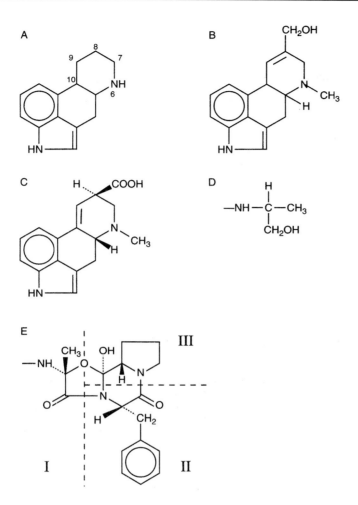

Figure 8.18   The ergot alkaloids. The common structural element of most ergot alkaloids is the tetracyclic ergoline ring system (A). In most naturally occurring ergot alkaloids the N atom at position 6 is methylated, there is a double bond at position 8–9 or 9–10 and a C atom is carried at position 8. For example, in the clavine alkaloid elymoclavine (B) the N at 6 is methylated, there is a double bond at C-8 to C-9 and C-8 carries a —CH$_2$OH group. In lysergic acid (C), the double bond is between C-9 and C-10 and C-8 carries a carboxylic acid group. In the lysergic acid amides the hydroxyl of the carboxylic acid is replaced by an amide group, for example propanolamide in ergometrine (D). In the peptide antibiotics the hydroxyl is replaced by a peptide composed of three amino acids. In ergotamine (E) the amino acid (I) linked to the lysergic acid group is alanine, but in other peptide alkaloids it can be valine or α-aminobutyric acid. In ergotamine the second amino acid (II) is phenylalanine but other possibilities are valine, leucine, isoleucine, homoleucine and α-aminobutyric acid. The third amino acid (III) in all peptide ergot alkaloids is proline.

such that heavy natural infections are common. In the past this could give rise to local outbreaks of 'ergotism', with the ergot alkaloids as mycotoxins (c.f. page 441). This phenomenon was known as 'St Anthony's fire' because of the burning sensation experienced by the victims. When the medical value of the ergot alkaloids became known, it was realized that the ergot could be worth more than the grain, and that heavy infections were to be welcomed. Deliberate infection was introduced in the 1920s, and subsequently mechanical methods of inoculation and harvesting were developed. The inoculum was produced in pure culture from strains of *C. purpurea* giving high yields of the required alkaloids in field experiments. The agricultural production of ergot remains economic in areas where little but rye can be grown and labour is relatively cheap. It is important in parts of Eastern Europe.

A very long period elapsed between the pure culture of *Claviceps* spp. and the achievement of alkaloid production by such cultures. Success was only obtained with *C. purpurea*, the species of greatest commercial interest, in the mid-1960s, but since then the fermenter production of peptide alkaloids by this species has become increasingly important. Alkaloids are synthesized only when the hyphae are composed of roughly spherical cells resembling those in the naturally occurring ergot. Such a morphology is obtained only in the presence of a high osmotic pressure, produced by including in the medium a high concentration of a slowly utilized carbon source, for example 30% sucrose. Alkaloid formation in fermenter culture was obtained with rather less difficulty with *C. paspali* and *C. fusiformis*, and these species are used for the production of lysergic acid amide and clavine alkaloids respectively.

Secondary Metabolites in Agriculture

Strobilurins
The major advance in agricultural fungicides in recent years has been the introduction of the strobilurins. This class of antifungal antibiotics was identified as metabolites of mushrooms. Early reports in the 1960s described 'mucidin', from *Oudemansiella mucida*, the 'porcelain fungus' growing on beech trees, which was marketed in Czechoslovakia. The full potential of this and related compounds was not widely realized for over 20 years, when the strobilurins were described from the related mushroom, *Strobilurus tenacellus*, a small mushroom found growing on buried pine cones. Strobilurin A was shown to be identical to mucidin (which had been incorrectly described). In contrast to many other biologically active secondary metabolites, strobilurin A, a derivative of methoxyacrylic acid (Fig. 8.14E), has a structure that is simple enough for easy chemical synthesis. The natural product proved to be unsuitable for use in the field, as it is too readily degraded by sunlight, and is too volatile, so has a very short life when sprayed on leaves. Synthetic strobilurins, with similar methoxyacrylate systems, were synthesized that have a much longer life in the field, and these were introduced as broad-spectrum systemic agricultural fungicides in Europe in the 1990s. They have very quickly established a large and rapidly growing market, accounting for nearly half of the UK market for cereal

fungicides in 1999, but unfortunately there is already a problem with the appearance of resistant strains of mildew. They give good protection against a range of diseases, and apparently in some cases actually increase yields by increasing plant photosynthetic capacity. They are very potent inhibitors of fungal respiration, acting by binding to the hydroquinone oxidation centre of the mitochondrial cytochrome $bc1$ segment of the respiratory chain. The fungi that produce them are resistant to their action by having amino acid substitutions in their cytochrome c polypeptides.

Gibberellins
The gibberellins are very potent plant hormones. They were, however, discovered as metabolites of the rice pathogen, *Gibberella fujikuroi*, which causes bakanae disease, literally 'foolish plant disease', characterized by excessive growth of the plants, so much so that they fall over and rot. In 1926, a Japanese plant pathologist reported that a substance excreted by the fungus in culture could cause increased growth of rice plants. It was not until many years later that their chemical nature was elucidated, and that they were identified as endogenous plant hormones, produced in tiny amounts. The different gibberellins are involved in a wide range of important development and growth processes in plants. At least 27 gibberellins (GAs) are produced by *G. fujikuroi*, of which 14 also occur in plants. They are diterpenoids with 19 or 20 carbon atoms (Fig. 8.14F), biosynthesized from mevalonic acid as are sterols and carotenoids. Industrial production is by fermentation of improved strains of *G. fujikuroi*, giving yields of several grams of GAs per litre in comparison to less than 100 mg l^{-1} for wild-type strains. Annual production throughout the world is probably between 20 and 30 tonnes, with a value of about 200 million dollars. GA_3 is used to improve fruit set and size in seedless grapes and citrus fruits, and in some countries to speed up malting of barley in brewing, while mixtures of GA_4 and GA_7 are used to control development and ripening of apples.

The Selection and Genetic Improvement of Fungal Strains

The success of any fermentation is dependent on obtaining a suitable microorganism and then, commonly, on improving its performance by genetic modification. The approaches used in achieving this will depend on whether a well-established process is being improved, or novel products being sought.

Sources of Fungal Strains

There are a large number of fungal species in culture collections, and for some species many strains are available. In nature, however, there is a far greater genetic diversity. Not only are there, for each species, many more strains than there are in culture collections, but in addition many yet undiscovered species.

The successful isolation of useful fungal strains from nature is dependent on well-planned and often ingenious screening programmes. Novel pharmacologically active compounds continue to be discovered in screens of fungal extracts. A recent example is a metabolite of an isolate of the Ascomycete, *Pseudomassaria* sp., that is a very specific, potent insulin mimetic that shows promise in the development of novel treatments of some types of diabetes. Unexpectedly it is a small aromatic derivative of quinone, demethylasterriquinone B-1, in contrast to the protein structure of insulin. It was discovered during a screen for insulin-like activity that involved over 50 000 mixtures of synthetic and natural products.

Screening for strains with improved performance

If a fungal strain produces a small amount of a metabolite of interest, then a programme of mutation and strain selection can increase the yield of the metabolite. Before undertaking such genetic modification, however, it is advisable to investigate other strains of the same and related species to see if any would make a better starting point for the programme. The early history of penicillin (page 516) illustrates the usefulness of such an approach. Penicillin was at first produced from surface cultures of *Penicillium notatum* with yields of about 2 mg l^{-1}. The testing of a large number of other strains and related species found a strain of *Penicillium chrysogenum* that not only gave higher yields, at 30 mg l^{-1}, but had the ability to do so in submerged culture, unlike *P. notatum*. There are other attributes, in addition to rapid growth and metabolite production in submerged culture, that are of importance in fermentation technology but which cannot be produced readily by breeding programmes. For example, strains should be stable yet respond well to mutagenic treatments (page 528). In addition, with mycelial fungi good sporulation is of value, since spores provide a convenient inoculum, survive well in storage and, if uninucleate and haploid, are appropriate for mutagenesis. Another useful attribute is the absence of toxic constituents or of materials that interfere with product purification.

The isolation of strains related to one already in culture is facilitated by a knowledge of the ecology and physiology of the species sought. Then diluted samples from the relevant habitat, such as soil or decaying plant materials, can be spread on an agar medium of suitable pH and the correct nutrients to encourage growth of the required fungal group. Bacterial growth is usually discouraged by the inclusion of a broad-spectrum antibacterial antibiotic. Some of the colonies arising are then likely to be of the desired species and can be isolated for further examination.

Screening for microorganisms giving new products

Sometimes new products may be obtained by screening strains and species related to an already important fermentation organism. The isolation of new strains of *Claviceps*, for example, has led to the discovery of new ergot alkaloids. More often the search for new products entails making dilutions of soil or decaying plant material and carrying out isolations in a way that indicates organisms possessing the properties that are being sought. The organisms that are obtained may be moulds or yeasts but can equally well be

Actinomycetes (hyphal bacteria superficially resembling moulds) or other bacteria. It is, however, the products that are made and not the place of the producer organisms in schemes of classification that are of interest to the fermentation technologist.

Sometimes what is sought, as in the production of single cell protein, is an organism capable of utilizing a particular type of substrate, such as starch or petroleum hydrocarbons. Although extensive tests will be needed to establish that the protein is of high quality and that toxic substances are absent, the first step is the isolation from appropriate sites of organisms able to use the selected substrate. The starch-utilizing *Fusarium* that is used to produce mycoprotein, for instance, was isolated from a starch-rich effluent. An enormous range of substances occur in small amounts in soils, and organisms capable of degrading them are also present, but in correspondingly small numbers. Their numbers can, however, be increased, prior to attempting isolation, by enriching the soil sample with an appropriate substrate and incubating. The agar medium on which diluted samples are spread will contain a properly chosen carbon source, for example starch, if starch-utilizing organisms are being sought. In this instance not only is the growth of starch-metabolizing organisms encouraged, but their presence is indicated by a clear circular zone around the colonies. This results from the diffusion of amylase from the colonies and the hydrolysis of the opaque starch. The *Aspergillus* species used in amylase production are isolated in this way. Organisms that produce enzymes such as pectinase, cellulase and protease that degrade other macromolecules can be detected and isolated by similar methods. Production of organic acids, such as citric acid, can be detected by the inclusion of calcium carbonate in the medium, which must not, however, contain salts such as ammonium chloride where the uptake of the cation causes the production of mineral acid and hence irrelevant zone formation. Organisms producing an excess of a vitamin can be detected through stimulation of the growth of a tester organism unable to synthesize the vitamin, the test organism being seeded into or on a layer of agar added after the growth of the organisms being screened. The production of an antibiotic by a microorganism can be recognized by the presence around a colony of a zone within which few other microorganisms grow. The colony can then be isolated and its ability to inhibit the growth of selected pathogens determined. Although in the past this procedure yielded many useful antibiotics, it now tends to result only in the rediscovery of antibiotics that are already known. Other procedures have hence to be used, which will not detect the production of antibiotics in general but only that small proportion that have a required property. For example, the inhibition of chitin synthase, essential for cell wall growth in fungi, has been used as a test for the detection of new antifungal antibiotics. An inhibition of β-lactamase activity implies the presence of a penicillin or cephalosporin that has bound to the enzyme molecule but has not been degraded, so the reaction has been used in the detection of new β-lactam antibiotics that resist enzymic destruction. A range of other ingenious tests, based on a knowledge of physiology and biochemistry, have been devised for the detection of antiviral agents, antitumour agents, immunosuppressants, hormones and other pharmacologically active substances.

Mutation and Strain Selection

Selection of naturally occurring strains is, as indicated above, often useful. The biotechnologist, however, usually seeks a strain that converts a substantial part of the carbon source into a single product that is then released into the medium. In nature this would be a waste of resources and an organism exhibiting such behaviour would not survive competition with other strains. Improved performance – from the viewpoint of the biotechnologist – has sometimes been obtained by the isolation of spontaneous mutants from a strain already in use. The rate of mutation can, however, be greatly increased by the use of mutagens, so artificial mutagenesis is the basis of most programmes of strain improvement. The labour involved in detecting mutant strains with improved yield is considerable, so the devising of efficient selection procedures is important.

Mutagenesis
A fungal strain for use in fermentations should be genetically stable, except when treated with mutagenic agents. It hence should not be an aneuploid, a heterokaryon or a heteroplasmon, all of which are genetically unstable. Instability is often indicated by the production of sectors of differing morphology or growth rate on agar media. A strain that is haploid in the vegetative phase is greatly to be preferred, since in a diploid or polyploid any new recessive mutation will be masked by the presence of the dominant allele.

A considerable number of genes are concerned in a biosynthetic pathway, some in specifying the enzymes involved and some in regulating overall activity in the pathway. Mutagenic agents act in a variety of ways, and the susceptibility of DNA to the different agents will depend on the nucleotide sequence. Hence with a given mutagenic agent some genes will be mutated readily and others less often. The improvement in product yield that can be obtained by mutating a single gene, either once or repeatedly, is limited, so in a strain improvement programme a series of mutagenic agents is used. Mutagenic agents in common use include X-rays, ultraviolet radiation, dimethylsulphonate and nitrosoguanidine. Mutagenic agents kill cells as well as mutating them, so investigations have to be carried out to determine the dose that gives the highest proportion of mutant cells among the survivors. Most mutagens are also carcinogens so proper precautions are needed in their use.

Mutations that constitute the steps in a strain improvement programme commonly do not affect morphology, and increase product yield by no more than 10–15%. Sometimes a mutation may result in a much larger increase in yield, but such a mutation is often accompanied by changes in morphology and behaviour that require extensive modification of the medium and fermentation process. Such infrequent 'major mutations' hence tend to be less useful than 'minor mutations', the cumulative effect of which can be impressive. An example of the power of these methods is given by the enormous increase in yield of penicillin production by *Penicillium chrysogenum* from from 150 mg l^{-1} in the 1940s to more than 30 g l^{-1} today.

Selection procedures

Following mutagenesis, strains giving improved performance have to be detected. The original approach was random screening. Samples of the cells that had been treated with the mutagen were spread on agar media and colonies arising from single cells were isolated. The isolates were then grown in liquid medium and the culture filtrate assayed for the required product. Such a procedure is very laborious, since only a small proportion of the isolates will show any improvement in yield. A reduction in labour could be achieved where it was possible to detect improved performance on agar media, employing methods similar to those used for the initial detection of products of interest (page 526). In more recent years, however, there has been much effort and ingenuity put into the development of high throughput screens ('HTS'), involving robotic handling and scoring of very large numbers of samples very quickly. Indirect screening methods are sometimes feasible, in which what is estimated is not the amount of the desired product, but some readily determined attribute likely to be correlated with product yield. For example, sensitivity to sodium monofluoracetate is correlated with accumulation of citric acid. This is because the compound inhibits the enzyme aconitate hydratase which converts citric acid into isocitric acid prior to its further metabolism. Hence a mutant with low amounts of this enzyme and so more sensitive to the poison will also be one in which citric acid tends to accumulate. Sometimes, in an indirect screening method, selection for the desired mutant will occur. Thus mutants of a *Geotrichum* with higher yields of azasterols, potential antifungal agents, have an enhanced resistance to pimaricin, and hence grow better than other strains in the presence of this polyene antibiotic.

Genetic Recombination

Strain improvement in the fermentation industry has been based largely on taking a single promising strain and then carrying out a programme of mutation and strain selection, as described in the previous section. In addition it is possible to take two strains with different features of value, to bring about genetic recombination between them, and to select from among the recombinants single strains that have the desirable features of both of the original strains. Such recombination can be brought about by sexual or parasexual hybridization.

Sexual hybridization

Mating occurs in few fungi of commercial interest, but these include mushrooms and some yeasts. With both there are problems in utilizing the sexual process for obtaining improved strains. Spore germination is poor in the cultivated mushroom *Agaricus bisporus*, and about 70% of the spores are self-fertile, due to the presence of nuclei of both mating types in the usually binucleate spores (page 64). In spite of these problems, homokaryons of opposite mating type can be obtained, and mated to yield new dikaryotic strains that give fruiting bodies. It has become clear, however, that there is very little genetic diversity among cultivated varieties – all may have come from a single dikaryotic strain brought into cultivation in France about 300 years ago (page 492). For this reason there is interest in the hybridization of cultivated strains with *A. bisporus* collected from nature. Knowledge of the genetics of *A. bisporus* remains sparse, but the genetics and life cycle of *Saccharomyces cerevisiae* have been intensively studied (page 72). Its entire genome has been sequenced and functional roles for all of these genes are rapidly being elucidated (page 88). However, the strains used in brewing often sporulate poorly or not at all, spores if obtained may prove difficult to germinate, and mating of the resultant haploids may not be easy. In spite of difficulties, mating has been used in strain improvement. For example, mating among haploids obtained from the sporulation of a diploid lager yeast gave some diploids with better flocculation and lower diacetyl production, both desirable features, than occurred with the parent strain. Here the beneficial results will have been obtained through the elimination of undesirable dominant genes in the original diploid; the introduction of new genetic material requires mating with the haploid progeny of a different diploid strain.

Parasexual hybridization

Many species of *Aspergillus*, *Penicillium* and other fungi of industrial importance lack a sexual cycle. They are, however, able to undergo a parasexual cycle (page 256). The first step in the cycle, heterokaryon formation, is now often accomplished by fusing the protoplasts of the two strains (page 105). The next step, vegetative diploidization, is normally a rare event, but its frequency can be increased greatly by treating cultures with camphor or UV radiation. Other treatments can be used to increase both the frequency of recombination between homologous chromosomes and of haploidization, the return to the haploid state. The parasexual cycle has proved useful in bringing about genetic recombination between closely related industrial strains, for example between strains of *P. chrysogenum* with high yields of penicillin and ancestral strains with lower yields but superior growth and sporulation. Vegetative incompatibility hinders heterokaryon formation between distantly related strains, and although protoplast fusion can to some extent bypass the barrier, there has been rather little success in obtaining recombinants between such strains.

Gene Cloning

The methods of strain improvement discussed in the previous sections (pages 526–529) are uncertain in their effects. A mutagen that has brought about the desired genetic change in a strain may at the same time have caused other mutations that are disadvantageous. The meiotic or mitotic recombination involved in the sexual and parasexual cycles will result in progeny that differ in a variety of ways from their parents. It is, however, possible to obtain very precise genetic changes in an organism by utilizing gene cloning. For example, a mutant gene can be used to replace its allele without bringing about any changes in the rest of the genome and hence any unintended effects. There is currently a great deal of research on the application of **gene cloning** to fungi and in future it will be widely used for the genetic improvement of industrial strains. Its relevance for the future of fungal biotechnology is, however, much wider and will be dealt with further below.

Gene Cloning and the Future of Fungal Biotechnology

During the past three decades procedures have been developed which together constitute a technology of great power. This technology, variously termed **genetic engineering**, **genetic manipulation**, **recombinant DNA technology** and **gene cloning**, has revolutionized both our knowledge of living organisms and our ability to change them. DNA is extracted from a **donor** – animal, plant or microbe – and treated in various ways. A desired portion of the donor DNA is then introduced into a **host** organism. Here it may merely be replicated, or may in addition be expressed – transcribed into RNA and translated to yield protein. At first the only host available was the bacterium *Escherichia coli*. With this host much information of value was obtained about the genes of many donors, including several fungi and humans. *E. coli* was also used to produce large quantities of protein and peptide hormones that are only produced in minute amounts in mammals. This made them available for detailed study and for therapeutic use. Subsequently, ways of using yeasts and moulds as hosts for gene cloning were devised. As a result, selected fungi can be used for the production of heterologous proteins, i.e. ones that would not normally be made by the species in question, including mammalian proteins.

The Production of Mammalian Proteins by Fungi

The bacterium, *Escherichia coli*, was not only the first host for gene cloning, but also the first microbe to be used for the commercial production of mammalian proteins. Human insulin, human growth hormone and the antiviral factor interferon have all been produced using *E. coli*. In spite of these successes, *E. coli* is not ideal for the production of heterologous proteins, especially those of mammalian origin. As a polypeptide chain is synthesized in a mammalian cell it folds and disulphide bridges are formed to give a protein with a precise three-dimensional structure. The folding and disulphide bridge formation can occur differently in a prokaryotic cell, giving a protein with the same amino acid sequence but different properties. In mammalian cells a peptide can be modified in a variety of ways following its synthesis. Hence the active product is sometimes a protein in which part of the original polypeptide chain has been removed by proteolytic cleavage. Alternatively, the active product may result from the enzymic addition of groups such as oligosaccharides, phosphates or sulphates to specific amino acids. Bacterial cells are usually incapable of these post-translational modifications. Furthermore *E. coli* usually retains the proteins that it makes within the cell instead of exporting them into the culture medium. This means that downstream processing is more difficult, especially since *E. coli* produces endotoxins (toxins associated with the cell) and pyrogens (fever-inducing materials) which must be eliminated completely during purification. Finally *E. coli* is a normal inhabitant of the human gut, so that there is a risk that an accident could cause infection – a potentially serious matter if the strain is producing pharmacologically active material.

These shortcomings have led to research aimed at finding ways of using other species as hosts for gene cloning. The first fungus with which success was achieved was the yeast *Saccharomyces cerevisiae*. Biochemists, geneticists and fermentation technologists are well versed in handling this species, and centuries of experience in baking and brewing make it clear that it does not produce toxins or infect humans. It has, however, the ability to secrete only a limited range of proteins. There are ways of overcoming this problem (page 357), but the proteins then secreted may not always have the expected properties. *S. cerevisiae* is hence not always the ideal host for the expression of mammalian genes and other fungal hosts are being sought. For example, another yeast, *Kluyveromyces lactis*, has been found to be an efficient host for synthesizing and secreting bovine prochymosin. This can be converted readily into the commercially important milk-clotting enzyme, chymosin, used in cheese manufacture (page 477). Another yeast, the methylotroph *Hansenula polymorpha*, is also used as a host for the production of heterologous proteins. Among the moulds several species of *Aspergillus*, a genus well known in fermentation technology for the efficient secretion of a wide variety of enzymes, are used as hosts for mammalian gene expression.

Obtaining DNA sequences that code for mammalian proteins

Many mammalian genes contain one or more introns, sequences of nucleotides that do not correspond to any part of the amino acid sequence of the protein specified by the gene. The nucleotide sequence of such a gene is transcribed into mRNA, but this mRNA is processed or 'edited' before being passed from the nucleus to the cytoplasm. During processing the introns are removed from the molecule of mRNA and the rest of the molecule rejoined or 'spliced', to give a molecule of processed mRNA. The nucleotide sequence of this processed mRNA is translated in the cytoplasm to give the amino acid sequence of the protein. Prokaryotes are unable to carry out the processing of mRNA described above. Hence, when *E. coli* is used for the production of a mammalian protein, it has to be provided with a DNA sequence corresponding to that of processed mRNA. Such a DNA sequence is known as **complementary DNA** or **cDNA**. Fungi, being eukaryotes, are able to process mRNA but introns are rather infrequent in the nuclear DNA of fungi, and the processing is not in all respects identical with that in mammals. Here too, therefore, the use of the appropriate cDNA is more certain to provide an active protein than is the mammalian gene.

Most of the mammalian proteins of therapeutic interest are produced by specialized cells. Such cells will also contain the relevant mRNA which will be absent from other types of cells. Sometimes only a single protein is made in large amounts, for example, the islet cells of the pancreas produce large amounts of insulin and very little other protein. In such cells nearly all the mRNA will be of the relevant type, simplifying isolation and purification. Ingenious methods have been developed for dealing with cells that contain more than one type of mRNA. For example, one procedure eliminates all the mRNAs common to two classes of cells, leaving only a few specialized types of mRNA for separation. The enzyme **reverse transcriptase** can be used to make a single-stranded DNA copy of the mRNA, and then a DNA polymerase to get double-stranded cDNA. As an alternative to the preparation of cDNA from mRNA, it may be possible, if the amino acid sequence of a protein is known, to prepare the corresponding DNA by chemical synthesis. Once a small amount of the required DNA sequence is available, as much as is required can be made by the **polymerase chain reaction** (page 326). Sometimes it is not necessary to prepare cDNA, the required sequence having already been cloned into *E. coli*.

The introduction of DNA into fungal cells

A prerequisite for gene cloning is the availability of a **vector**, a DNA molecule that will survive and replicate in the chosen host and into which the gene to be cloned can be inserted. A vector is usually a circular DNA molecule that can be opened with a **restriction enzyme**, an endonuclease that recognizes a specific nucleotide sequence and cuts the molecule at that point and at no other. A preparation of the opened vector circles is then mixed with the mammalian or other DNA to be inserted into the vector, and the mixture treated with the enzyme DNA ligase which rejoins cut DNA molecules. A proportion of the closed DNA circles that result will be vector molecules with the added DNA sequence. Host cells are then exposed to the preparation, after which cells that have acquired the vector have to be selected. Finally, it is necessary to establish that the desired DNA sequence has been introduced into the host along with the vector.

The yeast *Saccharomyces cerevisiae* has been used extensively as a host in gene cloning, and a variety of vectors have been employed, many of them, as in Fig. 8.19A, having been constructed from the 2 μm DNA plasmid of *S. cerevisiae* (page 92), using restriction enzymes and DNA ligase. The 2 μm plasmid has an origin for DNA replication and genes that are needed for replication and the maintenance of numerous copies, up to about 100, in the yeast cell. About half the plasmid, however, is not concerned with these functions and can be dispensed with in vector construction. A vector has to have a **cloning site** containing sequences that can be cut by restriction enzymes to permit the insertion of the DNA to be cloned. A polylinker cloning site is now often employed, a polylinker being a short synthetic double-stranded sequence that consists of a number of sites on which different restriction enzymes can act, giving experimental versatility. With an identical site at each end of the polylinker, the entire polylinker can be removed with a single restriction enzyme, and replaced with inserted DNA. The thick cell walls of yeasts and other fungi are a major obstacle to the entry of DNA into the cell. Hence initially cells of potential hosts had to be converted into protoplasts (page 105) prior to treatment with the vector and with calcium chloride and polyethylene glycol which facilitates DNA uptake. It was then found that lithium ions will render fungal walls permeable to DNA, so the use of protoplasts is not necessary. Recipes have been refined to give increasingly greater efficiency of transformation. A typical procedure is to incubate suitable cells with a mixture of lithium acetate, polyethylene glycol, single-stranded 'carrier' salmon sperm DNA (which increases efficiency) together with the plasmid, followed by a heat shock at 42°C. This should give a yield of more than 2×10^6 transformants per microgram of plasmid per 10^8 cells. When a vector based on the 2 μm plasmid is employed, a plasmid-free host strain is used, so that plasmids will not compete with the vector. Selection of the treated cells that have acquired the vector is achieved through the inclusion in the vector of a **selectable**

marker. The vector, for example, may have inserted within it the *S. cerevisiae* chromosomal gene *LEU2*, essential for leucine biosynthesis. Then if a host strain without *LEU2* is chosen, and a medium lacking leucine employed, only cells transformed by the vector will grow.

Alternatively the vector may include a gene giving resistance to an antibiotic or other factor to which the host is normally susceptible, and selection for transformed cells may be carried out on a medium containing the factor. Most vectors used with yeast are **shuttle vectors**, able to replicate in a second species, *Escherichia coli*. As already indicated, gene cloning was first carried out with *E. coli*, and the vectors used were able to replicate only in *E. coli*. This species remains a good one for the initial cloning of DNA and for the production of cloned DNA for structural analysis. However, the function of eukaryote genes is best studied with a eukaryote host, motivating the development of shuttle vectors enabling genes to be cloned in *E. coli* and studied in another host. One of the earliest shuttle vectors was developed from an *E. coli* plasmid vector, pMBa, and the 2 µm plasmid of *S. cerevisiae*. A shuttle vector able to survive in *E. coli* as well

Figure 8.19   Vectors for gene cloning in fungi. A, A yeast episomal plasmid (YEp) constructed from portions of the 2 µm DNA plasmid of *Saccharomyces cerevisiae* (light grey) and of an *Escherichia coli* plasmid vector such as pBR322 (dark grey). The YEp is a shuttle vector, incorporating sequences necessary for replication not only in *S. cerevisiae* but also in *E. coli*. The replication origin sites active in the yeast and the bacterium are indicated by O_y and O_b, respectively. The selective marker for survival in the yeast is a yeast chromosomal gene (HIS) needed for histidine biosynthesis, and for survival in the bacterium a bacterial plasmid gene (AMP) for resistance to the antibiotic ampicillin. The points at which four different restriction enzymes (R1–R4) act are indicated by R1–R4. Enzyme R1 was used to cut pBR322 for the insertion of the HIS gene, and its application to the vector would release the gene. Insertion of DNA at R2 would inactivate the HIS gene, but R3 and R4 are possible cloning sites. For the construction of an expression vector, able to express a range of heterologous genes, the cloning site would have to lie between a promoter and a terminator. B, The relationship between yeast vectors. Yeast integrating plasmids (YIps) can be constructed from bacterial plasmid vectors by the insertion of a yeast chromosomal gene such as LEU or HIS. A host with a deficiency with respect to the gene will then survive on a minimal medium only if the vector is integrated, with the active gene replacing the inactive homologous DNA. Vector replication in yeast can be obtained either by the inclusion of yeast 2 µm plasmid DNA to give a yeast episomal plasmid (YEp) or of an autonomously replicating sequence (ARS) from a yeast chromosome to give a yeast replicating plasmid (YRp). The addition of a yeast centromere (CEN) to a YRp yields a yeast centromere plasmid (YCp), and the further addition of telomeres (TEL) gives a yeast artificial chromosome (YAC). C, An integrating plasmid for transformation and gene expression in the filamentous fungus *Aspergillus nidulans*. The portion from an *E. coli* plasmid vector (dark grey) contains an origin (O_b) for replication in the bacterium, for which there is also a selective marker for ampicillin resistance (AMP). The selective marker for the fungal host is a gene for pyrimidine biosynthesis (PYR) from *Neurospora crassa*. However, increasing use is now being made of selective markers for resistance to antifungal antibiotics such as bleomycin or hygromycin, since it is then unnecessary to have a mutant host with a nutritional deficiency. The vector is designed for the expression of the mammalian prochymosin gene (PCH) which has upstream the *A. nidulans* glucoamylase promoter (PRM) and a signal sequence (S) for secretion, and downstream the glucoamylase terminator (TER). A sequence (ANS) that increases transformation frequency in *A. nidulans* is also included.

A

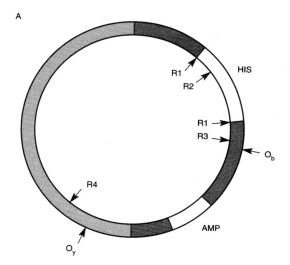

B

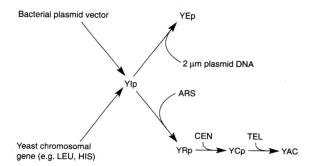

C

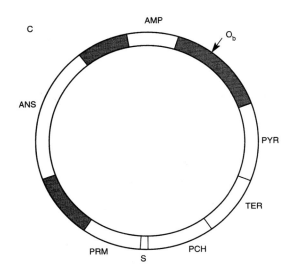

as in *S. cerevisiae* has to contain a site for the initiation of replication that will function in *E. coli* as well as a selectable marker that will be suitable for that host, for example a gene for ampicillin resistance. Where the production of a mammalian protein by yeast is envisaged, the relevant gene, after cloning in *E. coli* using a shuttle vector, is transferred to *S. cerevisiae*. However, for the expression of the gene it must be preceded (with respect to the direction of mRNA transcription) by a **promoter,** so that RNA polymerase is bound and transcription initiated, and followed by a **terminator**, which brings transcription to a close. Hence in an expression vector (one designed to bring about gene expression) the cloning site must be flanked 'upstream' by a promoter (the properties of which will be considered further, page 537) and 'downstream' by a terminator. Vectors based on the 2 μm plasmid and having as a selectable marker, a yeast chromosomal gene, are known as **yeast episomal plasmids** (YEps). An episome is a plasmid that can replicate independently but can also be integrated into a chromosome. YEps may do this as a result of sequence homologies between the selective marker gene of the vector and the mutant allele of the host, allowing recombination. YEps are very effective in bringing about transformation, and occur in high copy number. Hence there can be a high rate of mRNA transcription and a high output of the protein specified by the cloned gene. However, in the absence of selective conditions, YEps are gradually lost from the host.

Vectors other than YEps (Fig. 8.19B) have been developed for *S. cerevisiae*. **Yeast replicative plasmids** (YRps) carry a replication origin obtained from yeast chromosomal DNA, known as an autonomous replicating sequence (ARS), and hence YRps are sometimes termed ARS vectors. **Yeast centromere plasmids** (YCps) or minichromosome vectors carry a chromosomal centromere in addition to an ARS sequence. A centromere is the chromosomal region involved in the correct distribution of chromosomes to daughter cells during cell division. **Yeast integrating plasmids** (YIps) are vectors that must integrate into a host chromosome in order to replicate. The frequency of transformants obtained is low, but transformants are very stable. **Yeast artificial chromosomes** (YACs) resemble chromosomes in that they are linear with a centromere and at each end a telomere, a sequence that prevents attack on chromosome ends by exonucleases. YACs, which have to be large to be stable, are ideal for cloning long DNA sequences. The various types of vector now available for *S. cerevisiae*, and the extensive biochemical and genetic knowledge of the organism, make it very useful as a host for gene cloning.

Although the range of approaches available for gene cloning in fungi is well illustrated by reference to *S. cerevisiae*, vectors have now been developed for many other fungal hosts. In the yeast *Kluyveromyces lactis* vectors equivalent to *S. cerevisiae* YEps have been obtained from a 1.6 μm circular plasmid, and also from linear killer plasmids (page 92). In addition an integrating vector, equivalent to a YIp, has been developed as an expression vector for the production and secretion by *K. lactis* of the mammalian protein prochymosin, readily convertible to the commercially important enzyme, chymosin. Transformation has now been achieved with a wide variety of filamentous fungi. Few plasmids are known in filamentous fungi, and most of the vectors used are equivalent to YIps, being constructed from an *E. coli* vector and a selectable marker effective in the host fungus (Fig. 8.19C). The latter, with sequences homologous with those of the corresponding mutant gene in the host, facilitates integration into the host chromosome.

Recent experiments have shown the potential of further methods for the transformation of fungi with exogenous DNA. One is the discovery that the plant pathogenic bacterium, *Agrobacterium tumefaciens* can transfer its Ti DNA plasmid to protoplasts and spores of a range of yeasts and filamentous fungi, with integration into host chromosomal DNA. Another is the use of ballistic transformation, in which DNA-coated gold particles are fired from a 'gun' into the fungal cells. Both methods have been used widely for the introduction of heterologous genes into plant cells, but it is somewhat surprising that they have been successful with fungal cells.

The expression of heterologous DNA in fungi
Various factors determine the effectiveness with which a gene is expressed, but of special importance is the nature of the promoter. Transfer of promoters between species have shown that promoters are usually efficient only in the species in which they naturally occur. Hence, in the construction of expression vectors, the promoter that is included is almost always from the host species in which gene expression is required. However, the promoters occurring in a species vary greatly in their properties. Some proteins are required by a cell in small and some in large amounts. The genes that specify the latter have strong promoters that give a high transcription rate and high protein production. Some enzymes are constitutive, being produced at all times by an active cell, whereas others are inducible, being formed only when their substrate, or a structurally related inducer, is present. Inducible enzymes are commonly also repressible, ceasing to be produced in the presence of certain metabolites. Genes specifying inducible enzymes have regulatory sequences associated with their promoters, which respond to the presence of inducers and repressors and determine whether RNA polymerase is bound and transcription initiated by the promoter. In the construction of a vector for gene expression, a strong promoter is needed. It is also useful for the promoter to be one that can have its activity regulated, so that the required protein can be synthesized at the most appropriate stage in a fermentation. Although in nature genes have their own promoters, one well-chosen promoter can be used in a vector for the expression of a wide range of heterologous genes. Examples of promoters that have been used in expression vectors are those for galactose epimerase and for acid phosphatase in *S. cerevisiae*, lactase in *K. lactis*, methanol oxidase in the methylotrophic yeast *Hansenula polymorpha*, alcohol dehydrogenase and glucoamylase in *Aspergillus nidulans* and cellobiohydrolase in *Trichoderma reesei*. The activity of most of these promoters can be induced or repressed. For example, starch acts as an inducer and xylose as a repressor with the *A. nidulans* glucoamylase promoter, and cellulose as an inducer and glucose as a repressor with the *T. reesei* cellobiohydrolase promoter. Other factors that influence the amount of protein synthesized are the nature of the leader sequence – the nucleotides between the expressed sequence of a gene and the promoter, which are transcribed but not translated – and the copy number of a gene. A strong promoter, however, will give abundant protein production from a single

copy of a heterologous gene integrated into a host chromosome along with the promoter and terminator.

Some of the proteins that a cell synthesizes are retained within the cell and others are secreted – transported to the cell surface and released. Such secretion greatly facilitates purification and is hence a desirable feature from the commercial viewpoint. For secretion to occur, a polypeptide has to begin with a signal sequence of 15–30 mainly hydrophobic amino acids. The signal sequence interacts with the endoplasmic reticulum, invaginations of the plasma membrane, to initiate secretion, and in the course of secretion is cleaved from the rest of the protein molecule. Signal sequences vary considerably between different genes and between species. Where secretion of heterologous protein is desired, DNA specifying a signal sequence has to be included in the vector. The efficiency with which a protein is secreted and even its properties can be influenced by the choice of the signal sequence. For the secretion of heterologous proteins by *S. cerevisiae* the signal sequence for α-factor (page 207) is commonly used. However, *S. cerevisiae* normally secretes few polypeptides or proteins, so is not very efficient at secretion. *K. lactis* secretes proteins more effectively, and here a variety of signal sequences, including that for the *S. cerevisiae* α-factor, have been employed. Many filamentous fungi, such as *Aspergillus*, secrete proteins very efficiently, and this is increasingly leading to their use as hosts for heterologous protein production.

Other Applications of Gene Cloning in Fungi

So far fungal biotechnology has centred on the products of fungi that can be grown easily on a large scale in pure culture. There are, however, many fungi, such as lichen mycobionts (page 76), that are difficult to grow in pure culture but which produce unique metabolites of potential commercial interest. It should become possible to transfer gene clusters involved in the synthesis of metabolites of interest from such fungi into readily cultured species. A similar transfer has already been accomplished with the gene cluster responsible for penicillin biosynthesis in *Penicillium chrysogenum,* resulting in the experimental production of the antibiotic by *Aspergillus niger* and *Neurospora crassa*. Gene cloning may also be used to enable strains that grow well to use different nutrients. This could endow fermentation organisms with the ability not only to use different and abundant substrates, but also to provide fungi able to break down some of the refractory and toxic wastes associated with industrial society. Finally, knowledge obtained by the application of gene cloning in basic mycological research is likely to have important practical applications.

Acknowledgement

We are grateful to Professor Joan Bennett for reading the first edition of this book and advising on the updating needed, especially with reference to Chapter 8.

Further Reading and Reference

Biotechnology makes use of both bacteria and fungi, and the problems involved in their use are similar. Hence most biotechnology textbooks deal with both groups. With respect to traditional products of fermentation, quite old works remain valuable, but in areas such as gene cloning and enzyme technology, only very recent publications can give a realistic picture of current practice and possibilities. This makes articles on biotechnology in semi-popular monthly journals and annual review volumes very useful for keeping aware of current developments.

General Microbial Biotechnology

Bains, W. (1998). *Biotechnology from A to Z*. Oxford University Press, Oxford.

Demain, A. L. & Davies, J. E., eds. (1999). *Manual of Industrial Microbiology and Biotechnology*, 2nd edn. Blackwell Science, Oxford.

Glazer, A. N. & Nikaido, H. (1995). *Microbial Biotechnology: Fundamentals of Applied Microbiology*. W.H. Freeman, New York.

Glick, B. R. & Pasternak, J. J. (1998). *Molecular Biotechnology*. American Society for Microbiology, Washington, DC.

Kreuzer, H. (1996). *Recombinant DNA and Biotechnology: A Student Guide*. American Society for Microbiology, Washington, DC.

Moses, V. & Cape, R. E., eds. (1991). *Biotechnology: the Science and the Business*. Harwood Academic Publishers, Chur, Switzerland.

Primrose, S. B. (1991). *Molecular Biotechnology*, 2nd edn. Blackwell, Oxford.

Rehm, H. -B. & Reed, G., eds. (1981–89). *Biotechnology*, 8 vols. VCH, Weinheim.

Scientific American (1981). *Industrial Microbiology and the Advent of Genetic Engineering*. Freeman, San Francisco. (Excellent illustrations of industrial processes and plant).

Smith, J. E. (1996). *Biotechnology*, 3rd edn. Arnold, London.

Walker, J. M. & Cox, M. (1995). *The Language of Biotechnology: A Dictionary of Terms*, 2nd edn. American Chemical Society, Washington.

Weide, H., Páca, J. & Knorre, W. A., eds. (1991). *Biotechnologie*. Gustav Fischer, Jena.

Gene Cloning, General

Brown, T. A. (1990). *Gene Cloning: an Introduction*, 2nd edn. Chapman & Hall, London.

Drlica, K. (1996). *Understanding DNA and Gene Cloning: A Guide to the Curious*, 3rd edn. Wiley, New York.

Glover, D. M. & Hames, B. D. (1998). *DNA Cloning: Expression Systems, a Practical Approach*. IRL Press, Oxford.

Kingsman, S. M. & Kingsman, A. J. (1988). *Genetic Engineering: an Introduction to Gene Analysis and Exploitation in Eukaryotes*. Blackwell, Oxford.

Old, R. W. & Primrose, S. B. (1994). *Principles of Gene Manipulation: an Introduction to Genetic Engineering*, 5th edn. Blackwell, Oxford.

Oliver, S. G. & Ward, J. M. (1985). *A Dictionary of Genetic Engineering*. Cambridge University Press, Cambridge.

Rapley, R. & Walker, J. M. (1998). *Molecular Biomethods Handbook*. Humana Press, Totowa, NJ.

Tait, R. C. (1997). *An Introduction to Molecular Biology*. Horizon Scientific Press, Norfolk.

Fungal Biotechnology, General

Anke, T., ed. (1997) *Fungal Biotechnology*. Chapman & Hall, London.

Arora, D. K., Elander, R. P. & Mukerji, K. G., eds. (1992). *Handbook of Applied Mycology*, vol. 4. *Fungal Biotechnology*. Marcel Dekker, New York.

Berry, D. R., ed. (1988). *Physiology of Industrial Fungi*. Blackwell, Oxford.

Berry, D. R., Russell, I. & Stewart, G. G., eds. (1987) *Yeast Biotechnology*. Allen & Unwin, London.

Finkelstein, D. B. & Ball, C., eds. (1992). *Biotechnology of Filamentous Fungi: Technology and Products*. Butterworth-Heinemann, Boston.

Kück, U. (1994). *Genetics and Biotechnology*, Vol. II, *The Mycota*, K. Esser & P. A. Lemke, eds. Springer-Verlag, Berlin.

Leatham, G., ed. (1993). *Frontiers in Industrial Mycology*. Chapman & Hall, London.

Smith, J. E., Berry, D. R. & Kristiansen, B., eds. (1983). *The Filamentous Fungi*, vol. 4. *Fungal Technology*. Arnold, London.

Wainwright, M. (1992). *An Introduction to Fungal Biotechnology*. Wiley, Chichester.

Walker, G. M. (1998). *Yeast Physiology and Biotechnology*. Wiley, Chichester.

Wolf, K. (1996). *Nonconventional Yeasts in Biotechnology: A Handbook*. Springer-Verlag, Berlin.

Gene Cloning in Fungi

Bennett, J. W. & Lasure, L. L., eds. (1985). *Gene Manipulations in Fungi*. Academic Press, Orlando.

Bennett, J. W. & Lasure, L. L., eds. (1991). *More Gene Manipulations in Fungi*. Academic Press, New York.

Chaure, P., Gurr, S. J. & Spanu, P. (2000). Stable transformation of *Erysiphe graminis*, an obligate biotrophic pathogen of barley. *Nature Biotechnology* **18**, 205–207.

De Groot, M. J. A., Bundock, P., Hooykaas, P. J. J. & Beijersbergen, A. G. M. (1998) *Agrobacterium tumifaciens*-mediated transformation on filamentous fungi. *Nature Biotechnology* **16**, 839–842.

Kinghorn, J. R. & Turner, G., eds. (1992). *Applied Molecular Genetics of Filamentous Fungi*. Chapman & Hall, Andover.

Leong, S. A. & Berka, R. M., eds. (1991). *Molecular Industrial Mycology: Systems and Applications for Filamentous Fungi*. Marcel Dekker, New York.

Peberdy, J. F., Caten, C. E., Ogden, J. E. & Bennett, J. W., eds. (1991). *Applied Molecular Genetics of Fungi. Eighteenth Symposium of the British Mycological Society*. Cambridge University Press, Cambridge.

Fermentation and Enzyme Technology

Atkinson, B. & Mavituna, F. (1990). *Biochemical Engineering and Biotechnology Handbook*, 2nd edn. Macmillan, Basingstoke.

Bickerstaff, G. F. (1996). *Immobolization of Enzymes and Cells. Methods in Biotechnology*, Volume 1. Humana Press, Totawa, NJ.

Calam, C. T. (1987). *Process Development in Antibiotic Fermentations*. Cambridge University Press, Cambridge.

Chaplin, M. F. & Bucke, C. (1990). *Enzyme Technology*. Cambridge University Press, Cambridge.

Fogarty, W. M. & Kelly, C. T. (1990). *Microbial Enzymes and Biotechnology*. 2nd edn. Elsevier Applied Science, London.

McNeil, B. & Harvey, L. M. (1990). *Fermentation: a Practical Approach*. IRL Press, Oxford.

Nielsen, R. I. & Oxenbøll, K. (1998). Enzymes from fungi: their technology and uses. *Mycologist* **12**, 69–71.

Pyle, D. L. (1993). *An Introduction to Process Biotechnology*. Blackwell, Oxford.

Ward, O. P. (1989). *Fermentation: Principles, Processes and Products*. Wiley, Chichester.

Woodward, J., ed. (1985). *Immobilized Cells and Enzymes*. IRL Press, Oxford.

Food and Beverage Fermentations

Arora, D. K., Mukerji, K. G. & Marth, E. H., eds. (1991). *Handbook of Applied Mycology*, vol. 3. *Food and Feed*. Marcel Dekker, New York.

Bamforth, C. (1998). *Beer: Tap into the Art and Science of Brewing*. Plenum, New York.

Beuchat, L. R., ed. (1987). *Food and Beverage Mycology*, 2nd edn. Van Nostrand Reinhold, New York.

Board, R. G., Jones, D. & Jarvis, B., eds. (1995) Fermentation: foods, feeds and condiments. *Journal of Applied Bacteriology, Symposium Supplement* **79**, 96S–107S.

Chang, S. T. & Miles, P. G. (1992). Mushroom biology – a new discipline. *Mycologist* **6**, 64–65.

Chang, S. T., Buswell, J. A. & Miles, P. G., eds. (1993) *Mushroom Biology and Mushroom Products*. The Chinese University Press, Hong Kong.

Hesseltine, C. W. (1983). Microbiology of fermented foods. *Annual Review of Microbiology* **37**, 575–601.

Hornsey, L. (1999). *Brewing*. Royal Society of Chemistry, London.

McGee, H. (1991) *On Food and Cooking*. Harper Collins, London.

Norris, J. R. & Pettipher, G. L., eds. (1987). *Essays in Agricultural and Food Microbiology*. Wiley, Chichester.

Rose, A. H., ed. (1982). *Fermented Foods*. Academic Press, London.

Trinci, A. P. J. (1992). Mycoprotein: a twenty-year overnight success story. *Mycological Research* **96**, 1–13.

Special Topics

Baumberg, S., Hunter, I. & Rhodes, M., eds. (1989). *Microbial Products: New Approaches. Forty-fourth Symposium of the Society for General Microbiology*. Cambridge University Press, Cambridge.

Bull, A. T., Goodfellow, M. & Slater, H. J. (1992). Biodiversity as a source of innovation in biotechnology. *Annual Review of Microbiology* **46**, 219–252.

Hacking, A. J. (1986). *Economic Aspects of Biotechnology*. Cambridge University Press, Cambridge.

Harman, G. E. & Kubicek, C. P., eds. (1998). *Trichoderma* and *Gliocladium*, Vol. 2. *Enzymes, Biological Control and Commercial Applications*. Taylor & Francis, London.

Kristiansen, B., Linden, J. & Mattey, M., eds. (1999). *Citric Acid Biotechnology*. Taylor & Francis, London.

Langley, D. (1997). Exploiting the fungi. *Mycologist* **11**, 165–167.

Powell, K. A., Renwick, A. & Peberdy, J. F., eds. (1994). *The Genus Aspergillus*. Plenum Press, New York.

Smith. J. E., ed. (1994). *Aspergillus. Biotechnology Handbooks*, Vol. 7. Plenum Press, New York.

Tribe, H. (1998). The discovery and exploitation of cyclosporin. *Mycologist* **12**, 20–22.

Some Biotechnology Journals

Articles on biotechnology can be found in many of the journals listed at the end of Chapter 1. The journals given below, however, are wholly or largely devoted to biotechnology.

Advances in Applied Microbiology
Advances in Biochemical Engineering and Biotechnology
Bio/Technology
Biotechnology and Genetic Engineering Reviews
Current Opinion in Biotechnology
Nature Biotechnology
Trends in Biotechnology

Questions

With each question one statement is correct. Answers on page 562.

8.1 Mycoprotein for human consumption is grown in:
    a) a solid state fermentation
    b) a chemostat
    c) an airlift fermenter
    d) a stirred tank fermenter

8.2 The most important fungi for commercial enzyme production are species of:
    a) *Aspergillus*
    b) *Agaricus*
    c) *Neurospora*
    d) *Saccharomyces*

8.3 Most wine is made commercially with:
    a) a culture of *Saccharomyces cerevisiae*
    b) the natural fungal flora on the grapes
    c) both a) and b)
    d) an inoculum from the previous fermentation

8.4 A characteristic of fungal secondary metabolites is that:
    a) they are produced by exponentially growing cells
    b) most are produced as families of related molecules
    c) each is produced by just one specific strain of fungus
    d) they play essential roles in spore formation

8.5 Which of the following is not true about cyclosporin:
    a) it is a cyclic peptide
    b) it is produced by strains of several fungal species
    c) it has antifungal as well as immunosuppressive activity
    d) its amino acid sequence is encoded by a specific messenger RNA

Appendix 1

Glossary of common mycological terms

The categories used in classifying fungi and the names of major groups are given in Chapter 2 and Appendix 2. Terms widely used in science or occurring in most dictionaries are generally not included, nor are terms used only at one point in the text and there explained, and which can be easily located with the index. For terms from molecular biology and genetics the reader is referred to dictionaries such as Lackie, M. & Dow, J. (1999). *The Dictionary of Cell and Molecular Biology*, 3rd edn. Academic Press, London.

Aeciospores   Binucleate infective spores produced by a rust fungus.
Air spora   The population of spores found in the air.
Anamorph (Greek: *morphe*, shape)   The form of a fungus produced in its asexual sporing phase
Anastomosis   Fusion between hyphae.
Anemotropism (Greek: *anemos*, wind)   Tropism (qv) into a wind.
Antheridium (-a)   Male gametangium (qv).
Apothecium (-a)   Cup-shaped fruiting body (qv) of some Ascomycetes.
Appressorium   Adhesive pad formed by the hypha of a pathogenic fungus on the surface of its host to aid penetration.
Arthrospore (Greek: *arthron*, joint)   Spore formed by the breaking up of a length of mycelium into segments which become spores.
Ascocarp   A structure bearing asci, the fruit body (qv) of Ascomycetes.
Ascogenous hypha   The dikaryotic hypha (qv) emerging from an ascogonium (qv) after fertilization, which gives rise to the asci in Ascomycetes.
Ascogonium   Cell of Ascomycete protoperithecium (qv) taking part in fertilization.
Ascoma (plural, ascomata)   Synonym for ascocarp.
Ascospore   Sexual spore of Ascomycetes.

Ascus (Greek: *askos*, leather bag) The microscopic sac containing the ascospores in Ascomycetes.

Axenic (Greek: *a*, not; *xenos*, stranger) In the absence of contamination (qv).

Ballistospore The actively-discharged spore of Basidiomycete.

Basidiocarp The fruit body (qv) of Basidiomycetes.

Basidioma (plural, basidiomata) Synonym for basidiocarp.

Basidiospore Sexual spore of Basidiomycetes.

Basidium (-a) The enlarged terminal cell of a hypha which bears basidiospores, in Basidiomycetes.

Biomass The mass (in grams, etc.) of living material present.

Biotroph (Greek: *bios*, life; *trophe*, food) Fungus deriving its nutrients from living cells of a host.

Bipolar Bipolar heterothallism: requirement for different alleles at one mating type locus, for sexual compatibility.

Chemostat Continuous culture system in which the population size is held constant by control of the rate at which nutrients are supplied.

Chemotaxis Taxis (qv) towards or away from the source of a specific chemical.

Chemotropism Tropism (qv) to or from a specific chemical.

Chitin (Greek: *chiton*, coat of mail/garment) The main structural component of the walls of higher fungi, a polymer of N-acetylglucosamine.

Chlamydospore (Greek: *chlamys*, cloak) Thick-walled, usually asexual, resting spore (qv).

Clamp connection Short, backwardly-directed side branch found only in hyphae of Basidiomycetes, formed at the time of septum formation, and fusing with the cell to the rear.

Cleistothecium (-a) (Greek: *cleistos*, closed) Fruiting body (qv) of some Ascomycetes in which the asci are entirely surrounded by a wall of mycelium.

Colony An assemblage of hyphae which usually develops from a single source and grows in a coordinated way.

Conidiophore (Greek: *phoreo*, I bear) A hypha that gives rise to conidia.

Conidium (-a) An asexual spore produced on the external surface of mycelium, not in a sporangium (qv).

Contamination Growth of unwanted microbes in cultures that should contain a single, or sometimes a few, specified types of microbes.

Coprophilous (Greek: *copros*, dung; *phileo*, I love) Dung-inhabiting.

Coremium (-a) Synonym for synnema (qv).

Cornmeal Ground maize, used as a fungal growth medium.

Cyst A spherical cell, formed from the amoebae of slime moulds or from the zoospores of lower fungi, by wall formation (encystment, qv).

Dermatophyte A fungus infecting the skin (Greek: *derma*, skin; *phyton*, a plant).

Dikaryon (Greek: *dis*, two; *karyon*, nut) Mycelium containing two genetically different types of nuclei. Usually refers to Basidiomycetes with two nuclei of different mating type (qv), in each hyphal compartment (qv).

Dikaryotization Formation of a dikaryon by fusion and nuclear migration between monokaryons (qv).

Dolipore septum Septum with elaborate ultrastructure found in Basidiomycetes.

Elicitor Substance derived from a plant pathogenic fungus which induces a plant to resist infection.

Encystment Formation of a tough wall, as occurs round zoospores when they settle and start to grow.

Endophyte (Greek: *endon*, within; *phyton*, a plant) Fungus which inhabits a plant harmlessly.

Facultative Possible but not obligatory.

Fermentation Form of catabolism not requiring oxygen or other external electron acceptor.

Fermenter Vessel used for producing a microbial product such as alcohol, by fermentation.

Filamentous fungus A fungus with hyphae, not unicellular like yeast.

Fruit(ing) body The large spore-bearing structure in Ascomycetes and Basidiomycetes (e.g. mushrooms, truffles).

Gametangium Part of a hypha specialized for fusion in sexual reproduction.

Gene-for-gene hypothesis The theory that a plant's gene for resistance to its pathogen is matched by a gene in the pathogen, and both are involved in triggering plant resistance.

Geotaxis Taxis (qv) in response to gravity.

Geotropism Tropism (qv) in response to gravity.

Germ-tube The hypha that emerges from a spore.

Haustorium The part of a symbiotic fungus that lives within a host cell.

Heterokaryon Mycelium or hypha with nuclei of two or more different genotypes.

Heteroplasmon Mixture of protoplasm from more than one source.

Heterothallism (Greek: *heteros*, different) Requirement for two compatible mating types for the sexual process. Self-sterility.

Homokaryon A mycelium or hypha with nuclei of only one genotype.

Homothallism (Greek: *homo*, the same) A second mating type not required for the sexual process. Self-fertility.

Horizontal resistance A form of resistance by plants, giving some protection from attack by all strains of a fungal pathogen. Contrast vertical resistance (qv).

Host The larger organism in a symbiotic or parasitic relationship.

Hydrophobin (Greek: *hydro*, water; *phobos*, fear) Hydrophobic (i.e. unwettable) protein which is produced and added to the hyphal wall at the time of fruit body development, allowing hyphae to emerge from an aqueous environment (page 102).

Hymenium (Greek: *hymenaeos*, wedding) The surface of a fruit body (qv) on which sexually produced spores are borne.

Hypha (Greek: *hypha*, thread) The tubular cell growing at one end which is the developmental unit of mycelium.

Hyphal growth unit   The ratio between total length of hyphae in a mycelium and the number of hyphal tips.

Incubator   Cabinet for maintaining controlled conditions for growing microbes.

Isolate (noun, from verb, to isolate)   A strain (qv) of a fungus brought into pure culture (i.e. isolated) from nature.

Karyogamy (Greek: *karyon*, nut; *gamos*, wedding)   Fusion of nuclei preceding the production of sexually-produced spores.

Mating type   The factor determining whether a strain will or will not be able to mate with another strain.

Medium   A preparation used for culture of fungi or other microbes. Usually appropriate nutrients dissolved in water, used either in liquid form or gelled with agar.

Mildew   A plant disease with prominent surface growth of the fungus. Downy mildews are caused by Peronosporalceae and powdery mildews by Erysiphales.

Monkaryon (Greek: *karyon*, a nut; *monos*, alone)   Strain with nuclei of a single genotype. Often refers to a Basidiomycete mycelium with nuclei of one mating type, and which must encounter a mycelium of different mating type before fruit bodies can be formed.

Mycelial strand or cord   Linear aggregate of hyphae formed behind an advancing margin in which the hyphae are separate.

Mycelium (Greek: *mykes*, fungus)   The mass of hyphae, not in the form of large structures such as mushrooms, of which the fungi are mainly composed.

Mycorrhiza (pl., strictly mycorrhizae, now usually mycorrhizas)   Symbiosis between plant root and fungal mycelium.

Mycostasis   The dormancy imposed on fungal spores by soil and sometimes leaf surface microbes.

Necrotroph (Greek: *necros*, death; *trophe*, food)   Fungus that kills the cells of a living host and subsequently utilizes their remains for food.

Obligate   The opposite of facultative, a condition in which the fungus has no alternative state.

Oidium (-a)   A form of asexual spore. The term is most often used for those produced on monokaryons (qv) of some Basidiomycetes. It is thought that these are more important in germinating on, and bringing about dikaryotization of, other monokaryons of the same species, than in establishing new colonies.

Oogonium (-a)   Female gametangium (qv).

Oospore   Diploid spore produced by Oomycetes.

Parasexuality   Sequence of nuclear fusion and irregular division accompanied by genetic recombination found to occur in some otherwise asexual fungi.

Pellets   Multihyphal structures formed when some fungi are grown in fermenters.

Perithecium (-a)   Small bottle-shaped fruiting body of some Ascomycetes, from the neck of which one ascus (qv) at a time discharges.

Petri dish   Shallow transparent dish with lid for the culture and observation of fungi and other microbes.

Phototaxis (Greek: *photos*, light)   Taxis (qv) to or from a light source.

Phylloplane (Greek: *phyllon*, leaf)   The region on the surface of a leaf which provides a habitat for microscopic fungi.

Phytoalexin (Greek: *phyton*, plant; *alexo*, defence)   Substance produced in plants when damaged, and which inhibits fungal growth.

Pileus (Latin: *pileus*, felt hat)   The cap of a mushroom.

Plasmodium (-a)   Mass of protoplasm formed by slime moulds.

Plasmogamy (Greek: *plasma*, a thing moulded/formed; *gamos*, a wedding)   The fusion of cytoplasm from two different hyphae which precedes nuclear fusion during the sexual cycle. In Basidiomycetes plasmogamy and nuclear fusion can be separated by a long interval.

Polyphosphate   Form in which mycelium stores phosphate reserves.

Primordium   The earliest visible stage in the development of a structure.

Protoperithecium (-a)   The structure found in mycelium of Ascomycetes which is the site of fertilization and subsequent fruit body development.

Protoplast   Spherical protoplasmic structure which results when the fungal cell wall is removed.

Pycniospores   Very small spores produced by rust fungi whose sole function is to dikaryotize the mycelium by fusing with receptive hyphae.

Radial growth   Growth from a centre. The radial growth rate of a colony is the rate at which the hyphal margin advances.

Resource   A source of nutrients for a fungus.

Resting spore   A spore with prolonged survival as its main role, or a spore that is in a state of dormancy.

Rheotaxis   Taxis (qv) against the direction of a stream (of water).

Rhizoid   A fine, branched hypha which grows into material like a root, and anchors surface mycelium.

Rhizomorph   Multihyphal structure with a root-like apex and mode of growth.

Rhizosphere   Region surrounding a root in which conditions for microbial growth are enhanced.

Rust fungus   A member of the Uredinales, plant pathogenic fungi many of which produce rusty-red spores.

Saprotrophic (Greek: *sapros*, rotton; *trophe*, food)   Using remains of dead organisms as food.

Sclerotium (-a)   Mass of hyphae with protective rind and containing food reserves.

Smut fungus   A member of the Ustilaginales, plant pathogenic fungi many of which produce sooty-black spores.

Sp., spp.   Abbreviations, sing. and pl., for species, used with a generic name. For example, *Agaricus* sp. means 'a species of *Agaricus*, specific name uncertain', and *Agaricus* spp. means 'various species of *Agaricus*'.

Species   The basic unit of classification, distinguished by a Latin binomial, e.g. *Agaricus bisporus*. For fuller discussion, see page 247.

Sporangiophore   A sporophore (qv) which bears a sporangium.

Sporangiospore   Asexual spore borne in a sporangium.

Sporangium (-a)   Sac containing sporangiospores.

Sporophore   A structure that bears spores. It is used not only for massive sexually produced fruit bodies, but also for small structures that bear asexual spores in lower fungi and slime moulds.

Sporopollenin   A very resistant biological polymer, found in the walls of pollen grains and some fungal spores.

Sporulation   The process of forming spores.

Sterigma (-ata)   Microscopic points at the end of a basidium (qv), on each of which a basidiospore is produced.

Stipe   The stalk of a mushroom or toadstool.

Strain   A genetic variety of a fungus, either an isolate (qv) from nature or arising by mutation or recombination in the laboratory.

Stroma (-ata)   A mass of hyphae on which spores or fruit bodies are borne.

Substratum   The physical surface on or in which mycelium grows.

Synnema (-ata)   Structure of aggregated hyphae bearing spores.

Taxis   (Greek: *taxis*, arranging)   Movement of a free cell, such as a swimming cell, towards or away from the source of a physical (e.g. light) or chemical (e.g. sex attractant) stimulus.

Teleomorph (Greek: *teleos*, finished; *morphe*, form)   The form in which a fungus exists when producing sexual spores, seen as the end of the life cycle.

Teliospores (Syn. teleutospores)   Overwintering spores formed by Uredinales at the end of the host's growing season.

Tetrapolar   Tetrapolar heterothallism: requirement for different alleles at each of two genetic loci, for sexual compatibility.

Thigmotropism   Tropism (qv) in response to touch.

Translocation   Transport of nutrients within mycelium by processes other than those of growth.

Trichogyne   Receptive hypha in Ascomycete fertilization.

Tropism (Greek: *tropos*, turn)   The bending of a hypha, through differential growth on near and far side, or a reoriented apex, towards or away from the source of a physical (e.g. light) or chemical (e.g. sex attractant) stimulus.

Turbidostat   Continuous culture system in which the number of cells in the culture vessel is kept constant by monitoring the cell density and adjusting flow rate.

Uredospores (Latin: *uredo*, blight upon plants). Abundantly produced dikaryotic spores of rusts (Uredinales).

Vegetative incompatibility   Inability of different mycelia of the same species to fuse successfully and function as one.

Vertical resistance   A form of resistance by plants to fungal attack that gives resistance, often total, to some strains of a pathogen but not to others. Contrast horizontal resistance (qv).

Vesicular–arbuscular mycorrhiza (VAM)   A mycorrhiza produced by members of the Glomales, in which highly characteristic structures are formed within host cells.

Water activity, Water potential   Measures of water availability (page 158).

Yeast   Depending on context, can mean baker's and brewer's yeasts (*Saccharomyces cerevisiae*) or any unicellular fungus multiplying by budding, or, in a few instances, fission.

Zoosporangium (-a)   Sac in which zoospores develop, and from which they are released.

Zoospore (Greek: *zoos*, living)   Spore which can swim in water using one or two flagella.

Zygospore (Greek: *zygon*, yoke)   Spore formed by fusion of two gametangia in Zygomycetes.

Appendix 2

Classification of Fungi

The classification below is based on that in the 8th edn of *Ainsworth & Bisby's Dictionary of the Fungi* (Hawksworth, Kirk, Pegler, Sutton & Ainsworth, 1995). Species in the text are listed, but where more than one member of a genus has been mentioned in the text, only one, or sometimes a few, are given here. Where the names for both an anamorphic and a teleomorphic state, or a synonym, are widely used, these are given (ana., tel., syn.). Yeasts are assigned to Ascomycetes, Basidiomycetes and mitosporic fungi, but with their yeast status indicated (Y).

The organisms studied by mycologists occur in three kingdoms, the Protozoa, Chromista and Fungi, in the domain Eukaryota.

The Kingdom Protozoa

Four protozoan phyla have been studied by mycologists. Two of these, the Dictyosteliomycetes and Acrasiomycetes (not mentioned further) lack plasmodia and are informally known as cellular slime moulds and two, the Myxomycota and Plasmodiophoromycota, form plasmodia and hence are plasmodial slime moulds.

I. Phylum **DICTYOSTELIOMYCOTA**
Cellular slime moulds, with an amoeboid trophic (feeding) phase, and multicellular fruit bodies. Contains a single order, the Dictyosteliales.
Dictyostelium discoideum

II. Phylum **MYXOMYCOTA**
Plasmodial slime moulds, with an amoeboid trophic (feeding) phase, a plasmodial phase also able to ingest food particles, and multicellular fruit bodies. Contains

two classes, the Myxomycetes and Protosteliomycetes, only the former being dealt with in the present work.

Didymium iridis
Physarum polycephalum

III. Phylum **PLASMODIOPHOROMYCOTA**

Contains a single class, the Plasmodiophoromycetes, obligate parasites of plants having minute intracellular plasmodia.

Plasmodiophora brassicae

The Kingdom Chromista

Three Chromistan phyla, the Oomycota, the Hyphochytriomycota and the Labyrinthulomycota have been studied by mycologists, but only the Oomycota is considered in detail in the present work.

I. Phylum **OOMYCOTA**, the Oomycetes

Most species have a mycelial trophic phase resembling that of fungi, and many produce zoospores with an anterior tinsel and a posterior smooth flagellum. Nine orders are recognized, with members of four being mentioned in the present work.

(a)   Order Saprolegniales. The water moulds
   Achlya bisexualis
   Aphanomyces eutiches
   Dictyuchus
   Nematophthora gynophila
   Pythiopsis
   Saprolegnia
   Thraustotheca

(b)   Order Leptomitales
   Aqualinderella fermentans
   Leptomitus

(c)   Order Pythiales
   Lagenidium
   Phytophthora infestans
   Pythium ultimum

(d)   Order Peronosporales
   (i) Family Peronosporaceae. The downy mildews
      Bremia lactucae
      Peronospora tabacina
      Plasmopora viticola
      Pseudoperonospora humuli
   (ii) Family Albuginaceae. The white rusts
      Albugo candida

II. Phylum **LABYRINTHULOMYCOTA**
Two orders of freshwater and marine organisms, only one of which is mentioned in the present text.
(a) Order Thraustochytriales
     Thraustochytrium

The Kingdom Fungi

Heterotrophic, absorptive organisms, typically mycelial, but sometimes unicellular, as in yeasts. There are four phyla, the Chytridiomycota, the Zygomycota, the Ascomycota and the Basidiomycota, and an informal category, the mitosporic fungi, members of which lack a sexual phase and have not yet been assigned to taxa in the formal system of classification.

I. Phylum **CHYTRIDIOMYCOTA**, the Chytridiomycetes
Zoospores, and gametes if motile, have posterior smooth flagella – usually one, occasionally more. Five orders are recognized, and four of these are mentioned in the present work.
(a) Order Blastocladiales
     Allomyces macrogynus
     Blastocladia
     Blastocladiella emersonii
     Catenaria anguillulae
     Coelomomyces
(b) Order Chytridiales. The chytrids
     Batrachochytrum dendrobatides
     Chytriomyces
     Rhizophylyctis rosea
     Rhizophydium
     Synchytrium endobioticum
(c) Order Monoblepharidales
     Monoblepharis
(d) Order Neocallimastigales. The anaerobic rumen fungi
     Neocallimastix frontalis
     Piromonas
     Sphaeromonas
(e) Order Spizellomycetales
     Karlingia asterocyta

II. Phylum **ZYGOMYCOTA**
Zygospores, large resting spores, result from the sexual process.
1. Class **Trichomycetes**. Obligate parasites of arthropods.
2. Class **Zygomycetes**. The Zygomycetes. Seven orders are recognized, and five are mentioned in the present work.
(a) Order Mucorales
     Absidia glauca
     Actinomucor elegans (syn. A. repens).

Amylomyces rouxii
Blakeslea trispora
Chaetocladium jonesii
Cunninghamella
Dicranophora fulva
Mortierella
Mucor mucedo
Mycotypha
Phycomyes blakesleeanus
Pilaira
Pilobolus kleinii
Rhizopus oryzae
Spinellus
Syzygites
Zygorhynchus

(b) Order Dimargaritales. Obligate parasites, mainly of Mucorales
Dimargaris
Dispira
Tieghemomyces

(c) Order Zoopagales. Parasites of fungi (mycoparasites) and of small animals such as amoebae and nematodes
Piptocephalis

(d) Order Entomophthorales. Parasites of insects and other arthropods
Basidiobolus ranarum
Conidiobolus coronatus
Entomophaga grylli
Entomophthora muscae
Erynia
Massospora

(e) Order Glomales. Mycorrhiza-forming fungi, the vesicular–arbuscular endophytes (VAMs)
Acaulospora
Entrophospora
Gigaspora
Glomus macrocarpum
Sclerocystis
Scutellospora

III. Phylum **ASCOMYCOTA**, the Ascomycetes

Asci, containing ascospores, result from the sexual process. Forty-six orders are recognized, with members of nineteen being mentioned in the present work. Lichenized fungi are indicated (L).

(a) Order Saccharomycetales. The asci of members of this order and of the Schizosaccharomycetales are solitary, not in ascocarps. Hence at one time such fungi were included in the class Hemiascomycetes, 'half Ascomycetes'. In contrast, other orders have the asci in ascocarps and hence were regarded as Euascomycetes, 'true Ascomycetes'. Most of the Saccharomycetales are yeasts (Y), but a few do not have a yeast phase.

Ashbya gossypyii (Y)
Debaromyces hansenii
Dipodascus geotrichum (ana. *Geotrichum candidum*)
Endomycopsis fibuligera (syn. *Saccharomycopsis fibuligera*) (Y)
Hansenula anomala (Y)
Kluyveromyces lactis (Y)
Pichia saitoi (Y)
Saccharomyces carlsbergensis (lager strains of *S. cerevisiae*) (Y)
Saccharomyces cerevisiae (Y)
Saccharomyces ellipsoideus (wine strains of *S. cerevisiae*) (Y)
Saccharomyces rouxii (syn. *Zygosaccharomyces rouxii*) (Y)
Saccharomycodes ludwigii (Y)
Saccharomycopsis fibuligera (syn. *Endomycopsis fibuligera*) (Y)
Saccharomycopsis lipolytica (ana. *Candida lipolytica* (Y)
Zygosaccharomyces rouxii (syn. *Saccharomyces rouxii*) (Y)

(b) Order Schizosaccharomycetales
Schizosaccharomyces pombe (Y)

(c) Order Diaporthales
Cryphonectria parasitica (syn. *Endothia parasitica*)
Diaporthe (ana. *Phomopsis*)
Endothia parasitica (syn. *Cryphonectria parasitica*)
Gaeumannomyces graminis (syn. *Ophiobolus graminis*)
Ophiobolus graminis (syn. *Gaeumannomyces graminis*)

(d) Order Dothidiales
Cochliobolus miyabeanus (ana. *Drechslera oryzae*, *Helminthosporium oryzae*)
Didymella (ana. *Ascochyta*)
Eudarluca caricis (ana. *Sphaerellopsis filum*)
Leptosphaeria
Leptosphaerulina trifolii
Mycosphaerella tassiana (ana. *Cladosporium herbarum*)
Pleospora herbarum (ana. *Stemphylium botryosum*)
Venturia inaequalis (ana. *Spilocaea pomi*)

(e) Order Erysiphales. The powdery mildews
Blumeria graminis (syn. *Erysiphe graminis*)
Erysiphe graminis (syn. *Blumeria graminis*)

(f) Order Eurotiales
Emericella nidulans (ana. *Aspergillus nidulans*)
Eupenicillium brefeldianum (ana. *Penicillium dodgei*)
Eurotium herbariorum (ana. *Aspergillus glaucus*)
Talaromyces (ana. *Penicillium*)

(g) Order Halosphaeriales
Chadefaudia corallinum
Remispora maritima

(h) Order Hypocreales
Atkinsonella hypoxylon
Balansia cyperi
Claviceps purpurea
Cordyceps

Epichloë typhina
Gibberella zeae (ana. *Fusarium graminearum*)
Hypocrella
Nectria haematococca (ana. *Fusarium solani*)
Neocosmospora
Neotyphodium lolii

(i) Order Lecanorales
Cladonia arbuscula (L)
Rhizocarpon concentricum (L)

(j) Order Leotiales (formerly Helotiales)
Amorphotheca resinae (ana. *Hormoconis resinae*, *Cladosporium resinae*)
Botryotinia fuckeliana (ana. *Botrytis cinerea*)
Bulgaris inquinans
Helotium chloropodium
Hymenoscyphus ericae
Pezicula aurantiaca

(k) Order Onygenales (formerly Gymnoascales)
Ajellomyces capsulata (syn. *Emmonsiella capsulata*, ana. *Histoplasma capsulata*) (Y)
Ajellomyces dermatitidis (ana. *Blastomyces dermatitidis*) (Y)
Emmonsiella capsulata (syn. *Ajellomyces capsulata*, ana. *Histoplasma capsulata*) (Y)
Gymnoascus

(l) Order Ophiostomatales
Ceratocystis fagacearum
Ophiostoma novo-ulmi and *Ophiostoma ulmi* (syn. *Ceratocystis ulmi*)

(m) Order Peltigerales
Peltigera canina (L)
Pseudocyphellaria rufovirescens
Sticta (L)

(n) Order Pezizales
Aleuria vesiculosa
Ascobolus immersus
Gyromitra esculenta
Monascus bisporus (syn. *Xeromyces bisporus*)
Morchella hortensis
Pyronema confluens (syn. *P. omphalodes*)
Sclerotinia fructigena
Tuber melanosporum
Xeromyces bisporus (syn. *Monascus bisporus*)

(o) Order Phyllachorales
Glomerella cingulata (ana. *Colletotrichum gloeosporoides*)
Magnaporthe grisea (syn. *Pyricularia oryzae*)

(p) Order Rhytismatales
Rhytisma acerinum

(q) Order Sordariales
Chaetomium globosum
Neurospora crassa

Podospora anserina
Sordaria fimicola
(r) Order Teloschistales
Xanthoria parietina (L)
(s) Order Xylariales
Biscogniauxia nummularia
Daldinia concentrica
Pseudomassaria
Xylaria hypoxylon

IV. Phylum **BASIDIOMYCOTA**, the Basidiomycetes
A basidium bearing basidiospores results from the sexual process. Three classes are recognized.

1. Class **Basidiomycetes**. The Basidiomycetes that have fruit bodies (basidiocarps), plus a few yeasts. The class replaces two classes that formerly accommodated Basidiomycete macrofungi, the Hymenomycetes and the Gasteromycetes, the latter a very artificial assemblage. Thirty-two orders are recognized, with members of fourteen being mentioned in the present work.
(a) Order Agaricales. Most mushrooms and toadstools.
Agaricus bisporus
Agrocybe gibberosa
Amanita phalloides
Armillaria mellea
Bolbitius
Clitocybe
Collybia velutipes (syn. *Flammulina velutipes*)
Coprinus cinereus
Flammulina velutipes (syn. *Collybia velutipes*)
Hypsizygus marmoreus
Laccaria
Lepiota
Marasmius androsaceus
Mycena galopus
Oudemansiella mucida
Panellus
Pholiota nameko
Pluteus
Podaxis
Psilocybe cubensis
Stobilurus tenasellus
Termitomyces titanicus
Volvaria volvacia (syn. *Volvariella diplasia*)
(b) Order Auriculariales
Auricularia polytricha
(c) Order Boletales
Boletus subtomentosus
Coniophora puteana

Omphalotus
Rhizopogon roseolus
Serpula lachrymans
Suillus bovinus
(d) Order Cantharellales
   Cantharellus cibarius
   Clavaria
(e) Order Ceretobasidiales
   Thanatephorus cucumeris (ana. *Rhizoctonia solani*)
(f) Order Cortinariales
   Cortinarius speciosissimus
   Inocybe
(g) Order Ganodermatales
   Ganoderma adspersum
(h) Order Halospheriales
   Remispora maritima
(i) Order Hericiales
   Hericium erinaceus
(j) Order Lycoperdales. Puffballs
   Geastrum triplex
   Lycoperdon perlatum
(k) Order Nidulariales. Birds' nest fungi
   Crucibulum laeve
   Sphaerobolus
(l) Order Phallales. The stinkhorns
   Phallus impudicus
(m) Order Poriales
   Bjerkandera adusta
   Coriolus versicolor
   Fomes fomentarius
   Gloeophyllum trabeum
   Grifola frondosa
   Heterobasidion annosum
   Lentinus edodes
   Phaeolus schweinitzii
   Pleurotus ostreatus
   Polyporus brumalis
   Poria vaillantii
   Trametes gibbosa
(n) Russulales
   Russula
(o) Order Schizophyllales
   Schizophyllum commune
(p) Sclerodermatales
   Pisolithus tinctorius
(q) Order Stereales
   Phanerochaete chrysosporium (ana. *Sporotrichum pulverulentum*)
   Phlebia gigantea

  Phlebiopsis
  Stereum sanguinolentum
(r) Order Tremellales
  Filobasidiella neoformans (ana. *Cryptococcus neoformans*) (Y)
  Tremella mesenterica
2. Class **Ustomycetes**. The smuts. Seven orders are recognized, with members of two being mentioned in the present work.
(a) Order Ustilaginales
  Tilletia caries
  Ustilago maydis
(b) Order Sporidiales
  Rhodosporidium toruloides (ana. *Rhodotorula rubrum*)
3. Class **Teliomycetes**. Two orders are recognized.
(a) Order Uredinales. The rusts
  Hemileia vastatrix
  Melampsora lini
  Puccinia graminis
  Uromyces phaseoli
(b) Class Septobasidiales. Parasites of scale insects

V. Informal group **MITOSPORIC FUNGI**
Previously known as Deuteromycetes, Fungi Imperfecti or Imperfect Fungi, are classified on the basis of asexual morphology, although with some a sexual, teleomorphic phase is known. This is indicated (tel.) where the name is widely used.
1. **Coelomycetes**. Conidia are formed in a cavity composed of fungal or mixed host and fungal tissue.
  Ascochyta (tel. *Didymella*)
  Colletotrichum gloeosporoides (tel. *Glomerella cingulata*)
  Coniothyrium minitans
  Microsphaeropsis
  Phoma (tel. *Pleospora*)
  Phomopsis (tel. *Diaporthe*)
  Sphaeropsis filum (tel. *Eudarluca caricis*)
2. **Hyphomycetes**. Conidia not within cavities. Includes yeasts (Y), sometimes separated as Blastomycetes.
  Acremonium strictum
  Akanthomyces
  Alternaria tenuis
  Aphanocladium album
  Articulospora tetracladia
  Arthrobotrys oligospora
  Aspergillus glaucus (tel. *Eurotium herbariorum*)
  Aspergillus nidulans (tel. *Emericella nidulans*)
  Aspergillus niger
  Aureobasidium pullulans
  Beauveria bassiana
  Bipolaris (tel. *Cochliobolus*)

Blastomyces dermatitidis (tel. *Ajellomyces dermatitidis*) (Y)
Botrytis cinerea (tel. *Botryotinia fuckeliana*)
Candida albicans (Y)
Candida lipolytica (tel. *Saccharomycopsis lipolytica*) (Y)
Candida utilis (Y)
Cephalosporium
Cercospora (tel. *Mycosphaerella*)
Cercosporidium personatum
Cladosporium herbarum (tel. *Mycosphaerella tassiana*)
Cladosporium resinae (syn. *Hormoconis resinae*, tel. *Amorphotheca resinae*)
Coccidioides immitis (Y)
Cryptococcus neoformans (tel. *Filobasidiella neoformans*) (Y)
Curvularia lunata
Cylindrocarpon tonkinense
Deightoniella
Dendryphiella salina
Drechslera oryzae (syn. *Helmithosporium oryzea*, tel. *Cochliobolus miyabeanus*)
Epicoccum
Epidermophyton
Fusarium culmorum
Fusarium graminearum (tel. *Gibberella zeae*)
Fusarium solani (tel. *Nectria haematococca*)
Fusidium coccineum
Geotrichum candidum (tel. *Dipodascus geotrichum*)
Gliomastix
Graphium
Helicodendron glomeratum
Helminthosporium oryzea (ana. *Drechslera oryzae*, tel. *Cochliobolus miyabeanus*)
Hirsutella
Histoplasma capsulatum (tel. *Ajellomyces capsulata*, *Emmonsiella capsulata*) (Y)
Hormoconis resinae (syn. *Cladosporium resinae*, tel. *Amorphotheca resinae*)
Humicola lanuginosa
Malassezia furfur (Y)
Memnoniella
Meria coniospora
Metarhizium anisopliae
Myrothecium
Nigrospora
Oidiodendron oryzae
Paecilomyces lilacinus
Paracoccidioides brasiliensis (Y)
Penicillium chrysogenum
Penicillium dodgeii (tel. *Eupenicillium brefeldianum*)
Periconia abyssa
Phialophora fastigiata
Pithomyces (tel. *Leptosphaerulina*)
Pityrosporon (Y)
Pyricularia oryzae (tel. *Magnaporthe grisea*)

Retiarius
Rhizoctonia solani (tel. *Thanatephorus cucumeris*)
Rhodotorula rubra (tel. *Rhodosporidium toruloides*) (Y)
Rhynchosporium secalis
Spilocaea pomi (tel. *Venturia inaequalis*)
Sporobolomyces (Y)
Sporodesmium sclerotivorum
Sporotrichum pulverulentum (tel. *Phanerochaete chrysosporium*)
Stachybotrys
Stemphylium botryosum (tel. *Pleospora herbarum*)
Thermomyces lanuginosus
Tilletiopsis minor (Y)
Tolypocladium inflatum
Torulopsis (Y)
Trichoderma viride
Trichophyton
Trichosporon
Verticillium albo-atrum

3. **Agonomycetes** Also known as Mycelia Sterilia. Lacking spores, they are not strictly speaking mitosporic, but are included with mitosporic fungi for convenience. Some have morphological features such as sclerotia or traps for eelworms that facilitate identification.

   Sclerotium cepivorum

Answers to Questions

Chapter 2
2.1 (c) Oomycetes
2.2 (c) 70 000
2.3 (b) Basidiomycetes
2.4 (a) *Glomus*
2.5 (a) Ascomycetes or basidiomycetes with algae or bacteria

Chapter 3
3.1 (c) hyphal anastomosis
3.2 (c) a polysaccharide
3.3 (b) soon after the spindle plaque has been duplicated
3.4 (b) branching of its component hyphae
3.5 (a) the plasma membrane

Chapter 4
4.1 (c) conidia
4.2 (a) oospores
4.3 (b) sporangiospores
4.4 (d) a lipopeptide
4.5 (c) two dissimilar flagella
4.6 (c) in nutrient medium with added vitamins

Chapter 5
5.1 (d) 50%
5.2 (b) A_1B_2 with A_2B_1
5.3 (c) resists only strains of the fungus that carry a particular avirulence gene
5.4 (a) recombination without meiosis
5.5 (a) Chytridiomycetes

Chapter 6
6.1 (c) degrading wood and other plant remains to carbon dioxide
6.2 (d) an organic carbon source
6.3 (d) the soil
6.4 (a) oxidizing free radicals
6.5 (a) a specially raised monoclonal antibody

Chapter 7
7.1 (a) orchid mycorrhiza
7.2 (d) targeted gene disruption

7.3 (b) haustoria
7.4 (c) *Armillaria mellea*
7.5 (b) *Malassezia*

Chapter 8
8.1 (c) an airlift fermenter
8.2 (a) *Aspergillus*
8.3 (c) both (a) and (b)
8.4 (b) most are produced as families of related molecules
8.5 (d) its amino acid sequence is encoded by a specific messenger RNA

Index

Page references in *italics* refer to tables and figures; those in **bold** refer to main discussion